DYNAMICAL SYSTEMS FOR BIOLOGICAL MODELING

AN INTRODUCTION

Advances in Applied Mathematics

Series Editor: Daniel Zwillinger

Published Titles

Green's Functions with Applications, Second Edition *Dean G. Duffy*

Introduction to Financial Mathematics *Kevin J. Hastings*

Linear and Integer Optimization: Theory and Practice, Third Edition
Gerard Sierksma and Yori Zwols

Markov Processes *James R. Kirkwood*

Pocket Book of Integrals and Mathematical Formulas, 5th Edition
Ronald J. Tallarida

Stochastic Partial Differential Equations, Second Edition *Pao-Liu Chow*

Dynamical Systems for Biological Modeling: An Introduction
Fred Brauer and Christopher Kribs

Advances in Applied Mathematics

DYNAMICAL SYSTEMS FOR BIOLOGICAL MODELING

AN INTRODUCTION

Fred Brauer

University of British Columbia
Vancouver, British Columbia, Canada

Christopher Kribs

University of Texas at Arlington
Arlington, Texas, USA

CRC Press is an imprint of the
Taylor & Francis Group, an **informa** business
A CHAPMAN & HALL BOOK

Back cover image credit: (Photo of healthcare workers) Cleopatra Adedeji, CDC PHIL public domain collection.

CRC Press
Taylor & Francis Group
6000 Broken Sound Parkway NW, Suite 300
Boca Raton, FL 33487-2742

Printed by CPI UK on sustainably sourced paper
Version Date: 20151012

International Standard Book Number-13: 978-1-4200-6641-8 (Hardback)

Visit the Taylor & Francis Web site at
http://www.taylorandfrancis.com

and the CRC Press Web site at
http://www.crcpress.com

to our children and grandchildren

Contents

Preface

Some understanding of difference equations and differential equations is becoming essential for students in the biological sciences. There are several recent texts providing an introduction to mathematical biology through dynamical systems, but most of these are accessible primarily to students of mathematics with some interest in applications to biology rather than to biology students who have not (yet) developed a disposition to describe real systems in mathematical terms. Many of these texts present fully developed models of biological systems ready for mathematical analysis. Such models provide students with exercises in applying mathematical techniques but little experience in the (often iterative) translation of biological concepts into mathematical terms and vice versa, which is at the heart of modeling. It is our intention to try to address both of these issues, to prepare students of biology as well as of mathematics with the understanding and techniques necessary to undertake basic modeling of biological systems through the development and analysis of dynamical systems.

We propose to present an introduction to dynamical systems, covering both discrete (difference equation) and continuous (differential equation) types, for students who have had an introduction to calculus. While we assume that students have learned basic material including differentiation, integration, and exponential and logarithmic functions, we will recall topics which may not have made sufficient impression to have been absorbed completely and will describe some topics from a slightly different perspective. Our motivation will come from biological topics, including but not confined to population biology and epidemiology. Our approach will emphasize qualitative ideas rather than explicit computations; we feel that this approach is both easier for students whose primary interests are not in mathematics and more useful in many applications. This material is not, however, intended to serve as a differential equations course for students in the physical sciences as it does omit many techniques and topics that are essential for such students.

In presenting mathematical topics, we will attempt to tell the truth and nothing but the truth, but not necessarily the whole truth. We will try to emphasize the basic truths but not to overwhelm the student with precise technical detail. Some results will be stated without proof, while others will be accompanied by outlines of the reasons why they are true. We will normally avoid detailed, rigorous proofs.

Mathematics is not a spectator sport, and can be learned only by solving problems. We include examples of problems with solutions and some exercises which follow the examples quite closely. However, a library of solved examples used as templates will not be sufficient to meet the needs of a developing scientist. For this reason, we also include problems that go beyond the examples, both in mathematical analysis and in the development of mathematical models for biological problems, in order to encourage deeper understanding and an eagerness to use mathematics in learning about biology. We also include some problems, marked with an asterisk (*), which are more challenging.

We recommend the introduction to modeling in Chapter 1 as a beginning of any course on biological modeling using this text. The contents of Chapters 2 through 4 overlap considerably with the material which would be covered in a unified course (probably two semesters in length) which covers calculus and some difference equations and elementary differential equations. For students who have already taken such a course, a suitable course could review these topics as needed (after starting with Chapter 1) and then continue with Chapters 5, 6, and possibly 7. For students who have had a semester of calculus without difference or differential equations, a suitable course could consist of the first four chapters.

We provide some support for helping readers use software packages to generate numerical solutions and graphs (as we ourselves have done to make some of the figures in this book), primarily in the sections dealing with numerical analysis and in the appendices. These packages and the companies that produce them are listed below.

Maple is a registered trademark of Waterloo Maple, Inc., *www.maplesoft.com.*

Mathematica is a registered trademark of Wolfram Research, Inc., at *www.wolfram.com.*

MATLAB is a registered trademark of The Mathworks, Inc. For product information please contact The Mathworks, Inc., 3 Apple Hill Drive, Natick, MA 01760-2098 USA, tel. 508-647-7000, fax 508-647-7001, e-mail *info@mathworks.com, www.mathworks.com.*

Trademarked names may be used in this book without the inclusion of a trademark symbol. These names are used in an editorial context only; no infringement of trademark is intended.

Acknowledgments

We thank Bob Stern at CRC Press for planting the initial idea for this book, and both him and Bob Ross for their support and understanding in the face of long delays during the preparation of the manuscript. (Although we do not include delay equations in this book, there certainly were delays.) We also thank everyone at Taylor & Francis who helped with the production of this book, especially Shashi Kumar and Marcus Fontaine for critically useful help involving LaTeX, Karen Simon, and Kevin Craig for graphic design assistance.

The figures and photos accompanying discussion of the various biological systems studied and discussed play a crucial role in bringing the models (and their motivations) to life for the reader, and many appear in this work through the kind permission of others. We therefore acknowledge here all those who generously permitted us to print their photos, or helped us obtain permission: Donna Anstey, Francine Bérubé, Daniel Bowen, John Calambokidis, Tom Chrzanowski, Patricia Ernst, Carla Flores, Tim Gerrodette, Stefano Guerrieri, Ray Hamblett, Alan M. Hughes, Duncan Jackson, Russell S. Karow, Carolyn Kribs, Joel Michaelsen, Bernard E. Picton, Dave Powell, Francis Ratnieks, Helen Sarakinos, Howard Swatland, Michael Tildesley, and S. Bradleigh Vinson.

We also thank all those photographers who contributed indirectly, including photos in the public domain: Cleopatra Adedeji, Lennert B., David Burdick, Janice Haney Carr, Mark Conlin, Karen Couch, Jan Derk, Gary Fellers, Ryan Hagerty, William Chapman Hewitson, Steve Hillebrand, Tom Hodge, John & Karen Hollingsworth, Al Mare, Maureen Metcalfe, Benjamin Mills, A.J. Nicholson, Sergey Nivens, Zev Ross, Jeff Schmaltz, Greg Webster, Gary Zahm, and the following organizations: the Centers for Disease Control, Florida Keys National Marine Sanctuary, Ken Gray Image Collection at Oregon State University, MODIS/NASA, National Diabetes Information Clearinghouse, NOAA, the Pennsylvania Dept. of Conservation and Natural Resources, River Alliance of Wisconsin, the Sickle Cell Foundation of Georgia, the Southwest Fisheries Science Center of the NOAA Fisheries Service, U.S. Fish and Wildlife Service, U.S. Geological Survey, U.S. National Park Service, and the Wisconsin Dept. of Natural Resources.

Part I

Elementary Topics

Chapter 1

Introduction to Biological Modeling

1.1 The nature and purposes of biological modeling

Mathematical modeling has been used to explain biological systems for centuries, going back to work such as that of Thomas Malthus (1798) on the growth of populations, Daniel Bernouilli (1760) on smallpox vaccination and arguably Fibonacci (Leonardo of Pisa, 1202) on the well-known rabbit problem. Modern mathematical biology has been an ongoing scientific endeavor since the end of the nineteenth century, growing out of the work of individuals such as P.D. En'ko in epidemiology, D'Arcy Wentworth Thompson in change in biological organisms, and G. Stokes in fluid flows, and the technical developments of the late twentieth century have placed it in the forefront of scientific research. The use of theoretical models to describe and predict the behavior of biological systems is especially useful today in examining large-scale questions where controlled experiments are either impossible or unethical to carry out — the effects of large-scale alterations in environmental conditions on local ecology, or of slight decreases in efficiency of certain disease control measures. Even in situations where some experimental data is available, mathematical models may validate existing theories or provide new insights by describing the mechanisms through which biological processes occur, taking in a whole spectrum of possible measurements in a single step. More specifically, one might use modeling

- to predict unknown behavior (such as intervention effects) based on given factors,
- to generate virtual experiments that inform general biological theory,
- to account for observed behavior as simply as possible,
- to compare competing hypotheses — which better accounts for observed results?
- to evaluate or rule out hypotheses,
- to suggest conjectures and new hypotheses, or

- to guide empirical inquiry — does this process make a key difference in the behavior of the system?[1]

In short, modeling helps us consider the question, "What if ...?" The language of mathematics, meanwhile, provides powerfully compact descriptions of complicated ideas, and is capable, at its best, of offering equally simple insights, such as the idea of a single threshold quantity which determines the outcome of a complex biological process and brings together all the relevant factors in a way that clarifies the relative importance of each (cf. Section 4.4).

Mathematical models, however, are far from perfect: as intricate as they may appear at times, they are mere sketches or caricatures of the living world. Biological systems are complex, with many underlying forces and interactions, heterogeneity, and random variations and fluctuations. In modeling such a system, we can only hope to develop a good enough approximation to reality to accomplish our purposes. The engineering statistician George Box wrote, "All models are wrong, but some are useful."[2] We use models to give rough descriptions of reality in cases where rough is still accurate enough to give useful insights.

By its nature, modeling involves both *iteration* and *compromise*. As modeling is imperfect, the results of a given model may not agree with observations, or may have limitations that keep the model from being useful. In these cases, the modeler must make changes to the model, or even replace it altogether, and then analyze the new model to see if it describes the biology better (or well enough). One has to be willing to tinker and try again. Iteration thereby makes the modeling process cyclical rather than linear. The following section sketches this iterative modeling process.

Section 1.3 goes into more detail about the types of mathematical models one might use. The selection of a particular kind of mathematical structure as a model follows from one's research questions. A laboratory or field biologist, interested in the specific biology of the organisms in the system, may be most interested in quantitative or numerical results, and models which fit observed data closely. Theoretical biologists, on the other hand, may be more interested in qualitative, structural insights, which can suggest trends across models of a number of similar systems, such as the role of genetic diversity in the survival of populations. In each case, the researcher's perspective and interests will shape the research question and, through it, the type(s) of model and analysis most likely to provide fruitful answers. The types of models we shall explore in this text are called *dynamical systems*, which mean that they change as functions of some independent variable(s), usually time.

Facing up to the imperfection of any model means acknowledging its limitations. Modeling does not mean ignoring or brushing aside the assumptions

---

[1] cf. E. Smith, S. Haarer, J. Confrey (1997). Seeking diversity in mathematics education: mathematical modeling in the practice of biologists and mathematicians, *Science and Education* 6(5): 441–472.

[2] G.E.P. Box (1976). Science and statistics, *Journal of the American Statistical Association* 71: 791–799.

and simplifications we make in constructing and interpreting models; rather, it is an active exercise in compromise: including as much detail as necessary while keeping the model manageable in terms of analysis and (as applicable) data collection. A saying attributed to Albert Einstein runs, "Everything should be made as simple as possible, but not simpler." Section 1.4 discusses some of the most common compromises faced in mathematical modeling, as well as their consequences.

Finally, the modeling process should also include an explicit choice of scale and unit(s) of measurement for the quantities or populations being modeled, and Section 1.5 introduces two related issues which are discussed further as they recur later in the text. In each section, we shall speak of biological systems in general, as the focus here is on modeling, but the reader can substitute specific systems, such as the human circulatory system, the current generation of Monarch butterflies, or the population of Toronto at risk for SARS in March 2003, for the word "system" where it occurs.

1.2 The modeling process

Although mathematical modeling may look like a great variety of different things, there are certain commonalities in the process, in the "big picture," and as this text is aimed at introducing students to biological modeling, it is just as important to become familiar with the higher-order tools and elements of modeling as with the mathematical techniques specific to each type of model. As mentioned in the previous section, modeling is an iterative process. Figure 1.1 sketches the flow of this process.

In practice, we begin by developing our first try at a model (which process includes defining the problem at hand), translating the biological system into a mathematical system. We then apply mathematical tools to analyze the model. The mathematical results must then be interpreted back into biological terms, and finally we must evaluate the answers our model has given us, to decide whether they are satisfactory — biologically reasonable, and sufficient to meet our objectives. If they are not (which is the case often enough), we must adjust the model to try to remedy the observed shortcomings, and repeat the process until the results are satisfactory.

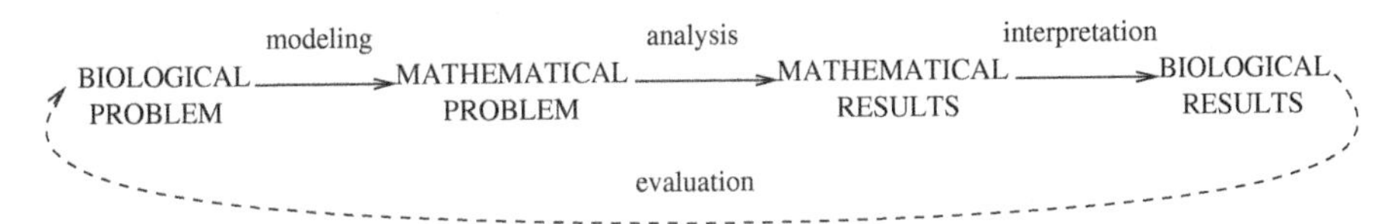

FIGURE 1.1: Mathematical modeling as an iterative process.

Mathematical biologists therefore need to be able to:

1. take biological features and hypotheses and turn them into mathematical structures;

2. apply relevant mathematical analysis tools;

3. interpret mathematical results in biological terms; and

4. evaluate biological results critically in order to evaluate the underlying mathematical model.

Let us consider each of these activities one by one.

(1) MODELING What does model construction look like? The details can vary widely, but the philosophy is to describe the essential characteristics of your system in mathematical terms. For us, this will typically mean a population or group of interacting populations, although we will apply the term *population* to collections of anything from molecules to households. One perspective on modeling is that there are two directions in which to go about it:

- *forward* — begin with observed characteristics, factors, rates, make your model reflect these, and see if the results match observations; or

- *backward* — use observations to develop known (desired) model behavior, and craft a model known to exhibit this behavior.

In other words, one might begin by assembling model elements which correspond to the biological features of the system and consider the results (a *predictive* model), or one might instead begin with a model of a type known to behave as desired, and modify the structure to fine-tune (or fit data) (a *descriptive* model). If we consider the three purposes suggested earlier for biological modeling, we see that a forward modeling approach suits the goal of predicting unknown behavior based on known forces, while a backward (or back-end) approach might best suit the goal of accounting for observed behavior. Evaluating competing hypotheses may require a combination of the two approaches. It is important, however, to acknowledge the distinction between *first principles* models derived directly from known laws (common in physics and chemistry, e.g., using Newton's Laws, but less so in biology) and *ad hoc* models which use terms, expressions or functions commonly used to represent a particular feature or process of the system based on past observations of their fit to data (e.g., using a logistic model to describe population growth with limited resources (see Sections 2.4 and 3.2), or a mass-action term to describe the interaction of two populations (see Section 4.3). Likewise there is a distinction between broadly descriptive models, which attempt to capture as much detail as possible in a biological system, and focused theoretical models which incorporate only those features relevant to a specific research question.[3] Of course, this perspective for developing models is by no means exclusive, and should be considered only as long as it is helpful in providing direction.

The elements of model development include identifying the problem or research question as specifically as possible (a complex task we shall not do more than illustrate), writing careful definitions of relevant terms (variables and parameters), specifying all the assumptions being made (see Section 1.4 for a fuller discussion), drawing diagrams (sometimes), choosing an appropriate mathematical structure (see Section 1.3) and scale (see Section 1.5), and assembling all these elements into a coherent mathematical description (for us, an equation or set of equations). It is also important to articulate one's objectives — what kind of answers one expects — as they will influence the types of model and analysis that one uses.

(2) ROBUSTNESS In the physical sciences, models often are based on specific assumptions about the terms included in the model, and these assumptions are based on mechanistic derivations. In the biological sciences, terms that are included in models do not necessarily have mechanistic derivations, but are included to give tractable problems. Thus, for example, the logistic equation often used to model the growth of a single population assumes a specific form for the growth rate of the population. Many of the results that can be deduced from a logistic equation model require only that the growth rate be zero when the population size is zero and when the population size has a positive value K called the carrying capacity, and that the growth rate be positive when the population is between zero and the carrying capacity. Robustness of a model is the property that conclusions drawn from the model are valid under a less specific set of assumptions. Often one does not know the details of a biological problem, and one has greater confidence in the conclusions drawn if changes in detail do not affect the conclusions drawn.

(3) ANALYSIS The mathematical analysis tools and techniques required vary by model type; for example, *stochastic* models, which incorporate random variations, typically require a numerical approach, in which one runs a large battery of computer simulations where one or more parameters (model inputs) vary according to a specified probability distribution. For the *deterministic* (non-probabilistic) dynamical systems which will form the basis for models considered in this text, we have three possible approaches: exact, qualitative, and quantitative. *Exact* solutions, in which we obtain an explicit formula for a biological quantity (usually as a function of time), can be obtained only for the simplest possible types of dynamical systems. We will find them useful mostly for *linearizations*, simplifications of complicated models that sketch model behavior near special points. *Qualitative* analysis provides general information on how a given system behaves, and is typically used to identify what affects the system in the long term. *Quantitative* methods use computers to approxi-

[3] Castillo-Chávez, as quoted in E. Smith, S. Haarer, J. Confrey, Seeking diversity in mathematics education: mathematical modeling in the practice of biologists and mathematicians, *Science and Education*, **6**: 441–472 (1997).

mate solutions numerically, and are useful for illustration purposes, or when a model is too complicated to complete a qualitative analysis, or when we have good estimates of the parameters involved. In this text we shall show how to apply all three types of analysis, although qualitative analysis will receive the most emphasis, as a way to prove how complicated models behave, regardless of the exact parameter values. Although the mathematical nature of such models can generate convincingly real data, it is often important to remember that, as gross oversimplifications of reality, mathematical models are just as much rough sketches as the nonmathematical theories with which biologists are familiar, and qualitative descriptions of behavior befit this notion.

(4) INTERPRETATION Interpreting mathematical results in biological terms requires a return to the original context of the problem, and is easier with a well-articulated research question. That is, rather than simply asking, "What happens if we model the following biological system?" one ought to ask a specific question that can be addressed with a model — for example, "What effects do seasonal temperature variations have on the size of this population?" or "How does behavior change based on observed infection levels affect the eradicability of this disease?" (A general statistical analysis of an experimental data set, however, may show up trends that shape subsequent research directions.) Interpretation is also facilitated by a close correspondence between each element of the mathematical model and the elements of the biological system. The conclusion "this population will go extinct if beta x is less than 1" is only useful if one can put "beta x" in biological terms — what do beta and x mean? can we measure them (or at least their product)? Which might we be able to control, and which are inherent properties of the biological system?

The basis for an interpretation may be numerical, qualitative, or graphical. When most parameters (fixed quantities) in a model can be estimated fairly well, numerical information can provide ranges within which some varying or controllable quantity causes the biological system to behave in a particular way, or critical values which cause that behavior to change. Identifying thresholds between survival and extinction, or between settling down to an equilibrium level and oscillating forever, or comparing the behavior of two related but slightly different models, can answer qualitative questions such as, "Which explanation for this phenomenon better accounts for our observations?" or "Does the system behave differently if we take into account this additional factor?" The ready availability of numerous computational tools also makes it possible to represent results graphically, and a graph can be an especially compelling, concise statement of a complex result. Graphs can illustrate both general trends and striking thresholds.

However, just as with algebraic expressions, graphical depictions of mathematical information can be so concise that the job of unpacking them into nonmathematical terms can be significant. The ability of mathematics in all its forms to represent complicated ideas compactly makes it both powerful and challenging, and if one considers all the time invested in becoming familiar with and manipulating algebraic expressions, it should not be surprising that unpacking the information contained in graphs requires practice and thought. For example, Roth[4] found that even as apparently simple a graph as that of population growth rates shown in Figure 1.2 created multiple interpretation challenges. While a definitive primer in graph interpretation is not the aim of this text, we shall discuss interpretation of figures we present in later chapters, and also here consider some particular questions regarding the graph in Figure 1.2 (cf. Figures 2.19 and 2.27 in Section 2.5).

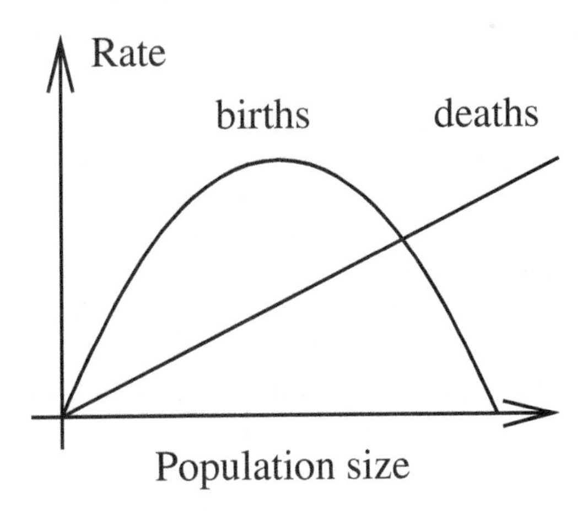

FIGURE 1.2: A graph comparing birth and death rates vs. population size, after Roth (2001).

- Note first that the *horizontal axis* in the graph is not time but population size. That is, the graph shows how birth and death rates vary depending on the population, rather than over time. Since many continuous graphs depict how some quantity changes over time, one must readjust one's reading of the graph to reflect what it is that is varying.
- Note second that the *vertical axis* in the graph gives birth and death *rates* (measured in individuals per day, week, or year), and not total births and deaths. That is, the graph shows how *fast* births and deaths occur, as population size grows (varies). The bending downward of the birth rate curve, for example, as population size increases does not address what has happened to previous births in the past of any population that reaches that size. The distinction (and relationship) between population

[4]W.-M. Roth (1998), Unspecified things, signs, and "natural objects": towards a phenomenological hermeneutic of graphing. In S. B. Berenson, K. R. Dawson, M. Blanton, W. N. Coulombe, J. Kolb, K. Norwood, and L. Stiff (Eds.), Proceedings of the Twentieth Annual Meeting of the North American Chapter of the International Group for the Psychology of Mathematics Education (Vol. I, pp. 291–297), ERIC Clearinghouse for Science, Mathematics, and Environmental Education, Columbus, OH, 1998.

sizes and their rates of change (mathematically, their derivatives with respect to time) is a fundamental prerequisite of the models we will develop in the chapters ahead.

- The simpler of the two curves to interpret is the death rate, which exhibits a *linear* rise, corresponding to a constant *per capita* death rate (the total death rate divided by the population size), visible in the graph as the slope of this line. Since constant per capita death rates due to natural causes are commonly assumed (see Section 2.1), one can interpret this line as indicating that there are no *density-dependent* causes of death in this system.

- The rise and fall of the birth rate curve, however, suggests that there *are* density-dependent forces influencing the birth rate (otherwise this graph would be a line, too). The decrease in birth rate for larger populations suggests a factor such as limited resources and/or competition which reduces reproduction in large populations. (The graph also suggests that for large enough populations there will be no births at all, which is not realistic, but see Section 2.4.)

- Ultimately, interpretation of this graph requires the reader to see the "big picture" depicted here, what one might call the *emergent properties* of the graph: what does it communicate? In this case, the superimposition of the two curves allows us to compare them, to see for what sizes the population grows and for what sizes it shrinks. Where the birth rate is greater, the population is growing; where the death rate is greater, it is shrinking. Our attention should therefore focus on the point(s) where the two curves intersect, in particular the point where the death rate catches up with the birth rate. Beyond all the individual details described above, the main conclusion this graph was drawn to suggest is this population level, where (for reasons we shall explore in the chapters ahead) the population will eventually settle.

Finally, it is also important to acknowledge explicitly the limitations on the results given: the range within which one's model is a good enough description of biological reality, and the extent to which the interpreted results may be expected to hold true. For example, one of the most frequent simplifications we shall undertake in our mathematical analyses in this text involves making nonlinear (complicated) models linear near certain key points. Almost none of our models will themselves be linear, however. Taylor's theorem (see Appendix B) guarantees that linear approximations are reasonable within a very small operating range, but to replace, say, the graph of the birth rate curve in Figure 1.2 above with the tangent line at any one of its points would give a very misleading idea indeed of the relationship between population size and birth rate. (Recall that the tangent line is the linear approximation to a curve at that point.)

(5) EVALUATION Evaluating the results given by a particular model requires asking questions such as the following: Are results reasonable? Do they agree with observations to the extent possible? Do the assumptions made impose too many limitations on the model's ability to explain or describe the biological system? The answers to these questions determine whether it is necessary to change or replace the model. Discussions in the text where evaluation receives special emphasis include those in Sections 2.4, 3.2, 5.1, 5.2, 6.1, and 7.1.

Note that many of the decisions which need to be made in modeling — like what features are relevant, what assumptions are reasonable — really become easier to make only with experience, and the examples in the chapters ahead provide a beginning. Keep at it!

1.3 Types of mathematical models

Once a research question has been defined and model development begins, one must decide what type of mathematical model is most appropriate for the given situation. In this text we shall discuss *dynamic* models, called in mathematics *dynamical systems*, which track the change of one or more variables (for us, usually populations) over time (and occasionally over other variables as well). This discussion therefore leaves out static models, such as statistical analyses of datasets, which are nevertheless useful for other purposes. We shall organize our discussion by considering, in turn, the various choices to be made.[5]

One recurring issue in modeling is the question, *discrete or continuous?* We shall consider this question in three different regards. The terms *discrete dynamical system* and *continuous dynamical system* generally refer to models that are discrete or continuous in *time*. Discrete-time models measure the size of a population at regular, fixed intervals, saying nothing about its size at moments in between those snapshots. This approach is appropriate for systems which change or reproduce in distinct generations (e.g., annually, like many species of fish), or for which data are available at regular intervals, and results in a model composed of *difference equations*, as we shall see in Chapter 2. Continuous-time models depict population growth continuously over time and are appropriate for systems with ongoing and/or rapid change, and manifest as *differential equations*, as we shall see in Chapters 3 and 4.[6] For

[5]Another accessible and thorough discussion of types and purposes of mathematical models applied specifically to the epidemiology of sexually transmitted diseases is given in G.P. Garnett (2002), An introduction to mathematical models in sexually transmitted disease epidemiology, *Sexually Transmitted Infections* 78: 7–12.

[6]A detailed comparison between a discrete-time model and the continuous-time model that corresponds to it is outlined in the review exercises at the end of Chapter 3.

example, suppose we are studying how a certain bird population is affected by fluctuations in its food source. If the birds are the swallows of Mission San Juan Capistrano, in California, around which recent development has seriously depleted nearby insect populations, we might consider a discrete-time model spanning several years and focus on the number of swallows which return to the mission on March 19, or the number still at the mission the day before they migrate south on October 23. However, if we are studying the effects of seasonal variation in insect availability on a non-migratory bird such as the Carolina wren, we might instead consider a continuous-time model.

The discrete/continuous question may also arise in two different ways in defining variables: *heterogeneity* and *population size*. If some property, such as age, spatial location or risk level, varies significantly enough within a population to warrant modeling that variation, we must decide whether to incorporate the heterogeneity discretely with a *stratification* or continuously with a *distribution*. Stratifications divide a population into groups, each of which follows similar rules or behaves similarly, and requires defining one state variable for each group. The simplest possible stratification is to define two groups, such as juveniles and adults, habitat 1 and habitat 2, or high-risk and low-risk. Continuous distributions introduce an additional independent variable like age or risk level, and typically lead to *partial differential equations*, which are mathematically more complicated. Many human censuses gather data by age ranges, such as 0–4 years, 5–9 years, 10–15 years of age, etc., and models built to make use of such data will usually stratify accordingly by age. With the injunction to simplify in mind, many models stratify into two or a few groups rather than introduce a second independent variable. For example, a landmark study on the transmission of gonorrhea in the United States[7] focused on variation in sexual activity levels. Although in practice there is a wide variation in such levels, the study was able to provide important insights by stratifying the population into two groups: a low-activity group called the non-core which includes most of the population, and a small, high-activity group called the core. However, many applications require detailed modeling of spatial or age information, and continuous distributions are important then. For instance, studies of pattern formation on growing animals (fish, cats, crocodiles) require keeping careful track of location on the animal in a continuous way that permits one to see the growth of waves in the patterns.

The state variables which keep track of the sizes of populations and other dynamic quantities in the model can also be discrete or continuous. At some level, real populations are always composed of a whole number of individuals (organisms, molecules, etc.) although of course measures of size such as mass and volume vary with each individual. Models which keep track of the status of each individual in a population must by nature have a discrete population size (even if that size changes with births, deaths, arrivals and departures), but

[7] H.W. Hethcote and J.A. Yorke (1984). *Gonorrhea transmission dynamics and control.* Lecture Notes in Biomathematics 56. New York: Springer-Verlag.

models which keep track only of the overall sizes of otherwise homogeneous groups may allow the variables representing them to vary continuously, especially if the unit (see Section 1.5) makes this meaningful: biomass measured in kilograms, or individuals measured in thousands or millions (where each individual is such a small quantity that the difference between discrete and continuous is minimal). In general, however, the choice between discrete and continuous state variables is linked to a more fundamental decision: whether to make the model deterministic or stochastic.

A *deterministic* model describes a system in terms of averages: every possible outcome — dying or not dying, reproducing or not reproducing, recovering from a disease at any moment from just after infection to years afterward — occurs, in the proportions corresponding to the relative probabilities of each event happening. If the average lifetime for a population under study is 30 years, then the corresponding average (per capita) death rate is its reciprocal, $\frac{1}{30}/\text{yr} \approx 0.03/\text{yr}$, and in a deterministic model precisely 1/30th, or $3\frac{1}{3}\%$, of the population will die each year. In some sense, it is more accurate to say that in a deterministic model, 1/30th of each individual dies per year. This type of model is appropriate for large populations, where we call on the so-called Law of Large Numbers, which says that as the number of trials of a random event (such as whether or not a given individual dies) increases, the actual experimental probability (here the proportion of the population that undergoes a certain change) approaches the theoretical probability of the event. Deterministic models also commonly use continuous state variables.

On the other hand, a *stochastic*, or probabilistic, model describes all events in terms of probabilities, and essentially flips a coin or tosses a die for each individual member of the population, to see whether that member dies or not, reproduces or not, etc. Consequently, these models use discrete state variables. A stochastic approach generally requires a little more mathematical machinery — in particular, familiarity with probability distributions — but is appropriate in cases where populations are small enough that the Law of Large Numbers does not apply, or where for some other reason it is important to couch the model in terms of probabilities: for example, to determine a possible probability distribution for the outcome of a certain event (like how many cases an outbreak of disease will cause). Recently, studies have been made of the structure of social contact networks,[8] and in some cases detailed data has been recorded for contact networks (e.g., SARS outbreaks, and sexual contact networks in small closed populations); studying how these networks affect the spread of contact-driven phenomena such as infectious diseases requires a stochastic approach. Likewise, studies of disease eradication must focus on the last few infected individuals, where probabilistic effects are important. For the large-scale study of gonorrhea mentioned earlier, however, a deterministic

[8]See, for instance, S.H. Strogatz (2001). Exploring complex networks, *Nature* 410: 268–276, and M.E.J. Newman (2003). The structure and function of complex networks, *SIAM Review* 45: 167–256.

approach is not only simpler but has provided significant enough insights to inform disease control strategies.

In addition, statistical analyses can be especially important for some models. Two particular calculations of relevance here are *sensitivity analysis* and *uncertainty analysis.* In a sensitivity analysis, a probability distribution of parameter values is chosen around the parameters' estimated averages, in order to determine how strongly the model results depend on each parameter; the outcome is a ranking of parameters to which some particular model output — say, the minimum influx of new individuals per year needed in order to sustain a population which reproduces slowly — is sensitive. For example, one might find this hypothetical minimum influx to be more sensitive to the existing population's death rate than to its reproductive rate. An uncertainty analysis studies the impact of measurement errors (how far off the estimated parameter values are); the outcome is a probability distribution for some model output, say, the spread of possible values for the required minimum influx given estimated probability distributions for the birth and death rates.

No single text can hope to make a reader conversant with all of these types of mathematical models. In this book we shall focus on deterministic models, with continuous state variables, stratified discretely when necessary to incorporate heterogeneity, the alternative being continuous variation using partial differential equations. Chapter 2 will consider models of biological systems appropriate for discrete dynamical systems (recall that by *discrete* we mean discrete in time), while the remaining chapters will consider models using continuous dynamical systems. In each case, we will begin our study with simple (linear) single equations, proceed to more complicated (nonlinear) single equations, and finally consider systems of interacting populations.

1.4 Assumptions, simplifications, and compromises

Modeling, as noted earlier, is an exercise in compromise: we include as much detail as we can (and must) without losing our ability to analyze the model. This means that, in developing a model and evaluating the results, we must identify which simplifying assumptions we make, why we make them, and how they influence (or limit) the conclusions we draw. It is worth mentioning some of the trade-offs we make most frequently (in general practice, as well as in the chapters ahead).

That we simplify at all means that to some extent we must be willing to give up biological accuracy in exchange for tractability or "analyzability." For example, we may assume that the total size of a given population is constant over time, in order not to have to keep track of it as a separate variable. Almost always in this text, we shall assume that the amount of time it takes for any biological process to happen is distributed exponentially, the way radioactive decay works (as opposed to, say, a fixed period of time); although some biological processes do work this way, the real reason we make this assumption is to make our models be differential equations rather than something more complicated, like delay or integral equations. Second, at the same time our willingness to consider reasonably complex models means giving up exact solutions to the equations in favor of qualitative information about the solutions: do they eventually taper off toward some equilibrium, or do they oscillate forever, or do they crash to zero in finite time, or do they do something stranger? In the case of a quantitative model, we give up exact solutions and broad, analytical guarantees of possible behaviors in exchange for quantitative approximations given specific parameter values.[9]

Third, typically we also make some kind of assumption of homogeneity of individuals — that, within a given group, they all behave the same way with regard to any events described in the model. Recall also the discussion of handling heterogeneity in the previous section; stratification typically includes this assumption of homogeneity within each class, in exchange for a simpler type of model. This assumption, as well as the first one above, arises from the motivation to minimize the number of state variables used (the *dimension* of the model).

Fourth, we sometimes give up immediate biological interpretability in return for tractability — that is, after initially defining a model in biological terms, we may further simplify it by redefining variables and parameters, in a way that makes the model mathematically simpler (usually reducing the number of different variables and/or parameters) but may complicate our ability to interpret the new quantities in biological terms. For example, one model for foraging ants[10] studied in Section 4.1 begins with five parameters:

N — the total number of ants;
α — the (per capita) rate at which ants find a food site;
β — the (per capita) rate at which ants find a pheromone trail left by other ants;
σ — the maximum rate at which ants retire from bringing food back from the site;
K — the number of ants at which the retirement rate reaches $\frac{1}{2}\sigma$.

It is determined, however, that the model can be simplified to one involving only three parameters:

$$a = \frac{\alpha N}{\sigma}, \quad b = \frac{K(\beta N - \alpha)}{\sigma}, \quad c = \frac{\beta K^2}{\sigma}.$$

[9] Computational schemes used in numerical analysis, however, are typically shown to provide approximations that converge to the exact solution as the time-steps involved get smaller. That is, one can make the approximate solutions as close as one likes to the exact solution, if one is willing to wait longer for the computer to run its calculations.

[10] D.J.T. Sumpter and S.C. Pratt (2003). A modelling framework for understanding social insect foraging, *Behav. Ecol. Sociobiol.* 53: 131–144. DOI 10.1007/s00265-002-0549-0.

The resulting model is simpler to analyze, but the new parameters are more difficult to interpret in biological terms because the biological information has been packed mathematically, and must be unpacked following analysis in order to be interpreted more easily. This process is called *rescaling* or *non-dimensionalization* and is discussed further in Section 1.5 below.

In articulating the simplifying assumptions involved in forming a model, one should consider the consequences of introducing these "errors." For example, what do we lose from the results of a model if we assume homogeneous mixing of sediments in a lake, or of a human population? The answers here, of course, may be very different: in the first case, we lose the ability to distinguish the effects of sediment build-up as a function of depth, or distance from the shore, including the resulting distribution of creatures in the lake that are affected by those sediments (small animals may hide from predators in the sediment, small animals and plants may even eat it; others, like the predators, may be hindered by it). In the second case we lose our ability to determine to what extent some individuals, who come into contact with more people than most, such as a supermarket checker or a delivery person, may be instrumental in the transmission of some contact-based process, such as an infectious disease or the spreading of news and rumors by word of mouth.

There is, finally, another type of simplification which is sometimes made without any intention of biological justification. In order to make inroads on a tough problem, we may deliberately oversimplify our model — for instance, by setting one or more quantities to zero — when the simpler model is easier to analyze, with the purpose of bringing those results back to the original problem to inform our intuition and expectations of how the full analysis will turn out. For example, if there are two processes taking place and one proceeds much more slowly than the other, we may assume temporarily that the "slow" variables remain constant. We mention this practice (discussed in Chapter 6) simply because it can be a useful tool for analyzing models at the edge of our ability to handle, and because it might not occur to modelers focused on justifying biologically all mathematical simplifications (the difference here being that they are only temporary simplifications).

In any case, since models are imperfect descriptions of reality, unless our main focus really is close fitting of data, our ultimate goal for conclusions drawn from modeling should be a notion called *robustness*. This means that the biological insights obtained from a mathematical model should be independent of the particular mathematical structure and functions used, to the extent that we are uncertain what the exact structure or function should be. For example, as a population grows to fill its habitat, it runs up against the limitations on the habitat's resources, which begins to limit the population's ability to reproduce. This is true on scales anywhere from a Petri dish to the planet Earth. This limitation on the birth rate can be described mathematically by any function which tapers off eventually toward zero after an initial increase (see, e.g., Figure 1.2). The most common function used for this purpose is the quadratic that leads to the logistic equation, analysis of which shows that the population will eventually level off at a size determined by the resources. However, other functions of similar shape (that described above) and similar properties can be used instead, with the same results. Likewise, we might hope that stratifying a population into three or four age or risk categories will provide results qualitatively similar to those of a model which stratifies the population into ten or twenty classes. (We certainly would not like to analyze the latter model!) Another example is that of threshold quantities, mentioned above and discussed in Section 4.4. In this way, we justify to a limited extent the imperfection of our model.

Sometimes models make predictions that are surprising or even alarming. Such predictions from simple models imply a need for further careful study of the phenomenon, in order to see if more detailed models make similar predictions. Predictions which are robust across multiple models are more likely to be accurate (but no more accurate than the models) and should be taken seriously.

1.5 Scale, and choosing units

Last in our discussion of what biological modeling is like is a discussion of scale and units. When one decides to model a certain problem, such as the variable infectivity observed with HIV, the human immunodeficiency virus, it is necessary to choose the scale on which to study the problem. Accordingly, the state variables involved should have appropriate units. At one extreme, we might choose to focus on the molecular level, in units of nanometers and seconds — the biochemistry of the immune system's initial response to the viral invasion. Moving up a level, we might instead focus on the cellular level, viewing the interactions of individual virus cells with helper T-cells and infected cells in units of microns, minutes and days. Moving up further, we might develop a model in the space of a single organ or individual, in units of centimeters and days to years. Beyond this, we might focus on the interactions of a small collective of individuals, in units of meters and weeks or months. Finally, at the upper extreme, we might make a model at the level of an entire population, in units of kilometers and weeks to years. Note that each jump in scale also effectively uses the lower level as the building block for the higher level: we model (and count) molecules interacting within a cell, cells interacting within an organ or individual, individuals interacting within a collective, and collectives (or individuals) interacting within a population. This spectrum is illustrated in Figure 1.3.

Arguably the least common, and to some extent newest, of the levels in this proposed spectrum is that of the *collective* — a small group within a population which functions effectively as a unit with regard to the entire population, and within which individuals interact much more with each other

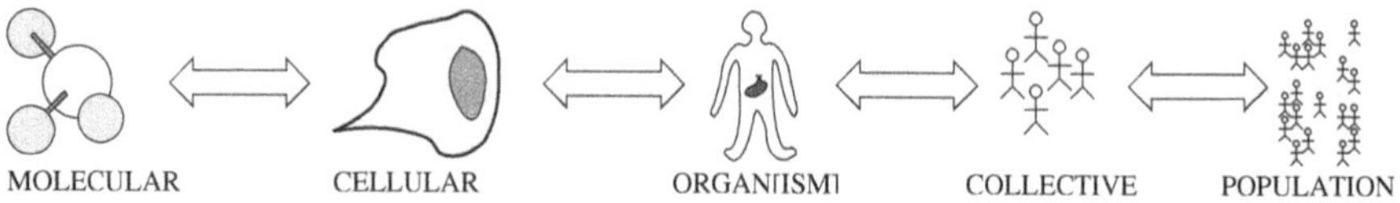

MOLECULAR CELLULAR ORGANISM COLLECTIVE POPULATION

FIGURE 1.3: A diagram showing the spectrum of scale levels on which modeling can take place.

than with others outside the collective. Recent research in mathematical biology has identified numerous situations in which collectives are important: for example, individual households in studying infectious diseases. Within a household, infections tend to spread faster among household members than between a household member and an individual outside the household. This is true of human respiratory diseases, as many families know, but also of pests such as mice, which carry hantaviruses, and insects, which act as vectors for diseases such as malaria (mosquitoes) and Chagas' disease (triatomines or "kissing bugs"). At the population level, it then also becomes reasonable to make households the unit of measurement, and count infected houses rather than infected individuals, with recovery coming only when no individuals in a given household remain infected, or when treatment has eliminated all the pests in the house.

Another issue related to scale and units is the rescaling procedure mentioned in the previous section. Sometimes we find it possible to reduce the mathematical complexity of a model by reducing the number of variables and/or parameters. Although the rescaled model can be more difficult to interpret immediately, the resulting simplifications in our calculations are typically worth it, especially when we can always undo the transformation afterward to facilitate interpretation. Probably the most common type of rescaling is to redefine variables relative to benchmark parameters. For example, in the model of foraging ants mentioned in the previous section (and explored more fully in Section 4.1), Sumpter and Pratt also rescaled both the state variable $E(t)$, representing the number of ants exploiting the food site, and the independent variable t for time, as follows:

$$x = \frac{E}{K}, \quad \tau = \frac{t}{K/\sigma},$$

so that both x and τ are dimensionless. That is, x gives the number of exploring ants as a multiple of the benchmark value K, and τ gives the time as a multiple of the benchmark value K/σ (which is how much time it takes, on average, for K ants to retire). This is why the process is also called nondimensionalization. (The transformation to x in the article also involves elimination of another state variable for the number of ants not exploiting the food source by assuming the total ant population constant, another simplifying technique mentioned in the previous section.)

Exercises

Consider the following biological systems from a modeling perspective. For each, determine which type of mathematical model and analysis you believe would be most appropriate with regard to the criteria developed in this chapter:

- discrete or continuous independent variable(s) (time, age, space),
- discrete or continuous dependent variable (population or quantity),
- type of stratification or continuous distribution of traits, if any,
- deterministic or stochastic,
- quantitative (numerical) or qualitative analysis,
- scale and units.

Defend or justify each of your choices with a sentence. (This exercise is most useful if discussed in a group setting after each person has written down his/her choices.)

1. the number of salmon that return to a particular pool to spawn each year
2. a population of pea aphids whose genetic resistance to infection by fungal spores varies inversely with their resistance to attack by parasitic wasps (studied by students in the paper by Smith, Haarer and Confrey)
3. the frequency of a particular allele within each generation of a population
4. the likelihood of extinction of a rare allele (for reasons other than genetic) over many generations of a population
5. the geographic dispersal (movement) of house sparrows within the United States, starting with the original eight pairs released in the spring of 1851 in Brooklyn, New York
6. the spread of tall grasses around the shore of a pond from an isolated cluster
7. seasonal competition for space among different species of trees in a rainforest
8. weekly harvesting (removal) of fish in a fish hatchery
9. the growth of a yeast culture in a Petri dish
10. the mixing and growth of bacteria and nutrients in a chemostat
11. the growth of a deer population subject to an annual hunting season

12. the spread of an epidemic in a large city
13. the role of "super-spreaders" in an epidemic's growth
14. the growth of a population of blue whales, which become sexually mature at age 10 years
15. the amount of a given drug remaining in the body over time
16. the kinetics of glucose, insulin and beta cells in the blood of diabetics
17. electrocardiac regulation
18. fluctuations in shark and fish populations in the Mediterranean (the sharks eat the fish)
19. population density control as a means of eradicating fox rabies
20. comparing two alternative explanations for the mechanisms underlying cell contraction
21. the concentrations of activator and inhibitor chemicals in pattern formation processes in the hydra, a multicellular water-based creature
22. uptake mechanisms for the diagnostic dye bromosulfophthalein into liver cells from the blood
23. competition between two species of bird for the same food source and nesting sites
24. the gradual degradation of the human immune system by HIV

The following three questions provide further practice in graph interpretation.

25. Answer the following questions about Figure 1.2.
    (a) What is the significance of the point where the birth rate curve crosses the x-axis?
    (b) What is the significance of the point where the birth rate curve has a horizontal tangent?
    (c) If the population begins at a level greater than that where the two curves cross, what will happen to its size?
    (d) If the population begins at a level less than that where the two curves cross, what will happen to its size?
26. In what important ways does Figure 1.4 below differ from Figure 1.2? Consider what will happen to the size of the population if it begins in each of the three regions marked (a), (b) and (c). (This graph exhibits something called an *Allee effect*, which is explored in Section 2.2.)

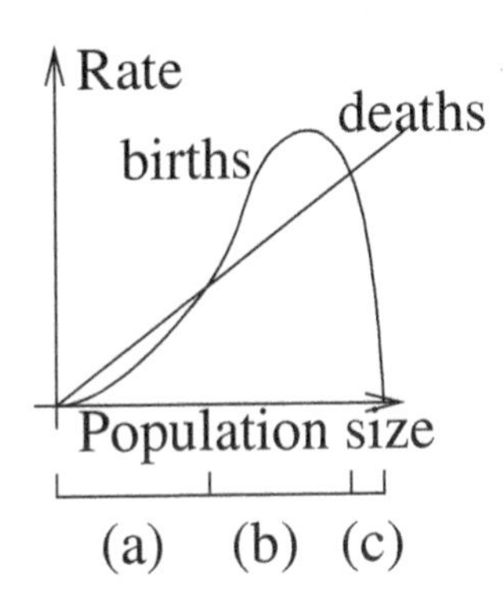

FIGURE 1.4: Graph for Exercise 26.

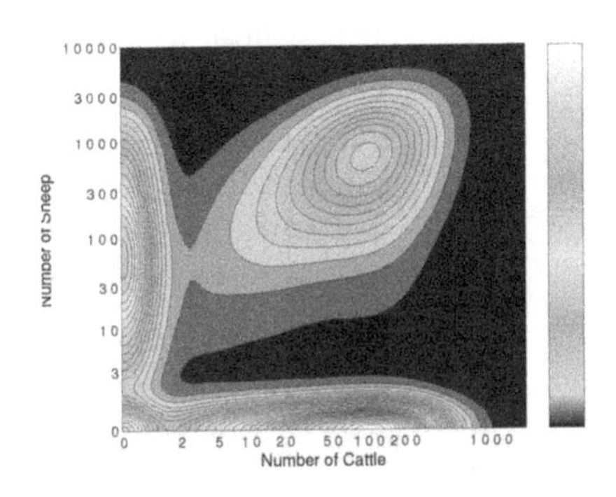

FIGURE 1.5: Graph for Exercise 27.

27. Figure 1.5 shows the distribution of farms in the U.K. during the 2001 outbreak of foot-and-mouth disease (FMD), in terms of the numbers of cows and sheep on each farm. Color intensity indicates the number of farms with the given numbers of animals.

    (a) Where, on this graph, are the farms with large numbers of cows but no sheep? Where are the farms with many sheep but no cows?

    (b) What does the graph indicate about the distinct types of farms that exist?

    (c) Since there is no treatment for FMD, the epidemic was brought under control by culling — slaughtering all susceptible animals on farms within a certain radius of any animal found to be infected. This wiped out numerous small farms altogether. How might the small farmers use this graph to argue for a change in control policy?

28. Smith, Haarer and Confrey (1997) studied a group of graduate students working together in a course on mathematical biology (see footnote 1). They found that the goals of students from different fields, as well as the course instructors, who had different backgrounds, differed at times. If we consider three perspectives from which students using mathematical models to represent biological systems might come, we might say that:

    Those who came with the perspective of an *experimental biologist* wanted models to incorporate all their data, and to fit and explain them.

    Those who came with the perspective of a *theoretical biologist* wanted to use the data obtained beforehand to suggest trends to explore in the model, to use the model as a means of exploring the effects of particular phenomena (like genetic variation) across particular systems, i.e., for many different species and habitats. They saw the model as a virtual experiment, whose results become a single datum in the identification of how general biological systems behave.

Those who came with the perspective of a *mathematical biologist* wanted their models to be analyzable, focused only on research-related features of the system rather than broadly descriptive, and were interested by mathematical results and their biological significance.

Compare these goals and purposes with each other. To what extent are they compatible? To what extent would the models produced by each differ (that is, the models themselves, as opposed to the uses to which they are put)?

29. Suppose you are investigating the effects of competing predators on the life cycle of the alewife, a freshwater fish that serves as prey for other fish such as salmon, trout and bass. Make a list of the five factors you expect would be most important to include in a mathematical model for this biological system, and another list of five factors which, while relevant to the life cycle of the alewife, you would exclude from this model. Justify both lists biologically.

30. Part of the homogeneity assumption discussed in this chapter is an assumption that when some particular type of contact occurs between members of two different groups — say, a species of predator and its prey, or an infectious individual and a susceptible individual — it is equally likely for the contact to involve any member of each group. In reality, individuals don't move about randomly. List five factors which affect which member of the first group is likely to be the one which interacts with a member of the second group, and indicate some way a model might take each factor into account.

31. One simplification that most models of biological systems make (and certainly those in this book) is to assume that events take place instantaneously. That is, infectious contacts between two individuals last only a moment, and predators and parasites locate and catch their prey immediately. (The alternative is to have the model remove the two individuals involved from their respective classes for the time necessary to make the contact, and after the given delay return the survivor(s) to those classes, which complicates the model.) What biological justification can be made for this simplification? In what cases might this assumption introduce important inaccuracies in the model and its behavior?

Chapter 2

Difference Equations (Discrete Dynamical Systems)

2.1 Introduction to discrete dynamical systems

Many organisms have births, deaths, and other demographic processes that occur at distinct, regular intervals (see Figure 2.1). Perhaps best-known among these are *semelparous* animals that reproduce only once before dying, such as the Pacific salmon (five species of the genus *Oncorhynchus*), which battles its way back upstream from the sea to spawn and die in the river where it was born. Likewise, the cicada family Cicadidae, which includes some 1500 species worldwide, includes the genus *Magicicada*, native to North America, whose members emerge from their underground hibernation every 13 or 17 years to mate and die. Analogues also exist in the plant kingdom, such as western North America's monument plant (also known as green gentian) *Frasera speciosa* and century plant (also known as Parry's agave) *Agavi americana*, both of which live for over twenty years before growing a tall, thin blooming spike which produces flowers and seeds but saps the plant's resources to the point that it dies afterward. Such plants are called *monocarpic*. Some plants, such as the carrot *Daucus carota*, also have fixed life cycles. The carrot and other similar plants are biennial: at the end of the first year, they store nutrients in the root that support flowering in the second year, at the end of which they die. On a smaller scale, there are processes such as ovulation and even neuron firing which occur discretely at certain set frequencies, rather than continuously. Finally, human interactions with some populations can also cause discrete cycles, such as harvesting or wildlife population management programs. This includes situations in which empirical studies necessitate a discrete perspective on time because the only data available have been gathered at regular intervals, making everything that happens in between observations equivalent to one demographic fell swoop.

In cases such as these where it is appropriate to model some biological quantity or process in terms of the number or amount present at discrete times, we shall use models which are *discrete dynamical systems*, that is, systems which evolve over time through equations (formulas) which tell how to figure the size of the next generation, based on the size of the present generation.

Photo courtesy U.S. Fish and Wildlife Service, dls.fws.gov

Photo courtesy Joel Michaelsen

FIGURE 2.1: The semelparous salmon (left) and the monocarpic century plant (right) reproduce in distinct generations. Note the different stages of the blooming spikes on the three century plants.

These equations are called *difference equations*, since they give the difference in size between one generation and the next, and *first-order* difference equations[1] have the form $y_{k+1} = f(y_k)$, which can be read as saying that the size of the next generation (generation number $k+1$ of population y) is a function f of the present generation (generation number k of population y). In the rest of this section, we will discuss the simplest type of difference equation and what its solutions look like. Following that, the next sections in this chapter develop some tools, both analytical and computational, for studying more complicated difference equations. Later sections will look at classes of biological systems that can be studied using discrete dynamical systems, including population genetics, competition, harvesting, and finally systems involving more than one quantity or population.

2.1.1 Linear difference equations

The rate of change of some quantity is often proportional to the amount of the quantity present. This may be true, for example, of the size of a population with unrestricted growth (say, lab bacteria in a petri dish). Let $y(t)$ represent the number of members of a population of simple organisms at time t. If we assume that these organisms reproduce by splitting, and that on average a fraction a of the members split into two members in unit time, then for a small period h of time

$$y(t+h) - y(t) \approx a\,h\,y(t), \tag{2.1}$$

where the symbol $\approx$ signifying approximate equality means that the error in this approximation is small for small h, in the sense that this error divided by

[1] First-order difference equations will be our primary focus in this chapter.

h approaches zero as $h \to 0$, i.e., that

$$\lim_{h\to 0} \frac{y(t+h) - y(t) - a\,h\,y(t)}{h} = 0.$$

For now, we neglect the small error and model the population size by the equation

$$y(t+h) = y(t) + a\,h\,y(t). \tag{2.2}$$

In this way we can predict $y(t+h)$ given $y(t)$ and a.

If a fraction b of the members reproduce by splitting and a fraction d of the members die in unit time, then (2.1) would be true with $a = b - d$. The constant a may be either positive or negative, depending on whether $b > d$ (more births than deaths) or $b < d$ (more deaths than births). We will therefore allow a to designate the constant of proportionality in (2.1) or (2.2), whether it is positive or negative.

We may rewrite (2.2) in the form

$$y(t+h) = (1 + ah)y(t), \tag{2.3}$$

which expresses $y(t+h)$ as a function of $y(t)$. In this chapter we shall think of h as a fixed time interval, so that if we start at time $t = 0$ the function y is defined not for all t but only for $t = kh$ $(k = 0, 1, ...)$. In other words, we are observing the value of y only at regular intervals, such as every hour, or every twenty-four hours. If we define $t_k = kh$, the kth observation time, and let $y_k = y(t_k)$, the value observed at that time, then the function y is described by the sequence of values $\{y_k\}$. Since $t_k + h = t_{k+1}$, (2.3) becomes the difference equation

$$y_{k+1} = (1 + ah)y_k \;\; (k = 0, 1, 2, \ldots). \tag{2.4}$$

In order to simplify the notation, we let $r = 1 + ah$, so that (2.4) becomes

$$y_{k+1} = ry_k \;\; (k = 0, 1, 2, \ldots). \tag{2.5}$$

By a *solution* of the difference equation (2.5) we mean an algebraic expression which gives us values for all the y_k $(k = 0, 1, \ldots)$. If we know the initial population size y_0 then we can calculate first $y_1 = ry_0$, then $y_2 = ry_1 = r^2 y_0$, etc. Thus we may solve the difference equation (2.5) *recursively.* The specification of the value y_0 is called an *initial condition.* The graph of the solution is the discrete set of points $\{(t_k, y_k), k = 0, 1, 2, \ldots\}$, but it is customary to connect these points with line segments to give a continuous graph.

If we return to the bacteria multiplying in the petri dish, and measure time in multiples of the bacteria's doubling time, and measure y in multiples of the original colony size, then Example 1 gives a model for this simple doubling process. Example 2 gives a variation on this theme, with time steps during which the colony size increases by 10%, beginning with a colony five times the size of that in Example 1.

EXAMPLE 1.
Solve the difference equation $y_{k+1} = 2y_k$, with $y_0 = 1$.

Solution: From $y_1 = 2y_0$ (the difference equation with $k = 0$) we see that $y_1 = 2y_0 = 2$. Then $y_2 = 2y_1 = 4$, $y_3 = 2y_2 = 8$, ... We may then guess (and prove by induction) that $y_k = 2^k$. This is the solution of the given problem. Graphing will show the familiar exponential curve. □

EXAMPLE 2.
Verify that $y_k = 5(1.1)^k$ is the solution of the difference equation $y_{k+1} = (1.1)y_k$, $y_0 = 5$.

Solution: First verify the initial condition: The expression $y_k = 5(1.1)^k$ with $k = 0$ gives $y_0 = 5(1.1)^0 = 5$, as desired. Now substitute the proposed solution into the difference equation: $y_{k+1} = (1.1)\left[5(1.1)^k\right] = 5(1.1)^{k+1}$. Thus the given y_k satisfies the given difference equation and initial condition. □

Equation (2.5) fits the general form $y_{k+1} = f(y_k)$ of a first-order difference equation; it also belongs to a more specific class of equations of the form

$$y_{k+1} = ry_k + b, \tag{2.6}$$

called *linear* difference equations because the right-hand side $f(y_k)$ is a linear function of y_k. The special case (2.5) where $b = 0$ is known as the *homogeneous* case. These simplest types of difference equation have special names for two reasons: first, as we shall see below, they are relatively easy to solve outright (in fact, they are the only kind of difference equation we shall attempt to solve explicitly); and, second, as we shall see in a later section, they turn out to be key to understanding the behavior of more complicated difference equations.

2.1.2 Solution of linear difference equations

It is easy to find an explicit formula for the solution of the linear homogeneous difference equation with constant coefficients (2.5) for any constant r. We merely observe that

$$y_1 = r\,y_0,\;\; y_2 = r\,y_1 = r^2\,y_0,\;\; y_3 = r\,y_2 = r^3\,y_0,\;\; \ldots$$

Then it is natural to guess that in general

$$y_k = r^k\,y_0, \tag{2.7}$$

and it is easy to verify (by induction) that this is correct.

EXAMPLE 3.
Find the solution of the difference equation $y_{k+1} = -y_k$, $y_0 = 1$.

Solution: From the formula (2.7) with $r = -1$ we have $y_k = (-1)^k$; notice that y_k oscillates between positive and negative values. □

EXAMPLE 4.
Find the solution of the non-homogeneous linear difference equation

$$y_{k+1} = -y_k + 1, \; y_0 = 1.$$

Solution: We begin by calculating

$$y_1 = -y_0 + 1 = 0, \;\; y_2 = -y_1 + 1 = 1, \;\; y_3 = -y_2 + 1 = 0,$$

and then conjecture that y_k alternates between 0 and 1. To verify the correctness of this conjecture, we need only note that if $y_k = 1$, then $y_{k+1} = 0$ and if $y_k = 0$, then $y_{k+1} = 1$. Thus y_k does alternate between 0 and 1, which we may express explicitly as $y_k = \frac{1}{2}[1 + (-1)^k]$. □

It is also not difficult to solve the general linear non-homogeneous difference equation (2.6) with constant coefficients r and b. We calculate

$$\begin{aligned} y_1 &= r\, y_0 + b \\ y_2 &= r\, y_1 + b = r(r\, y_0 + b) + b = r^2\, y_0 + r\, b + b \\ y_3 &= r\, y_2 + b = r(r^2\, y_0 + r\, b + b) = r^3\, y_0 + r^2\, b + r\, b + b, \quad \text{etc.} \end{aligned}$$

The formula for the general term is

$$y_k = r^k\, y_0 + b\left(1 + r + r^2 + \ldots + r^{k-1}\right). \tag{2.8}$$

By using the formula

$$1 + r + r^2 + \ldots + r^{k-1} = \frac{1 - r^k}{1 - r}$$

for the sum of a geometric series (with $r \neq 1$), we may write this solution in the form

$$y_k = r^k\, y_0 + b\frac{1 - r^k}{1 - r} = \left(y_0 - \frac{b}{1 - r}\right) r^k + \frac{b}{1 - r}. \tag{2.9}$$

We may also consider what happens to solutions of linear difference equations over long periods of time – in mathematical terms, as $k \to \infty$. Considering the solution (2.9), we see that if r is large in size, i.e., $r > 1$ or $r < -1$, then r^k grows unbounded as $k \to \infty$, and thus y_k grows unbounded too. The only exception occurs when $y_0 = \frac{b}{1-r}$, in which case y_k is a constant $\frac{b}{1-r}$ for all k (cf. (2.9)) and thus approaches $\frac{b}{1-r}$ as $k \to \infty$. Small differences, however, from this *equilibrium solution* will be magnified for large r (cf. again (2.9)).

If r is instead small in size (between -1 and 1), then r^k approaches zero as $k \to \infty$, and in view of (2.9) the solution y_k approaches the limit $\frac{b}{1-r}$ as $k \to \infty$ regardless of the initial value y_0.

If $r = -1$, r^k alternates between -1 and 1, and y_k does not have a limit (unless $y_0 = \frac{b}{1-r}$). If $r = 1$, the formula (2.9) is meaningless, but the formula (2.8) becomes $y_k = y_0 + rk$; thus y_k becomes unbounded as $k \to \infty$.

The way the solution of the linear homogeneous difference equation (2.5) depends on the value of the survival-and-growth term r will be important when we study qualitative behavior of solutions of difference equations in Section 2.3. From the formula (2.7) we see that $y_k \to 0$ as $k \to \infty$ if $|r| < 1$ and y_k grows unbounded as $k \to \infty$ if $|r| > 1$. More precisely, if $0 \leq r < 1$, y_k decreases monotonically to zero and if $-1 < r < 0$, y_k oscillates between positive and negative values in approaching zero. If $r > 1$, y_k increases to $+\infty$ and if $r < -1$, y_k oscillates unboundedly. The "boundary" cases are $r = -1$, in which y_k oscillates between $\pm y_0$ and does not approach a limit, and $r = 1$, in which y_k is the constant y_0. The essential property we shall need is the following result.

> The solution of the difference equation $y_{k+1} = r\ y_k$ approaches zero as $k \to \infty$ if and only if $-1 < r < 1$.

EXAMPLE 5.
Find for which values of a every solution of the difference equation $y_{k+1} = (1 + a)y_k$ approaches zero as $k \to \infty$.

Solution: Every solution approaches zero if and only if $|1 + a| < 1$, or $-1 < 1 + a < 1$, or $-2 < a < 0$. □

2.1.3 Nonlinear difference equations

Note that model (2.5) assumes a constant growth rate independent of population size. This assumption is unlikely to be reasonable for real populations, except possibly while the population is small enough in size not to be subject to the effects of overcrowding. Various nonlinear difference equation models have been proposed as more realistic in the general case, where resource limitations constrain growth. For example, the difference equations

$$y_{k+1} = \frac{ry_k}{y_k + A} \tag{2.10}$$

due to Verhulst,[2] and the second-order Hill function

$$y_{k+1} = \frac{ry_k^2}{y_k^2 + A^2} \tag{2.11}$$

have been suggested as descriptions for populations whose growth rates saturate for large population sizes. Here A is the population size at 50% saturation

[2] P. F. Verhulst, Récherches mathématiques sur la loi d'accroissement de la population, *Mem. Acad. Roy. Brussels* 18 (1845), 1–38.

(plug in A for y_k in (2.10) to see why). For small y_k (relative to A), the denominator of the right-hand side of (2.10) and (2.11) is essentially A, so that (2.10) has $y_k \approx \frac{r}{A}y_k$, similar to the unrestricted growth of the linear equation (2.5). However, for large y_k, A is relatively insignificant, making the right-hand sides of (2.10) and (2.11) (and thus y_{k+1}) approximately equal to r.

Another much-studied example is the logistic difference equation

$$y_{k+1} = ry_k\left(1 - \frac{y_k}{K}\right), \tag{2.12}$$

also introduced by Verhulst, with a growth rate which decreases to zero as y_k approaches the carrying capacity K and which becomes negative for $y_k > K$. The logistic difference equation should not be taken seriously as a model for large population sizes as y_{k+1} is negative if $y_k > K$. Other difference equations which have been used as models to fit field data are

$$y_{k+1} = ry_k(1 + \alpha y_k)^{-\beta} \tag{2.13}$$

and

$$y_{k+1} = \begin{cases} ry_k^{1-\beta}, & \text{for } y_k > \epsilon, \\ ry_k, & \text{for } y_k < \epsilon. \end{cases} \tag{2.14}$$

None of the difference equations (2.10), (2.11), (2.12), (2.13), (2.14) are derived from actual population growth laws. Rather, they are attempts to give quantitative expression to rough qualitative ideas about the biological laws governing population growth. For this reason, we should be skeptical of the biological significance of any deduction from a specific model which depends on the precise formula for the solution of that model. Our goal should be to formulate principles which are *robust*, that is, which are valid for a large class of models embodying some set of qualitative hypotheses. Therefore, we shall be more concerned with *qualitative* properties of solutions of difference equations than with formulae for solutions. In fact, although we have seen some examples of linear difference equations for which a solution formula is available, the solution of nonlinear difference equations like (2.10), (2.11), (2.12), (2.13), and (2.14) usually cannot be found in general as functions of the equation parameters. In Section 2.3 we will develop tools for analyzing the behavior of these more complicated equations.

The most that we can do *quantitatively* with many nonlinear difference equations is to calculate solutions numerically by iteration for particular choices of parameter values, as illustrated by the following two examples.

EXAMPLE 6

Find the first four terms of the solution of the difference equation $y_{k+1} = y_k(1 - y_k)$ with $y_0 = \frac{1}{2}$.

Solution: We have $y_1 = \frac{1}{2}(1 - \frac{1}{2}) = \frac{1}{4} = 0.25$, $y_2 = \frac{1}{4}(1 - \frac{1}{4}) = \frac{3}{16} = 0.1875$, $y_3 = \frac{3}{16}(1 - \frac{3}{16}) = \frac{39}{256} = 0.152344$, $y_4 = \frac{39}{256}(1 - \frac{39}{256}) = 0.129135$. □

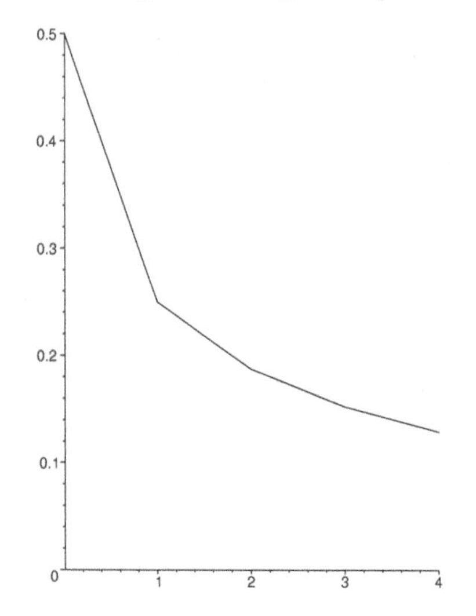

FIGURE 2.2: Solution to Example 6.

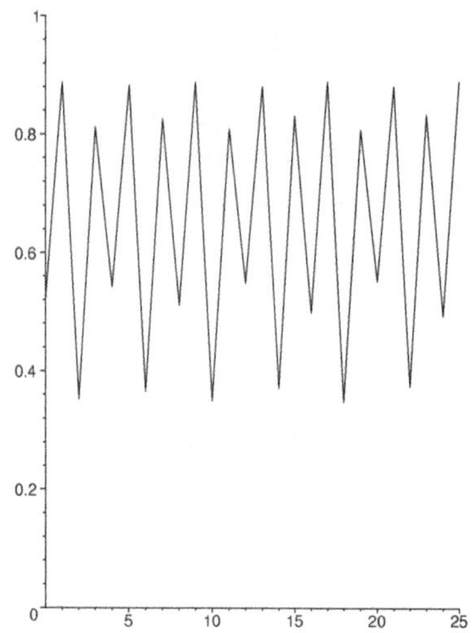

FIGURE 2.3: A solution to $y_{k+1} = 3.62y_k(1 - y_k)$.

EXAMPLE 7.
Verify that the constant sequence $y_k = \frac{1}{2}$ $(k = 0, 1, 2, \ldots)$ is a solution of the difference equation $y_{k+1} = 2y_k(1 - y_k)$.

Solution: If $y_k = \frac{1}{2}$, then $2y_k(1-y_k) = 2(\frac{1}{2})(1-\frac{1}{2}) = \frac{1}{2}$. Thus $y_k = \frac{1}{2}$ satisfies $y_{k+1} = 2y_k(1 - y_k)$. $\square$

Solutions of difference equations may be calculated and graphed easily using a computer algebra system such as Maple™, Mathematica®, or MATLAB®. Again, although a graph of a solution is technically a set of discrete points (k, y_k), we shall follow the customary procedure of connecting the points and presenting the graph as a continuous sequence of line segments. Figure 2.2, for example, shows the solution of the difference equation of Example 6 graphed with the aid of Maple; a program for this is given in Appendix A. Figure 2.3 gives an example which would be more difficult to calculate by hand, namely 25 terms of the solution of the difference equation $y_{k+1} = 3.62y_k(1 - y_k)$ with $y_0 = 0.52$. In the next section we shall present a graphical approach to analyzing difference equations.

Exercises

1. The bacteria *E. coli* doubles in a little over 20 minutes in good lab conditions. (For purposes of this exercise, we will assume it takes exactly 20 minutes.)

   (a) Write a difference equation that models the growth of an *E. coli* colony in a large nutrient dish, with an initial population of $y_0 = 10000$ bacteria, and time measured in increments of 20 minutes. What is the solution to this equation and initial condition?

(b) How would the equation change if the time between measurements was one hour instead of 20 minutes? (Consider what happens to the size of the colony in one hour.)

(c) In light of the discussion in the previous chapter, why would these models be inappropriate for a colony beginning with a single bacterium, or only a few?

(d) Why would these models be inappropriate for tracking the growth of the colony over a period of months?

(e) Suggest an alternate (nonlinear) model that might be more appropriate in tracking the growth of the colony from its inception until well after it fills the petri dish. Explain in biological terms any numbers or parameters.

2. Gressel and Segel[3] derived the following annual model for the density of a certain weed species (in weeds per square meter):

$$w_{n+1} = Fw_n + \mu,$$

where w_n is the density of the weed in year n, μ accounts for growth due to mutation from another, closely related species, and F is the ratio of the fitness of the weed species relative to that of its main competitor (so, e.g., $F > 1$ means the weed is more fit than its competitor).

(a) Solve this difference equation to obtain an expression for w_n in terms of μ, F, and the initial condition w_0.

(b) In light of the discussion in this section on difference equations of this form, what eventually (after many years) happens to the density of this particular weed if $F < 1$? Interpret this conclusion biologically.

(c) If instead $F > 1$, what happens to the weed density after many years? Interpret this conclusion biologically.

(d) Which of these conclusions are realistic consequences of the original assumptions, and which aren't? Evaluate this model, and give a range within which it gives a useful description of the weed's growth.

3. Find the first three terms of the solution to $y_{k+1} = y_k(1 - y_k)$, $y_0 = \frac{3}{4}$.

4. Find the first three terms of the solution to $y_{k+1} = y_k(1 - y_k)$, $y_0 = \frac{1}{4}$.

[3] J. Gressel and L.A. Segel, The paucity of plants evolving genetic resistance to herbicides: possible reasons and implications, *Journal of Theoretical Biology* 75: 349–371, 1978.

5. Find the first three terms of the solution to $y_{k+1} = \frac{y_k}{y_k+1}$, $y_0 = 1$.

6. Find the first three terms of the solution to $y_{k+1} = \frac{y_k^2}{y_k^2+1}$, $y_0 = 1$.

7. Verify that $y_k = 1$ is a constant solution of the difference equation $y_{k+1} = y_k\, e^{1-y_k}$.

8. Verify that $y_k = 1$ is a constant solution of the difference equation $y_{k+1} = \frac{2y_k}{y_k+1}$.

9. * Verify that if $r > 2$ the sequence given by

$$y_k = \frac{(r+2) + \sqrt{r^2-4}}{2r} \quad (k \text{ even})$$

$$y_k = \frac{(r+2) - \sqrt{r^2-4}}{2r} \quad (k \text{ odd})$$

is a solution of the difference equation $y_{k+1} = y_k + ry_k(1 - y_k)$.

10. * Show that a constant solution $y_k = \hat{y}$ $(k = 0, 1, 2, ...)$ of a difference equation $y_{k+1} = g(y_k)$ must satisfy the relation $g(\hat{y}) = \hat{y}$.

In the following exercises, find the solution of the given difference equation.

11. $y_{k+1} = (1.1)y_k$, $y_0 = 1$
12. $y_{k+1} = -(1.1)y_k$, $y_0 = 1$
13. $y_{k+1} = \frac{1}{2}y_k$, $y_0 = 1$
14. $y_{k+1} = \frac{1}{3}y_k$, $y_0 = 1$
15. $y_{k+1} = (1.1)y_k - 0.1$, $y_0 = 1$
16. $y_{k+1} = -(1.1)y_k + 0.1$, $y_0 = 0$
17. $y_{k+1} = \frac{1}{2}y_k - \frac{1}{2}$, $y_0 = 0$
18. $y_{k+1} = \frac{1}{3}y_k + 1$, $y_0 = 1$

In the following exercises, determine which difference equations have solutions that approach a limit as $k \to \infty$, and then find the limits in those cases.

19. $y_{k+1} = -0.2y_k$
20. $y_{k+1} = -\frac{1}{2}y_k$
21. $y_{k+1} = 0.2y_k + 1$
22. $y_{k+1} = -\frac{1}{2}y_k - 1$

2.2 Graphical analysis

There is a simple graphical method for solving difference equations, called the *cobwebbing method.* We shall illustrate the method first by applying it to the linear homogeneous difference equation

$$y_{k+1} = r\, y_k \tag{2.15}$$

which we have already solved analytically in Section 2.1. The method is also applicable to difference equations which cannot be solved analytically. It may be carried out on a computer with the aid of a computer algebra system such as Maple, Mathematica, or MATLAB. Some of the figures in this section were produced using Maple, with a program given in Appendix A.

This method begins by drawing in the $y - z$ plane the *reproduction curve,* which is the graph of the function on the right-hand side of our difference equation, and the line $z = y$, which we shall use for reflection purposes. For equation (2.15), the reproduction curve is the line $z = r\,y$ (taken from $r\,y_k$). Next, we mark y_0 on the y-axis, and go vertically to the reproduction curve (meeting it at height ry_0). The next step is to go horizontally to the line $z = y$, meeting it at the point (y_1, y_1). (Recall $y_1 = ry_0$ from (2.15).) Then we repeat the process, going vertically to the reproduction curve (at height ry_1), then horizontally to the line $z = y$ at the point (y_2, y_2), where $y_2 = ry_1$. Continuing in the same manner, we reach successively the values y_3, y_4, y_5, ... The graphic portrayal will have four different cases — $r > 1$ (Figure 2.4), $0 \le r < 1$ (Figure 2.5), $-1 < r < 0$ (Figure 2.6), $r < -1$ (Figure 2.7) — corresponding to different relative positions of the reproduction curve and the line $z = y$. In each case, the graphical solution illustrates the behavior already

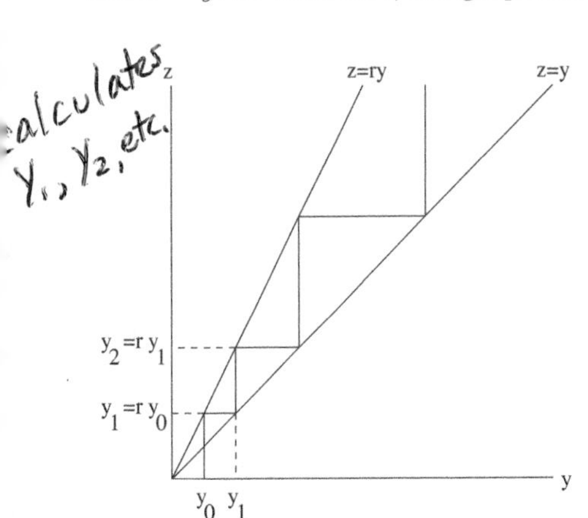

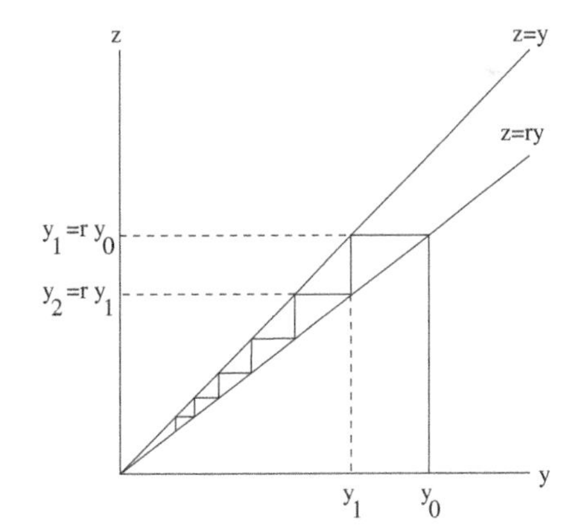

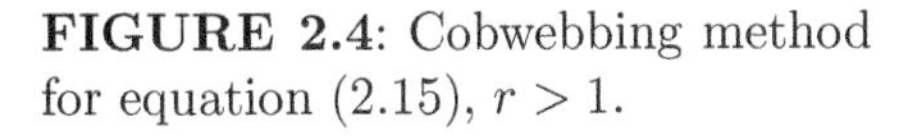
FIGURE 2.4: Cobwebbing method for equation (2.15), $r > 1$.

FIGURE 2.5: Cobwebbing method for equation (2.15), $0 < r < 1$.

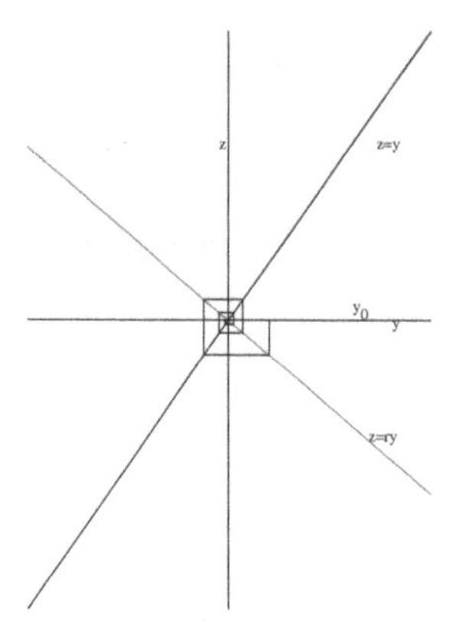

FIGURE 2.6: Cobwebbing method for equation (2.15), $-1 < r < 0$.

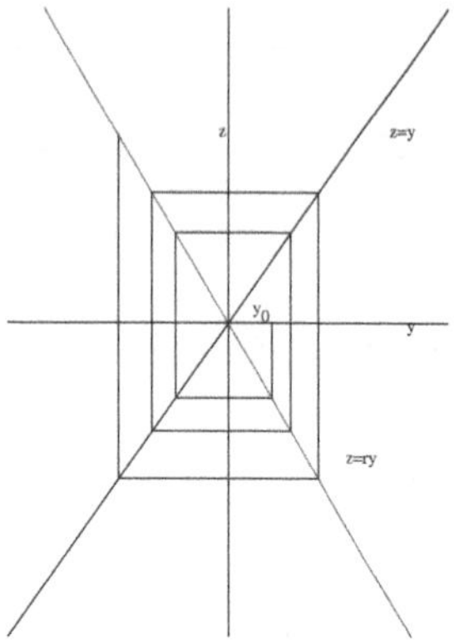

FIGURE 2.7: Cobwebbing method for equation (2.15), $r < -1$.

obtained analytically in Section 2.1. Note that we exclude the case $r = 1$, which would make (2.15) $y_{k+1} = y_k$.

The cobwebbing method may be applied to any difference equation of the form

$$y_{k+1} = g(y_k) \tag{2.16}$$

using the reproduction curve $z = g(y)$ and the line $z = y$. It gives information about the behavior of solutions and is particularly useful for difference equations whose analytic solution is complicated.

EXAMPLE 1.
Apply the cobwebbing method to describe the solutions of the Verhulst equation

$$y_{k+1} = \frac{r\,y_k}{y_k + A}.$$

Solution: The reproduction curve is $z = \frac{r\,y}{y+A}$ and its slope is

$$\frac{dz}{dy} = \frac{rA}{(y+A)^2}.$$

At $y = 0$ the slope is r/A. If $r < A$, this slope is less than 1, so the reproduction curve lies below the line $z = y$, while if $r > A$ this slope is greater than 1, so the reproduction curve begins above the line $z = y$ but comes down to intersect it at $y = r - A$. If $r > A$, every solution, regardless of the initial value y_0, approaches the limit $y_\infty = r - A$ (Figure 2.8), and if $r < A$, every solution approaches zero (Figure 2.9). In either case, the limit is an intersection of the reproduction curve and the line $z = y$. □ y_∞

Set $z = y$. $y = \frac{ry}{y+A}$ ← Solve for y.

$y(y+A) = ry$
$y^2 + Ay = ry$

$y^2 + Ay - ry = 0$
$y^2 + (A-r)y$
$y(y+A-r) = 0$

$y = 0$
$y = r - A$

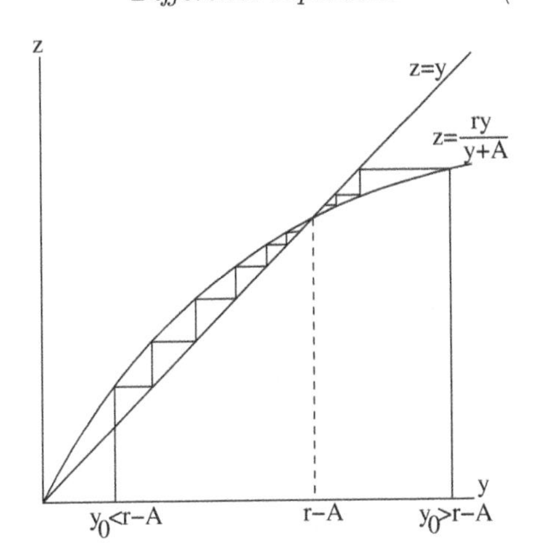

FIGURE 2.8: $r > A$ in Example 1.

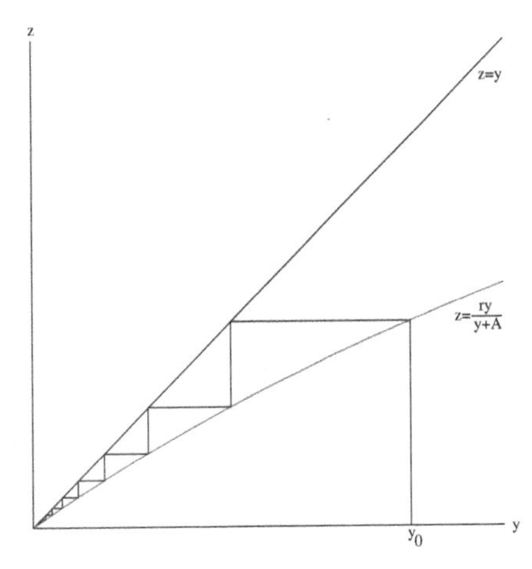

FIGURE 2.9: $r < A$ in Example 1.

EXAMPLE 2.
Apply the cobwebbing method to describe the solutions of the difference equation

$$y_{k+1} = \frac{r\, y_k^2}{y_k^2 + A}.$$

Solution: The reproduction curve is

$$z = \frac{ry^2}{y^2 + A},$$

which intersects the line $z = y$ when $y = 0$ and

$$y = \frac{r \pm \sqrt{r^2 - 4A}}{2}.$$

Thus if $r > 2\sqrt{A}$ there are three real intersections and if $r < 2\sqrt{A}$ the only real intersection is at $y = 0$. If $r < 2\sqrt{A}$, every solution approaches zero (Figure 2.10). If $r > 2\sqrt{A}$, every solution with y_0 less than the smaller positive intersection approaches zero (Figure 2.11), and every solution with y_0 greater than the smaller positive intersection approaches the limit $y_\infty = \frac{r+\sqrt{r^2-4A}}{2}$ (Figure 2.12). This model has been used to describe a population which collapses if its initial size is too small but survives if the initial population size exceeds a threshold, which in this example is $\frac{r-\sqrt{r^2-4A}}{2}$. Such behavior in a population is called an *Allee effect*. As in Example 1, limits of solutions are intersections of the reproduction curve and the line $z = y$. □

In Section 2.5 we shall see how we can identify limits of solutions (defined in the next section as *equilibria*) graphically for different parameter values, by this same superposition of two curves.

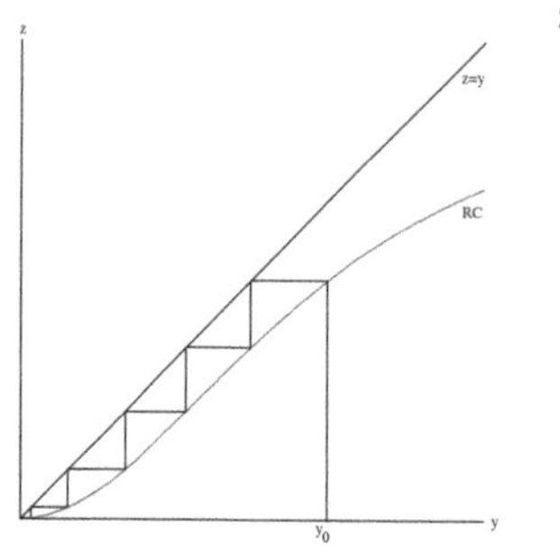

FIGURE 2.10: $r < 2\sqrt{A}$ in Example 2.

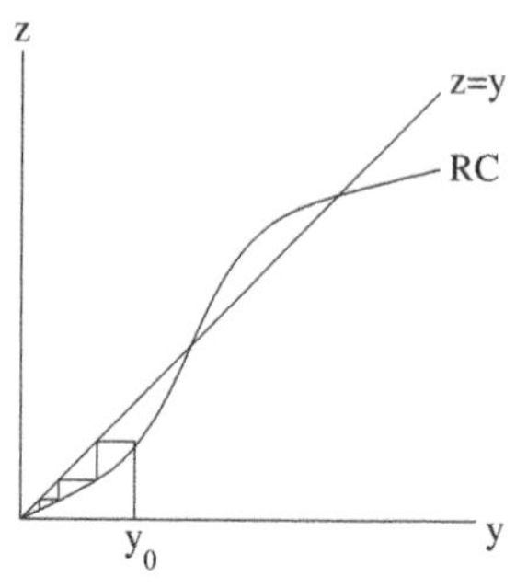

FIGURE 2.11: $r > 2\sqrt{A}$, $y_0 < y_-$ in Example 2.

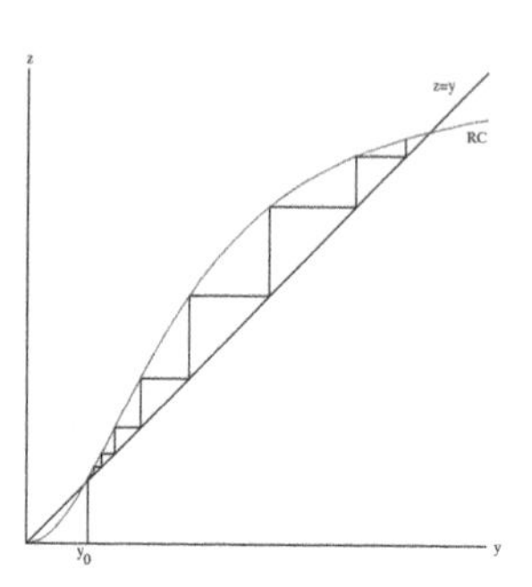

FIGURE 2.12: $r > 2\sqrt{A}$, $y_0 > y_-$ in Example 2.

Exercises

In Exercises 1-6, apply the cobwebbing method to describe the solutions of the given difference equation.

1. $y_{k+1} = 2y_k + 3$
2. $y_{k+1} = -3y_k + 1$
3. $y_{k+1} = \frac{1}{2}y_k(1 - y_k)$
4. $y_{k+1} = y_k + 2y_k(1 - y_k)$
5. $y_{k+1} = y_k e^{1-y_k}$
6. $y_{k+1} = \frac{2y_k}{4y_k+1}$

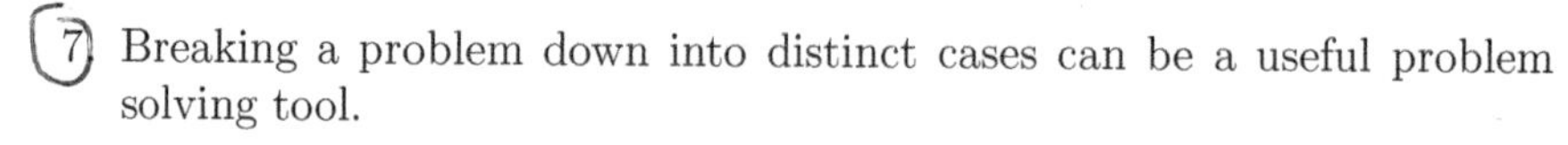

7. Breaking a problem down into distinct cases can be a useful problem solving tool.
   (a) Why are there four different cases to consider in cobwebbing the generic homogeneous linear difference equation (2.15)?
   (b) At what two points in Example 2 does the solution split into two cases or subcases?

8. (a) What happens mathematically to solutions of (2.15) when $r < 0$ (cf. Figures 2.6 and 2.7)?
   (b) What does this mean in biological terms?
   (c) Why do Examples 1 and 2 not consider the case $r < 0$?

9. Consider the nonlinear difference equation

$$y_{k+1} = 8y_k^3(1 - y_k),$$

which provides another way to model growth limitations for large populations (here large means closer to 1 than to 0, so consider the units of y to be the maximum population size at which reproduction can occur). Draw cobweb diagrams for three different initial conditions:

(a) $y_0 = 0.32$, (b) $y_0 = 0.64$, (c) $y_0 = 0.96$. Explain in biological terms what the model predicts will happen in each case. To what extent is this prediction realistic?

10. For most complicated models, we will typically apply qualitative analysis tools as developed in Section 2.3, to obtain information about the eventual behavior of solutions. However, there are cases in which even these methods are difficult to apply; in such cases, numerical exploration is again useful. Apply cobwebbing to the following two models of intraspecies competition (see Section 2.4) and try to describe qualitatively the behavior of solutions in each case.

    (a) The logistic model $y_{k+1} = ry_k(1 - y_k)$ with $r = 3.55$

    (b) Ricker's model $y_{k+1} = ry_ke^{-y_k}$ with $r = 3.55$

2.3 Qualitative analysis and population genetics

As remarked in the previous section, nearly all models of biological systems are nonlinear, and typically cannot be solved outright to obtain the population size (or other quantity) as a function of time (here, a discrete time index). We therefore need to develop tools that will allow us to describe the behavior of solutions qualitatively. In this section we introduce a technique for qualitative analysis based on a process called *linearization*, and apply it to some simple problems in population genetics.

2.3.1 Linearization and local stability

The motivation for linearization is fairly simple. Take any spot on a reasonably smooth curve (say, a curve with continuous first derivative) and zoom in on that spot on the graph. No matter how many zigs and zags the curve contains, if it is smooth then as you zoom in on the given point, restricting your attention to a smaller and smaller neighborhood of the chosen point, the graph will begin to straighten out, until it resembles — locally — a straight line. This suggests that, very close to any particular point, any smooth function behaves like a line (in particular, the line tangent to the curve at the given point). Figure 2.13 illustrates this concept for the graph of a particular curve. Consequently, within a very limited range, sufficiently smooth nonlinear dynamical systems behave like linear dynamical systems — and we already know how those behave. The utility of this approach turns out to depend on choosing the right points around which to linearize: a special kind of point called an equilibrium.

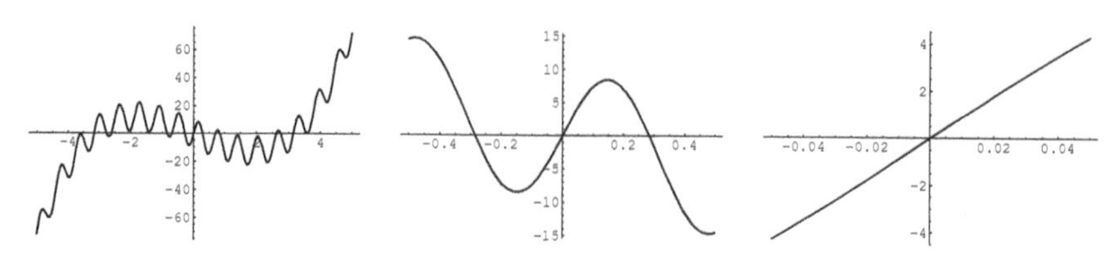

FIGURE 2.13: Three views of the function $y = x^3 - 10x + 10\sin 10x$, each centered at the origin. As we restrict our view to smaller x-intervals around 0, the graph appears increasingly flat.

The constant solutions of a difference equation

$$y_{k+1} = g(y_k), \tag{2.17}$$

if there are any, are particularly easy to find. They are simply the solutions of the equation $g(y) = y$. We define an *equilibrium* of the difference equation (2.17) to be a solution y_∞ of the equation $g(y) = y$. This is also referred to as a *fixed point* of the map g. Geometrically, an equilibrium is an intersection of the reproduction curve $z = g(y)$ and the line $z = y$. In the nonlinear difference equations used as examples in the previous section, we have observed a tendency for solutions to approach a limit as $k \to \infty$, and these limits are equilibria. Go back and look at the cobwebbing diagrams in the previous section to see this (note that in most cases, cobwebbing leads us to a point where the reproduction curve meets the line $z = y$). It will turn out that we can obtain a great deal of information about the solutions of a difference equation by studying the nature of its equilibria.

An essential fact in studying the behavior of solutions of a difference equation near an equilibrium is that a difference equation may be approximated by a linear difference equation near an equilibrium. If y_∞ is an equilibrium of the difference equation (2.17), so that $g(y_\infty) = y_\infty$, we make the change of variable

$$u_k = y_k - y_\infty$$

so that u_k represents the deviation from the equilibrium. Substitution into (2.17) gives

$$y_\infty + u_{k+1} = g(y_\infty + u_k).$$

By Taylor's Theorem (see Appendix B), we may write

$$g(y_\infty + u_k) = g(y_\infty) + g'(y_\infty)u_k + \frac{g''(c)}{2!}u_k^2$$

for some value c between y_∞ and $y_\infty + u_k$. Since $g(y_\infty) = y_\infty$, the difference equation (2.17) is equivalent to the difference equation

$$u_{k+1} = g'(y_\infty)u_k + \frac{g''(c)}{2!}u_k^2 \tag{2.18}$$

for u. The term $\frac{g''(c)}{2!}u_k^2$ is not really a quadratic function of u_k because the intermediate value c may depend on u_k, but it is small compared to u_k in the sense that this term divided by u_k approaches zero as $u_k \to 0$. The *linearization of* (2.17) *at the equilibrium* y_∞ is defined to be the homogeneous linear equation

$$u_{k+1} = g'(y_\infty)u_k \tag{2.19}$$

obtained from (2.18) by neglecting this "higher order" term. Note that the constant term drops out only because we are linearizing about an equilibrium, where $g(y_\infty) = y_\infty$.

EXAMPLE 1.
Find the linearization of the logistic difference equation

$$y_{k+1} = ry_k\left(1 - \frac{y_k}{K}\right)$$

at each of its equilibria.

Solution: The equilibria are the solutions of the equation

$$y = ry\left(1 - \frac{y}{K}\right),$$

namely $y = 0$ and $y = K(1 - \frac{1}{r})$ (the latter positive for $r > 1$). Here the function $g(y)$ is given by

$$g(y) = ry\left(1 - \frac{y}{K}\right),$$

and

$$g'(y) = r - \frac{2ry}{K}.$$

Thus $g'(0) = r$, and the linearization at the equilibrium $y = 0$ is $u_{k+1} = r\,u_k$. Also, $g'\left[K(1 - \frac{1}{r})\right] = r - 2r(1 - \frac{1}{r}) = 2 - r$, and so the linearization at $y = K(1 - \frac{1}{r})$ is $u_{k+1} = (2 - r)u_k$. □

The importance of the linearization lies in the fact that the behavior of solutions to the *linearization* of a difference equation at an equilibrium describes the behavior of solutions of the *original* difference equation *near* the equilibrium. In particular, what we want to know is, if a solution (the population size) is close to a particular equilibrium, will it approach the equilibrium, or will it go away from the equilibrium? This attractive-repulsive quality is called *stability*, and because it is based on linearizations, which are only good approximations of the original when we are very close to an equilibrium, the stability information we get from them is called *local stability*, as it does not (yet) tell us what happens when we start far away from an equilibrium.

An equilibrium of the difference equation (2.17) is said to be *asymptotically stable* if every solution whose initial value is sufficiently close to the equilibrium remains close to the equilibrium and approaches it as $k \to \infty$. An equilibrium is said to be *unstable* if there are solutions with initial values arbitrarily close to the equilibrium which fail to remain near the equilibrium. For an example, consider the generic homogeneous linear difference equation (2.15), $y_{k+1} = r\,y_k$, which has the unique equilibrium $y = 0$. As seen in the previous section, when $-1 < r < 1$ (the second and third cases, Figures 2.5 and 2.6), solutions approach the equilibrium at 0; in these cases, the equilibrium is asymptotically stable. However, when $r > 1$ or $r < -1$ (the first and fourth cases, Figures 2.4 and 2.7), solutions grow unbounded away from 0; in these cases, the equilibrium is unstable. This example actually provides the basis for the equilibrium stability theorem below, as all linearizations have this form. The two examples in the previous section can also be analyzed in this way (see, for instance, Example 3 below).

In applications to biological or other sciences, a difference equation model is only an approximation to reality. There are inevitably errors or approximations in the model, and there are experimental errors in the measurement of data to determine the parameters of the model. If an equilibrium is asymptotically stable, these errors are unimportant in the long-time behavior of the system, because a small movement away from the equilibrium will be wiped out over time. On the other hand, an unstable equilibrium is not experimentally observable because any small error will cause the solution to move away from it. Thus asymptotic stability of an equilibrium is an essential property for applications. (Sometimes errors or changes in parameter values can actually affect the stability of an equilibrium; these changes in stability, which occur for continuous-time systems also, are called *bifurcations* and are discussed in Section 4.4.)

It is now possible to establish the following important result, which we shall use often (we state it here without proof). It is valid not only for the first-order difference equations we have been studying here but also for systems of difference equations, and indeed for many other kinds of dynamical systems, including differential equations, which we shall meet in Chapter 3, and systems of differential equations, which we shall meet in Chapter 5.

LINEARIZATION THEOREM: If y_∞ is an equilibrium of the difference equation $y_{k+1} = g(y_k)$, and if all solutions of the linearization $u_{k+1} = g'(y_\infty)u_k$ at this equilibrium approach zero, then the equilibrium y_∞ is locally asymptotically stable. If the linearization has unbounded solutions, then the equilibrium y_∞ is unstable.

EXAMPLE 2.

Show that if $1 < r < 3$, the equilibrium $y = 0$ of the logistic difference equation

$$y_{k+1} = ry_k\left(1 - \frac{y_k}{K}\right)$$

is unstable and the equilibrium $y = K(1 - \frac{1}{r})$ is asymptotically stable. Show also that if $r > 3$ there is no asymptotically stable equilibrium.

Solution: As we have seen in Example 1, the linearization at the equilibrium $y = 0$ is $u_{k+1} = r\, u_k$. From Section 2.1 (see the boxed statement on page 28), we know that all solutions of this linearization approach zero if and only if $-1 < r < 1$. Thus if $r > 1$ the linearization has unbounded solutions, and by the linearization theorem the equilibrium $y = 0$ is unstable.

The linearization at the equilibrium $y = K(1 - \frac{1}{r})$ is $u_{k+1} = (2 - r)u_k$, and as we have seen in Section 2.1 all solutions approach zero if and only if $|2 - r| < 1$, or $-1 < 2 - r < 1$, or $1 < r < 3$. Thus the equilibrium $y = K(1 - \frac{1}{r})$ is asymptotically stable if $1 < r < 3$ and unstable if $r > 3$. Since both equilibria $y = 0$ and $y = K(1 - \frac{1}{r})$ are unstable if $r > 3$, there is no asymptotically stable equilibrium if $r > 3$. □

According to the linearization theorem, we can decide the asymptotic stability or instability of an equilibrium by examining the behavior of solutions of the linearization at the equilibrium. In Section 2.1 we analyzed completely the behavior of solutions of the linear homogeneous difference equation $u_{k+1} = r\, u_k$ (which every linearization resembles): if $|r| < 1$, every solution approaches zero, and if $|r| > 1$, there are unbounded solutions. This fact together with the linearization theorem gives the following result.

EQUILIBRIUM STABILITY THEOREM: Let y_∞ be an equilibrium of the difference equation $y_{k+1} = g(y_k)$. If $|g'(y_\infty)| < 1$, the equilibrium is asymptotically stable, and if $|g'(y_\infty)| > 1$, the equilibrium is unstable.

We have used the linearization theorem and our knowledge of how linear difference equations behave to develop the stability theorem, because this approach extends naturally to systems of difference equations. For the first-order difference equations we are studying here, however, there is also a simple direct approach given in Appendix D which uses the Mean Value Theorem.

It is also possible to refine this theorem and show that the approach to an asymptotically stable equilibrium y_∞ is monotone if $0 < g'(y_\infty) < 1$ and oscillatory if $-1 < g'(y_\infty) < 0$. The truth of this is suggested by the cobwebbing method (cf. Figures 2.5 and 2.6). For instance, we saw in Example 2 that the equilibrium $y_\infty = K(1 - \frac{1}{r})$ of the logistic difference equation is asymptotically stable if $1 < r < 3$. As $g'(y_\infty) = 2 - r$, it follows that solutions of the logistic difference equation approach y_∞ monotonically if $1 < r < 2$ and with oscillations if $2 < r < 3$. It is natural to conjecture that if $r > 3$, so that there is no asymptotically stable equilibrium, all solutions oscillate but without approaching an equilibrium.

This is a different sort of question than asking what happens near a particular equilibrium; it changes the scope of our study from local to global. Once the local stability of each equilibrium of a model has been determined, one can put the results together to form a picture of the overall model behavior.

In some cases, there is only one stable equilibrium, and within the *state space* (the set of possible values for our variable(s)) that equilibrium may then be not only locally stable but *globally stable*, that is, all solutions approach it no matter what the initial condition. For example, when $0 < r < 1$, the logistic difference equation has only one equilibrium, at 0, and it is locally stable. We can also see directly by inspection that $y_{k+1} < y_k$ when $r < 1$, and in fact in this case the equilibrium $y = 0$ is globally stable. In biological terms, the population's maximum reproductive ratio r is so low that the population cannot sustain itself. Then, when $1 < r < 3$, the equilibrium $y = 0$ is unstable, but the equilibrium $y = K\left(1 - \frac{1}{r}\right)$ exists and is locally stable. In this case, it can be shown that the latter equilibrium is globally stable.

There are other, more complicated possible scenarios for a model's global behavior. One possibility, which we shall explore further at the end of this section, is multiple locally stable equilibria. Another, which corresponds to the logistic difference equation with $r > 3$, is no stable equilibria at all. We shall return to this question in Section 2.6. For now, however, let us consider one more abstract example of local stability analysis before applying this technique to a biological problem.

EXAMPLE 3.
Determine the asymptotic stability of each equilibrium of the Verhulst difference equation

$$y_{k+1} = \frac{r\, y_k}{y_k + A}.$$

Solution: We have $g(y) = \frac{ry}{y+A}$, $g'(y) = \frac{rA}{(y+A)^2}$. For the equilibrium $y = 0$, since $g'(0) = r/A$, the equilibrium is locally asymptotically stable if $r < A$ and unstable if $r > A$. For the equilibrium $y_\infty = r - A$, which is biologically significant only if $r > A$, the equilibrium is locally asymptotically stable if $r > A$ since $g'(y_\infty) = A/r$. Thus there is always exactly one (globally) asymptotically stable equilibrium, $y = 0$ if $r < A$ and $y = r - A$ if $r > A$. The reader should compare this approach with the cobwebbing analysis given in Example 1 of Section 2.2, and Figures 2.8 and 2.9. □

2.3.2 A problem in population genetics

We will now use a difference equation to model how the distribution of two different alleles for the same gene changes from one generation to the next in a diploid organism (which has two copies of each gene). This discussion follows that in Segel,[4] Chapter 3, and Maynard Smith.[5]

We must first make some assumptions in order to define the problem more

---

[4]Lee A. Segel, *Modeling dynamic phenomena in molecular and cellular biology*, Cambridge University Press, Cambridge, 1984.

[5]J. Maynard Smith, *Mathematical ideas in biology*, Cambridge University Press, Cambridge, 1968.

precisely. There are several factors which might influence the distribution of alleles from one generation to the next, including: the number of alleles for the gene under study; the effect of *genotype* (which alleles a single individual has) on *viability*, the proportion[6] of individuals who survive to reproduce; the effect of genotype on mating preferences; the effect of genotype on *fertility*, the number of offspring an average individual produces; the (potentially genotype-dependent) mutation rate from one allele to another; and possible gender asymmetries. Let us examine here the role of genotype-dependent viability in shaping the distribution of alleles over many generations. To simplify our discussion, we will then assume in what follows that

(1) the gene under study has two distinct alleles, A and a.

There are therefore three possible *genotypes* a given individual might have: homozygous AA, heterozygous Aa, and homozygous aa. We further assume that

(2) genotype affects viability (some genotypes make individuals more likely to survive and reproduce than others), in the same manner for each generation.

In order to examine only the effects of this particular factor, we will also assume that the other potential influences do not play a role here; that is, that

(3) fertility is genotype-independent;

(4) mating preferences are genotype-independent;

(5) the mutation rate for all alleles is zero;

(6) all effects are gender-independent.

Note also that we are implicitly assuming distinct generations, with reproduction occurring only within a given generation (i.e., no mating between generations); this is appropriate only for some populations, although the general conclusions we draw can be applied more broadly.

Following assumption (2) we will incorporate differential viability by defining three parameters, ϕ_{AA}, ϕ_{Aa}, ϕ_{aa}, each of which measures the proportion of individuals of the given genotype who survive to reproduce. ϕ_{AA}, ϕ_{Aa}, ϕ_{aa} will take on values between 0 and 1 and will be assumed not to vary from one generation to the next. In order to measure the frequency of each allele, we define our variable p_n to be the proportion of alleles in the nth generation which are A, and $q_n = 1 - p_n$ to be the complementary proportion, the relative frequency of a alleles.

Now, in order to calculate the relative frequencies p_{n+1} and q_{n+1} in the next generation, we observe that each new individual receives one allele from

[6] Although we might more intuitively speak of the likelihood or probability of surviving to reproduce, this is a deterministic model, so we must instead speak of proportions.

each parent. From assumption (4) alleles are mixed at random, that is, in proportion to the relative frequencies of the gametes (eggs and sperm) produced (which are, in turn, the same as the relative frequencies of the alleles, by assumption (3)). Therefore a proportion $p_n p_n = p_n^2$ of the new individuals receive A alleles from both parents, a proportion $p_n q_n$ receive an A allele from the mother and an a allele from the father, a proportion $q_n p_n$ receive an a allele from the mother and an A allele from the mother, and a proportion $q_n q_n = q_n^2$ receive a alleles from both parents. (The reader can verify that these four terms sum to $(p_n + q_n)^2 = 1$, thereby accounting for the entire next generation.) Note that the relative frequency of heterozygotes Aa in the next generation is $2p_n q_n$, since the result (the zygote) is the same regardless of which allele came from which parent.

The genotype-dependent viability of assumption (2) means that of those AA-homozygotes born into generation $n+1$, only a proportion ϕ_{AA} of them survive to reproduce, that is, a proportion $\phi_{AA} p_n^2$ of the entire generation of newborns. Likewise, a proportion $\phi_{Aa} 2p_n q_n$ become mature heterozygotes, and a proportion $\phi_{aa} q_n^2$ become mature aa-homozygotes. Since each AA individual has two A alleles, and each Aa individual has one, the total frequency of A alleles is now

$$f_A = 2\left(\phi_{AA} p_n^2\right) + \left(\phi_{Aa} 2p_n q_n\right).$$

Likewise, the total frequency of a alleles in generation $n+1$ is

$$f_a = 2\left(\phi_{aa} q_n^2\right) + \left(\phi_{Aa} 2p_n q_n\right).$$

Therefore, finally, the relative frequency of the A allele in generation $n+1$ is

$$\begin{aligned} p_{n+1} &= \frac{f_A}{f_A + f_a} = \frac{2\left(\phi_{AA} p_n^2\right) + \left(\phi_{Aa} 2p_n q_n\right)}{2\left(\phi_{AA} p_n^2\right) + 2\left(\phi_{Aa} 2p_n q_n\right) + 2\left(\phi_{aa} q_n^2\right)} \\ &= \frac{\phi_{AA} p_n^2 + \phi_{Aa} p_n (1 - p_n)}{\phi_{AA} p_n^2 + 2\phi_{Aa} p_n (1 - p_n) + \phi_{aa} (1 - p_n)^2}, \end{aligned} \qquad (2.20)$$

and the relative frequency of the a allele in generation $n+1$ is $q_{n+1} = 1 - p_{n+1}$ (since $p + q = 1$ in each generation, we really only need keep track of one of the two).

We can simplify equation (2.20) by reducing the number of parameters to two. We do so by defining relative viabilities

$$F = \frac{\phi_{AA}}{\phi_{Aa}}, \quad G = \frac{\phi_{aa}}{\phi_{Aa}}.$$

F gives a ratio of AA viability to Aa viability; $0 < F < 1$ means a AA zygote has less viability than an Aa ($\phi_{AA} < \phi_{Aa}$); $F > 1$ means an AA zygote has higher viability than an Aa. Likewise G gives the viability of an aa zygote relative to an Aa. Effectively, the Aa viability has been made the yardstick by which others are measured. (The choice of Aa was arbitrary — we could instead choose one of the other two — but this choice preserves a certain symmetry.) If we now divide the numerator and denominator in (2.20) by ϕ_{Aa}, we obtain

$$p_{n+1} = \frac{Fp_n^2 + p_n(1-p_n)}{Fp_n^2 + 2p_n(1-p_n) + G(1-p_n)^2}. \tag{2.21}$$

We have now formulated a model that tracks the relative frequency of the two alleles across generations. As it is highly nonlinear, however, we cannot solve (2.21) for a non-recursive formula to determine p_n as a function of F, G, and the initial condition p_0. We will therefore resort to a qualitative analysis as developed earlier in this section.

We begin by finding equilibria. These are solutions p to the equation

$$p = \frac{Fp^2 + p(1-p)}{Fp^2 + 2p(1-p) + G(1-p)^2}. \tag{2.22}$$

Some algebra shows that the three equilibria are $p = 0$ (everyone is AA), $p = 1$ (everyone is aa), and $p = \frac{G-1}{F+G-2}$ (some of both). The last equilibrium, of course, only makes sense as long as it is in the interval [0,1]. Further algebra (Exercise 8) shows that the last equilibrium has a value between 0 and 1 only when F and G are either both greater than 1 or both less than 1.

To determine the local stability of each equilibrium, we linearize (2.21) about each one in turn. To apply the equilibrium stability theorem stated earlier in this section, we need the derivative $g'(p)$ of the right-hand side of the difference equation $p_{n+1} = g(p_n)$. It is

$$g'(p) = \frac{(F+G-2FG)p^2 + 2(F-1)Gp + G}{\left[Fp^2 + 2p(1-p) + G(1-p)^2\right]^2}.$$

This gives $g'(0) = 1/G$, $g'(1) = 1/F$, and $g'\left(\frac{G-1}{F+G-2}\right) = \frac{F+G-2FG}{1-FG}$. Therefore, the linearization about the equilibrium $p = 0$ is $u_{n+1} = \frac{1}{G}\,u_n$, and by our equilibrium stability theorem the equilibrium $p = 0$ is locally asymptotically stable if $1/G < 1$, that is, if $G > 1$. Likewise, the equilibrium $p = 1$ is locally asymptotically stable if $F > 1$. These are both biologically reasonable, as we shall explain shortly.

Determining the stability condition for the third equilibrium, $p_\infty = \frac{G-1}{F+G-2}$, is a little more complicated, but is simplified if we recall that this equilibrium only makes sense when either $F < 1$, $G < 1$ or $F > 1$, $G > 1$. The criterion for stability here is

$$\left|g'\left(\frac{G-1}{F+G-2}\right)\right| = \left|\frac{F+G-2FG}{1-FG}\right| < 1. \tag{2.23}$$

If $F < 1$, $G < 1$, then both the numerator and the denominator of $g'(p_\infty)$ are positive,[7] and (2.23) becomes $F + G - 2FG < 1 - FG$, which simplifies

[7] $FG < 1$, so the denominator is positive, and the numerator is $F + G - 2FG = (F - FG) + (G - FG) = F(1-G) + G(1-F)$, also positive.

to $G < 1$. Thus the third equilibrium is always locally asymptotically stable in this region. The other possibility is that $F > 1$, $G > 1$. In this case, both the numerator and the denominator of $g'(p_\infty)$ are negative, so (2.23) becomes $2FG - F - G < FG - 1$, which also simplifies to $G < 1$. Thus the third equilibrium is always unstable in this region.

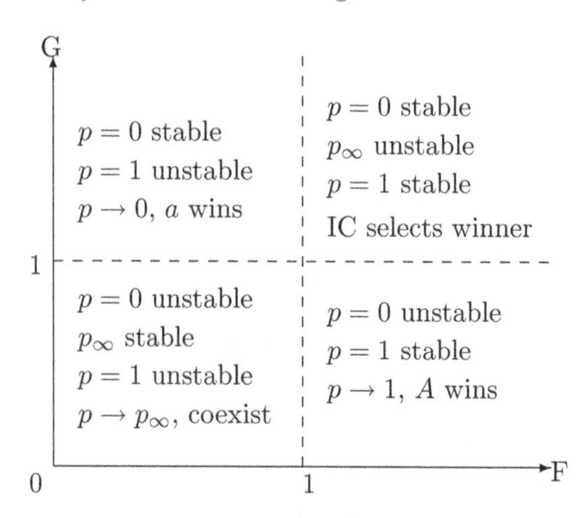

FIGURE 2.14: Stability of equilibria for equation (2.21), in terms of the parameters F and G.

We can now put these results together to form a complete picture of the model's behavior. There are four cases or regions in terms of F and G, as depicted in Figure 2.14. When $F < 1$ and $G > 1$ (which we can also write as $\phi_{AA} < \phi_{Aa} < \phi_{aa}$, so that AAs are less viable than Aas, which are in turn less viable than aas), the equilibrium at 0 is unstable, and the equilibrium at 1 is (globally) stable, so that $p_n \to 0$ as $n \to \infty$, that is, the proportion of A alleles dwindles to zero from one generation to the next, until eventually almost everyone is aa homozygous. Similarly, when $F > 1$ and $G < 1$, so that $\phi_{aa} < \phi_{Aa} < \phi_{AA}$ and the A allele contributes more to viability than the a allele, then the equilibrium at 0 is unstable, the equilibrium at 1 is (globally) stable, and $p_n \to 1$ as $n \to \infty$. In this case, eventually nearly everyone is AA homozygous. Note that in both of these two cases, one allele clearly contributes more to viability than the other, so it is not surprising that it comes to dominance. (Also note that in both cases, the coexistence equilibrium p_∞ is not present.)

When $F < 1$ and $G < 1$, both AA and aa genotypes are less viable than the heterozygous Aa genotype. This is the case, for example, in Africa with regard to the allele for sickle cell anemia. This allele provides some immunity against malaria, but homozygotes with two copies of the allele develop sickle cell anemia and often do not survive to maturity. Thus it is beneficial to have precisely one copy of the allele. In this scenario, the equilibria at 0 and 1 are both unstable, but the coexistence equilibrium is (globally) stable. Here all three genotypes persist, although both homozygotic genotypes are at a disadvantage.

Finally, when $F > 1$ and $G > 1$, both homozygotic genotypes are more viable than the heterozygote. Here the equilibria at 0 and 1 are both locally stable, but the coexistence equilibrium in between them is unstable: if $p_0 < p_\infty$, then p_1 will be even smaller, while if $p_0 > p_\infty$, then p_1 will be even larger. The unstable coexistence equilibrium acts as a boundary to separate those initial conditions (ICs) for which $p_n \to 0$ as $n \to \infty$ from those for which $p_n \to 1$ as $n \to \infty$. That is, if $p_0 < p_\infty$, then the a allele wins out, while if $p_0 > p_\infty$, then the A allele wins out. In either case, eventually nearly everyone is homozygous with two copies of the "winning" allele. This is our first instance of competition, which we shall study more in later sections. As Segel observes (p. 48), if we have multiple isolated habitats in which these same criteria hold ($F, G > 1$, so heterozygotes are at a disadvantage) but with different initial conditions, then we may expect to find the A allele "winning" in some of them, and the a allele dominating in others, so that this model also provides a possible explanation for genetic variability across similar isolated habitats.

Our global stability analysis here is not entirely rigorous, in that we have not proven conclusively that the simple assembly of local stability results provides a complete picture of the model's global behavior. Quantitative analysis, by having a computer solve the equation for particular values of the parameters F and G and particular initial conditions p_0, can further boost our intuitive understanding and confidence that this global picture is correct. We will forego further discussion of other possible global behaviors until Section 2.6 (but see Exercise 9).

Note, finally, how our decision to investigate the selective effects of genotype-dependent viability informed and shaped the assumptions that we made and the model we analyzed. This focused research question and the subsequent assumptions help us to frame and delimit the extent to which our conclusions (namely, that simple differential viability can account for either exclusive selection of one genotype or an equilibrium distribution of coexistence) hold. To complete the modeling cycle discussed in the previous chapter, we must evaluate the extent to which our interpreted conclusions provide a sufficient answer to our original question. Several of the exercises below investigate other factors for selection of alleles.

Exercises

In Exercises 1–6, find all positive equilibria of the given difference equation, and for each equilibrium determine whether it is asymptotically stable or unstable. If the difference equation involves one or more parameters, your answer may depend on the range of values of the parameters.

1. $y_{k+1} = 2y_k - y_k^2 + 2$
2. $y_{k+1} = \frac{1}{2}y_k + y_k^2 - \frac{7}{144}$
3. $y_{k+1} = \alpha y_k + \beta$
4. $y_{k+1} = ry_k(1 - y_k)$
5. $y_{k+1} = ry_k e^{1-y_k}$
6. * $y_{k+1} = ry_k(1 + y_k)^{-\frac{1}{2}}$

7. Find all positive equilibria of the difference equation $y_{k+1} = \frac{r\,y_k^2}{y_k^2+A}$ from Example 2 of Section 2.2, and for each equilibrium determine when (in terms of r and A) it is locally asymptotically stable or unstable. Compare your results to those given for the cobwebbing approach.

8. Solve the equilibrium condition (2.22) and show that the coexistence equilibrium $p_\infty = \frac{G-1}{F+G-2}$ can only exist when F and G are either both greater than 1 or both less than 1.

9. Suppose that, in the genetics discussion in this section, survival is genotype-independent, i.e., $\phi_{AA} = \phi_{Aa} = \phi_{aa}$. Show that the frequency of each allele is then constant across generations, establishing the *Hardy-Weinberg law of genetics* (so named for its independent discoveries by G.H. Hardy and Wilhelm Weinberg in 1908 [33, 104]), as follows: (a) Find the resulting values of the relative viabilities F and G. Where does this place the scenario relative to Figure 2.14? (b) Simplify equation (2.21).

10. Suppose that, in the genetics discussion in this section, we are studying a gene for which *aa* homozygotes never reproduce (they are either infertile or do not survive to maturity). The *a* allele may nevertheless survive among heterozygotes. In this case $\phi_{aa} = 0$, and thus $G = 0$. (Note also in this case we must have $p_0 > 0$, or else the entire population consists of *aa* homozygotes, and there is no next generation.)

    (a) How do equation (2.21) and its behavior change when $G = 0$?

    (b) Interpret biologically the criterion developed in (a) for survival of the *a* allele when $G = 0$.

    (c) Show that the $G = 0$ model in (a) guarantees $p_n > 1/2$ for $n \geq 1$. Explain biologically why this must be so.

    (d) Sketch (by hand or by computer) a graph of the remaining equilibria p as a function of the remaining parameter F. This type of graph is called a *bifurcation diagram*, and we shall discuss bifurcation diagrams further in Section 2.6, and in later chapters.

    (e) * Have a computer generate a three-dimensional bifurcation diagram for the model (2.21) in terms of both parameters F and G. (Segel[8], p. 50) If we reverse assumption (2) and suppose that all

[8] Lee A. Segel, *Modeling dynamic phenomena in molecular and cellular biology*, Cambridge University Press, Cambridge, 1984.

three genotypes are equally viable, what happens to the model (2.21)? Explain this result in biological terms.

11. Suppose that, in the discussion on population genetics in this section, we relax assumption (3) and allow fertility to be genotype-dependent. Again following Segel (pp. 49–50), define the average number of gametes produced by an adult of each genotype to be $2m_{AA}$, $2m_{Aa}$, and $2m_{aa}$, evenly divided among eggs and sperm, and in each case half of which correspond to each of the parent's two alleles.

    (a) Develop a difference equation for p_{n+1} in terms of p_n analogous to equation (2.21) which models the distribution of alleles under an assumption of genotype-dependent fertility.

    (b) By comparing this new model with equation (2.21), show that the behavior of the new model is qualitatively the same as that of (2.21), the only difference being different expressions for the fitness ratios K and L.

12. Segel also considers (p. 47) the possibility of mutation, rather than differential viability, driving the distribution of genotypes. In terms of the discussion in this section, assumptions (2) and (5) are reversed (and all the ϕ's are the same).

    (a) Segel suggests imagining mutations only in one direction, say from A to a. If a proportion μ of the A gametes in any generation mutate to a, then what proportion of the A gametes remain? If we assume all genotypes have the same viability, what is the corresponding equation for p_{n+1} in terms of p_n? (It should be linear.)

    (b) Solve the new model outright, and use the result to estimate how many generations it will take for this mutation to reduce the frequency of the A allele from 0.5 to 0.05, if μ has a value of 10^{-6} per generation.

    (c) Now suppose that back-mutations also occur, from a to A, with an average proportion μ' of the a gametes in any generation mutating to A. Write a difference equation that incorporates both mutations. Give a qualitative analysis of its behavior, assuming that $\mu > \mu'$.

    (d) Choose values of F and G which lead to the disappearance of the A allele (i.e., $F < 1 < G$), and have a computer calculate solutions of (2.21) using an initial condition of $p_0 = 0.5$. How many generations are necessary to reach a frequency of 0.05? Compare this number to that obtained in (b) above for mutation alone to accomplish the same result. Which selection mechanism appears to operate more quickly? How does this compare with currently held theory?

    (e) * Now write a difference equation that incorporates both differential viability and $A \to a$ mutation. To get you started, first reconsider the question, what proportion of new individuals receive an A allele from one parent? (It is now less than p_n.)

13. * If we reverse assumption (4) and instead assume extreme like-with-like mating preferences, so that individuals only mate with others of the same genotype, then we must keep track of the relative frequencies of each genotype, rather than those of alleles. If we define α_n and β_n, respectively, as the relative frequencies of AA homozygotes and aa homozygotes in generation n, then the relative frequency of Aa heterozygotes must be $(1 - \alpha_n - \beta_n)$.

    (a) Assuming random mixing of alleles within each genotype, write difference equations for α_{n+1} and β_{n+1}, each in terms of α_n, β_n, F, and G. (We will not, however, discuss the qualitative analysis of *systems* of two or more difference equations until Section 2.7.)

    (b) Show that the relative frequency of heterozygotes $(1-\alpha_n-\beta_n) \leq \frac{1}{2}$ for all $n \geq 1$, regardless of initial conditions. Explain biologically why this must be so.

2.4 Intraspecies competition

In discussing the linear homogeneous difference equation $y_{k+1} = ry_k$ as a model for population growth in Section 2.1, we observed that the model becomes inadequate as soon as the population grows enough to begin encountering the resource limitations of its habitat, whether on a cellular scale or a continental scale. Various examples were given of functions that have been proposed to address the effect of resource limitations on population growth for different populations. All these models must address, even if only implicitly, the way in which members of the population compete with each other when resources are scarce. This type of competition is called *intraspecies competition*, and is distinct from the *interspecies competition*we shall study when we introduce systems of interacting populations in Section 2.7 and in Chapter 6.

There are two general assumptions one might make about the way resources are distributed when there are not enough for everyone. *Scramble* competition, in which resources are divided evenly among all members, reflects an assumption that all individuals are equally capable of capturing resources for themselves, with the result that everyone suffers equally when resources are inadequate. In this case, reproduction may be severely curtailed, or even cease altogether, if no one is getting enough food to reproduce reliably. An example of a population observed to exhibit scramble competition is that of Nicholson's blowflies. In 1954 A.J. Nicholson conducted experiments on *Lucilia cuprina*, the Australian sheep blowfly, in which he observed large oscillations in the

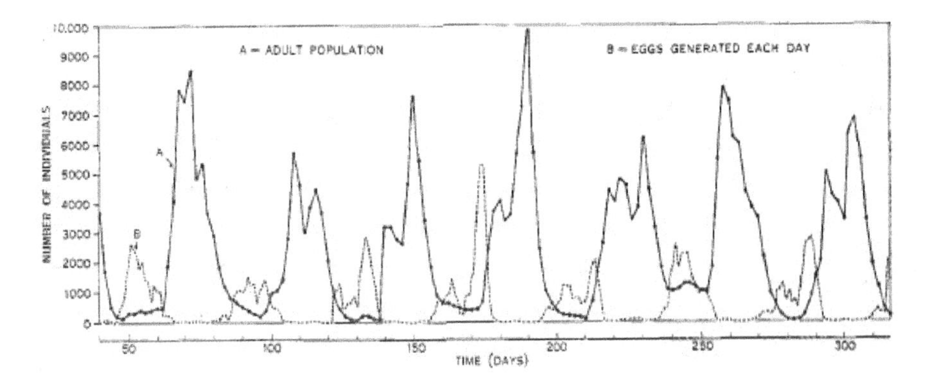

FIGURE 2.15: Sustained fluctuation in a blowfly (*Lucilia cuprina*) population with constant food supply, as observed by Nicholson (1954, Figure 3).

population size over time. Nicholson discovered[9] that the blowfly population, which emerged in distinct (or nearly distinct) generations, was controlled entirely by the rate at which food was provided, so that (when this rate was held constant) a small generation of blowflies, encountering no resource limitations, produced a large next generation. The next generation, however, was so large that each individual had a minuscule share of the food, and consequently very few were able to develop eggs, resulting in a very small next generation. These oscillations were sustained over a period of months (see Figure 2.15), and others have since observed similar fluctuations in fly populations.

Contest competition, on the other hand, assumes that some individuals — the most fit — are able to secure enough resources for themselves, while the less fit may end up with nothing. Here we may imagine a ranking in which resources are distributed to individuals in decreasing order of fitness, until resources are exhausted. Under contest competition, some number of individuals (as many as resources can support) will always be able to reproduce. For example, if we consider the number of trees of a given species in a certain patch of woods, often the tall trees spread their branches to catch as much sunlight as needed for growth through photosynthesis, crowding out smaller trees nearby (which then stop growing or may even die) until each tall tree's foliage meets that of its neighbors, allowing little sunlight to reach the ground. New trees then grow only when a large tree falls, creating an opening.

The differences between these two assumptions will lead naturally to differences in the mathematical models we develop to describe such populations. In particular, scramble competition is modeled with reproduction functions which rise to a maximum and then fall off, as excessive size reduces repro-

[9] A.J. Nicholson, An outline of the dynamics of animal populations, *Australian J. Zoology* 3: 9–65, 1954.

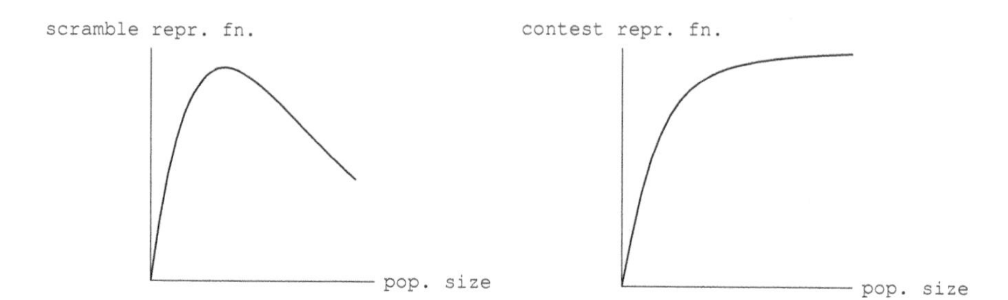

FIGURE 2.16: Scramble (left) and contest (right) competition reproduction functions.

duction. In some cases reproduction may reach zero for some finite carrying capacity, as with the logistic model (2.12), while in others it may taper off gradually, as with Ricker's model (first seen in the exercises at the end of Section 2.2 and discussed in more detail below). Contest competition, meanwhile, is modeled with reproduction functions which rise monotonically and flatten out as they reach the ceiling created by the resource limitations. The Verhulst model (2.10) incorporates contest competition. Figure 2.16 sketches the general form of these two types of functions.

We can consider the consequences of each type of intraspecies competition via the properties described above of the corresponding reproduction functions. In either case we will also require that the reproduction function $g(y)$ in the difference equation $y_{k+1} = g(y_k)$ have the following properties:

- For small populations y, reproduction should be roughly linear in y, corresponding to a constant (maximum) per capita ratio r called the *intrinsic growth rate*, i.e.,

$$\lim_{y \to 0} \frac{g(y)}{y} = r > 0.$$

- For large populations, reproduction should be constrained, i.e., $g(y)$ is bounded.

(The reader can verify that all models described in this section have these two properties.) Since, as we saw in Section 2.2, equilibria correspond graphically to intersections of the reproduction curve $z = g(y)$ (which is bounded by assumption) with the line $z = y$ (which grows unbounded), it follows that there must be a largest equilibrium y_∞, at which the line $z = y$ crosses the curve $z = g(y)$ for the last time, from below to above. At this point, the slope of $z = y$ (which is 1) must therefore be greater than the slope (derivative) of $z = g(y)$. That is, at this largest equilibrium y_∞, we have $g'(y_\infty) < 1$.

If, in addition, we assume contest competition, with monotone increasing reproduction, then we also have that $g'(y) \geq 0$ for all y, and more specifically $g'(y_\infty) \geq 0$. Therefore $0 \leq g'(y_\infty) < 1$, satisfying the stability condition $|g'(y_\infty)| < 1$, so that under contest competition there must always be at least one stable positive equilibrium (in a discrete-time model). In fact, if we consider that for any additional equilibria between 0 and y_∞, the crossing of the reproduction curve by the line $z = y$ alternates between below-to-above and above-to-below (as y increases), then the stability argument above can also be applied to every second equilibrium to the left of y_∞.

On the other hand, if we assume scramble competition, so that reproduction rises and then falls as y increases, then the last crossing of $z = g(y)$ by $z = y$ may occur on either the increasing or the decreasing part of the curve, depending on the value of the intrinsic reproduction ratio r. If r is small, then $z = y$ will outstrip $z = g(y)$ more quickly, and the last equilibrium y_∞ will occur on the increasing part of the curve, leading to an argument like that for contest competition in which $0 \leq g'(y_\infty) < 1$ and the equilibrium is stable. However, for large r the last equilibrium y_∞ will occur on the decreasing part of the curve $z = g(y)$, so that $g'(y_\infty) < 0$. In this case the previous conclusion that $g'(y_\infty) < 1$ no longer helps bound the derivative, and it is quite possible that $g'(y_\infty) < -1$, which implies instability for y_∞.

The logistic equation $y_{k+1} = ry_k(1 - y_k)$, introduced in Section 2.1, is an example of a population model with scramble competition. (Another form sometimes used for the logistic equation is $y_{k+1} = y_k + r\, y_k(1 - y_k)$, but we shall use that given above throughout this chapter.) In Section 2.6 we will describe the behavior of a difference equation such as the logistic equation with $r > 3$ (cf. Section 2.3, Example 2) modeling scramble competition and having no asymptotically stable equilibrium.

We can therefore conclude that populations with contest competition always reach an asymptotically stable [nonzero] equilibrium, but populations with scramble competition have an asymptotically stable [nonzero] equilibrium only if the intrinsic growth rate is sufficiently small (but still large enough for the equilibrium to exist in the first place, cf. the linear equations of Section 2.1). We can interpret this idea intuitively as saying that in contest competition, those who obtain enough resources to reproduce will ensure a subsequent generation which is as large as possible with the given resources, while in scramble competition too high an intrinsic reproduction rate would result in the division of available resources so many ways that reproduction would be severely curtailed, thereby setting the stage for sustained oscillations. This principle is *robust*: it does not depend on the specific form of the reproduction function but only on the qualitative assumptions on the nature of the competition.

We will now consider two models that have been proposed for different fish populations, one representing each type of intraspecies competition.

2.4.1 Two metered fish models

Many species of fish have an annual cycle in which all new fish are born in a short time span. It is reasonable to assume that the number of newborn fish is a function of the adult stock which has survived to the birth season and the newly matured fish, and this total will depend on the adult population from the previous year. Thus a difference equation model is appropriate. However, the survivorship from one year to the next is a continuous process, and the form of the model will depend on this process. This type of model, in which continuous short-term dynamics shape the long-term dynamics of discrete generations, is called a *metered* model. We shall now describe two frequently used metered models for different species of fish, although we will not be able to establish the appropriate survivorship until Section 3.4.

Some fish, such as haddock and North Atlantic plaice, have very high fertility rates but low survivorship to adulthood. For such fish, it is appropriate to assume a linear (rather than constant) per capita mortality rate between spawning seasons, and this assumption leads to the *Beverton-Holt model*[10]

$$y_{k+1} = \frac{ay_k}{1 + by_k} \tag{2.24}$$

where a and b are positive constants. This equation, which corresponds to contest competition, is the same as the Verhulst equation introduced in Section 2.1 with a replacing $\frac{r}{A}$ and b replacing $\frac{1}{A}$. From our study of the Verhulst equation (Section 2.3, Example 3), we now see that there is one asymptotically stable equilibrium, namely $y = 0$ if $a < 1$ and $y = \frac{a-1}{b}$ if $a > 1$. In this case, the rise in per capita mortality with population density creates contest competition, with the result that no matter how many fish are born in a given year, only a certain number survive to the next year. (In addition, the reproductive ratio $a = g'(0)$ must be high enough — $a > 1$ — to keep the population from dying out altogether. The parameter b, which gives a measure of the size of the density-dependent per capita mortality rate relative to the density-independent per capita mortality rate, only affects the size of the eventual [positive] equilibrium.)

Some species of fish, notably salmon, habitually cannibalize their eggs and young. Now the appropriate assumption is instead that the birth rate is proportional to the adult population size and that there is a per capita death rate proportional to the initial size of the young population. This assumption leads to the *Ricker model*[11]

$$y_{k+1} = ay_k e^{-by_k} \tag{2.25}$$

with a and b again positive constants. Here a is the constant of proportionality in the birth rate, while b measures the cannibalism rate due to adults.

---

[10] R.J.H. Beverton and S.J. Holt, The theory of fishing, *Sea fisheries: their investigation in the United Kingdom* (M. Graham, ed.), Edward Arnold, London, 1956, pp. 372–441.

[11] W.E. Ricker, Stock and recruitment, *J. Fisheries Research Board Canada* 11: 559–623, 1954.

Photo by Bernard E. Picton and Christine C. Morrow

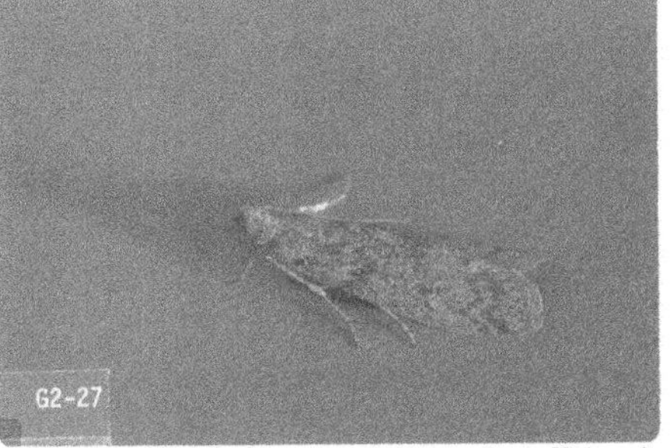

Ken Gray Image Courtesy of Oregon State University

FIGURE 2.17: The bottom-feeding plaice (left), hard to see against the sea floor except when they move, have high fertility and low survivorship to maturity, leading to contest competition. The Mediterranean flour moth or mill moth (right), seen in its characteristic resting pose, undergoes different types of intraspecies competition during its life cycle as a result of parasitism.

This model, corresponding to scramble competition (all young have a reduced chance of survival), exhibits behavior similar to that of the logistic model. Equilibria of the Ricker model are given by $y = 0$ and $ae^{-by} = 1$. Thus there is an equilibrium $y = 0$ describing extinction of the population and possibly an equilibrium describing survival given by $e^{by} = a$, or $y_\infty = \frac{\log a}{b}$ (by log we mean the natural logarithm). This possible survival equilibrium is positive only if $a > 1$. For the Ricker model, $g(y) = aye^{-by}$, $g'(y) = ae^{-by}(1 - by)$. Thus $g'(0) = a$, and the equilibrium $y = 0$ therefore is asymptotically stable if and only if $a < 1$. Also, if $a > 1$, so that the survival equilibrium y_∞ makes biological sense, then the condition $ae^{-by_\infty} = 1$ makes

$$g'(y_\infty) = 1 - by_\infty = 1 - \log a.$$

Thus the equilibrium y_∞ is asymptotically stable if and only if $-1 < 1 - \log a < 1$. This condition reduces to $\log a < 2$, or $a < e^2$. The result suggests that fish which cannibalize their young and have a high birth rate ($a > e^2$) may have an unstable equilibrium. We shall see in Section 2.6 that this may indicate oscillations in population size. Oscillations in population size can lead to times at which the population size is very small, so that a small perturbation could lead to wiping out the population. Fish populations which satisfy a Ricker model may be quite vulnerable to external influences.

In this and later sections, we will use the Ricker form $g(y) = aye^{-by}$ as a prototypical scramble recruitment function and the Verhulst form $g(y) = \frac{ry}{y+A}$ as a prototypical contest competition recruitment function.

2.4.2 Between contest and scramble

The equation

$$y_{k+1} = a y_k (1 + \frac{b}{n} y_k)^{-n} \tag{2.26}$$

with $a, b > 0$ and $n > 1$, introduced by Hassell[12] to model intraspecific larval competition in insects, is intended to describe a situation between the extremes of contest competition — $n = 1$, which gives the Beverton-Holt model — and scramble competition — $n \to \infty$, which gives the Ricker model since $\lim_{n\to\infty}(1+\frac{b}{n}y_k)^{-n} = e^{-by}$. Exercises 9–12 below develop some of the analysis of this model. Maynard Smith and Slatkin[13] proposed a similar model to fill in the spectrum between contest and scramble competition:

$$y_{k+1} = \frac{a y_k}{1 + (b y_k)^n} \tag{2.27}$$

with $a, b > 0$ as before. Here, as in the Hassell model, $n = 0$ corresponds to no competition ($y_{k+1} = \frac{a}{2} y_k$), $n = 1$ corresponds to contest competition (Beverton-Holt), and $n \to \infty$ corresponds to scramble competition. Some researchers have used statistical techniques such as least-squares regression to fit observed data to such models and take the best-fit value for n as a measure of the amount and type of intraspecies competition exhibited by the population under study.

In fact, some species of insects have been found to exhibit different types of competition in different parts of their life cycles, or depending on interactions with other species or the environment. Lane and Mills found[14] that parasitism can actually change the nature of intraspecific competition from scramble to contest in some insects — in particular, the Mediterranean flour moth *Ephesia kuehniella* shifts from scramble competition prior to parasitization to contest competition following it, since parasitized larvae are not only poor competitors but also more vulnerable to cannibalism than unparasitized larvae. Reeve, Rhodes and Turchin, meanwhile, found[15] that the southern pine beetle *Dendroctonus frontalis* shifts from contest competition early in its life cycle (when adults attack a tree to lay eggs in galleries) to scramble competition later on (brood survivorship), with the latter dominating. Flexible models such as those of Hassell (1975) and Maynard Smith and Slatkin (1973) can therefore be important tools for modeling this variability.

---

[12] M.P. Hassell, Density dependence in single species populations, *J. Animal Ecology* 44: 283–295, 1975.

[13] J. Maynard Smith and M. Slatkin, The stability of predator-prey systems, *Ecology* 54: 384–391, 1973.

[14] S.D. Lane and N.J. Mills, Intraspecific competition and density dependence in an *Ephesia kuehniella–Venturia canescens* laboratory system, *OIKOS* 101: 578–590, 2003.

[15] J.D. Reeve, D.J. Rhodes and P. Turchin, Scramble competition in the southern pine beetle, *Dendroctonus frontalis*, *Ecological Entomology* 23(4): 433–443, 1998.

Exercises

In Exercises 1–4 determine the values of r for which there is an asymptotically stable equilibrium.

1. $y_{k+1} = \frac{ry_k^2}{y_k^2+A}$

2. $y_{k+1} = ry(1+\alpha y_k)^{-\beta}, \ \ (0 < \beta < 1)$

3. $y_{k+1} = ry_k e^{1-\frac{y_k}{K}}$

4. $y_{k+1} = ry_k(e^{1-y_k} - d)$

   In Exercises 5–8 determine the range of values of H for which there is an asymptotically stable equilibrium. [The range of values may depend on r.]

5. $y_{k+1} = \frac{ry_k}{y_k+A} - H$

6. $y_{k+1} = y_k + ry_k(1 - \frac{y_k}{K}) - H$

7. $y_{k+1} = (0.9)y_k + 0.1 - H$

8. $y_{k+1} = \alpha y_k + \beta$ with $0 < \alpha < 1$.

The following exercises deal with the Hassell model (2.26).

9. * Show that $\lim_{n\to\infty}(1+\frac{b}{n}y_k)^{-n} = e^{-by}$, so that the limiting case of the Hassell model as $n \to \infty$ is the Ricker model. [HINT: Use the relation $\lim_{n\to\infty}(1+\frac{x}{n})^n = e^x$.]

10. Show that (2.26) has an equilibrium $y = 0$ which is asymptotically stable if $a < 1$.

11. Show that (2.26) has a positive equilibrium $y = \frac{n}{b}(a^{\frac{1}{n}} - 1)$ if $a > 1$.

12. * Show that the positive equilibrium of (2.26) is asymptotically stable if $n(1-a^{-1/n}) < 2$, and in particular for all a if $n < 2$.

13. Take the example given in this section of crowding of trees in a forest and develop a difference equation which models the contest competition that regulates the density of trees in a given patch. Explain the parameters in your model, and say how you could measure them.

14. Find the expression for the limiting (scramble competition) case of the Maynard Smith and Slatkin model (2.27) where $n \to \infty$, as a function of y_k. (You will need to consider three cases, depending on the value of by_k.)

15. The model in Exercise 3 is an alternate form of the Ricker model (2.25). Show that they are equivalent.

2.5 Harvesting

Another mechanism besides competition which regulates population sizes is the removal of individuals by another species, usually for food. Fish eat plankton, and aquarium owners often buy algae-eating fish to keep their tanks clean. Wolves and coyotes hunt deer and smaller mammals in the woods of North America; lions hunt gazelles on the plains of Africa. Humans also cut down trees to make paper and wood for construction. In areas where humans have removed natural predators, the prey populations often grow unchecked and can cause damage to the environment or become a nuisance. Deer populations, for instance, in areas where wolves have been shot by farmers, rise to the point that they run onto roads, causing traffic accidents, and wander into residential neighborhoods and eat flowers, bushes and gardens bare. Hunting seasons have been developed and managed as a response to this overpopulation, as a means of population control. Human colonization of wild areas has also led to other, more accidental removals of natural predator or competitor species, necessitating further population management programs based on careful study. All these types of removal and control are known as *harvesting*, and harvesting has become a major concern in wildlife and environmental management. Rempel and Kaufman, for instance, studied[16] the effects of both spatial and temporal tree harvesting patterns in the Nakina Forest of northwestern Ontario, Canada, on the preservation of wildlife habitat for animals such as caribou, moose, and martens, balancing commercial timber production objectives with a concern not to fragment forest habitats.

We shall consider here only one type of harvesting, that is *constant-rate* or *constant-yield* harvesting, by which we mean removing a fixed number of members just before the annual birth period. Thus a population described in the absence of harvesting by a difference equation $y_{k+1} = g(y_k)$ would be described under harvesting by a difference equation $y_{k+1} = g(y_k) - H$, where H is the number of individuals harvested per year. The questions of interest to investigate involve the relationship between the harvesting rate and the equilibrium population size. What is the maximum harvest rate that will allow the population to persist? What is the minimum harvest rate necessary to prevent excessive population growth? To what extent does harvesting make a population more vulnerable to extinction from a catastrophic event such as drought or disease? Can harvesting make a population *less* vulnerable to such events? We begin by considering a simple example.

EXAMPLE 1.
Consider a population governed by a logistic difference equation from which

[16] R.S. Rempel and C.K. Kaufman, Spatial modeling of harvest constraints on wood supply versus wildlife habitat objectives, *Environmental Management* 32(5): 646–659, 2003. doi: 10.1007/s00267-003-0056-8.

H members are removed in each generation. Determine the maximum value of H for which the population will persist (i.e., will have a positive equilibrium), and show that if $H > 0$ there can be an asymptotically stable equilibrium for values of r greater than 3.

Solution: The population is modeled by

$$y_{k+1} = ry_k\left(1 - \frac{y_k}{K}\right) - H. \tag{2.28}$$

An equilibrium of (2.28) is a solution of $y = ry\left(1 - \frac{y}{K}\right) - H$, i.e.,

$$y^2 - K\left(1 - \frac{1}{r}\right)y + \frac{HK}{r} = 0. \tag{2.29}$$

The quadratic equation (2.29) has two positive real roots

$$y = \frac{1}{2}\left[K\left(1 - \frac{1}{r}\right) \pm \sqrt{K^2\left(1 - \frac{1}{r}\right)^2 - \frac{4HK}{r}}\right] \tag{2.30}$$

if and only if

$$K^2\left(1 - \frac{1}{r}\right)^2 - \frac{4HK}{r} \geq 0,$$

which we can rewrite as

$$H \leq H_{max} = \frac{(r-1)^2}{4r}K.$$

This gives the range of values of H (for a given value of r) for which the population will persist.

To determine the (local) stability of the two equilibria found in (2.30), we recall the condition $|g'(y_\infty)| < 1$ for the stability of an equilibrium y_∞ of the difference equation $y_{k+1} = g(y_k)$. Here we find that for (2.28)

$$g(y) = ry\left(1 - \frac{y}{K}\right) - H, \quad g'(y) = r\left(1 - \frac{2y}{K}\right),$$

and substituting (2.30) for y_∞ we obtain

$$g'(y_\infty) = 1 \mp \sqrt{(r-1)^2 - \frac{4rH}{K}}.$$

Since $g'(y_\infty) > 1$ for the smaller of the two equilibria, only the larger equilibrium in (2.30) can be asymptotically stable, and the condition $-1 < g'(y_\infty) < 1$ for its asymptotic stability simplifies to $g'(y_\infty) > -1$, or

$$\sqrt{(r-1)^2 - \frac{4rH}{K}} < 2,$$

which reduces to

$$(r-1)^2 - \frac{4rH}{K} < 4.$$

We can rewrite this in terms of H as

$$H > H_{min} = \frac{(r-1)^2 - 4}{4r}K = \frac{(r-3)(r+1)}{4r}K,$$

which is satisfied for all $H \geq 0$ when $r < 3$. Therefore, when $r < 3$, all $H < H_{max}$ allow the population to persist at a (stable) positive equilibrium, provided that the initial population is above the unstable (smaller) equilibrium. (Too small an initial population will go extinct from even a little harvesting.)

The perhaps surprising result comes from addressing the second half of the original question. Recall (Section 2.3, Example 2) that for $r > 3$ the logistic equation without harvesting has no stable equilibria. When $r > 3$, however, (2.28) has a stable positive equilibrium for H in the interval $H_{min} < H < H_{max}$. For $H > H_{max}$, harvesting is too great, and the population is wiped out in a finite number of generations (since there is no equilibrium). For $H < H_{min}$, harvesting is not strong enough to temper the high reproductive rate, the equilibria remain unstable, and the population oscillates, as it does in the absence of harvesting. Therefore, we see that for populations that exhibit scramble competition (as in the logistic model) and a high fertility rate ($r > 3$, measured in appropriate units), it is actually possible to use harvesting as a means of stabilizing the population, which leaves it less vulnerable to fluctuations caused by other external or environmental factors. □

2.5.1 Fishery harvesting and graphical equilibrium analysis

Perhaps the greatest commercial application of harvesting to wildlife management is in fishery management. Fisheries around the world have often been wiped out or reduced to extremely low levels by overfishing (over-harvesting) in the past. An understanding of the effect of harvesting on a fishery population model may be helpful in avoiding such problems. Here again, the most important question is for what harvesting rates H there will be a positive asymptotically stable equilibrium — in other words, how much harvesting can the population sustain without dying out? In the remainder of this section, we shall apply constant-yield harvesting to the Beverton-Holt and Ricker models developed in the previous section. As the models become more complex, we will use a graphical approach to equilibrium analysis which resembles the setup for the cobwebbing approach developed in Section 2.2.

The harvested Beverton-Holt model is

$$y_{k+1} = \frac{ay_k}{1+by_k} - H, \tag{2.31}$$

and its equilibria are the solutions of

$$\frac{ay}{1+by} = y + H, \tag{2.32}$$

Photos courtesy U.S. Fish and Wildlife Service, dls.fws.gov

FIGURE 2.18: At left, fishery biologists pumping eggs from the Tenor River, Alaska, 1966. At right, a fishery biologist returns fish to their habitat. Fishery biologists play a key role in restoring such important species as Atlantic and Pacific salmon, American shad, striped bass, and Great Lakes trout.

given by

$$y_\infty = \frac{1}{2}\left[\left(\frac{a-1}{b} - H\right) \pm \sqrt{\left(\frac{a-1}{b} - H\right)^2 - 4\frac{H}{b}}\right]. \qquad (2.33)$$

With some algebra (see Exercise 4) one can show that these equilibria exist and are positive for $H \le H_c = (\sqrt{a}-1)^2/b$, $a > 1$. (For $H = 0$ these are the equilibria 0 and $\frac{a-1}{b}$ found in the previous section for the Beverton-Holt model without harvesting.) H_c, which is the greatest harvesting yield that allows the population to persist, is called the *critical harvesting yield.* Here $g(y) = \frac{ay}{1+by} - H$ and $g'(y) = \frac{a}{(1+by)^2}$, so the criterion $|g'(y_\infty)| < 1$ for stability becomes $y_\infty > \frac{\sqrt{a}-1}{b}$. With, again, some algebra, one can show (see Exercise 5) that $\frac{\sqrt{a}-1}{b}$ lies between the two equilibria (2.33) when $H \le H_c$, $a > 1$, so that only the larger equilibrium is stable. Thus, as with the logistic model (Example 2.5), any amount of harvesting ($H > 0$) will drive a population with low fertility ($a < 1$) to extinction in finite time (since there are no equilibria). Populations with high fertility ($a > 1$) can survive harvesting as long as the initial population is large enough (greater than the smaller, unstable equilibrium) and there is no over-harvesting (i.e., as long as $H < H_c$).

If we compare these results with those for the Beverton-Holt model without harvesting derived in the previous section, we see that the effects of harvesting on species exhibiting contest competition are threefold:

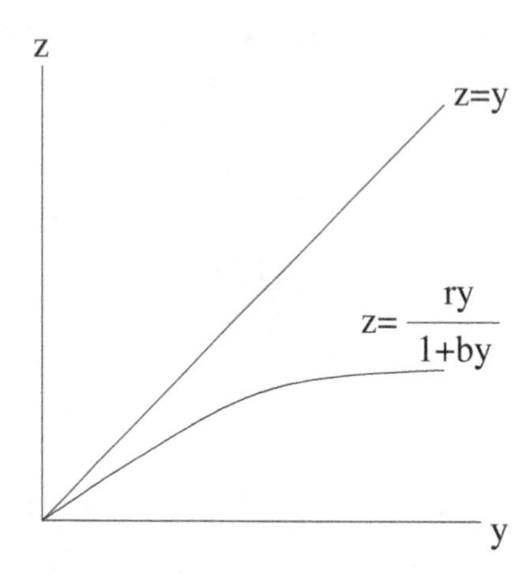

FIGURE 2.19: Beverton-Holt model, $a < 1$, $H = 0$.

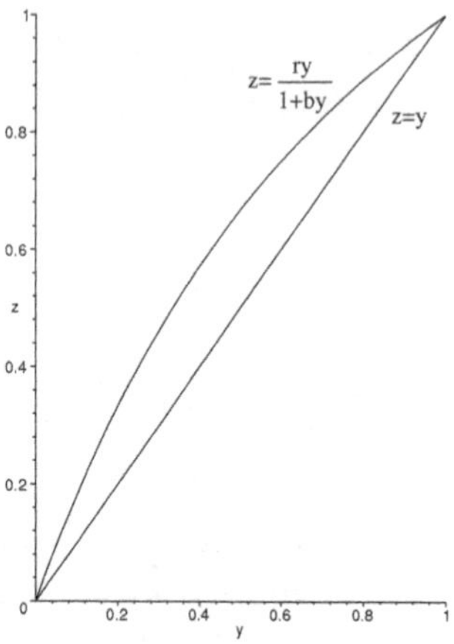

FIGURE 2.20: Beverton-Holt model, $a > 1$, $H = 0$.

- harvesting small populations causes extinction;
- over-harvesting causes extinction;
- extinctions occur in finite time rather than asymptotically.

Exercises 6 and 7 suggest further reflections on these issues.

Alternatively, we can use graphical methods to analyze this model. As observed in Section 2.3, equilibria of the difference equation $y_{k+1} = g(y_k)$ can be identified graphically as intersections of the graphs of the reproduction function $z = g(y)$ and the line $z = y$. These graphs will give us a visual approach to the analysis of this model. For (constant-yield) harvesting models, which take the form $y_{k+1} = g(y_k) - H$, equilibria obey the equation $y + H = g(y)$, so we can instead view them as intersections of the graphs of $z = g(y)$ and $z = y + H$.

Equilibria of the harvested Beverton-Holt model (2.31) are intersections of the graphs of $z = \frac{ay}{1+by}$ and $z = y + H$. The curve $z = \frac{ay}{1+by}$ starts at the origin with slope a and is monotone increasing, approaching the limit $\frac{a}{b}$ as $y \to \infty$. If $a \leq 1$, the line $z = y + H$ lies entirely above this curve, except for an intersection with the curve at $y = 0$ if $H = 0$ (Figure 2.19). Thus if $a < 1$ there is no equilibrium under harvesting and the population dies out. Since this is also what happens for $H = 0$, it is hardly surprising that the population cannot sustain any harvesting.

The case $a > 1$, in which the unharvested population has an asymptotically stable positive equilibrium, is more interesting. In this case, the line $z = y$ starts below the curve at $y = 0$ and intersects it again at a positive value of y, $y = \frac{a-1}{b}$ (Figure 2.20). Increasing H from 0 means moving the line $z = y + H$ up parallel to itself (its slope remains 1, but its y-intercept H increases). For

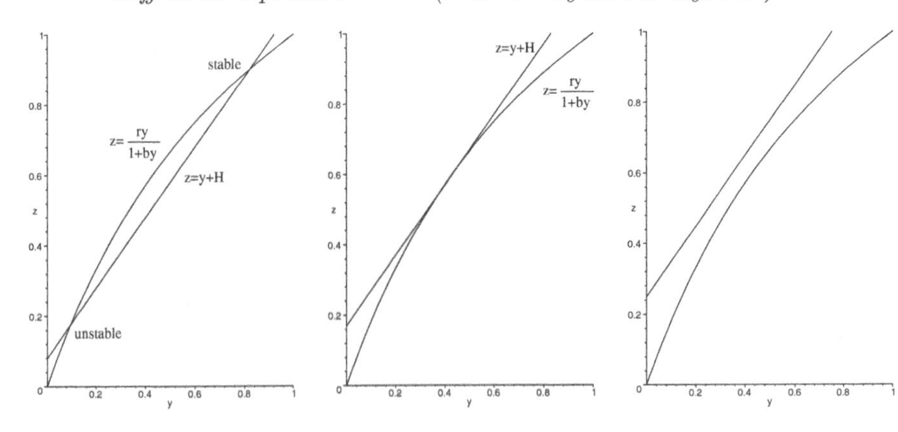

FIGURE 2.21: Beverton-Holt model, $a > 1$, $0 < H < H_c$; two equilibria.

FIGURE 2.22: Beverton-Holt model, $a > 1$, $H = H_c$; single equilibrium.

FIGURE 2.23: Beverton-Holt model, $a > 1$, $H > H_c$; no equilibrium.

sufficiently small H, there are two intersections, with one starting at 0 and increasing and the other starting at $\frac{a-1}{b}$ and decreasing (Figure 2.21). At the smaller intersection, the slope of the curve is greater than the slope of the line (which is 1) and thus by the stability theorem of Section 2.4 this equilibrium is unstable. At the larger intersection, the slope of the curve is positive but less than 1 and so by the stability theorem this equilibrium is asymptotically stable. This means that the limiting population size is the second equilibrium, as long as the initial population size is greater than the first equilibrium (otherwise the population will die out).

However, as H increases, it will reach a value where the two equilibria come together and curve and line are tangent (Figure 2.22). If H increases beyond this value, there will be no equilibrium (Figure 2.23). Thus the qualitative behavior of the system will be that there is an asymptotically stable positive equilibrium until H reaches this critical value, which we denote by H_c, and then the equilibrium jumps to zero. A jump in equilibrium value is called a *catastrophe*, because biologically it corresponds to the disaster of wiping out the population.

It is also possible to use graphs to show that the critical harvest rate is $\frac{(\sqrt{a}-1)^2}{b}$ and that the equilibrium population size at this critical harvest rate is $\frac{\sqrt{a}-1}{b}$. This may be shown by recognizing that, since the graphs of $z = \frac{ay}{1+by}$ and $z = y + H$ are tangent when $H = H_c$ (Figure 2.23), the equilibrium population size at the critical harvest rate is the point on the reproduction curve at which the slope is 1. Substituting this value of y into equation (2.32) and then solving for H will give us the critical harvest rate H_c (see Exercise 3 below).

We now consider a particular application.

EXAMPLE 2.
As reported in the previous section, Reeve, Rhodes and Turchin (1998) found that the southern pine beetle *Dendroctonus frontalis* experiences contest competition early in its life cycle, during oviposition. Their study measured the numbers of eggs deposited in galleries in each attack site on a tree, and, matching their data to the flexible competition model (2.27) proposed by Maynard Smith and Slatkin (1973), gave an estimate of the exponent n which (as also described in the previous section) measures the type of competition exhibited. The estimate $n = 1.5$ corresponds roughly to the Beverton-Holt model. They also gave parameter estimates of $a = 54$ and $b = 0.226$/egg.

Use these estimates to find the expected average (equilibrium) number of eggs per attack site, and the critical harvest rate H_c, the maximum number of eggs which could be removed or parasitized per attack site without driving the population to extinction.

Solution: In the absence of harvesting, the equilibrium value is $(a-1)/b$ as long as $a > 1$ (which it is here). This gives $(54-1)/0.226 \approx 235$ eggs per attack site, with a critical harvest yield

$$H_c = \frac{(\sqrt{a}-1)^2}{b} = \frac{(\sqrt{54}-1)^2}{0.226} \approx 178 \text{ eggs. } \square$$

The harvested Ricker model is

$$y_{k+1} = ay_k e^{-by_k} - H, \tag{2.34}$$

and its equilibria are the solutions of

$$aye^{-by} = y + H, \tag{2.35}$$

the intersections of the graphs of the two expressions. Since this equation is transcendental and cannot be solved explicitly, we will proceed with a graphical analysis. The curve $z = aye^{-by}$ starts at the origin with slope a, increases to a maximum value of $\frac{a}{be}$ at $y = \frac{1}{b}$ and then decreases, approaching a limit of zero as $y \to \infty$. If $a < 1$, the line $z = y + H$ lies above this curve, except for an intersection with the curve at $y = 0$ if $H = 0$, as was the case for the Beverton-Holt model. Thus if $a < 1$ there is again no equilibrium under harvesting, and the population dies out.

If $a > 1$, the analysis proceeds very similarly to the harvested Beverton-Holt model. For $H = 0$ (no harvesting), the line $z = y$ starts below the curve at $y = 0$ and intersects it again at a positive value of y (Figure 2.24). Again, increasing H from zero means moving the line $z = y + H$ up, parallel to itself. For sufficiently small H there are two intersections, with one starting at 0 and increasing, and the other starting at $\frac{\log a}{b}$ and decreasing (Figure 2.25). At the smaller intersection, the slope of the curve is greater than the slope of the line (which is 1) and thus by the stability theorem of Section 2.4 this equilibrium is unstable. At the larger intersection, stability depends on the slope of the

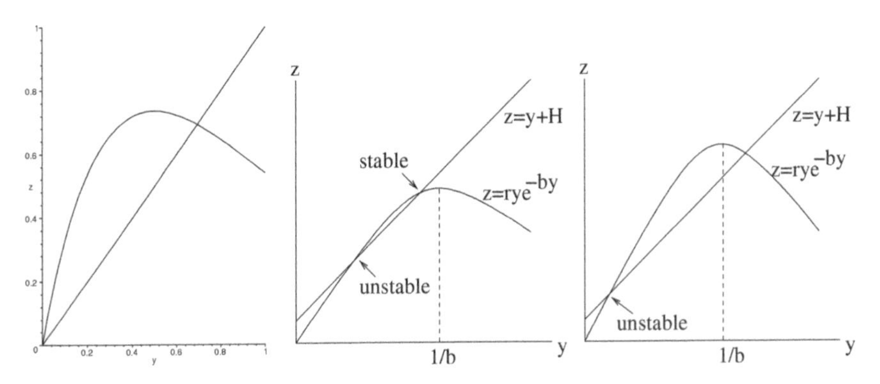

FIGURE 2.24: Ricker model, $r > 1$, $H = 0$.

FIGURE 2.25: Ricker model, $r > 1$, $0 < H < H_c$; second equilibrium has (left) $y < 1/b$, (right) $y > 1/b$.

curve. If the curve and line intersect to the left of $y = 1/b$ (Figure 2.25a), the slope of the curve is positive but less than 1 and by the stability theorem this equilibrium is asymptotically stable. To the right of $y = 1/b$ the curve's slope is negative (Figure 2.25b). If this slope is between 0 and -1 (which is guaranteed if $a < e^2$), the equilibrium will be asymptotically stable, but it is possible for the curve's slope to be less than -1 here, in which case neither equilibrium is stable.

As H increases, it will reach a value H_c where the two equilibria come together and curve and line are tangent (Figure 2.26). If H increases beyond this value, there will be no equilibrium (Figure 2.27). Thus the qualitative behavior of the system will be that for H near H_c there is an asymptotically stable positive equilibrium until H reaches this critical value, past which the population size falls to zero. Again, we have a catastrophe, because this situation corresponds biologically to the disaster of wiping out the population.

By matching the derivatives of the two graphs at the point of tangency when $H = H_c$ (Figure 2.26), we see that the equilibrium population size at this point is the solution y_∞ of the transcendental equation

$$ae^{-by_\infty}(1 - by_\infty) = 1$$

(the left-hand side is the derivative of the reproduction curve; the slope of the line is 1). From the equilibrium condition (2.35) we also see that the critical harvest rate is $aye^{-by} - y$ with this value of y_∞. However, it is not possible to solve the equation (2.35) explicitly for y. In practice, one would use a numerical approximation method to solve (2.35) and then one could calculate the critical harvest rate. It would also be necessary to use a numerical approximation method to find the equilibrium population size for a given harvest rate.

The stability analysis of the harvested Ricker equation is rather more complicated, and we shall not attempt to carry it out. In fact, as with the logistic

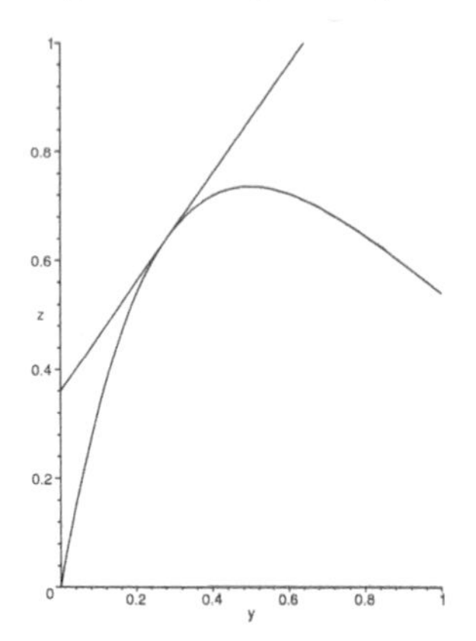

FIGURE 2.26: Ricker model, $r > 1$, $H = H_c$; single equilibrium.

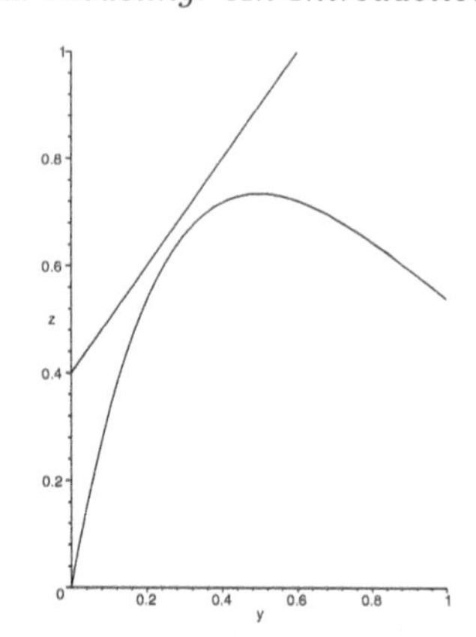

FIGURE 2.27: Ricker model, $r > 1$, $H > H_c$; no equilibrium.

equation (Example 1), it is possible to choose the parameters of the model so that the survival equilibrium with no harvesting is unstable but harvesting tends to stabilize this equilibrium. We can summarize the effects of harvesting on the Ricker model, our representative scramble competition model, by observing that while over-harvesting can of course drive a population to extinction, there is an interval leading up to the critical harvesting rate within which the population tends toward a stable equilibrium, even if without harvesting there would be no stable equilibrium. This shows how careful harvesting can be a tool for stabilizing a population that exhibits scramble competition. In the following section we shall investigate in more detail what happens to a population that reproduces in distinct generations when there is no stable equilibrium.

Exercises

1. Reeve, Rhodes and Turchin (1998) also fit the number of pine beetle oviposition (egg-laying) galleries per attack site on a tree to the Maynard Smith and Slatkin model (2.27) and found that it too exhibits contest competition, with an estimated $n = 0.828$ for naturally occurring attack sites and $n = 0.914$ for field-experiment sites (recall $n = 1$ corresponds to the Beverton-Holt model). For naturally occurring sites, the other parameter estimates were $a = 27.596$ and $b = 0.114$/gallery; for the field-experiment trees, they estimated $a = 25.639$ and $b = 0.133$/gallery. Assume a Beverton-Holt model in both cases.

   (a) Find the equilibrium gallery densities for both types of sites.

   (b) Calculate the critical harvest rate for both types of sites.

(c) Find the limiting numbers of galleries per attack if a certain number H of galleries are "harvested" (removed or prevented by experimenters) for $H = 75$ and 150. Which of the two types of site appears more sensitive to harvesting?

2. Suppose we now apply the parameter estimates from the previous problem ($a = 27.596$ and $b = 0.114$/gallery for naturally occurring attack sites) to a Ricker model with harvesting.

   (a) Estimate the critical harvest rate numerically.

   (b)* Estimate numerically the minimum harvest rate required to stabilize the population (i.e., the positive equilibrium). *[Hint: Plot $g'(y_\infty(H))$ as a function of H and see where it falls inside the interval $(-1, 1)$.]*

3. (a) Show that the point on the curve $z = \frac{ay}{1+by}$ which has slope 1 is $y = \frac{\sqrt{a}-1}{b}$.

   (b) Solve the equation $\frac{ay}{1+by} = y + H$ with $y = \frac{\sqrt{a}-1}{b}$ for H.

4. Show that the equilibria (2.33) of the harvested Beverton-Holt model exist and are positive precisely when $H \leq (\sqrt{a} - 1)^2/b$, $a > 1$. *[Hint: First find conditions on H under which y_∞ is real; then find conditions on H under which y_∞ is positive; finally, show that when $a > 1$ the positivity threshold lies between the two thresholds for y_∞ being real.]*

5. Complete the stability analysis for the harvested Beverton-Holt model by showing that $\frac{\sqrt{a}-1}{b}$ lies between the two equilibria (2.33) if and only if $H \leq (\sqrt{a} - 1)^2/b$, $a > 1$. *[Hint: Begin with the desired inequality and work backward.]*

6. In this section we saw that over-harvesting (harvesting beyond the critical rate) can create a situation in which no nonnegative equilibria exist. Choose one of the models studied in this section, and select parameter values that cause over-harvesting. Select an initial condition y_0 and calculate the next ten generations, $y_1, ..., y_{10}$. What happens to the predicted population size? What do the results tell you about the valid operating range of the model?

7. In this section we saw cases in which harvesting below the critical rate can create a situation in which the smallest positive equilibrium is unstable and there is no zero equilibrium. In such a case, what happens to a population which falls below the level of the smallest equilibrium? (Try some calculations if necessary, to see what happens.) What is a reasonable way to interpret this result biologically?

8. If $y_k < H$ for some k, then it is not possible to harvest H individuals from every generation. To make the harvesting yield more realistic (and the

model well-posed), suggest a way to make yield a function of population size for small populations.

9. Another type of harvesting model is *constant-effort harvesting*, in which a constant "effort" E is applied, and the yield Ey is proportional to the population size — the more fish there are, for example, in a lake, the easier it should be to catch one. The Beverton-Holt model with constant-effort harvesting is

$$y_{k+1} = \frac{ay_k}{1 + by_k} - Ey_k.$$

   (a) Show that this model has a zero equilibrium which is stable when $E - 1 < a < E + 1$.

   (b) Show that this model has a positive equilibrium $\left(\frac{a}{E+1} - 1\right)/b$ which is stable when $E < 1$ or $a < \frac{(E+1)^2}{E-1}$.

   (c) What happens when $a > \frac{(E+1)^2}{E-1}$? when $a < E - 1$? It may help to try some numerical examples or graph these criteria in the $E - a$ plane.

   (d) Based on this analysis, summarize the effects of constant-effort harvesting on a population undergoing contest competition.

10*. Write the difference equation for the logistic model with constant-effort harvesting, and derive the stability criterion for its zero equilibrium.

2.6 Period doubling and chaos

We now return to the question first raised in Section 2.3 of what behavior is possible for a system modeled by a difference equation which has no stable equilibria. We exclude the possibility of unrestricted growth first seen in the simple linear difference equation $y_{k+1} = ry_k$ where $r > 1$ since, as we have discussed, every biological system exists within an environment which has limited (even if very great) resources. We will begin by also returning to the context in which this question first arose.

A population undergoing scramble competition in discrete generations may be described by the logistic equation

$$y_{k+1} = rg(y) = ry_k\left(1 - \frac{y_k}{K}\right), \tag{2.36}$$

which we have seen in Section 2.3 has an asymptotically stable equilibrium at zero for $0 \leq r < 1$ (i.e., the population dies out if the intrinsic growth rate is too low) and an asymptotically stable positive equilibrium $K(1 - \frac{1}{r})$ for $1 < r < 3$ (i.e., the population approaches a steady level for intermediate growth rates). Figures 2.28 and 2.29 show graphical analyses for the latter case, using the initial value $y_0 = K/2$. As the growth rate parameter r passes through the value 3, there must be a fundamental change in the behavior of solutions of (2.36). While the equilibrium $y = K(1 - \frac{1}{r})$ exists for all values of r greater than 1, *no* solution other than the constant solution $y = K(1 - \frac{1}{r})$ approaches it if $r > 3$.

What could happen to a population whose intrinsic growth rate is too high from one generation to the next? The first generation, each of whose members

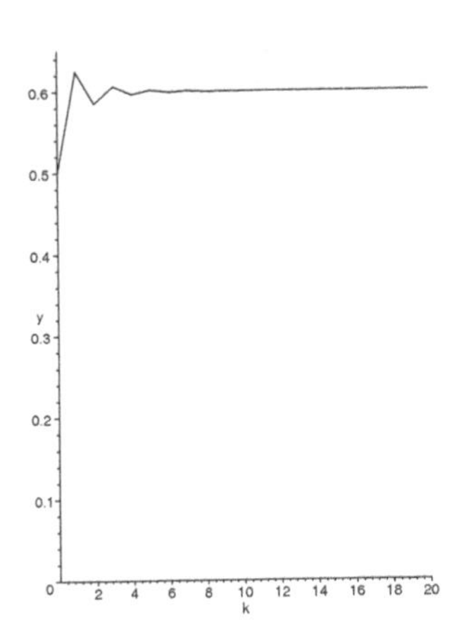

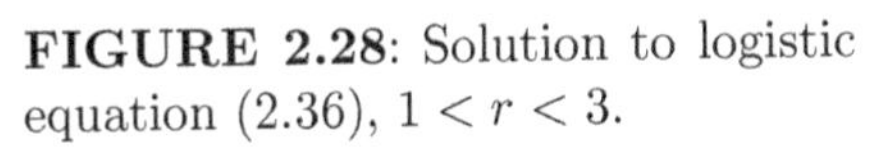

FIGURE 2.28: Solution to logistic equation (2.36), $1 < r < 3$.

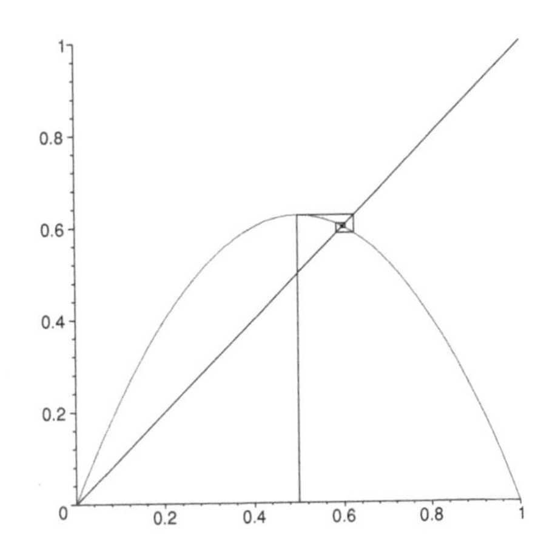

FIGURE 2.29: Cobweb diagram for logistic equation (2.36), $1 < r < 3$.

may have plenty of resources, will produce a second generation which is too great for the local resources to support. If the population exhibits scramble competition as in the logistic model, then the second generation will then be nearly unable to reproduce — that is, the third generation will be quite small, as very few members of the second generation will have had children. The third generation, however, being few in number, will have all the resources needed to reproduce at their intrinsic capacity, and the fourth generation will again be quite large. We observe an oscillation or alternation here between small and large generations, based on three factors: discrete generations (reproduction occurs simultaneously, rather than continually until environmental resources are all in use), scramble competition, and a very high intrinsic reproduction rate. In fact, we see this kind of oscillation in the logistic model when $2 < r < 3$ — for instance, in Figures 2.28 and 2.29, where $r = 2.5$ — but the oscillation dies down, and the population level eventually approaches a steady intermediate state. We might therefore conjecture that populations with an extremely high intrinsic reproduction rate may cause such wild fluctuations that the population size never settles down. Indeed, as we shall see, this is precisely what happens in the logistic model with $r > 3$.

When r increases beyond 3, a solution of period 2 appears in the logistic model. By this we mean that there is a pair of values y^+ and y^- such that $rg(y^+) = y^-$ and $rg(y^-) = y^+$. Thus the alternating sequence $\{y^+, y^-, y^+, y^-, \ldots\}$ is a solution of the difference equation (2.36). Since this solution repeats after two terms, we call it a solution of period 2, or a 2-cycle. We can verify this behavior and find the values of y^+ and y^- either numerically (choose values for the parameters, say $K = 1$ and r just beyond 3, and then iterate the solution many times) or analytically. In order to find the values analytically, we must solve the equation $y_{k+2} = y_k$, i.e., $rg[rg(y)] = y$. As the function $g(y)$ is a quadratic polynomial, this is a fourth degree polynomial equation. We already know two of the roots of this polynomial equation, namely the equilibria $y = 0$ and $y = K(1 - \frac{1}{r})$. By dividing out the factors corresponding to these two roots, we may reduce the fourth-degree equation to the quadratic equation

$$y^2 - K\left(\frac{r+1}{r}\right)y + K^2\left(\frac{r+1}{r^2}\right) = 0$$

whose roots are

$$y_+ = \frac{K}{2r}\left[r + 1 + \sqrt{r^2 - 2r - 3}\right], \; y_- = \frac{K}{2r}\left[r + 1 - \sqrt{r^2 - 2r - 3}\right].$$

These roots are real, and therefore meaningful, provided $r^2 - 2r - 3 = (r - 3)(r + 1) \geq 0$, that is, if $r > 3$ (since we know $r > 0$). Thus, if $r > 3$ the logistic difference equation (2.36) does indeed have a periodic solution of period 2. Figure 2.30 shows the graph of the solution and Figure 2.31 shows the cobweb diagram, again using $y_0 = K/2$.

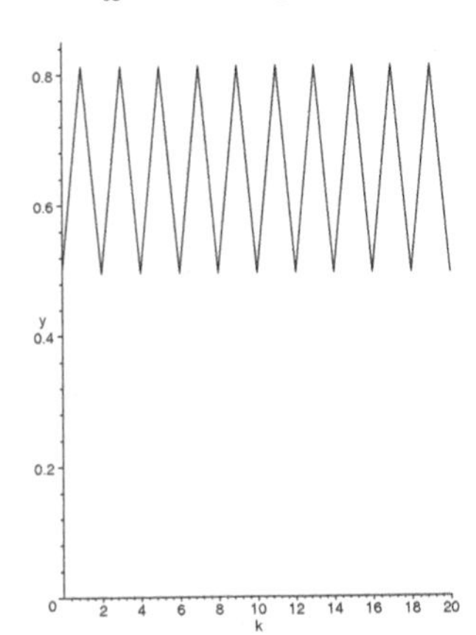

FIGURE 2.30: Solution to logistic equation (2.36), $3 < r < 1 + \sqrt{6}$.

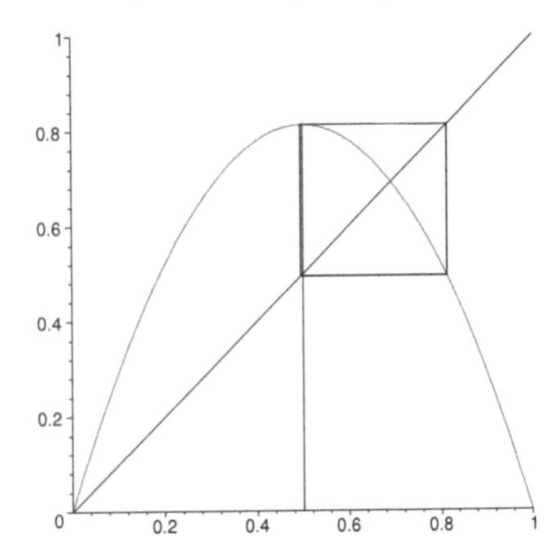

FIGURE 2.31: Cobweb diagram for logistic equation (2.36), $3 < r < 1 + \sqrt{6}$.

This periodic solution can be considered as an equilibrium of the second-order difference equation

$$y_{k+2} = rg[rg(y_k)], \tag{2.37}$$

and more generally the study of 2-cycles of first-order difference equations can be treated in terms of equilibria of second-order difference equations

$$y_{k+2} = g_2(y_k). \tag{2.38}$$

We may find equilibria of such equations by looking for constant solutions; for (2.38) the condition is $y_\infty = g_2(y_\infty)$. The linearization about each equilibrium is obtained by replacing any nonlinear functions by the linear part of their Taylor expansion and forming the corresponding linear difference equation; for the second-order equation (2.38) we replace $g_2(y)$ by $g_2'(y_\infty)u$ and obtain the linearization $u_{k+2} = g_2'(y_\infty)u_k$. Finally, the stability criterion is then obtained by recalling from Section 2.1 that solutions of linear difference equations have the form $u_k = u_0\lambda^k$ for some number λ, and substituting $u_0\lambda^{k+2}$ for u_{k+2} and $u_0\lambda^k$ for u_k; for (2.38) this gives $\lambda^2 = g_2'(y_\infty)$. The criterion $|\lambda| < 1$ for stability then becomes $|g_2'(y_\infty)| < 1$.

For the logistic function $g(y) = y(1 - \frac{y}{K})$, these computations are complicated. Exercises 3, 4 and 5 at the end of this section suggest one way to manage these calculations. It is possible to show that the equilibria y_+^* and y_-^* found above for (2.37) are asymptotically stable (and hence the corresponding 2-cycle of (2.36) is stable) if $3 < r < 1+\sqrt{6} \approx 3.4495$. That is, for r between 3 and 3.4495, every solution of the logistic equation (2.36) approaches a solution of period 2.

For $r > 3.4495$, the solution of period 2 is unstable, but it is possible to show (in the same way) that a solution of period 4 appears, and that this

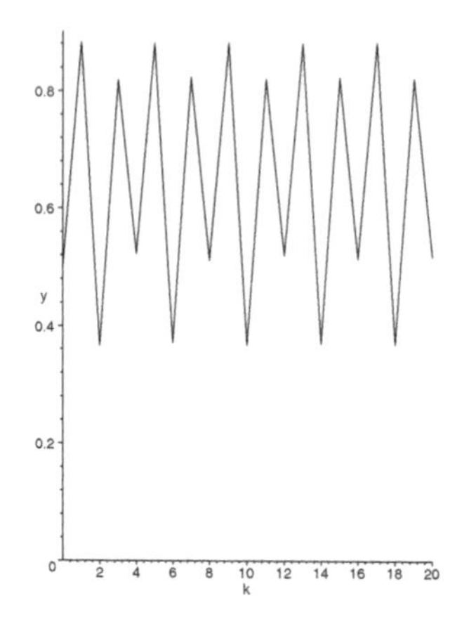

FIGURE 2.32: Solution to logistic equation (2.36), $3.4495 < r < 3.544$.

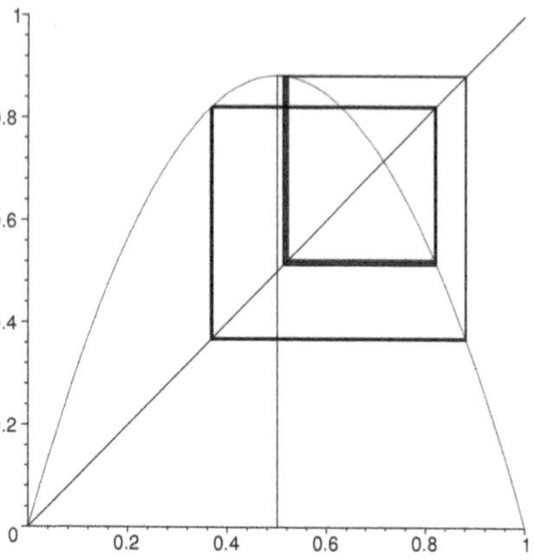

FIGURE 2.33: Cobweb diagram for logistic equation (2.36), $3.4495 < r < 3.544$.

solution is asymptotically stable if $3.4495 < r < 3.544$. Figure 2.32 shows the graph of the solution, and Figure 2.33 shows the cobweb diagram. When the solution of period 4 becomes unstable, a solution of period 8 appears, and this solution is asymptotically stable for $3.544 < r < 3.564$ (see Figures 2.34 and 2.35).

This *period doubling* phenomenon continues until $r = 3.570$, when periodic solutions whose periods are not powers of 2 begin to appear. Figure 2.36 shows what appears to be close to a solution of period 5 for $r = 3.75$, and Figure 2.37 shows the corresponding cobweb diagram. If we iterate more terms, the cobweb diagram looks less like a periodic solution. (Figures 2.38 and 2.39 show a solution for $r = 3.9$.) For any positive integer p there is some value of $r > 3.828$ for which the logistic equation has a periodic solution with period p, but different initial values give solutions whose behaviors are quite different. Such behavior is called *chaotic*.

These facts, whose proofs are difficult and require a close examination of the properties of continuous functions and of the fixed points of iterates of continuous functions, are not restricted to the logistic difference equation. It is a remarkably robust fact that for every difference equation of the form $y_{k+1} = g(y_k)$, where $g(y)$ is a function which increases from zero to a unique maximum and then decreases and approaches zero as $y \to \infty$ (possibly hitting zero for finite y), the period doubling phenomenon and the onset of chaos occur. The Ricker model introduced in Section 2.4 is another example of such a $g(y)$. (The Beverton-Holt model is not, because its $g(y)$ never decreases.)

In fact, the period doubling begins to happen at the same rate: the intervals of r values during which a given solution of period 2^n is stable begin to decrease in size by the same factor, as n increases. More specifically, if r_n is the value of r for which the asymptotically stable solution of period 2^n appears, then

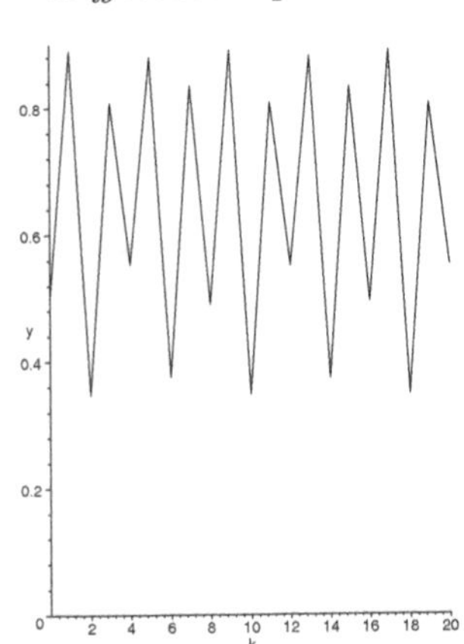

FIGURE 2.34: Solution to logistic equation (2.36), $3.544 < r < 3.564$.

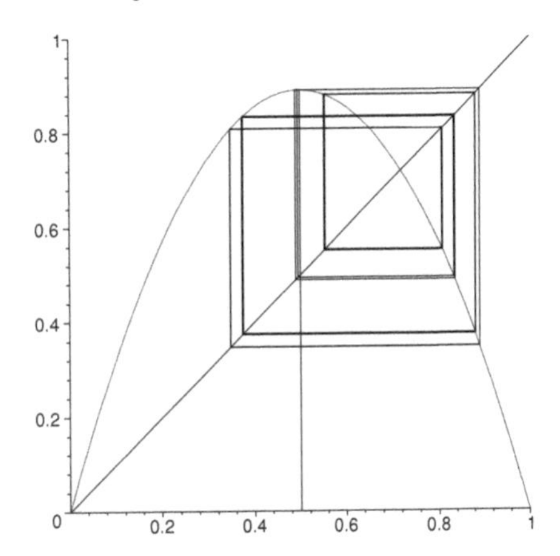

FIGURE 2.35: Cobweb diagram for logistic equation (2.36), $3.544 < r < 3.564$.

for the logistic equation, $r_1 = 3$, $r_2 \approx 3.4495$, $r_3 \approx 3.544$, and $r_4 \approx 3.564$. The period 2 solution is stable for an interval of length $r_2 - r_1 \approx 0.4495$, the period 4 solution for an interval of length $r_3 - r_2 \approx 0.0945$, and the period 8 solution for $r_4 - r_3 \approx 0.020$. Comparing these periods,

$$\frac{r_2 - r_1}{r_3 - r_2} \approx 4.73, \quad \frac{r_3 - r_2}{r_4 - r_3} \approx 4.725.$$

The periods appear to be shortening by a factor of about 4.7 each time. For any model of the form described above (logistic or otherwise), it has been shown that this factor approaches a limit,

$$\lim_{n \to \infty} \frac{r_{n+1} - r_n}{r_{n+2} - r_{n+1}} = 4.66920176\ldots,$$

a number which is called the *Feigenbaum constant.* Usually, the limiting value is approached very rapidly. This means that the period doubling values of r occur closer and closer together.

One way to illustrate period doubling and chaotic behavior is by means of a *bifurcation diagram*, which we shall discuss in greater detail in Section 4.4. A bifurcation diagram in general plots equilibria as a function of the parameter r. If there is an orbit of period 2 for a given value of r, then the bifurcation diagram will have two points for that value of r, and similarly an orbit of any period will show in the bifurcation diagram as a number of points equal to the period of the orbit for that value of r. To draw a bifurcation diagram for, say, the logistic difference equation, we iterate the logistic function for a sequence of values of r and plot the results obtained starting after enough iterations for the orbit to have approached the periodic orbit. This may be done using

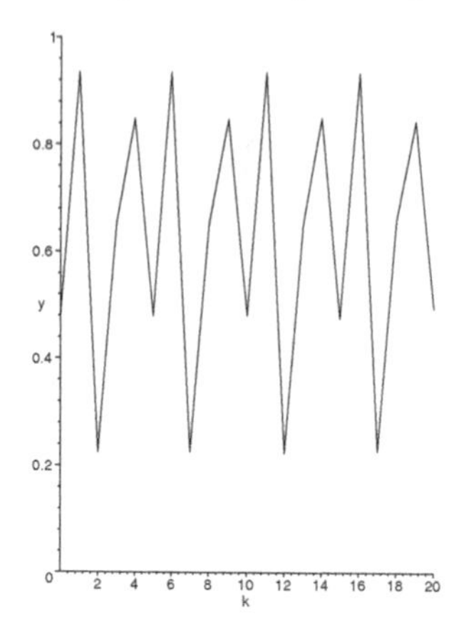

FIGURE 2.36: Solution to logistic equation (2.36), $r = 3.75$.

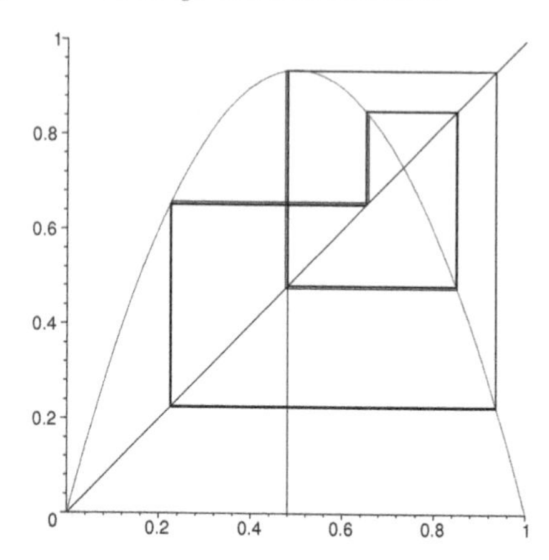

FIGURE 2.37: Cobweb diagram for logistic equation (2.36), $r = 3.75$.

a computer algebra system such as Maple (for which a program is given in Appendix A), and the result is displayed in Figure 2.40.

From a biological point of view, these results are also remarkable. Chaotic behavior in a population model is a very disturbing phenomenon, as it challenges our notion that if we know the initial state of a system and the law governing its development we should be able to predict the behavior. In chaotic behavior, two solutions whose initial values are very close together — so close that their difference is smaller than we can measure experimentally — may behave very differently. This means that knowing initial values to within as low a tolerance as we like is still not enough to allow us to predict the long-term behavior of the system.

One interpretation of the presence of chaos in the logistic equation is that even very simple models can give rise to apparently unpredictable behavior. This suggests that complicated behavior may arise even under simple governing laws. There is experimental data on insect populations which appears to support the possibility of chaotic behavior. Chaotic behavior would mean that experimental results may not be repeatable. In the range of values of the parameter r for which there is an asymptotically stable solution, we may predict behavior. In the chaotic range we will have to be satisfied with less information; perhaps upper and lower bounds can be obtained for solutions even if it is not possible to describe behavior more precisely.

It should also be noted, however, that (as will be seen in the following chapters) chaotic behavior occurs for single-population models only in discrete-time (difference) equations, where the built-in time lags due to having discrete time steps allow overshooting an equilibrium. Biologically this suggests that species which reproduce in distinct generations may be more subject to complicated fluctuations from one generation to the next. The foregoing discussion also highlights intraspecies scramble competition as a biological mechanism which

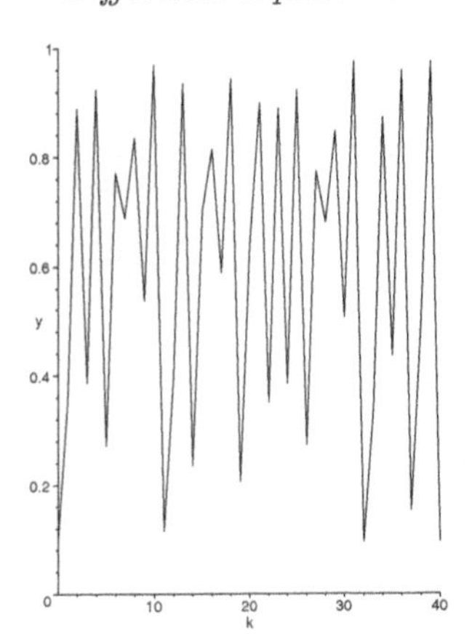

FIGURE 2.38: Solution to logistic equation (2.36), $r = 3.9$.

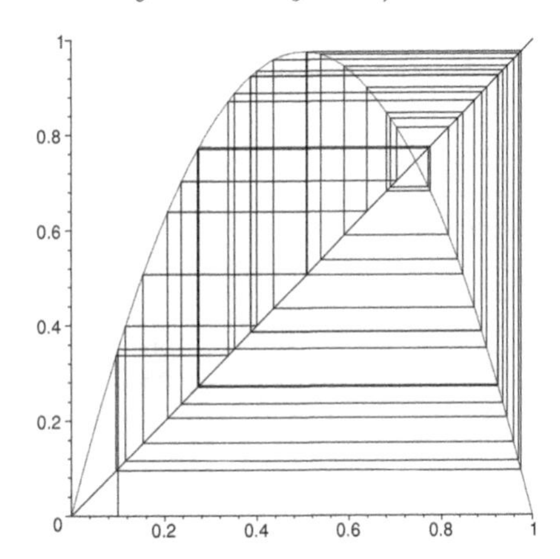

FIGURE 2.39: Cobweb diagram for logistic equation (2.36), $r = 3.9$.

opens the door to chaos for such populations. We may wonder what other biological mechanisms may produce chaotic behavior in populations with distinct generations. Although we cannot provide an exhaustive answer to this question, in the remainder of this section we shall consider two other examples of biological phenomena which have been associated with chaotic behavior.

2.6.1 Dispersal

The term *dispersal* refers to spatial redistribution of some or all of a population, often in response to population pressures (intraspecies competition as population density builds in a single area). While models for biological dispersal typically keep track of population sizes in several different habitats or patches (as we shall see in the following section), some simple studies have been made on the effects of dispersal in a single patch.

Since dispersal from a single source tends to reduce the population density in that location, we might suspect that it will also counter the chaotic tendencies of populations with high reproduction rates. We can see this very simply if we consider the usual form of dispersal, in which a given proportion d of individuals leave the patch (habitat) under study following reproduction. In this case, the model $y_{k+1} = g(y_k)$ becomes

$$y_{k+1} = (1 - d)g(y_k),$$

and the effective growth ratio r as in the logistic equation is reduced to $(1 - d)r < r$. For the logistic model, the criterion for avoiding period doubling and chaos then becomes $r < 3/(1 - d)$, which is higher (less restrictive) than 3 for $0 < d < 1$.

We might also consider migration into the patch, that is, dispersal from

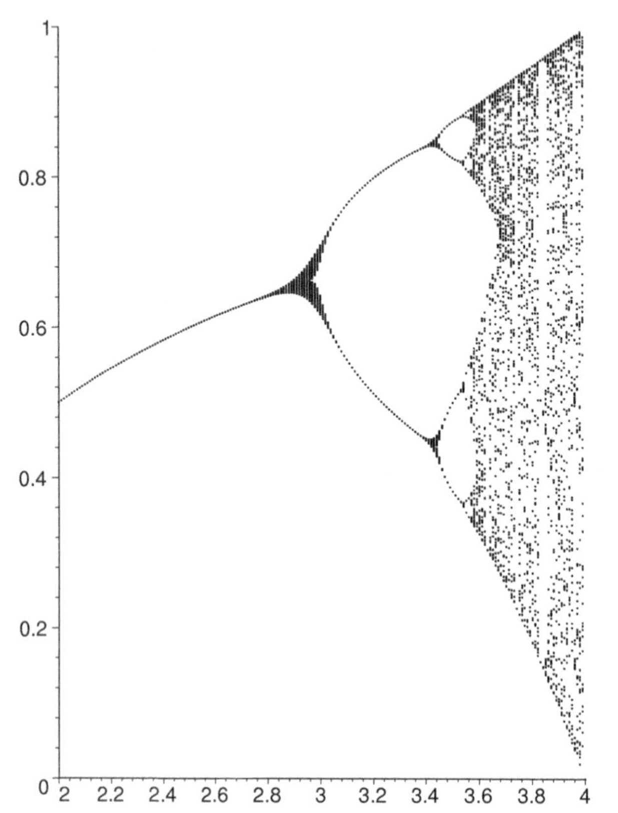

FIGURE 2.40: Bifurcation diagram for the logistic difference equation, with the parameter r on the horizontal axis.

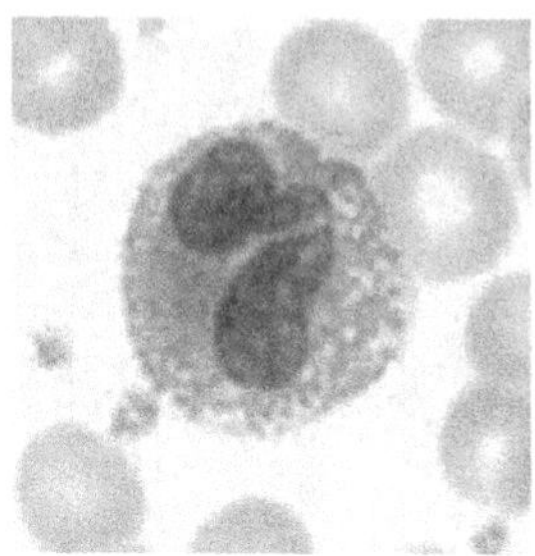

Photo courtesy Lennert B.

FIGURE 2.41: An eosinophil, a type of granulocyte, a specialized white blood cell, surrounded here by erythrocytes (red blood cells). Granulocytes fight infection by phagocytosis; eosinophils are identifiable by their multi-lobed nuclei. Their production exhibits regular oscillation patterns but can be affected by diseases such as leukemia.

other nearby patches into this one. Since we are not modeling the population densities in the other patches, we incorporate immigration as a constant c, making the basic form

$$y_{k+1} = g(y_k) + c. \tag{2.39}$$

The result of this immigration is to increase the equilibrium value y_∞ above what it would be without immigration, but the effect upon $|g'(y_\infty)|$ is difficult to predict, and so in terms of simplicity of dynamics, constant migration may either stabilize or destabilize a population's dynamics. By the same token, then, a constant dispersal (i.e., $c < 0$, as opposed to the proportional dispersal d discussed above) can also have either effect, in addition to reducing the value of the positive equilibrium. Note that constant dispersal is equivalent to constant-yield harvesting, so a high enough dispersal rate will drive a population to extinction in finite time (as seen in Section 2.5). Doebeli discusses[17] a way to use variable dispersal to stabilize a population's dynamics by making the magnitude of the dispersal dependent on population density.

We shall now look briefly at examples where immigration and dispersal can either stabilize or destabilize a population. We shall follow Doebeli's lead in using the flexible model (2.27) proposed by Maynard Smith and Slatkin (1973) to describe a population's natural growth. Constant migration applied to this model gives us the equation

$$y_{k+1} = \frac{ay_k}{1+(by_k)^n} + c. \tag{2.40}$$

EXAMPLE 1.
Describe the behavior of the Maynard Smith and Slatkin model with constant immigration (2.40) and parameter values $a = \frac{3}{2}$, $b = 1$/pop. unit, and $n = 5$, for migration values of $c = 0$, $c = \frac{1}{2}$, and $c = 1$ pop. units.

Solution: The parameter values given correspond to scramble competition ($n = 5$) with an intermediate intrinsic growth ratio($a = 3/2$). We may then expect the positive equilibrium to be stable, and it is. We calculate

$$y_\infty = \frac{(a-1)^{1/n}}{b} = \left(\frac{1}{2}\right)^{1/5} \approx 0.870551, \quad g'(y_\infty) = \frac{1-(a-1)(n-1)}{a} = -\frac{2}{3},$$

so without immigration ($c = 0$) the population density will stabilize around 0.87. For immigration of $c = 1/2$ per generation, we calculate $y_\infty \approx 1.1146$, $g'(y_\infty) \approx -1.19213$; the immigration has destabilized the equilibrium, causing periodic behavior. For immigration of $c = 1$ per generation, however, we find $y_\infty \approx 1.36049$, $g'(y_\infty) \approx -0.825852$, signaling damped oscillations and a return to stability. □

What happens here is that a little immigration increases the first generation just enough to make the second generation considerably smaller — that is, the decrease in natural population size outweighs the boost given by immigration — but a lot of immigration will end up dominating the population's natural growth, so that each new generation will consist almost entirely of newly arrived immigrants, and very few descendants of the previous generation (which was so numerous it was practically unable to reproduce).

EXAMPLE 2.
Describe the behavior of the Maynard Smith and Slatkin model with constant dispersal (2.40) and parameter values $a = 10$, $b = 1$/pop. unit, and $n = 2$, for dispersal values of $c = 0$ and $c = -1$ pop. units.

Solution: The parameter values correspond to intermediate or very weak

[17]M. Doebeli, Dispersal and dynamics, *Theoretical Population Biology* 47: 82–106, 1995.

scramble competition ($n = 2$) with an extremely high intrinsic growth ratio ($a = 10$). Here the competition is so weak that despite the high growth ratio, the positive equilibrium is stable in the absence of dispersal. We compute

$$y_\infty = \frac{(a-1)^{1/n}}{b} = 3, \quad g'(y_\infty) = \frac{1-(a-1)(n-1)}{a} = -\frac{4}{5},$$

verifying stability for $c = 0$. With a dispersal of 1 pop. unit per generation, however, we compute $y_\infty \approx 2.47419$, $g'(y_\infty) \approx -1.00983$, which is just enough to destabilize the equilibrium. As the dispersal amount increases, things get worse; for $c < -4.085$, there are no positive equilibria at all, and the population must "crash" to zero in finite time. Here, in some sense, dispersal magnifies the effect of the competition, so that instability occurs. □

Of course, a full study of the effects of dispersal should include multi-patch models in order to see how dispersal between two or more neighboring habitats affects population densities in each location, and how the single-patch dynamics interact. We shall not attempt a complete discussion of dispersal in this text, but will return to the idea in the next section, when we take up systems of interacting populations with discrete generations.

2.6.2 Dynamical diseases and physiological control systems

Many physiological systems normally display predictable patterns, either remaining almost constant or with regular oscillations. However, there are diseases, called *dynamical diseases*, which exhibit changes in these patterns such as changes in the nature of the oscillations. Generally, these diseases occur in physiological control systems. One area in which dynamical diseases have been observed is the blood system. Some forms of anemia and leukemia have been identified as dynamical diseases.

Except for lymphocytes, a type of white blood cell produced in lymphatic tissues, blood cells are formed from primitive stem cells in the bone marrow. The development process takes about six days. There is also a steady elimination process whose rate depends on the type of cell. For granulocytes (a kind of white blood cell, see Figure 2.41), the cell destruction rate is about 10% per day. While the cell production process is not well understood, it appears that the production rate should be small for small cell density levels, increasing to a maximum, and then decreasing to zero as cell density increases (a form similar to scramble competition reproduction functions). One form which Mackey and Glass suggested and used successfully in fitting data[18] is a production

[18] M.C. Mackey and L. Glass, Oscillation and chaos in physiological control systems, *Science* 197: 287–289, 1977; L. Glass and M.C. Mackey, Pathological conditions resulting from instabilities in physiological control systems, *Ann. N.Y. Acad. Science* 316: 214–235, 1979.

rate in terms of the cell density y, of the form

$$p(y) = \frac{b\theta^n y}{\theta^n + y^n}$$

with positive constants b, θ, n. We assume that the elimination rate has the form cy, and this suggests that if the production and elimination processes occurred at discrete intervals, we might model the blood cell production by a difference equation

$$y_{k+1} = y_k + \frac{b\theta^n y_k}{\theta^n + y_k^n} - cy_k = \frac{b\theta^n y_k}{\theta^n + y_k^n} + (1-c)y_k. \tag{2.41}$$

However, the actual process is continuous but with a time lag corresponding to the length of the production process. If there were no time lag, we would model the process by a differential equation using the balance relation that the rate of change of the blood cell density is equal to the production rate minus the elimination rate. This would give a continuous model

$$y' = \frac{b\theta^n y}{\theta^n + y^n} - cy.$$

We will see in Chapter 4 how to analyze such models, but in overlooking the time lag in the production process we actually lose an essential feature of the system. To incorporate a time lag τ in the production process, we would have to use as model a *differential-difference equation*

$$y'(t) = \frac{b\theta^n y(t-\tau)}{\theta^n + [y(t-\tau)]^n} - cy(t). \tag{2.42}$$

The analysis of differential-difference equations is complicated, well beyond the scope of this book. Instead, we shall use a difference equation model as a way of simulating the time lag. We think of the production time lag as the unit of time in a discrete model, and then we obtain (2.41) as a very simplified model.

Some parameter values which have been estimated for the model (2.42) are $b = 0.2/\text{day}$, $c = 0.1/\text{day}$, $\theta = 1$, $n = 10$. In order to use these for (2.41), we must change the time scale. In (2.41) we take 6 days (the time lag) as the unit of time, and this means we should use $b = 6(0.2) = 1.2$. Over 6 days, an elimination of 10% per day ($c = 0.1/\text{day}$) leads to a remainder of $(1-c)^6 = (0.9)^6 = 0.53$. Thus we should use $c = 0.47$ in (2.41).

Equilibria of (2.41) are given by $y = 0$ and $c = \frac{b\theta^n}{\theta^n + y^n}$. Thus there is a positive equilibrium $y = \theta\left(\frac{b-c}{c}\right)^{1/n}$ provided $b > c$. With the function $g(y) = \frac{b\theta^n y}{\theta^n + y^n} + (1-c)y$, we may calculate $g'(0) = b + 1 - c$, and from this the stability condition $-1 < g'(0) < 1$ reduces to $0 < c - b < 2$. The determination of the stability condition for the positive equilibrium is more complicated, but it can be calculated that this equilibrium is asymptotically stable if

$$0 < c\left[1 - \frac{nc - (n-1)b}{b^2}\right] < 2.$$

Neither of these conditions is satisfied with the parameter values $c = 0.47$, $b = 1.2$ obtained above. Simulations with these parameter values indicate oscillations with a period of about 20 days (3.33 time units). In chronic myelogenous leukemia, it is thought that the production time of blood cells increases. An increase in the time units would cause changes in the parameters b and c in the model (2.41) and would lead to more irregular behavior. This is not incompatible with observations, although the state of knowledge of the mechanisms and of experimental data precludes real comparisons.

Exercises

1. Find the value of r for which a solution of period 2 appears for the difference equation
$$y_{k+1} = ry_k e^{1-y_k}.$$

2. Consider the Hassell model $y_{k+1} = ay_k(1+\frac{b}{n}y_k)^{-n}$ with $a, b > 0$, $n > 1$. Show that this model does not exhibit period doubling if $n < 2$ and find the value of a for which a solution of period 2 appears if $n > 2$. [Hint: See Exercise 12, Section 2.4.]

3. Let $\{y_+, y_-\}$ be a solution of period 2 of the difference equation $y_{k+1} = g(y_k)$. Show that both y_+ and y_- are equilibria of the second-order difference equation $y_{k+2} = g[g(y_k)]$.

4. * Define a new index $m = \frac{k}{2}$ for k even and define the iterated function $g_2(y) = g[g(y)]$. Show that the solutions y_+, y_- of period 2 from Exercise 3 are equilibria of the first-order difference equation $y_{m+1} = g_2(y_m)$. [*Remark:* Exercise 4 and the stability theorem of Section 2.3 show that an equilibrium $y*$ of this second-order difference equation is asymptotically stable if $|g_2'(y*)| < 1$. Exercise 5 below gives another criterion for the asymptotic stability of a solution of period 2 of the difference equation $y_{k+1} = g(y_k)$.]

5. * (a) Let $\{y_+, y_-\}$ be a solution of period 2 of the difference equation $y_{k+1} = g(y_k)$. Use the chain rule of calculus to show that if $g_2(y) = g[g(y)]$, then
$$g_2'(y_+) = \left[\frac{d}{dy}\{g[g(y)]\}\right]_{y=y_+} = g'(y_-)g'(y_+).$$

   (b) Deduce from part (a) that the solution of period 2 is asymptotically stable if $|g'(y_-)g'(y_+)| < 1$.

6. * Doebeli (1995) writes regarding constant dispersal (or immigration) applied to the Maynard Smith and Slatkin (1973) model (2.27), "one can choose the parameters in [this equation] so that ... dispersal destabilizes the system. For example, [the equation] can have a stable equilibrium

while dispersal leads to a 2-cycle." Show that the possibility of dispersal or immigration destabilizing the positive equilibrium of (2.27) depends on a and n but not on b. More specifically, show that for the equation (2.40)

(a) there is an interval of negative values for c (i.e., dispersal) which cause a stable equilibrium to become unstable if and only if $1 < n < 3$ and the value of a lies between the curves $a = \frac{4n}{(n-1)^2}$ and $a = \frac{n}{n-2}$ (for $2 < n < 3$);

(b) there is an interval of positive values for c (i.e., immigration) which cause a stable equilibrium to become unstable if and only if $n > 3$ and

$$\max\left(1, \frac{4n}{(n-1)^2}\right) < a < \frac{n}{n-2}.$$

[Hint: Note that in both cases we need $|g'(y_\infty)| < 1$ and $|g'(y_*)| > 1$, where y_∞ is the equilibrium of the original model ($c = 0$) and y_* is the point where $g'(y)$ reaches a minimum. Since the equilibrium of the model (2.39) with dispersal or immigration increases with c, a third and final condition is to place y_* on whichever side of y_∞ will make the equilibrium approach y_* as c is moved from 0.]

7. Use a graphical argument similar to those made in Section 2.5 to show that a constant immigration term as in (2.39) increases the value of the fixed point y_∞.

8. The proportional dispersal model $y_{k+1} = (1 - d)g(y_k)$ discussed in this section assumes that individuals disperse *after* reproduction, so that in order to calculate the size of the next generation, we first apply the reproduction function, and then dispersal, to the previous generation y_k. Suppose instead that dispersal occurs *before* reproduction, or, equivalently, that we measure the size of each generation immediately after reproduction occurs.

   (a) What is the general form of the resulting model?

   (b) How does the effective intrinsic growth rate compare to that of the original proportional dispersal model if $g(y)$ is logistic?

9. Identify and discuss at least three modeling issues raised in the discussion of the blood cell model in this section.

10. What practical difference is there between constant-yield harvesting and constant-rate dispersal?

2.7 Structured populations

The simplifying assumption of homogeneity within a population is sometimes unrealistic because of important differences among its members. In many such cases, one can capture the essential variations among members by breaking the population into subgroups, called *compartments*, with movement often possible from one compartment to another. Two of the most common ways to structure a population are by age and by spatial distribution . In this section we shall consider simple examples of both of these kinds of structure.

The modeling of structured population systems requires a system of two or more equations. For simplicity, we shall consider here only systems of two equations, which may also be referred to as *two-dimensional* systems, but the methods apply equally to systems of higher dimension than two. First, however, we must consider how the mathematical tools we developed for single difference equations extend to systems of difference equations.

A general system of two first-order difference equations has the form

$$\begin{aligned} y_{k+1} &= f(y_k, z_k), \\ z_{k+1} &= g(y_k, z_k), \end{aligned} \tag{2.43}$$

where the growth of each new generation depends on the sizes of both populations in the previous generation. Algebraically, an equilibrium of this system is a solution of the pair of equations

$$\begin{aligned} y &= f(y, z), \\ z &= g(y, z). \end{aligned} \tag{2.44}$$

Geometrically, an equilibrium is an intersection of the two curves $y = f(y, z)$ and $z = g(y, z)$ in the $y-z$ plane. If (y_∞, z_∞) is an equilibrium of (2.43), then the system has a constant solution $y_k = y_\infty$, $z_k = z_\infty$ $[k = 0, 1, 2, \ldots]$.

The description of the behavior of solutions of systems near an equilibrium parallels the description given in Section 2.3 for a single first-order difference equation. If (y_∞, z_∞) is an equilibrium of (2.43), we make the change of variables $u_k = y_k - y_\infty$, $v_k = z_k - z_\infty$ $[k = 0, 1, 2, \ldots]$, so that (u_k, v_k) represents deviation from the equilibrium. We then have the system

$$\begin{aligned} u_{k+1} &= f(y_\infty + u_k, z_\infty + v_k) - y_\infty = f(y_\infty + u_k, z_\infty + v_k) - f(y_\infty, z_\infty), \\ v_{k+1} &= g(y_\infty + u_k, z_\infty + v_k) - z_\infty = g(y_\infty + u_k, z_\infty + v_k) - g(y_\infty, z_\infty). \end{aligned} \tag{2.45}$$

If we use Taylor's theorem to approximate $f(y_\infty + u_k, z_\infty + v_k)$ and $g(y_\infty + u_k, z_\infty + v_k)$ by their linear parts and neglect the remainder terms,

$$\begin{aligned} f(y_\infty + u_k, z_\infty + v_k) &\approx f(y_\infty, z_\infty) + f_y(y_\infty, z_\infty)u_k + f_z(y_\infty, z_\infty)v_k, \\ g(y_\infty + u_k, z_\infty + v_k) &\approx g(y_\infty, z_\infty) + g_y(y_\infty, z_\infty)u_k + g_z(y_\infty, z_\infty)v_k, \end{aligned}$$

where f_y, f_z, g_y, g_z denote the partial derivatives of the functions f and g with respect to the variables y and z, respectively, then we obtain a linear system

$$\begin{aligned} u_{k+1} &= f_y(y_\infty, z_\infty)u_k + f_z(y_\infty, z_\infty)v_k, \\ v_{k+1} &= g_y(y_\infty, z_\infty)u_k + g_z(y_\infty, z_\infty)v_k \end{aligned} \tag{2.46}$$

called the *linearization of the system* (2.43) *at the equilibrium* (y_∞, z_∞), which approximates the true system (2.43) near the equilibrium (y_∞, z_∞). The analogue of the linearization theorem of Section 2.3 is true for systems: If all solutions of the linearization at an equilibrium approach zero, then the equilibrium is asymptotically stable.

The condition that all solutions of the linearization approach zero is more complicated for systems than it was for a single equation, however. The idea behind solving this problem begins by supposing that solutions of the linearization have the form $u_k = u_0\lambda^k$, $v_k = v_0\lambda^k$ for some number λ, since this is the form that solutions to single linearized difference equations take (cf. equation (2.7) of Section 2.1). We then find the conditions that make $|\lambda| < 1$. If $|\lambda| < 1$, then $\lim_{k\to\infty} \lambda^k = 0$, so solutions $u_k = u_0\lambda^k$, $v_k = v_0\lambda^k$ of the linearization approach zero (that is, deviations from the equilibrium approach zero, so the equilibrium must be locally asymptotically stable).

In order to find solutions to (2.46) of the form given above, it must be possible to have nontrivial solutions (u_0, v_0) to the system of linear equations

$$\begin{aligned} u_1 &= f_y(y_\infty, z_\infty)u_0 + f_z(y_\infty, z_\infty)v_0 = \lambda u_0, \\ v_1 &= g_y(y_\infty, z_\infty)u_0 + g_z(y_\infty, z_\infty)v_0 = \lambda v_0. \end{aligned}$$

If we rewrite these equations in matrix form,

$$\begin{bmatrix} u_1 \\ v_1 \end{bmatrix} = \begin{bmatrix} f_y(y_\infty, z_\infty) & f_z(y_\infty, z_\infty) \\ g_y(y_\infty, z_\infty) & g_z(y_\infty, z_\infty) \end{bmatrix} \begin{bmatrix} u_0 \\ v_0 \end{bmatrix} = \lambda \begin{bmatrix} u_0 \\ v_0 \end{bmatrix},$$

and denote the 2×2 matrix above as $A(y_\infty, z_\infty)$, and the vector $[u_0 \ v_0]^T$ as $\vec{u}_0$, then we can write the equation in the form $A\vec{u}_0 = \lambda\vec{u}_0$, or

$$(A - \lambda I)\vec{u}_0 = 0, \tag{2.47}$$

where I is the 2×2 identity matrix. From linear algebra, we know that the only way for (2.47) to have nonzero solutions $\vec{u}_0$ is for the matrix $(A - \lambda I)$ to be singular. That is, this system has nontrivial solutions if λ satisfies

$$\det[A(y_\infty, z_\infty) - \lambda I] = 0. \tag{2.48}$$

(Students of linear algebra will recognize λ as the eigenvalue(s) of A.) Equation (2.48) is called the *characteristic equation* of this system, and for two-dimensional systems can be written in scalar form as

$$\lambda^2 - \operatorname{tr} A(y_\infty, z_\infty)\lambda + \det A(y_\infty, z_\infty) = 0, \tag{2.49}$$

where tr A and det A denote the trace and determinant of A, namely

$$\begin{aligned}\text{tr } A(y_\infty, z_\infty) &= f_y(y_\infty, z_\infty) + g_z(y_\infty, z_\infty),\\ \det A(y_\infty, z_\infty) &= f_y(y_\infty, z_\infty)\, g_z(y_\infty, z_\infty) - f_z(y_\infty, z_\infty)\, g_y(y_\infty, z_\infty).\end{aligned}$$

Our result, therefore, is that all solutions of the linearization (2.46) approach zero if all roots λ of the characteristic equation (2.48) or (2.49) satisfy $|\lambda| < 1$. In this matrix form, the stability theorem of Section 2.3 generalizes to two-dimensional systems, and indeed to systems of any dimension. (Note that for single difference equations, A and I are simply scalars, and the characteristic equation for $y_{k+1} = g(y_k)$ at y_∞ has the unique solution $\lambda = A = g'(y_\infty)$.) We shall state this result first for a system of any dimension, and then more specifically for two-dimensional systems.

EQUILIBRIUM STABILITY THEOREM (I): Let

$$\vec{y}_\infty = (y_{1_\infty}, y_2, \ldots, y_{n_\infty})$$

be an equilibrium of the system of difference equations

$$\begin{aligned} y_{1_{k+1}} &= f_1(y_{1_k}, y_{2_k}, \ldots, y_{n_k}),\\ y_{2_{k+1}} &= f_2(y_{1_k}, y_{2_k}, \ldots, y_{n_k}),\\ &\cdots\\ y_{n_{k+1}} &= f_n(y_{1_k}, y_{2_k}, \ldots, y_{n_k}),\end{aligned}$$

and let $A(\vec{y}_\infty)$ be the matrix of the linearization of the system at this equilibrium. If all roots of the characteristic equation (2.48) satisfy $|\lambda| < 1$, then the equilibrium is asymptotically stable, and if the characteristic equation has a root with $|\lambda| > 1$, the equilibrium is unstable.

There is a set of conditions, known as the *Jury criterion*, which gives necessary and sufficient conditions that all roots of a polynomial equation satisfy $|\lambda| < 1$. (In general, for an equation of degree n, the Jury criterion consists of n different inequalities.) The characteristic equation at an equilibrium of a two-dimensional system of difference equations is a quadratic equation. The Jury criterion for two-dimensional systems is as follows: Both roots of the quadratic equation

$$\lambda^2 + a_1\lambda + a_2 = 0$$

satisfy $|\lambda| < 1$ if and only if

$$|a_1| < a_2 + 1 < 2.$$

Appendix C offers a proof of the Jury criterion for two-dimensional systems.

For the characteristic equation (2.48), since $a_1 = -\text{tr } A(y_\infty, z_\infty)$, $a_2 = \det A(y_\infty, z_\infty)$, we obtain the following stability criterion:

EQUILIBRIUM STABILITY THEOREM (II): Let (y_∞, z_∞) be an equilibrium of the two-dimensional system of difference equations (2.43). If $A(y_\infty, z_\infty)$ is the matrix of the linearization of the system at the equilibrium, and if

$$0 < |\text{tr } A(y_\infty, z_\infty)| < \det A(y_\infty, z_\infty) + 1 < 2, \tag{2.50}$$

then the equilibrium is asymptotically stable.

EXAMPLE 1.
For each equilibrium of the system $y_{k+1} = 2z_k$, $z_{k+1} = y_k^2$, determine whether it is asymptotically stable.

Solution: Equilibria are solutions of $y = 2z$, $z = y^2$. Substituting the first of these equations into the second, we obtain $z = 4z^2$, which has two solutions, $z = 0$ and $z = \frac{1}{4}$. If $z = 0$, then $y = 0$, and if $z = \frac{1}{4}$, then $y = \frac{1}{2}$. Thus there are two equilibria: (0,0) and $(\frac{1}{2}, \frac{1}{4})$. The matrix of the linearization at (y, z) is

$$\begin{bmatrix} 0 & 2 \\ 2y & 0 \end{bmatrix}.$$

If $y = 0$, this matrix has trace 0 and determinant 0 and the conditions (2.50) are satisfied. Thus the equilibrium (0,0) is asymptotically stable. If $y = \frac{1}{2}$, the matrix has trace 0 and determinant -2, and the condition $|\text{tr } A| < \det A + 1$ is violated. The equilibrium $(\frac{1}{2}, \frac{1}{4})$ is therefore unstable. □

In our study of period doubling in Section 2.6, we were obliged to study a *second-order* difference equation, of the form $y_{k+2} = g_2(y_k)$. An alternative method for studying higher-order difference equations is to convert them to systems of first-order difference equations. For example, we can rewrite the second-order difference equation $y_{k+2} = g_2(y_k)$ by defining $z_k = y_{k+1}$, so that $y_{k+2} = z_{k+1} = g_2(y_k)$. This gives the system $y_{k+1} = z_k$, $z_{k+1} = g_2(y_k)$.

EXAMPLE 2.
Determine the condition for asymptotic stability of an equilibrium y_∞ of the second-order difference equation $y_{k+2} = g(y_k)$ by rewriting it as a first-order system.

Solution: We rewrite the equation as the system $y_{k+1} = z_k$, $z_{k+1} = g(y_k)$. An equilibrium is a solution of $y = z = g(y)$, and the matrix of the linearization at an equilibrium y is

$$\begin{bmatrix} 0 & 1 \\ g'(y) & 0 \end{bmatrix}.$$

The characteristic equation is $\lambda^2 - g'(y) = 0$, with roots $\pm\sqrt{g'(y)}$. Thus the condition for asymptotic stability is $|g'(y)| < 1$. □

It is, of course, not necessary to transform a second-order difference equation to a system in order to study equilibria and their stability. This approach merely gives another proof for the stability of solutions of period 2.

We now consider some common biological systems which must be modeled by systems of [two] first-order difference equations.

2.7.1 Spatial dispersal

In Section 2.6 we considered spatial dispersal of a species through the lens of a single patch (habitat), with constant dispersal either in or out of the patch. Since in reality dispersal is a function of population density, it is more realistic to consider this phenomenon as an interaction among all the patches between which dispersal occurs. We will limit ourselves here to two patches for simplicity.

In studying the spread of disease in an amphibian population, Stock and Emmert considered[19] a population of amphibians which disperses between an aquatic habitat and a land-based habitat (see Figure 2.42). In the 1990s, unexpectedly high mortality rates were observed in many amphibian species. In many cases, the deaths were found to be linked to pesticides and other contaminants found in the environment, even in relatively remote areas; in other cases the deaths were caused by viral or fungal infections. These die-offs are of great concern because amphibians are good barometers of significant environmental changes that may initially go undetected by humans.

We will suppose that, of the adults of a given generation in either patch, a proportion p of them survive to the next generation and remain in the same patch, a proportion q of them survive to the next generation and disperse to the other patch, and the remaining $\mu = 1 - p - q$ of them do not survive to the next generation. In addition, all reproduction occurs in the aquatic patch, with a density-dependent per capita reproduction ratio $B(y)$ (which we assume to be differentiable and monotone decreasing, and hence invertible on the interval $(0, B(0))$).

If we denote the amphibian population in the aquatic (playa) habitat by y_k and the population on land by z_k, and let the proportions p and q vary by patch, we can describe this system with the equations

$$\begin{aligned} y_{k+1} &= p_y y_k + q_z z_k + y_k B(y_k), \\ z_{k+1} &= q_y y_k + p_z z_k. \end{aligned} \tag{2.51}$$

The system (2.51) has two equilibria: extinction (0,0), which always exists,

[19]E.M. Stock, Deterministic discrete-time epidemic models with applications to amphibians, master's thesis, Tarleton State University, 2006.

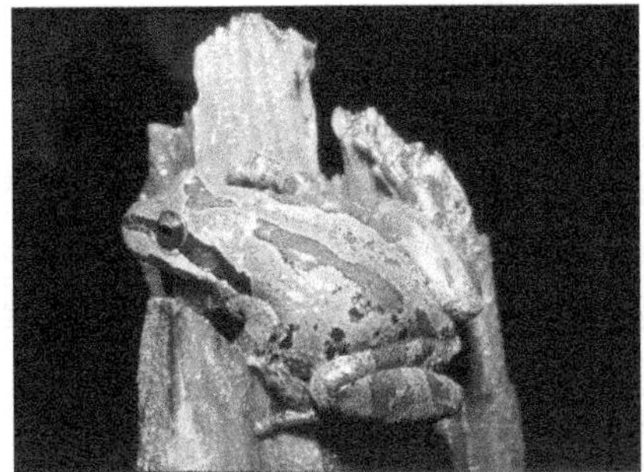

Photos by Gary Fellers, courtesy U.S. Geological Survey, www.usgs.gov

FIGURE 2.42: Left: a Pacific tree frog. Right: a high sierran lake with mountain yellow-legged frogs. Amphibians such as these frogs move between lakes and the surrounding land; the lakes, which serve as breeding sites, also act as focal points for viral and fungal infections and pesticide runoffs.

and survival $\left(B^{-1}(m), \frac{q_y}{1-p_z}B^{-1}(m)\right)$, which exists when $B(0) > m$, where m is an average effective mortality ratio for the playa patch,

$$m = (1 - p_y) - \frac{q_y q_z}{1 - p_z} = \mu_y + q_y \frac{\mu_z}{q_z + \mu_z}$$

(in every generation of reproductive adults y_k, a proportion q_y of them leave for land; a proportion $\frac{\mu_z}{q_z+\mu_z}$ of those die before returning to the water).

The asymptotic stability of the extinction equilibrium can be analyzed via the matrix for the linearization of (2.51) about (0,0),

$$A = \begin{bmatrix} p_y + B(0) & q_z \\ q_y & p_z \end{bmatrix}.$$

Applying the Jury criterion, we see that the extinction equilibrium is stable if (and only if)

$$0 < p_y + p_z + B(0) < 1 + (p_y + B(0))p_z - q_y q_z < 2.$$

Rewriting these inequalities in terms of $B(0)$, the stability conditions become

$$B(0) < m \text{ and } B(0) < k = \frac{1 - p_y p_z + q_y q_z}{p_z}.$$

A little algebra shows that $m < k$ ($mp_z(1 - p_z) < kp_z(1 - p_z)$ simplifies to $0 < (1 - p_z)^2 + q_y q_z$), so the extinction equilibrium is stable if $B(0) < m$, that is, if the effective mortality ratio exceeds the maximum reproductive ratio.

The asymptotic stability of the survival equilibrium depends on the form of $B(y)$. The matrix A for the linearization of (2.51) about the survival equilibrium is

$$\begin{bmatrix} p_y + m + B^{-1}(m)B'(B^{-1}(m)) & q_z \\ q_y & p_z \end{bmatrix}.$$

The second half of the Jury criterion, $\det A < 1$, becomes

$$\left(p_y + m + B^{-1}(m)B'(B^{-1}(m))\right) p_z - q_y q_z < 1,$$

which can be rewritten as $m + B^{-1}(m)B'(B^{-1}(m)) < k$. Since by assumption B is monotone decreasing, $B'(B^{-1}(m)) < 0$, so $m + B^{-1}(m)B'(B^{-1}(m)) < m < k$, and the condition is met. The first half of the Jury criterion, $|\text{tr}\, A| < 1 + \det A$, becomes

$$\begin{aligned} &|p_y + p_z + m + B^{-1}(m)B'(B^{-1}(m))| \\ &\qquad < 1 + p_y p_z - q_y q_z + p_z \left(m + B^{-1}(m)B'(B^{-1}(m))\right). \end{aligned} \tag{2.52}$$

If $\text{tr}\, A > 0$, then we drop the absolute value bars and rewrite the condition (with some algebra[20]) as $m + B^{-1}(m)B'(B^{-1}(m)) < m$, which is again always true because B is decreasing. However, if $\text{tr}\, A < 0$, the condition can be rewritten as

$$m + B^{-1}(m)B'(B^{-1}(m)) > \frac{q_y q_z}{1 + p_z} - (1 + p_y), \tag{2.53}$$

which does not hold if $B'(B^{-1}(m))$ (which is negative) is large enough.

If we assume that the reproduction ratio is determined by contest competition, then the total reproduction is an increasing function of population density, i.e., $\frac{d}{dy}\left(yB(y)\right) = B(y) + yB'(y) > 0$ for all values of y including the equilibria, so that $\text{tr}\, A = p_y + p_z + B(y_\infty) + y_\infty B'(y_\infty) > 0$, and the survival equilibrium is stable whenever it exists (i.e., for $B(0) > m$). If, on the other hand, the reproduction ratio is determined by scramble competition, then it is possible for B to decline so fast that (2.53) does not hold.

EXAMPLE 3.
Show that the survival equilibrium of (2.51) is always stable when it exists, if the fertility is described by the Beverton-Holt model.

Solution: We have $yB(y) = \frac{ay}{1+by}$, so that $B(y) = \frac{a}{1+by}$. Thus $B^{-1}(m) = \frac{1}{b}\left(\frac{a}{m} - 1\right)$, and the matrix for the linearization of (2.51) about the survival equilibrium is

$$A = \begin{bmatrix} p_y + \frac{m^2}{a} & q_z \\ q_y & p_z \end{bmatrix}.$$

Here $\text{tr}\, A = p_y + p_z + \frac{m^2}{a} > 0$, so the Jury criterion holds as noted above. □

EXAMPLE 4.
Determine the stability condition for the survival equilibrium of (2.51), if the fertility is described by the Ricker model.

[20] Move $p_y + p_z$ to the right-hand side, move the last term on the right to the left, and divide through by $1 - p_z$. The right-hand side becomes m, and the given inequality is obtained.

Solution: We have $yB(y) = aye^{-by}$, so that $B(y) = ae^{-by}$. Now $y_\infty = \frac{1}{b}\log\frac{a}{m}$ exists for $a > m$, and the case tr $A < 0$ becomes

$$a > m\exp\left(1 + \frac{p_y + p_z}{m}\right).$$

Also, in this case (2.53) becomes

$$m\left(1 - \log\frac{a}{m}\right) > \frac{q_y q_z}{1 + p_z} - (1 + p_y),$$

which, after some algebra, can be rewritten

$$a < m\exp\left(2 + \frac{2}{m}P\right), \quad \text{where } P = p_y + q_y q_z \frac{p_z}{1 - p_z^2} < p_y + q_y < 1. \quad (2.54)$$

Since

$$0 < 1 + \frac{p_y + p_z}{m} < 2 + \frac{2}{m}P$$

(the second inequality requires some algebra), the threshold value for a in (2.54) gives $a > m$ and tr $A < 0$, so (2.54) (together with $a > m$) is the stability criterion for the survival equilibrium. □

The general behavior of the models in the two examples above is similar to that of the simpler Beverton-Holt and Ricker models analyzed in Section 2.4. So in what way do these dispersal models give a better understanding of the effects of dispersal on the stability of the underlying populations?

In order to answer this question, we should compare the behaviors of the two two-patch dispersal models with the behaviors of the one-patch models that most closely correspond to them: models with the given reproductive functions, mortality ratios μ, and no dispersal (i.e., $q = 0$, so that $p + \mu = 1$ and we consider the entire population as a whole, without separating those in the lake or playa from those on land). For Beverton-Holt fertility, this is

$$y_{k+1} = \frac{ay_k}{1 + by_k} + (1 - \mu)y_k;$$

for Ricker fertility, it is

$$y_{k+1} = ay_k e^{-by_k} + (1 - \mu)y_k.$$

The analyses of these models are left as an exercise for the reader. The one-patch model with Beverton-Holt fertility has a survival equilibrium $\frac{1}{b}\left(\frac{a}{\mu} - 1\right)$, which exists and is stable when $a/\mu > 1$. The corresponding two-patch model replaces μ with m. The one-patch model with Ricker fertility has a survival equilibrium $\frac{1}{b}\log\frac{a}{\mu}$ which is stable for $1 < a/\mu < e^{2/\mu}$. The two-patch Ricker model has a survival equilibrium with $y_\infty = \frac{1}{b}\log\frac{a}{m}$ which is stable when $a > m$ and (2.54) holds. Therefore the effects of water-land dispersal for amphibians are twofold: in increasing the effective mortality ratio from μ_x to m due to deaths of nonreproducing adults on land, and, in the case of scramble (Ricker) competition, in making the survival equilibrium less likely to be stable, by reducing the upper bound for stability (it takes some algebra to show that $\frac{2}{\mu_y} > 2 + \frac{2}{m}P$). The model also gives a way to estimate the number of adults on land relative to the number in the water, although it is important to note that the model implicitly assumes that adults remain on land for an average period of time longer than that between breeding times (which is used as the time step).[21]

2.7.2 Two-stage populations

Insect populations normally pass through at least two developmental stages, larval and adult. Often there is a pupal stage as well. We shall construct a simple two-stage model covering a larval stage and an adult stage to give an idea how one might go about devising and analyzing more accurate models. We let y_k denote the larval population size in the kth generation and z_k the adult population size in the kth generation. We assume that the birth rate of larvae depends only on the adult population size and is rze^{-bz} when the adult population size is z (scramble competition). We assume that a fraction p of the larvae survive to become adults in the next generation and that a fraction μ of the adults die from one generation to the next. These assumptions lead to the model

$$\begin{aligned} y_{k+1} &= rz_k e^{-bz_k}, \\ z_{k+1} &= py_k + (1-\mu)z_k. \end{aligned} \tag{2.55}$$

(Note that this model also implicitly assumes that the time required for larvae to mature is approximately the same as the time between reproductive cycles.)

Equilibria of the model (2.55) are solutions of the equations $y = rze^{-bz}$ and $z = py + (1-\mu)z$ or $\mu z = py$. We substitute $y = \frac{\mu}{p}z$ into the second equation, obtaining $\mu z = rpze^{-bz}$. Thus either $z = 0$, which implies $y = 0$, or $rpe^{-bz} = \mu$, which implies

$$z = \frac{1}{b}\log\frac{rp}{\mu}, \quad y = \frac{\mu}{pb}\log\frac{rp}{\mu}. \tag{2.56}$$

This equilibrium has positive values of y and z, corresponding to survival of the species, only if $\frac{rp}{\mu} > 1$, or $rp > \mu$. The matrix of the linearization of the system (2.55) at an equilibrium (y, z) is

$$\begin{bmatrix} 0 & re^{-bz}(1-bz) \\ p & 1-\mu \end{bmatrix}.$$

At the equilibrium (0,0), this is

[21] It can be shown that the average number of generations that an individual spends in the land patch before dying or returning to water is $1/(1-p_z)$ in this model.

$$\begin{bmatrix} 0 & r \\ p & 1-\mu \end{bmatrix},$$

with trace $1-\mu$ and determinant $-rp$. The equilibrium is asymptotically stable if $1-\mu < 1-rp < 2$. Thus the asymptotic stability condition is $rp < \mu$, which is just the condition that (0,0) is the only equilibrium. At the survival equilibrium, the matrix is

$$\begin{bmatrix} 0 & \frac{\mu}{p}(1-bz) \\ p & 1-\mu \end{bmatrix},$$

with trace $1-\mu$ and determinant $\mu(bz-1)$. The asymptotic stability condition is

$$1-\mu < \mu(bz-1)+1 < 2.$$

The first inequality is satisfied automatically and the second condition is $\mu(bz-1) < 1$. Using the equilibrium value of z, we reduce this to $\mu \log \frac{rp}{\mu} < 1$. Taking exponentials, we have $\frac{rp}{\mu} < e^{1/\mu}$, or $rp < \mu e^{1/\mu}$. Since the existence of this equilibrium requires $rp > \mu$, we have an asymptotically stable survival equilibrium if and only if

$$\mu < rp < \mu e^{\frac{1}{\mu}}.$$

This result suggests that populations described by this model will tend to reach a (more or less) constant level when the effective reproductive ratio rp (the reproductive ratio multiplied by the proportion of larvae that survive to reproductive age) outpaces natural mortality μ but not by too much. Populations undergoing intraspecific scramble competition which have an excessively high growth ratio will tend to suffer fluctuations, just as observed for the simple logistic model in Section 2.6.

In evaluating the utility of this model, we should consider what additional detail it gives over the single-stage model with Ricker-function fertility and proportional mortality described in the subsection on dispersal above. The general behavior is the same; in the simpler model, the population tends to the same equilibrium value (with $a = rp$) as long as the effective reproductive ratio outpaces natural mortality μ, but not by too much. In the simpler model, the ratio a/μ should fall in the interval $(1, e^{\frac{2}{\mu}})$, while for the two-stage model it should fall within the interval $(1, e^{\frac{1}{\mu}})$ in order for the population density to stabilize. This suggests that the added delay incurred by newborns requiring a full generation of time to reach reproductive age makes it more difficult to maintain stability; this conclusion is consistent with our previous observation that delays (as represented by the discrete time steps implicit in difference equations) make fluctuations more likely. The other difference, of course, is that the two-stage model allows us to estimate the size of the larval population relative to the adult population at any time.

Emmert and Allen used a similar model (with an additional term qy_k in the first equation denoting those juveniles who survive without having reached maturation) as a basis for studying the fungal infection chytridiomycosis in amphibian juveniles and adults;[22] they first studied the model without disease in order to better understand the behavior of the model when the disease is introduced. Many of the other models we have studied in this chapter can likewise serve as baseline models to understand the behavior of a given species in the absence of some phenomenon (like disease) which can alter that behavior.

Exercises

In each of the following exercises, find all equilibria and determine which of them are asymptotically stable.

1. $y_{k+1} = 3y_k - 2z_k,$
   $z_{k+1} = 4y_k + z_k$
2. $y_{k+1} = z_k + z_k^2,\ z_{k+1} = y_k + 1$
3. $y_{k+1} = 2y_k,\ z_{k+1} = 4e^{-y_k}$
4. $y_{k+1} = ry_k e^{-by_k},$
   $z_{k+1} = ay_k(1 - e^{-by_k})$
5. $y_{k+2} + 4y_{k+1} + 4y_k = 0$
6. $y_{k+2} = e^{-y_k}$
7. For the model with Beverton-Holt fertility and proportional mortality μ

$$y_{k+1} = \frac{ay_k}{1 + by_k} + (1 - \mu)y_k,$$

   find all equilibria and analyze their asymptotic stability.

8. For the model with Ricker fertility and proportional mortality μ

$$y_{k+1} = ay_k e^{-by_k} + (1 - \mu)y_k,$$

   find all equilibria and analyze their asymptotic stability.

9. (a) Rewrite the third-order difference equation $y_{k+3} = g_3(y_k)$ as a first-order system of dimension three.
   (b^*) Linearize the resulting system and obtain the criterion for stability of an equilibrium y_∞ from the characteristic equation.
10. Write a model describing the dispersal of a species among three patches in which the second patch is connected to the first and third, but the first and third are not connected directly to each other. This situation is appropriate for three patches arranged in a more or less linear fashion, or for an amphibian population where the end patches are bodies of water and the center patch is the patch of land between them.
11. Write a model describing the dispersal of a species among three patches in which dispersal occurs directly among all three patches.

[22]K.E. Emmert and L.J.S. Allen, Population persistence and extinction in a discrete-time, stage-structured epidemic model, *Journal of Difference Equations and Applications* 10(13–15): 1177–1199, 2004.

2.8 Predator-prey systems

In the previous section we studied single populations structured by age or spatial distribution. Another type of biological system which requires multiple equations to model is one in which two or more species interact. One of the most common is a *predator-prey* system, in which one population benefits at the expense of the other. For instance, in modeling a population with discrete generations, it may be important to include in the model the resources on which the species being studied feeds, as well as the species of interest. This leads to a system of two equations, one each for the species and the resource. Another kind of species interaction is a host-parasitoid situation, most common among insects, in which one species lays its eggs in a host individual. Both of these systems are special cases of *predator-prey* relationships.

2.8.1 A plant-herbivore model

The cinnabar moth *Tyria jacobaeae* lives for one year, feeding on the perennial plant ragwort (see Figure 2.43). At the end of the year it lays eggs that hatch the following spring, and then dies. We let y_k denote the total biomass of ragwort and z_k the number of insect eggs at the start of the kth year. We assume that the number of eggs laid is proportional to the amount of ragwort at the start of the previous year, so that $z_{k+1} = ay_k$. We assume that, without the moth feeding, the ragwort would have a total biomass of B, but consumption reduces the stock of ragwort so that its biomass falls exponentially with the ratio of insect eggs to plant biomass the previous year, so that $y_{k+1} = Be^{-bz_k/y_k}$. This gives the model[23]

$$y_{k+1} = Be^{-\frac{bz_k}{y_k}}, \quad z_{k+1} = ay_k. \tag{2.57}$$

Equilibria of the system (2.57) are solutions of the pair of equations

$$z = ay, \quad y = Be^{-\frac{bz}{y}},$$

and we may eliminate z to give $y = Be^{-ab}$, $z = aBe^{-ab}$. The matrix of the linearization at an equilibrium (y, z) is

$$\begin{bmatrix} B\left(\frac{bz}{y^2}\right)e^{-bz/y} & -B\left(\frac{b}{y}\right)e^{-bz/y} \\ a & 0 \end{bmatrix}.$$

[23] E. van der Meijden, M.J. Crawley, and R.M. Nisbet, The dynamics of a herbivore-plant interaction, *Insect Populations: in Theory and Practice* (J.P. Dempster and I.F.G. McLean, eds.), Chapman and Hall, London, 1998.

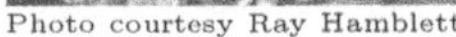
Photo courtesy Ray Hamblett

FIGURE 2.43: A cinnabar moth caterpillar eating ragwort. This tall, yellow flowering weed can be fatal if eaten by horses and cattle, and cinnabar caterpillars are used in some places as biocontrol. In Leicestershire, England, the caterpillars have reduced the ragwort population by up to 80%.

Photo courtesy S. Bradleigh Vinson

FIGURE 2.44: A parasitoid wasp attacking a host larva.

Because of the equilibrium conditions $e^{-bz/y} = e^{-ab}$, $\frac{bz}{y^2} = \frac{ab}{Be^{-ab}}$, $\frac{b}{y} = \frac{b}{Be^{-ab}}$, this matrix reduces to

$$\begin{bmatrix} ab & -b \\ a & 0 \end{bmatrix}.$$

The trace of this matrix is ab and the determinant is also ab; thus the stability condition is $ab < ab+1 < 2$, or $ab < 1$. Thus the equilibrium is asymptotically stable if and only if $ab < 1$ (note that ab is a dimensionless measure of the amount by which the moth egg density reduces the ragwort biomass). Observations indicate both asymptotically stable equilibria and unstable equilibria with oscillations are possible.

2.8.2 A host-parasitoid model

Insect parasitoids lay eggs in the larvae of a host species, causing the death of the host larvae and the survival to adulthood of the parasitoid (see Figure 2.44). We let y_k denote the host population size in generation k and z_k denote the parasitoid population size in generation k. We suppose that each host lays enough eggs to produce r larvae. The assumptions that (i) the number of encounters between host larvae and parasitoids is proportional to host population size and parasitoid population size, and that (ii) these encounters are distributed randomly among the available hosts, lead to the conclusion that the fraction of host larvae surviving to adulthood is e^{-bz_k} for some b. Both hosts and parasitoids live for only one generation of time. Then the host equation is $y_{k+1} = ry_k e^{-bz_k}$. Since each parasitized host larva leads to one adult parasitoid, the parasitoid equation is $z_{k+1} = ry_k(1 - e^{-bz_k})$. This gives the Nicholson-Bailey model for host-parasitoid dynamics,[24]

$$\begin{aligned} y_{k+1} &= ry_k e^{-bz_k}, \\ z_{k+1} &= ry_k(1 - e^{-bz_k}). \end{aligned} \tag{2.58}$$

Equilibria of (2.58) are solutions of the pair of equations $y = rye^{-bz}$, $z = ry(1 - e^{-bz})$. One equilibrium is $y = 0$, $z = 0$, corresponding to extinction of both populations. A second equilibrium is found by solving $re^{-bz} = 1$, which gives $z = \frac{1}{b}\log r$, and substituting into $z = ry(1 - e^{-bz})$, which then gives $\frac{1}{b}\log r = ry(1 - \frac{1}{r}) = (r-1)y$, or $y = \frac{1}{b(r-1)}\log r$ provided $r > 1$. Thus there is a survival equilibrium

$$y_\infty = \frac{1}{b(r-1)}\log r, \quad z_\infty = \frac{1}{b}\log r$$

if $r > 1$. The matrix of the linearization of (2.58) at an equilibrium (y, z) is

$$\begin{bmatrix} re^{-bz} & -rbye^{-bz} \\ r(1 - e^{-bz}) & rbye^{-bz} \end{bmatrix}.$$

At the extinction equilibrium (0,0), this matrix is

$$\begin{bmatrix} r & 0 \\ 0 & 0 \end{bmatrix},$$

with trace r and determinant zero. The stability conditions are $r < 1 < 2$, or $r < 1$. Thus if $r < 1$, the extinction equilibrium is the only equilibrium and it is asymptotically stable. At the survival equilibrium, the matrix of the linearization is

$$\begin{bmatrix} 1 & -by_\infty \\ r-1 & by_\infty \end{bmatrix},$$

with trace $1 + by_\infty$ and determinant rby_∞. The stability conditions are $1 + by_\infty < 1 + rby_\infty < 2$ or $by_\infty < rby_\infty < 1$. The first of these conditions is satisfied if and only if $r > 1$, and the second is satisfied if $rby_\infty = \frac{r}{r-1}\log r < 1$. However, this condition is not satisfied for any value of $r > 1$. Thus the survival equilibrium of the Nicholson-Bailey model can not be asymptotically stable. This means that either both host and parasitoid populations will be wiped out (if $r < 1$) or both populations will oscillate. These oscillations may have large amplitude and may bring one of the populations to a very low level from which a small perturbation could wipe it out.

Thus this rather simple model supplies an explanation for outbreaks and subsequent crashes of an insect population with a parasitoid. However, host-parasitoid populations which persist for many generations have been observed,

[24] A.J. Nicholson and V.A. Bailey, The balance of animal populations, *Proc. Zoological Soc. London* 3: 551–598, 1935.

and this can not be explained with a Nicholson-Bailey model. In order to allow the possibility of an asymptotically stable survival equilibrium we must make some different assumptions. One assumption which would lead to this conclusion is that adult hosts may live longer than one generation. If we assume that a fraction p of hosts survive to the next generation, the model (2.58) would be replaced by

$$\begin{aligned} y_{k+1} =& r y_k e^{-b z_k} + p y_k, \\ z_{k+1} =& r y_k (1 - e^{-b z_k}). \end{aligned}$$

This model does allow for the possibility of an asymptotically stable survival equilibrium, but the computations to show this are somewhat complicated (see Exercise 5). The result is that most of the adult hosts must survive from one generation to the next in order to overcome a high reproductive ratio r and stabilize the equilibrium.

The biological systems described in this section share the property that the equilibrium is unstable if the natural growth rate is too large. As discussed in previous sections, many discrete models have this property. In the next chapter, we shall consider continuous models, or ordinary differential equations, which do not share this property. Discrete models have a built-in time-lag, and this is what makes instability possible.

Exercises

In each of the following exercises, find all equilibria and determine which of them are asymptotically stable.

1. Determine which of the systems in Exercises 1–6 of Section 2.7 represent predator-prey systems, by considering y_k and z_k as two separate species in each case, and examining the effect that each population has on the other (via the sign of the associated coefficients).

2. In developing the model (2.57), van der Meijden, Crawley, and Nisbet assumed that the number of new eggs laid is proportional only to the biomass of ragwort, and not to the number of adult moths. What implicit assumption is being made about the relevant factors? That is, why might this be a reasonable assumption?

3. Suppose that the assumption discussed in the previous problem is not valid, and that in fact the number of new eggs laid is proportional to both the biomass of ragwort *and* the number of adult moths, i.e., $z_{k+1} = a y_k z_k$. Find the equilibria of this system, and derive criteria for their asymptotic stability.

4. Compare the behavior of the plant-herbivore model in the main text with that of the modified system proposed in the previous problem. Explain

in biological terms what difference it makes whether the number of new eggs laid depends upon the number of adult moths.

5. (a) Show that the revised Nicholson-Bailey host-parasitoid model

$$\begin{aligned} y_{k+1} =& r y_k e^{-b z_k} + p y_k, \\ z_{k+1} =& r y_k (1 - e^{-b z_k}). \end{aligned}$$

has an extinction equilibrium (0,0) which always exists, and a survival equilibrium

$$\left(\frac{1}{b(r+p-1)} \log \frac{r}{1-p}, \frac{1}{b} \log \frac{r}{1-p} \right)$$

which exists if and only if $r + p > 1$.

(b) Show that the extinction equilibrium is stable if and only if $r+p < 1$, and that the survival equilibrium is stable if and only if

$$\frac{(r+p)(1-p)}{(r+p-1)} \log \frac{r}{1-p} < 1.$$

(c^*) Use a computer to sketch a graph of the region in the $p-r$ parameter plane for which the survival equilibrium is stable.

Miscellaneous exercises

For each of the difference equations in Exercises 1–4, verify that the given function is a solution of the given difference equation.

1. $y_{k+1} = y_k(1 - y_k^2), \quad y_k = -1$

2. $y_{k+1} = 1/y_k, \quad y_k = c$, ($k$ odd), $y_k = 1/c$ (k even)

3. $y_{k+1} = y_k e^{r(1-y_k)}, \quad y_k = 1$

4. $y_{k+1} = y_k/(1 + y_k), \quad y_k = 0$

Solve each of the difference equations in Exercises 5–8 with initial value $y_0 = c$.

5. $y_{k+1} = (k+1)y_k$
6. $y_{k+1} = 2^k y_k$
7. $y_{k+1} = y_k/(k+1)$
8. $y_{k+1} = y_k + 1$

For each of the difference equations in Exercises 9–12, find the equilibria, and determine which equilibria are asymptotically stable.

9. $y_{k+1} = 2(y_k - y_k^2)$
10. $y_{k+1} = y_k^2 + 1/4$
11. $y_{k+1} = y_k e^{ry_k}$
12. $y_{k+1} = 4(y_k - y_k^3)/3$

For each of the difference equations in Exercises 13–16, determine for which values of H (possibly depending on r) there is a positive equilibrium which is asymptotically stable.

13. $y_{k+1} = ry_k - H$
14. $y_{k+1} = ry_k^2 - H$
15. $y_{k+1} = r\frac{1}{y_k} - H$
16. $y_{k+1} = -\frac{1}{2}y_k^2 - y_k + \frac{1}{2} - H$

17. Show that the difference equation $y_{k+1} = \dfrac{y_k}{1+y_k^2}$ has no non-zero constant solution.

18. Define the function $\Gamma(x) = \int_0^\infty t^{x-1}e^{-t}dt$ for $x \geq 0$.
    (a) Show that $\Gamma(1) = 1$ and $\Gamma(n+1) = n\Gamma(n)$ if n is a positive integer.
    (b) Deduce that $\Gamma(n+1) = (n+1)!$.

19. Consider the difference equation $y_{k+1} = \dfrac{1}{2}(y_k + \dfrac{a}{y_k})$.
    (a) Find all equilibria.
    (b) For which choices of y_0 does y_k approach $\sqrt{a}$?

20. Consider the difference equation $y_{k+1} = -\dfrac{1}{2}y_k^2 - y_k + \dfrac{1}{2}$.
    (a) Show that there is no asymptotically stable equilibrium.
    (b) Show that the initial value $y_0 = 1$ leads to a solution of period 3.
    (c) Use the cobwebbing method to draw several solutions corresponding to different initial values.

Solve each of the following systems.

21. $y_{n+1} = y_n + z_n, \quad z_{n+1} = -2y_n + 4z_n, \quad y_0 = 1, \quad z_0 = 1$
22. $y_{n+1} = -y_n + z_n, \quad z_{n+1} = y_n, \quad y_0 = 1, \quad z_0 = 2$
23. $y_{n+1} = -y_n + 2z_n, \quad z_{n+1} = 3y_n, \quad y_0 = 1, \quad z_0 = 0$
24. $y_{n+1} = 2y_n - z_n, \quad z_{n+1} = y_n + 3z_n, \quad y_0 = 0, \quad z_0 = 1$

In Exercises 25–28 find all equilibria of the given system and determine which are asymptotically stable.

25. $y_{n+1} = (1 - y_n) - z_n, \quad z_{n+1} = -y_n$
26. $y_{n+1} = \frac{ay_n}{1+z_n}, \quad z_{n+1} = \frac{bz_n - y_n}{1+y_n}$
27. $y_{n+1} = z_n, \quad z_{n+1} = -y_n + 1$
28. $y_{n+1} = y_n/2 - z_n^2 + w_n, \quad z_{n+1} = y_n - z_n + w_n, \quad w_{n+1} = y_n - z_n + w_n/2$

Chapter 3

First-Order Differential Equations (Continuous Dynamical Systems)

3.1 Continuous-time models and exponential growth

Most biological systems can be described on some scale in terms of continuous change over time, whether it is the gradual accumulation of algae on the edges of a stagnant pond or the rapid ebb and flow of species such as the mayfly *Dolania americana*, the females of which generally live less than five minutes as adults (Figure 3.1). For populations which do not reproduce in distinct, synchronized generations, or for which frequent data is available, considering changes to occur continuously in time may allow a model to capture important features of growth. Although populations of discrete individuals should increase or decrease by discrete numbers, rather than continuous amounts, for large populations the inaccuracies incurred by treating the population size as a continuous quantity are small. In cases where changes are continuous in time, we can describe those changes by describing the rate of change in terms of the time-derivative of the quantity, typically a population, rather than in terms of the difference between population sizes in two consecutive generations, as is done for discrete-time models. Doing so results in models composed of *differential equations*, rather than difference equations.

The models we shall consider in this chapter involve a particular sort of differential equation, and it is worth delineating the territory we shall explore. First, in this chapter we shall consider population sizes to vary as a function of time only, and not other continuous variables such as age, spatial location, temperature, etc. Differential equations which involve only functions of one independent variable are called *ordinary differential equations*, or sometimes simply ODEs, in order to distinguish them from *partial differential equations*, or PDEs, which involve two or more independent variables. Equations which involve only the first derivative of the population size (or other quantity under study) are again called *first-order* equations, to distinguish them from *higher-order* equations (e.g., second-order equations, which involve second derivatives, the most common physical application of which is acceleration as the second derivative of location). In this chapter and the next, we shall restrict

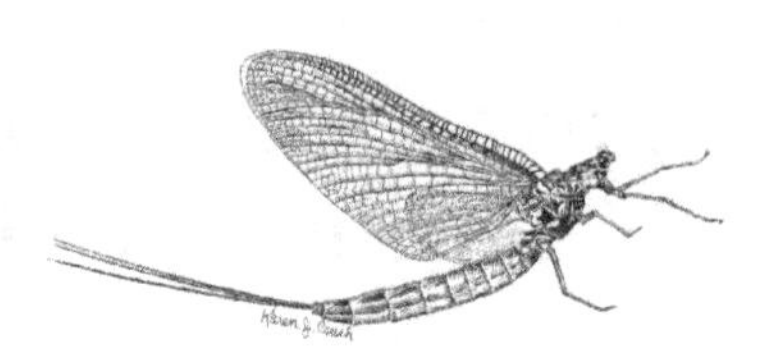

Photo and illustration courtesy U.S. Fish and Wildlife Service, dls.fws.gov

FIGURE 3.1: Left, algae accumulation, Kesterson National Wildlife Refuge, Alaska. Photo by Gary Zahm, USFWS. Right, the mayfly, some species of which live only minutes as adults after a longer nymph stage spent underwater. Illustration by Karen Couch, USFWS.

our study to biological systems that can be modeled with first-order ordinary differential equations.

We saw in the previous chapter that the simplest sort of change other than a constant-valued derivative is one in which the rate of change is a constant multiple of the quantity under study. This type of equation is called *linear*, and, although many linear models are far too simple to capture the complicated behavior of real biological systems, we also saw that the behavior of more complex, *nonlinear* equations can be analyzed in terms of the behavior of related linear equations. We thus begin our study of biological systems that change continuously in time with a look at the simplest sort of first-order ODE: the linear equation. Later sections will develop the tools necessary to examine nonlinear models, and consider some of the types of models and systems that have contributed most to our understanding of the biological world, from chemical and neural functions within a single organism to the growth and management of entire populations.

3.1.1 Exponential growth

The rate of change of some quantity is often proportional to the amount of the quantity present. This may be true, for example, of the size of a population with enough resources that its growth is unrestricted, and depends only on an inherent per capita reproductive rate. The idea can also apply to a decaying population — for example, the mass of a piece of a radioactive substance. In such a case, if $y(t)$ is the quantity at time t, then $y(t)$ satisfies the first-order homogeneous linear differential equation

$$\frac{dy}{dt} = ay, \tag{3.1}$$

where a is a constant which represents the proportional growth or decay rate. a is positive if the quantity is increasing and negative if the quantity is decreasing. The idea that populations grow in this way is old, used most famously by Thomas Robert Malthus to hypothesize geometric (exponential) growth of human populations.[1]

We can develop simple linear models like this from basic principles. For instance, let $y(t)$ represent the number of members of a population of simple organisms at time t. If we assume that these organisms reproduce by splitting, and that a fraction a of the members split into two members in unit time, then

$$y(t+h) - y(t) \approx ah\, y(t), \tag{3.2}$$

with the approximate equality ($\approx$) signifying that the difference between the two sides of the relation is small compared to h, in the sense that this difference divided by h approaches zero as $h \to 0$, i.e., that

$$\lim_{h\to 0} \frac{y(t+h) - y(t) - ahy(t)}{h} = 0. \tag{3.3}$$

So far, this derivation is exactly the same as the one which began Section 2.1. Now, however, instead of thinking of h as a fixed time interval, we allow h to approach zero. It follows from (3.3) that

$$y'(t) = \lim_{h\to 0} \frac{y(t+h) - y(t)}{h} = ay(t).$$

Thus the population size $y(t)$ satisfies the differential equation (3.1). If a fraction b of the members reproduce by splitting and a fraction d of the members die in unit time, then (3.2) would be true with $a = b - d$. The constant a may be either positive or negative, depending on whether $b > d$ or $b < d$. We will use a to designate the constant of proportionality in (3.1); this constant may be either positive or negative.

Although linear models may appear too simple to provide useful predictions, the hypotheses of constant per capita birth rates in the absence of resource limitations, and constant per capita death rates, do have some experimental support, even for relatively complex organisms. For instance, Hutchinson cited[2] the growth of the collared dove *Streptopelia decaocto* in Great Britain, which it invaded in 1954 after spreading across much of western Europe. In the absence of initial competition among the first pioneer birds to settle throughout Great Britain, the data (reproduced in Figure 3.2) show a remarkably close fit to exponential growth (the figure is on a logarithmic scale, making exponential curves appear linear) for several years, until population density slowed the growth. Hutchinson also observed a constant per

[1] T.R. Malthus, An essay on the principle of population, Harmondsworth, Middlesex (1798) [republished by Penguin, New York (1970)].

[2] G. Evelyn Hutchinson, *An introduction to population ecology*, Yale University Press, New Haven, 1978.

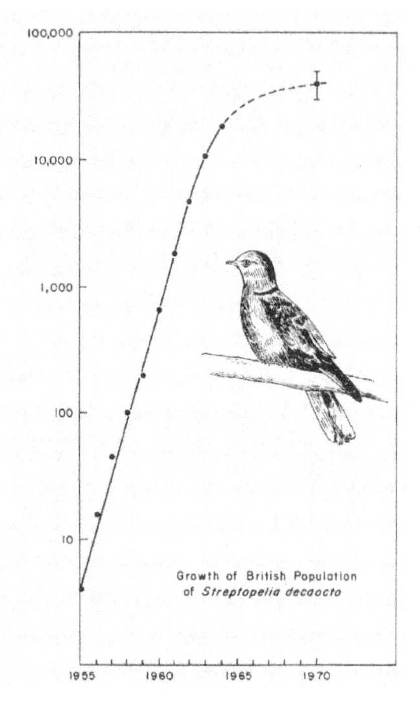

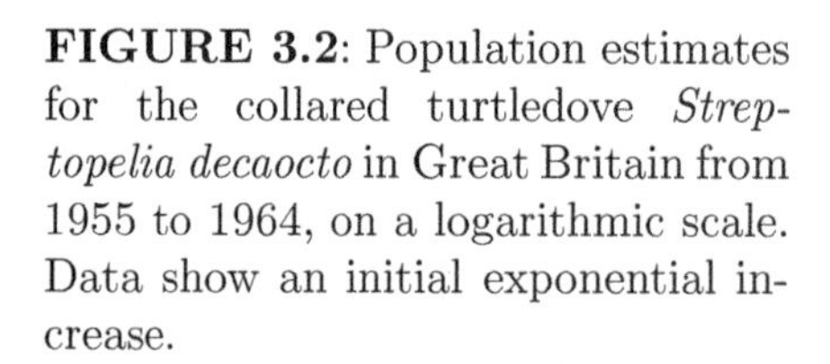

FIGURE 3.2: Population estimates for the collared turtledove *Streptopelia decaocto* in Great Britain from 1955 to 1964, on a logarithmic scale. Data show an initial exponential increase.

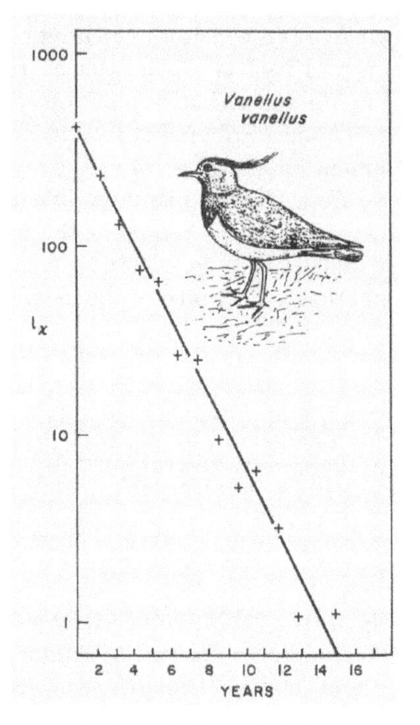

FIGURE 3.3: Composite age-specific survivorship curve for the lapwing *Vanellus vanellus*, based on ringed birds found dead in Europe. Data are presented from the end of the first year of life onward.

capita mortality rate reflected in data gathered by several researchers on some species of birds. Figure 3.3, also plotted on a logarithmic scale, shows an exponential decay describing mortality for the lapwing *Vanellus vanellus* following an initial high juvenile mortality period (omitted).

Functions (such as population sizes) which obey the linear differential equation (3.1) are exponential in form. To see, first, that exponential functions obey (3.1), we need only verify that $y = c\,e^{at}$ is a solution of the differential equation (3.1) for every choice of the constant c. By this we mean that if we substitute the function $y = c\,e^{at}$ into the differential equation (3.1) it becomes an identity. If $y = ce^{at}$, then $y' = ace^{at} = ay$, and this is the necessary verification. Thus the differential equation (3.1) has an infinite family of solutions (one for every choice of the constant c, including $c = 0$),

$$y = c\,e^{at}. \tag{3.4}$$

In order to show that *all* solutions to the differential equation (3.1) have the form (3.4), suppose that $y(t)$ is a solution of the differential equation $y' = ay$, that is, that $y'(t) = ay(t)$ for every value of t. If $y(t) \neq 0$, division of this equation by $y(t)$ gives

$$\frac{y'(t)}{y(t)} = \frac{d}{dt} \log |y(t)| = a. \tag{3.5}$$

We will always use *log* to denote the natural logarithm, rather than *ln*, which may have been used in your calculus course. Integration of both sides of (3.5) gives $\log |y(t)| = at + k$ for some constant of integration k. Then

$$|y(t)| = e^{at+k} = e^k e^{at}.$$

Because e^{at} and e^k are positive for every value of t, $|y(t)|$ cannot be zero, and thus $y(t)$ cannot change sign. We may remove the absolute value and conclude that $y(t)$ is a constant multiple of e^{at}, $y = ce^{at}$. We note also that if $y(t)$ is different from zero for one value of t then $y(t)$ is different from zero for every value of t. Thus the division by $y(t)$ at the beginning of the proof is legitimate unless the solution $y(t)$ is identically zero. The identically zero function is a solution of the differential equation, as is easily verified by substitution, and it is contained in the family of solutions $y = ce^{at}$ with $c = 0$.

The absolute value which appears in the integration produces some complications which may be avoided if we know that the solution must be nonnegative, so that $|y(t)| = y(t)$. This is the case in many applications. If we know that a solution $y(t)$ of the differential equation $y' = ay$ is positive for all t, we could replace (3.5) by

$$\frac{d}{dt} \log y(t) = a$$

and then integrate to obtain $\log y(t) = at + k$, $y(t) = e^{at+k} = e^{at}e^k = ce^{at}$.

The logical argument in the above proof is that if the differential equation has a solution, then that solution must have a certain form. However, it also derives the form and thus serves as a method of determining the solution.

In order for a mathematical problem to be a plausible description of a scientific situation, the mathematical problem must have only one solution; if there were multiple solutions we would not know which solution represents the situation. This suggests that the differential equation (3.1) by itself is not enough to specify a description of a physical situation. We must also specify the value of the function y for some initial time when we may measure the quantity y and then allow the system to start running. For example, suppose we impose the additional requirement, called an *initial condition*, that

$$y(0) = y_0. \tag{3.6}$$

A problem consisting of a differential equation together with an initial condition is called an *initial value problem.* We may determine the value of c for which the solution (3.4) of the differential equation (3.1) also satisfies the initial condition (3.6) by substituting $t = 0$, $y = y_0$ into the form (3.4). This gives the equation

$$y_0 = c\,e^0 = c$$

and thus $c = y_0$. We now use this value of c to give the solution of the differential equation (3.1) which also satisfies the initial condition (3.6), namely $y = y_0 e^{at}$. This procedure may be followed in any situation described by an initial value problem, including population growth and radioactive decay.

EXAMPLE 1.
Suppose that a given population of protozoa develops according to a simple growth law with a growth rate of 0.7944 per member per day, that there are no deaths, and that on day zero the population consists of two members. Find the population size after 6 days.

Solution: The population size satisfies the differential equation (3.1) with $a = 0.7944$, and is therefore given by $y(t) = ce^{0.7944t}$. Since $y(0) = 2$, we substitute $t = 0$, $y = 2$, and we obtain $2 = c$. Thus the solution satisfying the given initial condition is $y(t) = 2\,e^{0.7944t}$, and the population size after 6 days is $y(6) = 2\,e^{(0.7944)(6)} = 235$ (rounding the population size to the nearest integer). □

If we know that a population grows exponentially according to a simple growth law but do not know the rate of growth we view the solution $y = ce^{at}$ as containing two parameters (c and a) which must be determined. This requires knowledge of the population size at two different times to provide two equations which may be solved for these two parameters.

EXAMPLE 2. Suppose that a population which follows a simple growth law has 100 members at a starting time and 150 members at the end of 100 days. Find the population at the end of 150 days.

Solution: The population size at time t satisfies $y(t) = ce^{at}$ and $y(0) = 100$, $y(100) = 150$. Thus $y(0) = 100 = ce^0$, $y(100) = ce^{100a} = 150$. It follows that $c = 100$ and $150 = 100e^{100a}$. We obtain $e^{100a} = 1.5$, $a = \frac{\log 1.5}{100} = 4.05465 \times 10^{-3}$. Finally, we obtain

$$y(150) = 100\,e^{150(4.05465\times 10^{-3})} = 183.71.$$

Rounding off to the nearest integer, we obtain the population size 184 after 150 days. □

3.1.2 Radioactive decay

Radioactive materials decay because a fraction of their atoms decompose into other substances. If $y(t)$ represents the mass of a sample of a radioactive substance at time t, and a fraction k of its atoms decompose in unit time, then $y(t+h)-y(t)$ is approximately $-ky(t)$, and we are led to the differential equation (3.1) with a replaced by $-k$. If it is clear from the nature of the problem that the constant of proportionality must be negative, we will use $-k$ for the constant of proportionality, giving a differential equation

$$y' = -ky \tag{3.7}$$

with $k > 0$.

EXAMPLE 3.
The radioactive element strontium 90 has a decay constant 2.48×10^{-2} years^{-1}. How long will it take for a quantity of strontium 90 to decrease to half of its original mass?

Solution: The mass $y(t)$ of strontium 90 at time t satisfies the differential equation (3.7) with $k = 2.48 \times 10^{-2}$. If we denote the mass at time $t = 0$ by y_0, then $y(t) = y_0\, e^{-(2.48\times 10^{-2})t}$. The value of t for which $y(t) = y_0/2$ is the solution of

$$\frac{y_0}{2} = y_0\, e^{-(2.48\times 10^{-2})t}.$$

If we divide both sides of this equation by y_0 and then take natural logarithms, we have

$$-(2.48 \times 10^{-2})t = \log\frac{1}{2} = -\log 2$$

so that $t = (\log 2)/(2.48 \times 10^{-2}) = 27.9$ years. □

The time required for the mass of a radioactive substance to decrease to half of its starting value is called the *half-life* of the substance. The half-life T is related to the decay constant k by the equation

$$T = \frac{\log 2}{k}$$

because if $y(t) = y_0 e^{-kt}$ and (by definition) $y(T) = \frac{y_0}{2}$, then $e^{-kT} = \frac{1}{2}$, so that $-kT = \log\frac{1}{2} = -\log 2$. For radioactive substances it is common to give the half-life rather than the decay constant.

EXAMPLE 4.
Radium 226 is known to have a half-life of 1620 years. Find the length of time required for a sample of radium 226 to be reduced to three fourths of its original size.

Solution: The decay constant for radium 226 is $k = \frac{\log 2}{1620} = 4.28\times 10^{-4}$ years^{-1}.

In terms of k, the mass of a sample at time t is y_0e^{-kt} if the starting mass is y_0. The time τ at which the mass is $\frac{3y_0}{4}$ is obtained by solving the equation

$$\frac{3y_0}{4} = y_0\, e^{-k\tau}$$

or $\frac{3}{4} = e^{-k\tau}$. Taking natural logarithms we obtain $-k\tau = \log\frac{3}{4}$, which gives

$$\tau = -\frac{\log\frac{3}{4}}{k} = \frac{1620(\log\frac{3}{4})}{\log 2} = 672\,\text{years.} \ \square$$

The radioactive element carbon 14 decays to ordinary carbon (carbon 12) with a decay constant 1.244×10^{-4} years^{-1}, and thus the half-life of carbon 14 is 5570 years. This has an important application, called carbon dating, for determining the approximate age of fossil materials. The carbon in living matter contains a small proportion of carbon 14 absorbed from the atmosphere. When a plant or animal dies, it no longer absorbs carbon 14 and the proportion of carbon 14 decreases because of radioactive decay. By comparing the proportion of carbon 14 in a fossil with the proportion assumed to have been present before death, it is possible to calculate the time since absorption of carbon 14 ceased.

EXAMPLE 5.
Living tissue contains approximately 6×10^{10} atoms of carbon 14 per gram of carbon. A wooden beam in an ancient Egyptian tomb from the First Dynasty contained approximately 3.33×10^{10} atoms of carbon 14 per gram of carbon. How old is the tomb?

Solution: The number of atoms of carbon 14 per gram of carbon, $y(t)$, is given by $y(t) = y_0e^{-kt}$, with $y_0 = 6 \times 10^{10}$, $k = 1.244 \times 10^{-4}$, and $y(t) = 3.33 \times 10^{10}$ for this particular t value. Thus the age of the tomb is given by the solution of the equation

$$e^{-(1.244\times 10^{-4})t} = \frac{3.33 \times 10^{10}}{6 \times 10^{10}} = \frac{3.33}{6},$$

and if we take natural logarithms this reduces to

$$t = -\frac{\log 3.33 - \log 6}{1.244 \times 10^{-4}} = 4733\,\text{years.} \ \square$$

Exercises

In Exercises 1–6, assume the population size satisfies a simple growth law.

1. Suppose that the birth rate of a given population is 0.36 per member per day with no deaths. If the population size on day zero is 50, what is the population size 10 days later?

2. Suppose that the birth rate of a given population of protozoa is 0.2 per member per day with no deaths. If the population size on day zero is 10, find the population size 20 days later.

3. Suppose a population has 173 members at $t = 0$ and 262 members at $t = 10$. Estimate the population size at $t = 5$.

4. Suppose that a population has 87 members at $t = 0$ and 125 members at $t = 4$. Estimate the population size at $t = 6$.

5. Suppose that a population has 12 members at $t = 3$ and 5 members at $t = 10$. What was the population size at $t = 0$?

6. Suppose that a population has 13 members at $t = 4$ and 20 members at $t = 6$. What was the population size at $t = 0$?

7. If the half-life of a radioactive substance is 30 days, how long would it take until 99 % of the substance decays?

8. How long does it take for a piece of carbon 14 to decrease to 20 % of its original size?

9. In a sample of uranium 238, it is found that 0.0000154 % of the mass disintegrates in 1000 years. Find the half-life of uranium 238.

10. How old is a fossil in which 85 % of the carbon 14 has disintegrated?

11*. Show that the solution of the initial value problem $y' = ay$, $y(t_0) = c$ is $y(t) = ce^{a(t-t_0)}$.

12*. Use the graph in Figure 3.2 to estimate (a) the per capita birth rate of the collared turtledove, in units of per individual per year, and (b) how long it would take for a population of this bird to double in an environment with enough resources to prevent competition. *Hint: What form does* $\log y$ *have if* $dy/dt = ay$*? How can you read* a *from the graph?*

13. Explain why the graphs of solutions to the linear differential equation $dy/dt = ry$ appear linear on a logarithmic scale. Calculate the slope of such a line (supposing an initial condition $y(0) = y_0$) and interpret it.

14. Suppose that a given population exhibits exponential growth when it is small. How and why would you expect that to change when the population is large? How would the graph of the population over time differ from the graph of a purely exponential function?

3.2 Logistic population models

As has been observed earlier in this text, linear models are useful either for very simple situations such as those described in the previous section (e.g., unrestricted population growth) or as indicators of local behavior for more complicated systems near equilibria. However, most of the biological systems we may wish to model have essentially nonlinear features which are crucial in determining how the system behaves. In particular, any expanding population will eventually begin to feel growth restrictions as it runs up against the resource limitations of its environment.[3] At such a point in time, the per capita growth rate must become *density-dependent*, that is, a function (typically decreasing) of the population density in the local area. We have already seen in Chapter 2 how different assumptions on the nature of this dependence can produce important differences in the growth of populations which reproduce in discrete generations. In this section we will look at the continuous-time version of the simplest, most common model proposed to incorporate density-dependent growth limitations: the logistic model.

The discrete-time logistic equation was introduced in Chapter 2 as a means of accounting for restricted growth. The continuous-time logistic equation

$$y' = ry\left(1 - \frac{y}{K}\right) \tag{3.8}$$

can be derived as follows. Let $y(t)$ be the size of a population at time t. As we wish to consider $y(t)$ as a differentiable function of t, it is not quite appropriate to consider y to be the number of members of a population, although for large populations it may be a reasonable approximation. We will think of y as representing the *biomass*, that is, the total mass of the members of the population, and we think of this biomass as increasing due to the conversion of nutrients to increased mass of the members of the population. We let $x(t)$ be the amount of nutrient available at time t and we assume that consumption of one unit of nutrient leads to an increase of a units of population biomass, so that

$$y'(t) = -ax'(t). \tag{3.9}$$

We assume also that the per capita rate of population growth $\frac{y'(t)}{y(t)}$ is proportional to the amount of nutrient available, so that

$$\frac{y'(t)}{y(t)} = bx(t) \tag{3.10}$$

for some constant b. Integration of (3.9) gives

$$y(t) = -ax(t) + c \tag{3.11}$$

[3]For a more extensive discussion of the limitations of exponential growth models, see Mark Kot, *Elements of mathematical ecology*, Cambridge University Press, Cambridge, 2001, pp. 5–6.

with c a constant of integration; in fact, $c = y(0) + ax(0)$. Now, substitution of (3.11) into (3.10) gives

$$y'(t) = bx(t)y(t) = b\left(\frac{c - y(t)}{a}\right) y(t) = \frac{bc}{a}\, y(t)\left(1 - \frac{y(t)}{c}\right).$$

We now let $r = bc/a$, $K = c$, and we obtain the logistic differential equation (3.8) for the total biomass of the population. We again consider r to be the maximum intrinsic per capita growth rate (in units of per time) and K to be the carrying capacity determined by the resources of the environment and the needs of the population.

This equation was originally proposed by Verhulst[4] to model the assumption that the per capita growth rate should decrease as population size increases. Verhulst chose a decreasing linear function as the per capita growth rate for simplicity. Solutions of the logistic equation have been fitted to many different population models with considerable predictive success; although the notion of biomass may make one think of simple organisms, Verhulst's original application was to humans.

In Section 3.4, we will show how to solve the logistic differential equation (3.8) explicitly, using a technique called separation of variables. For now, we may verify that for every constant c the function

$$y = \frac{K}{1 + ce^{-rt}} \tag{3.12}$$

is a solution of this differential equation.[5] To see this, note that for the given function y,

$$y' = \frac{Kcr\,e^{-rt}}{(1 + ce^{-rt})^2}$$

and

$$1 - \frac{y}{K} = \frac{K - y}{K} = \frac{ce^{-rt}}{(1 + ce^{-rt})}.$$

Thus

$$ry\left(1 - \frac{y}{K}\right) = \frac{Krc\,e^{-rt}}{(1 + ce^{-rt})^2} = y',$$

and the given function satisfies the logistic differential equation for every choice of c.

[4] P.F. Verhulst, Notice sur la loi que la population suit dans son accroissement, *Corr. Math. et Phys.* 10(1838), 113–121.

[5] We observe that the family of solutions (3.12) of the logistic differential equation (3.8) includes the constant solution $y = K$ (with $c = 0$) but not the constant solution $y = 0$. The existence and uniqueness theorem of Section 3.3 shows that, since we have now obtained a solution corresponding to each possible initial condition, we have obtained all solutions of the logistic differential equation.

To find the solution which obeys the initial condition $y(0) = y_0$, we substitute $t = 0$, $y = y_0$ into the form (3.12), obtaining $\frac{K}{1+c} = y_0$ which implies $c = \frac{K-y_0}{y_0}$ as long as $y_0 \neq 0$ and gives the solution

$$y = \frac{K}{1 + \left(\frac{K-y_0}{y_0}\right) e^{-rt}} = \frac{K y_0}{y_0 + (K - y_0)\, e^{-rt}} \qquad (3.13)$$

to the initial value problem with $y_0 \neq 0$. Note that the denominator begins at K (for $t = 0$) and moves toward y_0 as $t \to \infty$.

Now suppose that K represents some physical quantity such that $K > 0$. One can see from the form (3.13) that if $y_0 > 0$, then the solution $y(t)$ exists for all $t > 0$, and $\lim_{t\to\infty} y(t) = K$. If $y_0 < 0$, then this solution does not exist for all $t > 0$, because $y(t) \to -\infty$ where the denominator changes sign: as $y_0 + (K - y_0)e^{-rt} \to 0$, or $t \to \log\left(\frac{y_0-K}{y_0}\right)$. If $y_0 = 0$, the solution of the initial value problem is not given by (3.13), but is the identically zero function $y = 0$. If, instead, $K < 0$, as will occur in some examples presented in this chapter, then the solution exists for all time if $y_0 < 0$, but approaches positive infinity in finite time if $y_0 > 0$, for the same reason as above.

Note also that (for $K > 0$, $y_0 > 0$) if $r > 0$ then $y \to K$, while if $r < 0$ (as also occurs for some examples in this chapter) then $y \to 0$. In either case a given solution approaches a limit monotonically, regardless of the magnitude of r, in contrast to the behavior of the logistic difference equation of Chapter 2, where more complicated behavior occurs for large values of the growth constant r.

3.2.1 All creatures great and small?

Although the simple reproduction process involved in the original derivation (and that above) for the logistic model would appear to provide biological justifications only for simple organisms which grow via cell division, in fact the logistic model has been found to fit data for the growth of many more complex organisms remarkably well. We shall now consider several such examples in order to get a feel for how this equation can be used to describe population growth.

One notable application of the logistic equation to simple biological populations is an experiment on the protozoa *Paramecium* by the Soviet biologist G.F. Gause.[6] Gause fit his experimental data to the solution (3.12) using the values of the constants given in Example 1, Section 3.1. Table 3.1 lists the observations, the solution of the logistic model with these constants, and the solution of the simple exponential population model with the same intrinsic growth rate. The calculated values are rounded off to the nearest integer. The reader will note that for the first 4 days the exponential model yields results

[6] G.F. Gause, *The Struggle for Existence*, Williams and Wilkins, Baltimore (1934).

TABLE 3.1: Gause's data for protozoan (*Paramecium*) growth, compared to logistic and exponential models' predictions.

t (days)	0	1	2	3	4	5	6	7	
y (observed)	2	3	22	16	39	52	54	47	
y (logistic)	2	4	9	17	28	40	51	57	
y (simple model)	2	4	10	22	48	106	...		
t (days)	8	9	10	11	12	13	14	15	16
y (observed)	50	76	69	51	57	70	53	59	57
y (logistic)	61	62	63	64	64	64	64	64	64

TABLE 3.2: Yeast population y vs. time t in hours, from Carlson (1913).

t	0	1	2	3	4	5	6	7	8	9
$y(t)$	9.6	18.3	29.0	47.2	71.1	119.1	174.6	257.3	350.7	441.0
t	10	11	12	13	14	15	16	17	18	
$y(t)$	513.3	559.7	594.8	629.4	640.8	651.1	655.9	659.6	661.8	

comparable to those of the more sophisticated logistic model. However, for $t \geq 5$ the simple model is hopelessly inaccurate while the logistic model fits the observations reasonably well.

Another example of the ability of the logistic model to predict the growth of simple organisms is an experiment of Carlson[7] on yeast cultures, later analyzed by Pearl[8] and Renshaw.[9] Carlson measured the amount of yeast in a particular culture every hour (see Table 3.2). Pearl and Renshaw (separately) fit this data to a logistic model by rewriting the culture size y in terms of the logarithm of relative growth

$$z(t) = \log \frac{K - y(t)}{y(t)},$$

which we can rewrite (solving (3.12) for $(K - y)/y$ and substituting) as

$$z = \log \frac{K - y_0}{y_0} - rt,$$

the graph of which (versus time) is a line with slope $-r$ and y-intercept $\log(K - y_0)/y_0$. Thus a modified linear regression[10] can provide estimates

[7] T. Carlson, Über Geschwindigkeit und Grösse der Hefevermehrung in Würze, *Biochemische Zeitschrift* 57: 313–334, 1913.

[8] R. Pearl, *Introduction of medical biometry and statistics*, Saunders, Philadelphia, 1930.

[9] Eric Renshaw, *Modelling biological populations in space and time*, Cambridge Studies in Mathematical Biology 11, Cambridge University Press, Cambridge, 1995, pp. 53–55.

[10] In this case, the usual linear regression calculation must be modified slightly since the data z_n are functions of the parameter K (otherwise we must first fix a value for K and then find the best r for that K). In searching for a minimum (with respect to K and r) of

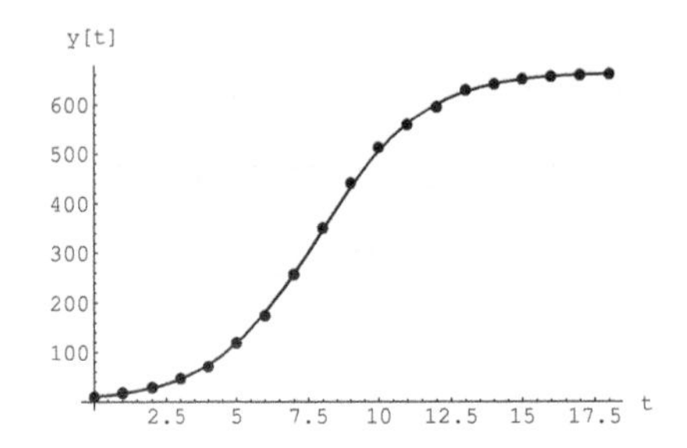

FIGURE 3.4: Yeast culture growth data from Carlson (1913) and its best-fit logistic curve.

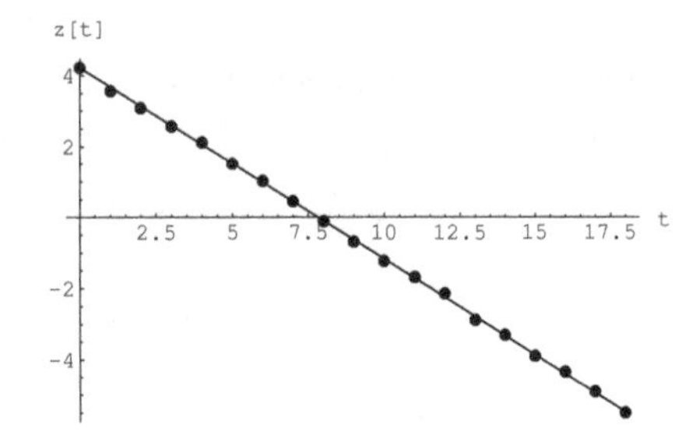

FIGURE 3.5: Logarithmic (z) data from Carlson (1913) and its best-fit line.

for the parameters r and K (given y_0). The best-fit values for the data in Table 3.2 are $K \approx 664.5$, $r \approx 0.539/\text{hr}$. Graphs of $y(t)$ and $z(t)$ comparing data and best-fit solution (similar to those in Renshaw, but using the parameter values given here) are shown in Figures 3.4 and 3.5. The fit is clearly quite good.

The logistic model has also been used with some success to fit data on growth of larger, more complex organisms, as the following example illustrates.

EXAMPLE 1.
Census figures (in millions) for the United States, $P(t)$ in the table below, fit the solution

$$y(t) = \frac{265}{1 + 69\,e^{-0.03t}}$$

of the logistic differential equation, with $t = 0$ corresponding to the year 1790, reasonably well. Predict the population in the years 1990 and 2000, and the limiting population size.

Year	$P(t)$	$y(t)$
1790	4	3.8
1800	5	5.1
1810	7	6.8
1820	10	9.1
1830	13	12.2
1840	17	16.2
1850	23	21.4

Year	$P(t)$	$y(t)$
1860	31	28.0
1870	39	36.5
1880	50	47.0
1890	63	59.7
1900	76	74.8
1910	92	91.8
1920	106	110.6

Year	$P(t)$	$y(t)$
1930	123	130.2
1940	132	150.0
1950	152	169.0
1960	180	186.5
1970	204	202.0
1980	226	215.3

the sum of squared errors

$$\sum_n \left[z_n - \left(\log \frac{K - y_0}{y_0} - rt_n \right) \right]^2,$$

substituting $z_n = \log[(K - y_n)/y_n]$ transforms the problem into one in which a computer can search simultaneously for the best-fit K and r.

Solution: The limiting population size as seen from the expression for $y(t)$ is 265. The population size in 1990 ($t = 200$) would be predicted from $P(t)$ as

$$\frac{265}{1 + 69\, e^{-(0.03)(200)}} = 226.3,$$

and the prediction for 2000 ($t = 210$) would be

$$\frac{265}{1 + 69\, e^{-(0.03)(210)}} = 235.2.$$

The actual population found in the 1990 census was approximately 250 million. The actual population given in the 2000 census was about 281.4 million. □

Another notable application of the continuous-time logistic model to the growth of more complex organisms provides a helpful perspective from which to evaluate this equation's overall utility in predicting population growth. Davidson made use of the extensive records of the sheep populations imported to South Australia and Tasmania in the 19th century — over a century of annual records in each case — to study their growth in habitats (albeit artificial — sheep cannot persist unaided in Australia) with a carrying capacity determined principally by the amount of land available for pasture. Davidson found[11] that the data for both populations matched logistic curves closely during the initial "sigmoid" (S-shaped) period of growth. Here again, the logistic parameters were fit via the transformation $z = \log[(K - y)/y]$ which makes the curve linear, here first choosing a value for K and then using linear regression to find r. The parameters c and r for (3.12) for the two populations are, in fact, remarkably close, differing only in the carrying capacities K. However, as the populations reached their carrying capacities, they exhibited ongoing, significant fluctuations, gaining or losing over 25% in a handful of years (see Figures 3.6 and 3.7).

Davidson's work is not the only instance where a logistic model fits data well only during the initial rapid growth period. Renshaw discusses (e.g., p. 55) several other situations in which lab cultures and other populations were observed to grow more or less logistically toward carrying capacity, but thereafter exhibited either sustained fluctuations or a gradual decline. The likely explanation for this phenomenon is that when a population has not yet run up against resource limitations, its reproductive capacity may often be the strongest force affecting its growth, whereas other factors such as stochastic effects, spatial heterogeneity, or changes in the environment (the fluctuations observed in the Australian sheep populations in the graphs were attributed to fluctuations in carrying capacity while pastures recovered from overgrazing) dominate when the logistic growth is small. In any case, the conclusion to draw

[11] J. Davidson, On the ecology of the growth of the sheep population in South Australia, *Transactions of the Royal Society of South Australia* 62: 141–148, 1938.

J. Davidson, On the growth of the sheep population in Tasmania, *Transactions of the Royal Society of South Australia* 62: 342–346, 1938.

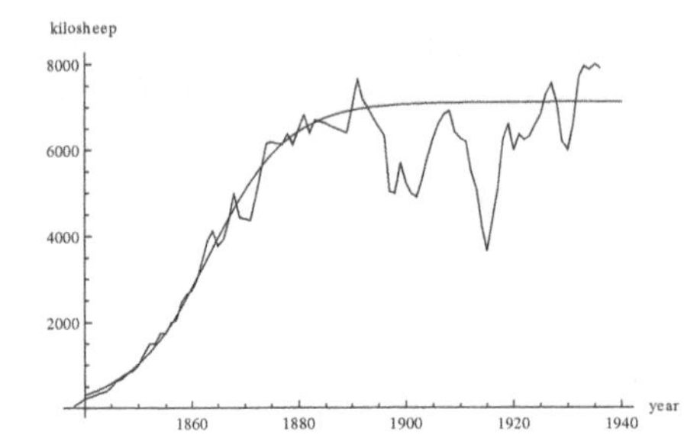

FIGURE 3.6: South Australian sheep population (in thousands) from 1838 to 1936 with the logistic curve $y(t) = 7115/[1 + \exp(249.11 - 0.13369\,t)]$ superimposed, after Renshaw (1995).

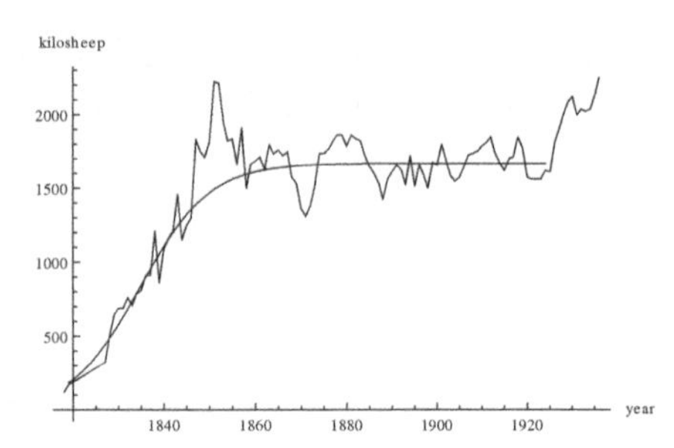

FIGURE 3.7: Tasmanian sheep population (in thousands) from 1818 to 1936 with the logistic curve $y(t) = 1670/[1+\exp(240.81 - 0.13125\,t)]$ superimposed, after Renshaw (1995).

seems to be that the continuous logistic model often does a remarkably good job of describing the rapid growth of even complex populations approaching carrying capacity, but is of at best limited use in describing the growth of populations at or near that capacity.

We finish this section with two extensions of the logistic model to multiple populations, which provide a glimpse of the modeling of multi-population biological systems undertaken in Chapters 5, 6, and 7.

3.2.2 Competition among plants

Bampfylde reviews[12] a set of simple deterministic models for hierarchical competition among plants in a habitat composed of multiple patches or sites, developed by Tilman[13] to account for the observed trade-offs between colonization and competitive ability (as measured by longevity) as plants allocate their energies between developing root systems and reproduction. One example cited by Tilman involves a prairie habitat in Minnesota where primary competition is for nitrogen in the soil; superior competitors such as bluestem grass develop extensive root systems often running two meters or more below the surface but spread slowly, whereas inferior competitors such as rough bentgrass proliferate quickly but have relatively shallow root systems which prevent it from invading areas where bluestem is established (see Figure 3.8 for photos).

If we take the habitat sites to be of sufficient size for one adult to survive, we can describe the proportion of sites occupied by a given species in the

[12]Caroline Bampfylde, Modelling rainforests, M.Sc. thesis, Oxford University, 1999, Section 2.2, pp. 6–9.

[13]David Tilman, Competition and biodiversity in spatially structured habitats, *Ecology* 75: 2–16, 1994.

Photo courtesy U.S. National Park Service, www.nps.gov
Photo by Dave Powell, USDA Forestry Service

FIGURE 3.8: Left, big bluestem (*Andropogon gerardii*), a competitive prairie grass with a deep root system. Right, rough bentgrass (*Agrostis scabra*), a perennial clump grass whose wide inflorescence breaks off and floats like a tumbleweed, enabling it to colonize quickly despite a relatively shallow root system.

rainforest as a function of time (i.e., without being spatially explicit). Suppose now that the proportion p_1 of sites occupied by species 1 is determined by (a) the per site rate c_1 at which seeds or other propagules disperse and proliferate, and (b) the per-site mortality rate m_1. Let us also assume that these rates are unaffected by the presence of other species, i.e., that species 1 dominates any competition with other species, and that colonization sites are determined at random. Then the rate of production of new sites is given by the product of the rate of seedling production by the occupied sites c_1p_1 and the proportion of sites available for colonization $(1 - p_1)$, while the rate at which sites are vacated is given by m_1p_1. This gives the model

$$\frac{dp_1}{dt} = c_1p_1(1 - p_1) - m_1p_1 \tag{3.14}$$

originated by Levins,[14] which can be rewritten in logistic form as

$$\frac{dp_1}{dt} = (c_1 - m_1)p_1\left(1 - \frac{p_1}{1 - \frac{m_1}{c_1}}\right). \tag{3.15}$$

We observe that $r = c_1 - m_1$ and $K = 1 - \frac{m_1}{c_1}$ are positive if and only if $c > m$. From our analysis earlier in this section, we know that (3.15) has the solution

$$p_1 = \frac{1 - \frac{m_1}{c_1}}{1 + Ce^{-(c_1 - m_1)t}},$$

[14]R. Levins, Some demographic and genetic consequences of environmental heterogeneity for biological control, *Bulletin of the Entomological Society of America* 15: 237–240, 1969.

where $C = \left(1 - \frac{m_1}{c_1} - p_1(0)\right)/p_1(0)$, as long as $p_1(0) \neq 0$. Therefore, if $c_1 > m_1$ then $p_1 \to 1 - \frac{m_1}{c_1}$, whereas if $c_1 < m_1$ then $p_1 \to 0$. This result agrees with intuition, in that the criterion for the species to survive is that it should reproduce faster than it dies out.

Now suppose that there is another species, species 2, which competes with species 1 for resources (and space) in the rainforest. Suppose that species 1 dominates species 2 in this competition, so that species 2 can only grow in those sites not occupied by species 1. Then this inferior competitor colonizes the proportion $(1-p_1-p_2)$ of sites occupied by neither the superior competitor nor itself at a base rate of c_2p_2, and vacates sites due to natural mortality at a rate m_2p_2, and due to displacement by the superior competitor at a rate $c_1p_1p_2$ (species 1 colonizes at a rate c_1p_1; a proportion p_2 of those sites are occupied by species 2, causing displacement). These assumptions, together with those on species 1, give us a system of equations

$$\begin{aligned}\frac{dp_1}{dt} &= c_1p_1(1-p_1) - m_1p_1,\\ \frac{dp_2}{dt} &= c_2p_2(1-p_1-p_2) - m_2p_2 - c_1p_1p_2.\end{aligned} \tag{3.16}$$

As species 1 remains unaffected by species 2, this system is said to *decouple*, that is, we can analyze the growth of the two species separately. Species 1 will approach its carrying capacity p_1^* as discussed in the previous paragraph, $p_1 \to p_1^*$, at which point (3.16) becomes

$$\frac{dp_2}{dt} = c_2p_2(1-p_1^*-p_2) - m_2p_2 - c_1p_1^*p_2,$$

which we can rewrite in explicit logistic form as

$$\frac{dp_2}{dt} = r_2p_2\left(1 - \frac{p_2}{r_2/c_2}\right),$$

where

$$r_2 = c_2 - m_2 - (c_1+c_2)p_1^* = \begin{cases} c_2\frac{m_1}{c_1} - m_2 - (c_1 - m_1), & c_1 > m_1;\\ c_2 - m_2, & c_1 < m_1.\end{cases}$$

The behavior of p_2 thus again depends on the sign of the coefficient r_2, approaching r_2/c_2 if $r_2 > 0$ and 0 if $r_2 < 0$. Under the (reasonable) assumption that for each species i found in the rainforest, $c_i > m_i$, that is, that in the absence of competition the species would persist, we take the corresponding form for r_2 and write the persistence condition for species 2 as

$$c_2 > \frac{c_1}{m_1}(m_2 + c_1 - m_1).$$

This system can be extended to a complete hierarchy of n species, ranked by dominance in competition (with species i dominating and displacing species j whenever $i < j$), so that, for instance,

$$\frac{dp_3}{dt} = c_3 p_3 (1 - p_1 - p_2 - p_3) - m_3 p_3 - (c_1 p_1 + c_2 p_2) p_3,$$

and more generally

$$\frac{dp_i}{dt} = c_i p_i (1 - \sum_{j=1}^{i} p_j) - \left(m_i + \sum_{j=1}^{i-1} c_j p_j \right) p_i. \tag{3.17}$$

Since the competition is completely ordered, each equation in the system can be analyzed independently (beginning with species 1), rewritten as a logistic equation, and shown to approach a unique equilibrium value, which is the greater of 0 and

$$p_i^* = 1 - \frac{m_i}{c_i} - \sum_{j=1}^{i-1} \left[p_j^* \left(1 + \frac{c_j}{c_i} \right) \right]. \tag{3.18}$$

Note that the last term in the expression for p_i^* represents the reduction in distribution of species i due to displacement and exclusion by its superior competitors. Tilman observed that depending on parameter values (i.e., a species's ability to colonize and survive), an inferior competitor may actually occupy a greater proportion of sites than a superior competitor. For instance, suppose species 1 has a high mortality rate, so that p_1^* is small. Then if species 2 has a low mortality rate,

$$p_2^* = 1 - \frac{m_2}{c_2} - p_1^* \left(1 + \frac{c_1}{c_2} \right) \approx 1 - \frac{m_2}{c_2} > p_1^*.$$

This model can be extended to accommodate, for example, habitat destruction and fragmentation (see Exercises 8 and 9 below). Bampfylde, who was interested in modeling competition in rainforests, observed that displacement does not occur among rainforest species, so competition is purely for sites opened through natural mortality. In this case, the displacement term in (3.17) is omitted, and the colonization term is adjusted to exclude displacement. Thus the colonization rate $c_i p_i$ is reduced by the proportion of open sites, $(1 - \sum_{j=1}^{n} p_j)$, rather than merely by the proportion of sites not occupied by superior competitors, $(1 - \sum_{j=1}^{i} p_j)$. The resulting model does not decouple, however, so we shall reserve its further study for Chapter 6.

3.2.3 The spread of infectious diseases

Another biological science to which mathematics has made important contributions is epidemiology. Mathematical models for the spread of infectious diseases typically divide populations into distinct classes based on epidemiological or demographic factors relevant to the transmission of the disease:

susceptible and infected individuals, high-risk and low-risk individuals, highly infectious and less infectious individuals, etc. For this reason they are often referred to as *compartmental* models. Most compartmental models involve keeping track of several populations, an activity which requires as many state variables (and equations) as compartments. The simplest possible epidemic model, however, can be rewritten as a single logistic equation, and so we include it here.

Consider a population of constant size N in which an infectious disease is introduced. We divide the population into two classes: susceptibles (not infected) and infectives (infected and contagious). Let $S(t)$ denote the number of susceptibles at time t and let $I(t)$ denote the number of infectives, so that $S(t)+I(t)=N$. We assume that the disease is spread from infectives to susceptibles through contact. Suppose that an "average" infective makes potentially infective contacts with a constant number βN of individuals in unit time. Then, presumably $\beta I(t)$ of these individuals are already infected, and the number of new infections caused by an "average" infective in unit time is $\beta S(t)$. Thus the total number of new infections in unit time is $\beta S(t)I(t)$. We assume also that the disease is never fatal, and that in unit time a fraction γ of the infectives recover, so that the number of recoveries in unit time is $\gamma I(t)$. This assumption is equivalent to the assumption that for each $s \geq 0$ the fraction of the infectives who remain infective for a time interval s is $e^{-\gamma s}$, and that the average length of the infective period is $1/\gamma$. Finally, we assume that on recovery infectives have no immunity against re-infection and return to the susceptible class.

We may formulate a model to describe $S(t)$ and $I(t)$ by thinking of a flow rate βSI from the susceptible class to the infective class and a flow rate γI from the infective class to the susceptible class. As noted with some previous population models, S and I should, properly speaking, take on only integer values, to count individuals. However, we will allow them to be real-valued in this model, and consider the model either as an approximation valid for large populations N, or (if we set $N=1$) as a representation of *proportions* of the population susceptible and infective. (Models for small populations are usually written as discrete processes, and stochasticity also plays an important role on this scale.)

We now have a pair of differential equations

$$S' = -\beta SI + \gamma I, \quad I' = \beta SI - \gamma I,$$

a model first described by W. O. Kermack and A. G. McKendrick.[15] Since $S(t)+I(t)=N$ for all t, we may replace S by $N-I$ and describe the situation by a single differential equation:

$$I' = \beta I(N-I) - \gamma I = (\beta N - \gamma)I\left[1 - \frac{I}{\frac{\beta N-\gamma}{\beta}}\right]. \tag{3.19}$$

[15] W.O. Kermack and A.G. McKendrick, Contributions to the mathematical theory of epidemics, Part II, *Proc. Royal Soc. London* 138 (1932), 55–83.

The differential equation (3.19) has the same form as the logistic equation (3.8), with r replaced by $\beta N - \gamma$) and K replaced by $\frac{\beta N-\gamma}{\beta}$. It is important to note that, as was true for the plant competition model studied earlier in this section, in (3.19) it is possible for $r = \beta N - \gamma$ to be either positive or negative. Recalling the solution (3.12) of (3.8), we again observe that if $r > 0$ (and $K > 0$), then every solution $y(t)$ with $y(0) > 0$ has $\lim_{t\to\infty} y(t) = K$, while if $r < 0$, then every solution $y(t) > 0$ has $\lim_{t\to\infty} y(t) = 0$. If we translate this result to (3.19), we see that if $\beta N/\gamma < 1$, then $\lim_{t\to\infty} I(t) = 0$, while if $\beta N/\gamma > 1$, then $\lim_{t\to\infty} I(t) = N - \frac{\gamma}{\beta} > 0$.

This result is the famous threshold theorem of Kermack and McKendrick: If the quantity $\beta N/\gamma$, called the *basic reproductive number*, is less than 1, then the number of infectives approaches zero, and the number of susceptibles approaches N. In epidemiological terms this means that the infection dies out. On the other hand, if the basic reproductive number $\beta N/\gamma$ exceeds 1, then the number of infectives remains positive and tends to a positive limit. In epidemiological terms, this means that the infection remains *endemic.* The quantity $\beta N/\gamma$ represents the number of secondary infections caused by each infective over the duration of the infection. Since βN is a number of contacts per infective in unit time, the dimensions of βN are time^{-1}. Thus the basic reproductive number $\beta N/\gamma$ is dimensionless, and does not depend on the units used in describing the model.

One infectious disease which has been studied mathematically by models such as (3.19), as well as by considerably more detailed models, is gonorrhea. For gonorrhea the average infective period $(1/\gamma)$ is known to be about one month. However, the transmission coefficient β must be estimated by some indirect means. One way to estimate β is to use the fact that for small values of I a solution of (3.19) grows exponentially with exponent $r = (\beta N - \gamma)$. Thus if an infection in some region begins with a small number of infectives, the time $\hat{t}$ for the number of infectives to double ($e^{r\hat{t}} = 2$) is approximately

$$\frac{\log 2}{r} = \frac{\log 2}{\beta N - \gamma}.$$

In one reported gonorrhea outbreak this doubling time was approximately 1.7 months, which leads to the estimate $\beta N/\gamma = 1.4$, which indicates a basic reproductive number of 1.4.

EXAMPLE 2.

A new strain of influenza is introduced into a town with 1200 inhabitants by two visitors. Assume that the average infective is in contact with 0.4 inhabitants per day and that the average duration of the infective period is 6 days. Will the infection die out or will the flu persist?

Solution: The number of potentially infective contacts per infective per day is $\beta N = 0.4$ days^{-1}. Thus the basic reproductive number is $\beta N/\gamma = (0.4 \text{ days}^{-1})(6 \text{ days}) = 2.4$; since this exceeds 1, the flu will persist. □

There are other problems such as the spread of a rumor and rates of chemical reactions which may also lead to a logistic differential equation model (see Exercises 17 and 18 below, and Exercise 1 in Section 4.3).

Exercises

1*. Suppose that a population satisfies a logistic model (3.8). Show that its growth rate y' is at a maximum when $y = \frac{K}{2}$ and that its maximum growth rate is $\frac{rK}{4}$. Hint: Maximize y' by setting

$$(y')' = \frac{d}{dt}\left(ry(1 - \frac{y}{K})\right) = 0.$$

2. The population of the United States in 1970 was 202.0 (millions) and the per capita rate of growth was approximately 0.7% per year. Use these data and an exponential growth model to predict the population in the years 1990, 2000, and 2100.

3. Suppose that the population of the United States satisfies a logistic model with carrying capacity 300 (millions). (a) Estimate the value of r in the logistic model by using the observed population size and per capita growth rate in 1970 as given in Exercise 2. (b) Predict the population in the years 1990, 2000, and 2100.

4. (a) Compare the predictions of the exponential and logistic models in Exercises 2 and 3 for the years 1990 and 2000 with each other and with actual data. (b) Compare their predictions for the year 2100. Which seems more realistic? Explain.

5*. Evaluate the logistic model as a predictor of the growth of the U.S. population (Example 1, this section).

   (a) The actual population ($P(t)$ in the table) is generally slightly ahead of the prediction $y(t)$ through 1910. Beginning in 1920, however, it is significantly behind the prediction. Why?

   (b) At what point, if any, does the model stop being an effective predictor? Speculate on possible causes for this change, and how they might be incorporated into the model.

6. Suppose a population satisfies the logistic model (3.8) with $r = 0.4$, $K = 100$, $y(0) = 5$. Find the population size for $t = 10$.

7. The Pacific halibut fishery has been modeled by the logistic equation (3.8) with parameters estimated as $r = 0.71\,\text{yr}^{-1}$, $K = 80.5 \times 10^6$ kilograms, where $y(t)$ is the total biomass at time t measured in kilograms. If $y(0) = \frac{K}{4}$, find the biomass 1 year later. Also, find how long it will take for the total biomass to increase to $\frac{K}{2}$.

8. Tilman et al.[16] generalized Tilman's multi-species competition model to include habitat destruction. If a proportion D of the sites are destroyed, only the remaining proportion $1 - D$ can be occupied. Rewrite (3.17) and the corresponding equilibrium values (3.18) to accommodate this assumption.

9. For the one-species version of the habitat destruction model of Tilman et al. discussed in Exercise 8, find the persistence criterion for the species (which previously was simply $c_1 > m_1$) given D.

10. Find parameter values c_i, m_i for a three-species version of Tilman's multi-species competition model (3.17) such that $0 = p_2^* < p_1^* < p_3^*$, that is, the intermediate competitor is driven to extinction by competition with species 1, while the inferior competitor occupies more sites than the superior competitor by virtue of longevity. (Recall the assumption that $c_i > m_i$.)

11. Calculate the net maximum growth rate r_i for species i in Tilman's multi-species competition model (3.17) by rewriting the equation in logistic form, $dp_i/dt = r_i p_i \left(1 - \frac{p_i}{p_i^*}\right)$, with p_i^* as given in (3.18).

12. Show that Bampfylde's displacement-free rainforest competition model as described in this section does not decouple, by writing out the equations for the two-species version: alter (3.17) by omitting the final term and replacing $(1 - \sum_{j=1}^{i} p_j)$ with $(1 - \sum_{j=1}^{n} p_j)$ in the colonization term.

13. A disease begins to spread in a population of 800. The infective period has an average duration of 14 days and the average infective is in contact with 0.1 person per day. What is the basic reproductive number? To what level must the average rate of contact be reduced so that the disease will die out?

14* European fox rabies is estimated to have a transmission coefficient β of 80 km^2 years^{-1} and an average infective period of 5 days. There is a critical carrying capacity K_c measured in foxes per km^2, such that in regions with fox density less than K_c rabies tends to die out, while in regions with fox density greater than K_c rabies tends to be endemic. Estimate K_c.

    [*Remark*: It has been suggested in Great Britain that hunting to reduce the density of foxes below the critical carrying capacity would be a way to control the spread of rabies.]

15*. A communicable disease from which infectives do not recover may be modeled by the pair of differential equations

$$S' = -\beta SI,\ I' = \beta SI.$$

---

[16] David Tilman, Robert M. May, Clarence M. Lehman, and Martin A. Nowak, Habitat destruction and the extinction debt, *Nature* 371: 65–66, 1994.

(a) Show that in a population of fixed size K, such a disease will eventually spread to the entire population.

(b) Given the meaning in practical terms of "eventually," what modeling assumption made implicitly above accounts for the fact that there always remain some individuals who never get infected, even if the population remains closed and of fixed size, and infected individuals really never recover.

16*. Consider a disease spread by carriers who transmit the disease without exhibiting symptoms themselves. Let $C(t)$ be the number of carriers and suppose that carriers are identified and isolated from contact with others at a constant per capita rate α, so that $C' = -\alpha C$. The rate at which susceptibles become infected is proportional to the number of carriers and to the number of susceptibles, so that $S' = -\beta SC$. Let C_0 and S_0 be the number of carriers and susceptibles, respectively, at the time $t = 0$.

(a) Determine the number of carriers at time t from the first equation.

(b) Substitute the solution to part (a) into the second equation and determine the number of susceptibles at time t.

(c) Find the number of members of the population who escape the disease, $\lim_{t\to\infty} S(t)$.

17*. Consider a population of fixed size K in which a rumor is being spread by word of mouth. Let $y(t)$ be the number of people who have heard the rumor at time t and assume that everyone who has heard the rumor passes it on to r others in unit time. Thus from time t to time $t+h$ the rumor is passed on $rh\,y(t)$ times, but a fraction $\frac{y(t)}{K}$ of the people who hear it have already heard it and thus there are only $rh\,y(t)\left(\frac{K-y(t)}{K}\right)$ people who hear the rumor for the first time. Use these assumptions to obtain an expression for $y(t+h) - y(t)$, divide by h and take the limit as $t \to \infty$ to obtain a differential equation satisfied by $y(t)$.

18. At 9 AM, 1 person in a village of 100 inhabitants has heard a rumor. Suppose that everyone who has heard the rumor tells one other person per hour. Using the model of Exercise 17, determine how long it will take until half the village has heard the rumor.

3.3 Graphical analysis

In Chapter 2 we explored cobwebbing, a method for graphical analysis of difference equations which yields both exact numerical solutions for discrete-time initial value problems (difference equations with initial conditions) and qualitative information about the nature of solutions to the given difference equation. In this section we will explore the use of *direction fields*, a method for graphical analysis of differential equations which yields similar information for continuous-time initial value problems (differential equations with initial conditions). First, however, we revisit the notions of solutions and families of solutions introduced in Section 3.1 in order to present an important result about when solutions are certain to exist, which will inform our use of direction fields.

3.3.1 Solutions

Recall that by a differential equation we mean simply a relation between an unknown function and its derivatives. The general form of a first-order differential equation is

$$y' = \frac{dy}{dt} = f(t, y), \tag{3.20}$$

with f a given function of the two variables t and y. By *a solution* of the differential equation (3.20) we mean a differentiable function y of t on some t-interval I such that, for every t in the interval I,

$$y'(t) = f\{t, y(t)\}.$$

In other words, differentiating the function $y(t)$ results in the function $f(t, y)$. For example, as we saw in Example 1 of Section 3.1, the function $y = 2e^{at}$ is a solution of the differential equation $y' = ay$ on every t-interval. We see this by differentiating $2e^{at}$ to get $a\,2e^{at}$, which we can rewrite as ay. To verify whether a given function is a solution of a given differential equation, we need only substitute into the differential equation and check whether it then reduces to an identity.

EXAMPLE 1.
Show that the function $y = \frac{1}{t+1}$ is a solution of the differential equation $y' = -y^2$.

Solution: For the given function,

$$\frac{dy}{dt} = -\frac{1}{(t+1)^2} = -y^2$$

and this shows that it is indeed a solution. □

In the same way we can verify that a family of functions satisfies a given differential equation. By *a family of functions* we mean a function which includes an arbitrary constant, so that each value of the constant defines a distinct function. The family ce^{5t}, for instance, includes the functions e^{5t}, $-4e^{5t}$, $12e^{5t}$, and $\sqrt{3}\,e^{5t}$, among others. When we say that a family of functions satisfies a differential equation, we mean that substitution of the family (i.e., the general form) into the differential equation gives an identity satisfied for every choice of the constant.

EXAMPLE 2.
Show that for every c the function $y = \frac{1}{t+c}$ is a solution of the differential equation $y' = -y^2$. (This equation models a decaying population whose per capita mortality rate is linear in population size, rather than constant.)

Solution: For the given function,

$$\frac{dy}{dt} = -\frac{1}{(t+c)^2} = -y^2,$$

and this shows that each member of the given family of functions is a solution. □

EXAMPLE 3.
Show that the family of functions

$$y = \frac{1+ce^t}{1-ce^t}$$

is a solution of the differential equation

$$y' = \frac{y^2-1}{2}$$

for every value of the constant c.

Solution: For the given function y,

$$y' = \frac{ce^t(1-ce^t) - (-ce^t)(1+ce^t)}{(1-ce^t)^2},$$

$$\frac{y^2-1}{2} = \frac{1}{2}\left[\frac{(1+ce^t)^2-(1-ce^t)^2}{(1-ce^t)^2}\right],$$

and algebraic simplification shows that these two expressions are equal. □

In applications we are usually interested in finding not a family of solutions of a differential equation but a single solution which satisfies some additional requirement, usually an initial or boundary condition. Geometrically, an initial condition picks out the solution from a family of solutions which passes through the point (t_0, y_0) in the t-y plane. Physically, this corresponds to measuring the state of a system at the time t_0 and using the solution of the initial value problem to predict the future behavior of the system.

EXAMPLE 4.
Find the solution of the differential equation $y' = -y^2$ of the form $y = \frac{1}{t+c}$ which satisfies the initial condition $y(0) = 1$.

Solution: We substitute the values $t = 0$, $y = 1$ into the equation $y = \frac{1}{t+c}$, and we obtain a condition on c, namely $1 = \frac{1}{c}$, whose solution is $c = 1$. The required solution is the function in the given family with $c = 1$, namely $y = \frac{1}{t+1}$. □

EXAMPLE 5.
Find the solution of the differential equation $y' = -y^2$ which satisfies the general initial condition $y(0) = y_0$, where y_0 is arbitrary.

Solution: We substitute the values $t = 0$, $y = y_0$ into the equation $y = \frac{1}{t+c}$ and solve the resulting equation $y_0 = \frac{1}{c}$ for c, obtaining $c = \frac{1}{y_0}$ provided $y_0 \neq 0$. Thus the solution of the initial value problem is

$$y = \frac{1}{(t + \frac{1}{y_0})^2}$$

except if $y_0 = 0$. If $y_0 = 0$, there is no solution of the initial value problem of the given form; in this case the identically zero function, $y = 0$, is a solution. We have now obtained a solution of the initial value problem with arbitrary initial value at $y = 0$ for the differential equation $y' = -y^2$. □

A family of solutions may arise if we are considering a differential equation with no initial condition imposed, and we will then also be concerned with the question of whether the given family contains all solutions of the differential equation. To answer this question, we will need to make use of a theorem which guarantees that each initial value problem for the given differential equation has exactly one solution. More specifically, if an initial value problem is to be a usable mathematical description of a scientific problem, it must have a solution, for otherwise it would be of no use in predicting behavior. Furthermore, it should have only one solution, for otherwise we would not know which solution describes the system. Thus for applications it is vital that there be a mathematical theory telling us that an initial value problem has exactly one solution. This was not an issue in Chapter 2; for a difference equation we can always iterate (in theory) to form a solution. For differential equations, however, solutions are not obtained by a direct construction, and it is necessary to have some result telling us that a problem really does have a solution. Fortunately, there is a very general theorem which tells us that this is true for the initial value problem (3.20), $y(0) = y_0$ provided the function

f is reasonably smooth. We will state this result and ask the reader to accept it without proof because the proof requires more advanced mathematical knowledge than we have at present.

EXISTENCE AND UNIQUENESS THEOREM: If the function $f(t, y)$ is continuous and differentiable with respect to y in some region of the plane which contains the point (t_0, y_0), then the initial value problem consisting of the differential equation $y' = f(t, y)$ and the initial condition $y(t_0) = y_0$ has a unique solution which is defined on some t-interval containing t_0 in its interior.

Even though the function $f(t, y)$ may be well-behaved in the whole t-y plane, there is no assurance that a solution will be defined for all t. As we have seen in Example 1, the solution $y = \frac{1}{t+1}$ of $y' = -y^2$, $y(0) = 1$ exists only for $-1 < t < \infty$. As we have seen in Example 2, each solution of the family of solutions $y = \frac{1}{t+c}$ has a different interval of existence. In Example 5 we have shown how to rewrite a family of solutions for a differential equation in terms of an arbitrary initial condition — that is, as a solution of an initial value problem for that differential equation. We have also seen how to identify those initial conditions which cannot be satisfied by a member of the given family. Often there are constant functions which are not members of the given family but which are solutions and satisfy initial conditions that cannot be satisfied by a member of the family. The existence and uniqueness theorem tells us that if we can find a family of solutions, possibly supplemented by some additional solutions, so that we can find this collection contains a solution corresponding to each possible initial condition, then we have found the set of all solutions of the differential equation.

3.3.2 Direction fields

The geometric interpretation of a solution $y(t)$ to a differential equation (3.20) is that the curve $y = y(t)$ has slope $f(t, y)$ at each point (t, y) along its length. Thus we might think of approximating the solution curve by piecing together short line segments whose slope at each point (t, y) is $f(t, y)$. To realize this idea, we construct at each point (t, y) in some region of the plane a short line segment with slope $f(t, y)$. The collection of line segments is called the *direction field* of the differential equation (3.20). The direction field can help us to visualize solutions of the differential equation since at each point on its graph a solution curve is tangent to the line segment at that point. We may sketch the solutions of a differential equation by connecting these line segments by smooth curves.

Drawing direction fields by hand is a difficult and time-consuming task.

There are computer programs, both self-contained and portions of more elaborate computational systems such as Mathematica, MATLAB and Maple, which can generate direction fields for a differential equation and can also sketch solution curves corresponding to these direction fields. We give some examples here which have been produced by Maple (see Appendix A); the reader with access to a facility which is capable of drawing direction fields is urged to reproduce these examples before trying to produce other direction fields.

EXAMPLE 6.
Draw a direction field and some solutions of the differential equation $y' = y$. (This is the prototypical linear homogeneous differential equation, with the time unit rescaled so that the growth rate a is equal to 1; see Section 3.1 for applications.)

Solution: See Figure 3.9. The direction field suggests exponential solutions, which we know from Section 3.1 to be correct. □

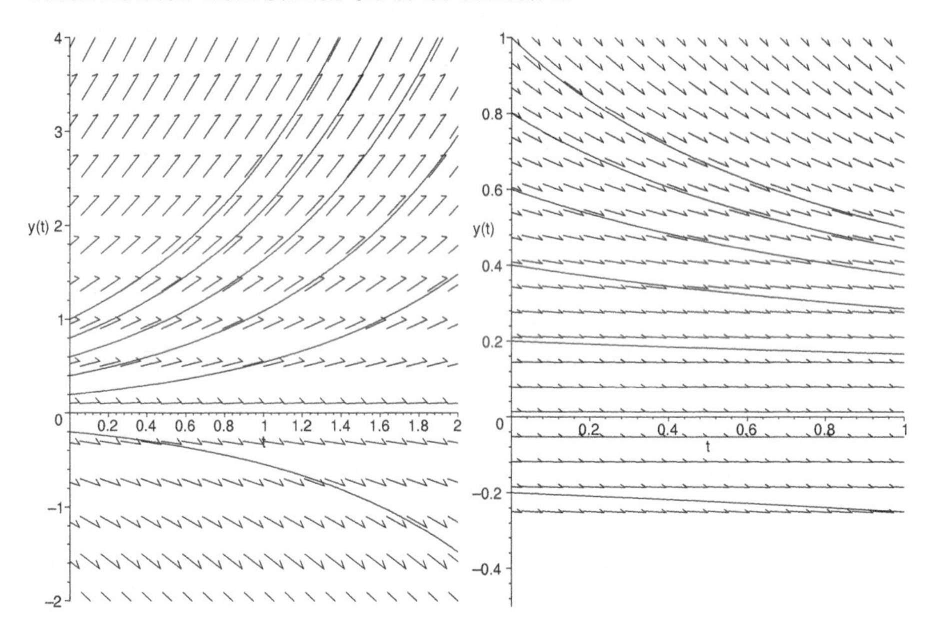

FIGURE 3.9: Direction field and solutions for $y' = y$.

FIGURE 3.10: Direction field and solutions for $y' = -y^2$.

EXAMPLE 7.
Draw a direction field and some solutions of the differential equation $y' = -y^2$.

Solution: See Figure 3.10. The direction field indicates that solutions below the t-axis become unbounded below while solutions above the t-axis tend to zero as $t \to \infty$, as could be seen from Example 2. □

EXAMPLE 8.

Draw a direction field and some solutions of the differential equation $y' = y(1-y)$. This is the prototypical logistic differential equation, in which both time and population units have been rescaled so that r and K are both equal to 1.

Solution: See Figure 3.11. The direction field indicates that solutions below the t-axis are unbounded below while solutions above the t-axis tend to 1 as $t \to \infty$, consistent with what we have established for this logistic differential equation. □

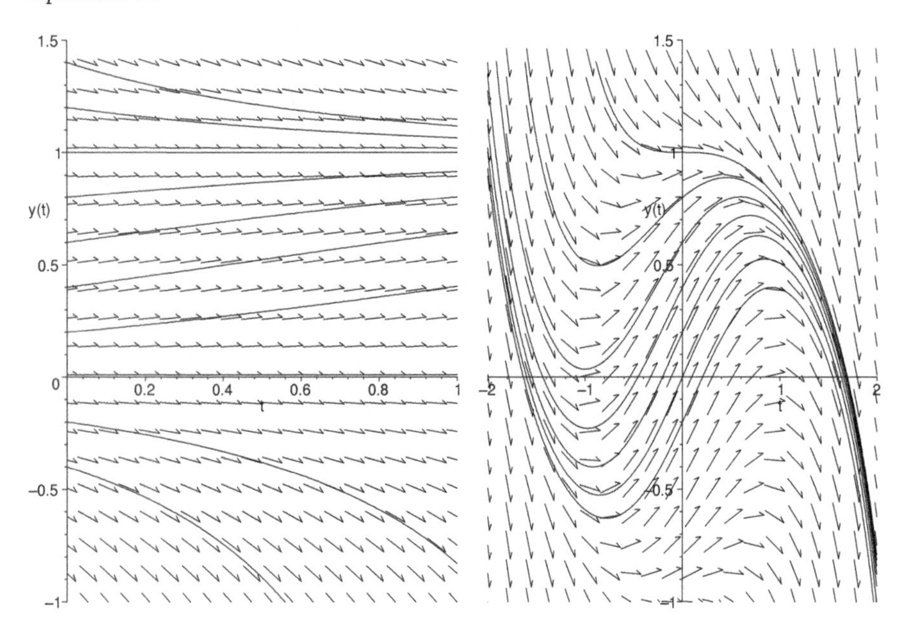

FIGURE 3.11: Direction field and solutions for $y' = y(1-y)$.

FIGURE 3.12: Direction field and solutions for $y' = 1 - t^2 - y^2$.

EXAMPLE 9.

The differential equation $y' = 1 - y^2$ is a slight variation on the logistic equation $y' = y - y^2$; the important difference between the two for purposes of modeling biological systems is that when $y = 0$ the new model has $y' = 1$, rather than $y' = 0$ as the logistic model has. This allows the population to grow to its asymptotic value of 1 even when measurement errors in the initial population size, which may be very small (given the rescaling of units so that $K = 1$), yield a value of 0 or even a small negative number. Although the new model is not quite biologically sound at $y = 0$, it may be used with the given caveat.

Suppose now that we wish to incorporate an additional feature into the model: namely, an eventual decline in population once it has reached its carrying capacity (as mentioned in Section 3.2, such declines have been observed following the initial growth period). A simple modification might involve adding a negative term dependent on t, say $-t^2$, to y'. Draw a direction field and some solutions of the differential equation $y' = 1 - y^2 - t^2$, to evaluate the effects of this modification.

Solution: See Figure 3.12. It can be seen that regardless of initial conditions (i.e., above or below the carrying capacity of 1), solutions eventually grow negative and unbounded. The $-t^2$ term does produce an eventual decline from $y = 1$, but the decline is too strong; the term needs to be modified to restrict its maximum size, so that solutions remain positive. □

As the reader can see, direction fields provide a useful visual indication of system behavior for a range of initial conditions. Although the solutions obtained by connecting the segments are approximate rather than exact, and approximate numerical solutions for specific initial conditions are found more commonly via numerical analysis, direction fields are useful tools in suggesting qualitative behavior that varies by initial condition in cases when it may be difficult to produce a complete analysis using the usual qualitative tools (which will be developed in Section 3.5).

The geometric view of differential equations presented by the direction field will appear again when we examine some qualitative properties in Section 4.1, and is also the basis of the numerical approximation of solutions in Section 4.5.

Exercises

1. Show that $y = (1 - t^2)^{-\frac{1}{2}}$ is a solution of the differential equation $y' = ty^3$.

2. Show that $y = 3t$ is a solution of the differential equation $y' = y/t$.

3. Show that $y = (c - t^2)^{-\frac{1}{2}}$ is a solution of the differential equation $y' = ty^3$ for every choice of the constant c.

4. Show that $y = ct$ is a solution of the differential equation $y' = y/t$ for every choice of the constant c.

5. Among the family of solutions $y = (c - t^2)^{-\frac{1}{2}}$ of $y' = ty^3$, find the solution such that $y(0) = 1$.

6. Find the solution of $y' = ty^3$ such that $y(0) = 0$.

7. Show that the solution of the initial value problem $y' = ty^3$, $y(0) = -1$ is $y = -(1 - t^2)^{-\frac{1}{2}}$.

8. Among the solutions $y = \frac{1+ce^t}{1-ce^t}$ of the differential equation $y' = \frac{1}{2}(y^2 - 1)$, find the ones which satisfy the initial conditions $y(0) = 1$, $y(0) = 0$, and $y(2) = 0$.

9. Show that if $\hat{y}$ is a constant such that $f(t,\hat{y}) = 0$ for all t, then $y = \hat{y}$ is a constant solution of $y' = f(t,y)$.

10*. * If $y(t)$ is a solution of $y' = f(t,y)$ which has a local maximum at the point $(\hat{t},\, \hat{y})$, show that $f(\hat{t},\, \hat{y}) = 0$.

11. Draw a direction field and some solutions of the differential equation $y' = ty$.

12. Draw a direction field and some solutions of the differential equation $y' = t^2 + y^2$.

13. Draw a direction field and some solutions of the differential equation $y' = 1 + y^2$.

14. Draw a direction field and some solutions of the differential equation $y' = y - t$.

15*. For the differential equation $y' = -t + \sqrt{2y + t^2}$,

    (a) Show that $y = ct + \frac{1}{2}c^2$ is a solution for every choice of the constant c.

    (b) Show that $y = -\frac{t^2}{2}$ is a solution.

    (c) Find the tangent line to the curve $y = -\frac{t^2}{2}$ at the point $(-c,\, -\frac{c^2}{2})$.

16*. Improve the modification to the model $y' = 1 - y^2$ presented in Example 9 so that it does not cause all solutions to grow unbounded and negative.

3.4 Equations and models with variables separable

In this section we shall learn a method for finding solutions of a class of differential equations. Although exact solutions are not necessary in order to use models to answer questions about biological systems, they provide additional information that can lead to helpful insights for many of the most common model equations.

The method of *separation of variables* is based on the approach we used in Section 3.1 to solve the differential equation of exponential growth or decay, and is applicable to differential equations with *variables separable.* A differential equation is called *separable*, or is said to have *variables separable*, if the function f which describes the rate of change of the population y over time can be expressed in the form

$$f(t,y) = p(t)q(y); \tag{3.21}$$

that is, if the changes in population size due to population density y can be factored out from those due to the time index t (such as seasonal factors). For example, $y' = \frac{y}{1+t^2}$ and $y' = y^2$ are both separable, whereas $y' = \sin t - 2ty$ is not separable. The examples discussed in Section 3.1 are also separable, and the method of solution described for equation (3.1) of Section 3.1 is a special case of the general method of solution to be developed for separable equations.

The reason for the name *separable* is that the differential equation can then be written as

$$\frac{y'}{q(y)} = p(t),$$

with all the dependence on y on the left and all the dependence on t on the right-hand side of the equation, provided that $q(y) \neq 0$ (indeed, this is the first step in the method). An important special case, which includes most of the examples in this chapter, is when f is completely independent of t and is a function of y only (i.e., $p(t) = 1$); in this case, the equation is said to be *autonomous.* The method of separation of variables applies to many relatively complex differential equations, and is worth learning for this reason.

Although the models and systems explored in this section are thus united by a mathematical, rather than biological, feature, two classes of models which allow separation of variables are the linear and logistic differential equations, which form the basis for many models of biological systems. We shall begin our discussion with some applications of these models, and continue with some other, more specialized separable models.

3.4.1 A linear model for the cardiac pacemaker

The firing of neurons and the regulation of cardiac activity have long been the subjects of intense study, including numerous attempts to capture via modeling the nuances of the cardiac pacemaker, the sinoatrial node in the heart whose cells provide the impulses that regulate heartbeats. Most of these complex models are well beyond the scope of this chapter, but we shall use this context to adapt a simpler, linear model that has been proposed for the periodic firing of neurons in general, and of the cardiac pacemaker wave in particular.[17]

The sinoatrial node consists of a cluster of cells which depolarize, fire, and then reset periodically to send a wave of electrical signals that cause the atria of the heart to contract. After reaching the atrioventricular node, the pulses, distributed through the His-Purkinje system, then cause the ventricles to contract a fraction of a second later, pumping blood from the heart to the rest of the body (see Figure 3.13). The mechanism that produces these impulses in the sinoatrial node involves differences in concentrations of sodium, calcium, and potassium ions inside and outside the cells. Small channels in the cell

[17] See, e.g., B.W. Knight, Dynamics of encoding in a population of neurons, *J. General Physiology* 59: 734–766, 1972, and D.S. Jones and B.D. Sleeman, *Differential equations and mathematical biology*, Boca Raton, FL: CRC Press, 2003.

membranes open and close to change the electrical potential via an exchange of particular ions.

Single autonomous differential equations typically do not lead to sustained oscillations like a heartbeat, but we can obtain periodic behavior by introducing a control variable u that is sensitive to the cellular voltage levels, and switches on and off when the voltage crosses set thresholds. In this way we may think of the control as a regulating force like that of an artificial pacemaker (although such devices' controls are typically calibrated to times, rather than voltages). We shall assume, then, that we know four voltage levels: the control values u_L and u_H and the threshold values y_L and y_H, with $u_L < y_L < y_H < u_H$. We shall also assume that the control signal u works as follows: when the cellular voltage level is low, $y(t) \leq y_L$, the control switches on, $u = u_H$; when the voltage level rises above the upper threshold, $y(t) \geq y_H$, the neuron fires and the control switches off, $u = u_L$. While the voltage level is rising or falling between the two thresholds, $y_L < y < y_H$, the control remains at its present level. Finally, we shall assume that, apart from the influence of the control signal, the pacemaker signal $y(t)$ naturally decays at a rate proportional to its present level.

These assumptions lead to a linear heterogeneous model for the pacemaker signal voltage,

$$y' = au - ay = a(u - y), \tag{3.22}$$

where $a > 0$ is a rate constant which determines how quickly the signal decays from its previous values (such constants are sometimes said to measure the "forgetfulness" of the system). Although the control signal u may change from time to time as the voltage y crosses the upper or lower threshold, it is said to be *piecewise constant*—that is, constant on short periods of time—and so we will study how this system behaves on these short periods, later connecting the solutions during each period to form a long-term picture.

Equation (3.22) is separable, and we shall solve it using the method of separation of variables, which we will explain more generally later in this section. (Another method for solving linear differential equations will be presented in Section 3.5.) To separate variables, we first divide both sides of (3.22) by $a(u-y)$, which is permissible as long as $y \neq u$ (and remember that the model, if successful, will constrain y within $[y_L, y_H]$, so that it never reaches u_L or u_H). This yields

$$\frac{1}{a(u-y)}\frac{dy}{dt} = 1.$$

Integrating both sides of this equation with respect to t, considering y as a function $y(t)$ of t,

$$\int \frac{1}{a(u-y)}\frac{dy}{dt}\,dt = \int 1\,dt,$$

we simplify to

$$\int \frac{dy}{a(u-y)} = \int dt. \tag{3.23}$$

Alternatively, and more colloquially, we can treat the derivative dy/dt as a ratio of differentials dy and dt, and multiply equation (3.22) by $dt/a(u-y)$, to obtain

$$\frac{dy}{a(u-y)} = dt.$$

This form of the equation is meaningless without an interpretation of differentials, but integration then yields the meaningful form (3.23) above.

Integration of (3.23) gives

$$-\frac{1}{a}\log|y-u| = t + c,$$

with c an arbitrary constant of integration. Algebraic solution now gives

$$\begin{aligned}
\log|y-u| &= -a(t+c) \\
|y-u| &= e^{-a(t+c)} = e^{-ac}\,e^{-at} \\
y-u &= \pm e^{-ac}e^{-at} \\
y &= u \pm e^{-ac}\,e^{-at}.
\end{aligned}$$

We now rename the arbitrary constant $\pm e^{-ac}$ as a new arbitrary constant, C, which measures the initial distance of the cellular voltage from the control. We thus obtain the family of solutions

$$y = u + Ce^{-at}. \tag{3.24}$$

Note that although our separation of variables explicitly excluded the constant solution $y = u$, this solution is contained in the family (3.24).

In order to find the solution of (3.22) which satisfies the initial condition

$$y(0) = y_0 \tag{3.25}$$

we substitute $t = 0$, $y = y_0$ into (3.24), obtaining $y_0 = u + C$ or $C = y_0 - u$. This value of C gives the solution

$$y = u + (y_0 - u)e^{-at} = u(1 - e^{-at}) + y_0 e^{-at} \tag{3.26}$$

of the initial value problem (3.24), (3.25).

In applications, the specific form of the solution of a differential equation is often less important than the behavior of the solution for large values of t. Because $e^{-at} \to 0$ as $t \to \infty$ when $a > 0$, we see from (3.24) that every solution of (3.22) tends to the limit u as $t \to \infty$ if $a > 0$. This is an example of *qualitative* information about the behavior of solutions which will be useful in applications. In Section 4.1 we shall examine other qualitative questions—information about the behavior of solutions of a differential equation which may be obtained indirectly rather than by explicit solution.

If we now set time to zero at the moment when the control switches on,

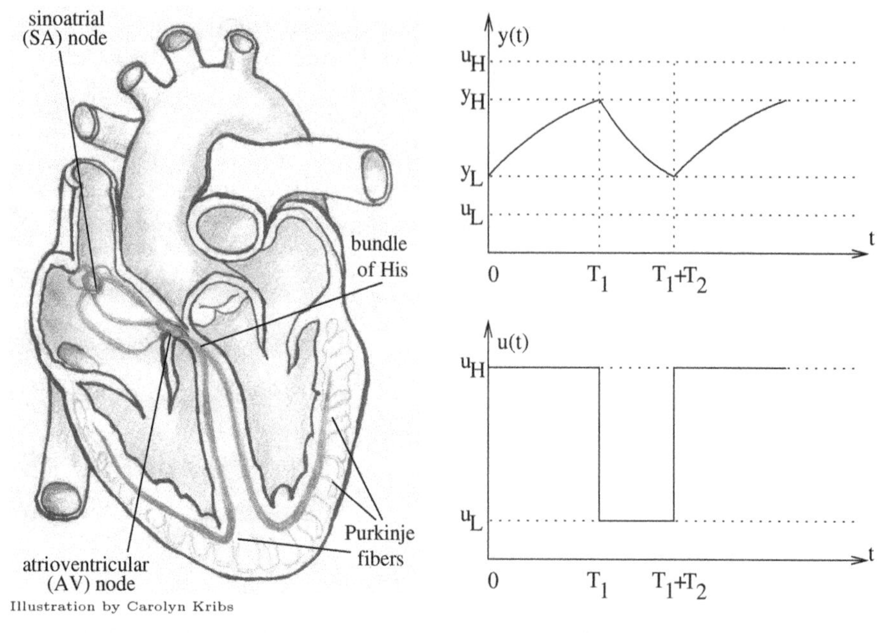

FIGURE 3.13: The cardiac electrical system.

FIGURE 3.14: Pacemaker and control voltage over one period.

we have $y_0 = y_L$, and $u = u_H$ for t on an interval $[0, T_1]$, where T_1 is the time required for the voltage to reach the upper threshold and cause the neuron to fire. During this interval, we have, from (3.26),

$$y(t) = u_H + (y_L - u_H)e^{-at}.$$

Thus the voltage rises from y_L toward u_H, until it crosses the upper threshold y_H, at which point the neuron fires and the control resets to u_L. At this moment we have $y(T_1) = y_H$, which we use as a new initial condition in (3.26) to obtain a description of the voltage as it falls back toward u_L,

$$y(t) = u_L + (y_H - u_L)e^{-a(t-T_1)}$$

($t - T_1$ is time since T_1). This description of y holds until the voltage falls below the lower threshold y_L to begin the cycle over again T_2 time units later (i.e., for $T_1 < t \leq T_1 + T_2$). Figure 3.14 connects the graphs of y on these two intervals to show a complete period, alongside the corresponding graph of the control variable u. The graphs repeat every $T_1 + T_2$ time units.

In practice, we might use such a model to predict the behavior of the cardiac rhythm under different conditions (such as an alteration of the threshold voltages, or of the control signal—see Exercise 18 below), or to calculate information that is not directly observable, based on known data. For instance, if we know three of the voltage parameters and the two time periods T_1 and T_2, we can determine the remaining voltage parameter and the rate constant a. Conversely, if we know the four voltage parameters and a, we can determine T_1 and T_2. These relationships come from the continuity of the voltage y at the points where the control signal changes:

$$y(T_1) = u_H + (y_L - u_H)e^{-aT_1} = y_H \tag{3.27}$$

and

$$y(T_1 + T_2) = u_L + (y_H - u_L)e^{-a(T_2)} = y_L. \tag{3.28}$$

Experimental observations have shown that the voltage level ranges from $y_L = -70$ mV to $y_H = +30$ mV during one period, which lasts about one second. If observations show $T_1 \approx 600$ msec and $T_2 \approx 400$ msec, and we suppose a lower control voltage of $u_L = -100$ mV, then we can calculate

$$e^{-aT_1} = \frac{y_H - u_H}{y_L - u_H}, \quad e^{-aT_2} = \frac{y_L - u_L}{y_H - u_L},$$

so that

$$aT_1 = \log\left(\frac{y_L - u_H}{y_H - u_H}\right), \tag{3.29}$$

$$aT_2 = \log\left(\frac{y_H - u_L}{y_L - u_L}\right). \tag{3.30}$$

To find the rate constant a, we substitute into (3.30):

$$a(400\,\text{msec}) = \log\left(\frac{30\,\text{mV} - (-100\,\text{mV})}{-70\,\text{mV} - (-100\,\text{mV})}\right) = \log\frac{130}{30},$$

and find that $a = 0.003666\,\text{msec}^{-1} \approx 0.004\,\text{msec}^{-1}$, or $4\,\text{sec}^{-1}$. To find u_H, we can then substitute into (3.29) and solve, to find $u_H = +42.5\,\text{mV} \approx +40$ mV.

While this model uses an external control variable rather than incorporating details of the ion channel biochemistry to generate periodic behavior, the exercise above does illustrate the utility of even simple models, as well as the procedure of separation by variables. Before considering further applications, we shall next give a general description of this procedure.

3.4.2 General procedure

Suppose that we wish to solve the differential equation

$$y' = p(t)q(y) \tag{3.31}$$

where p is continuous on some interval $a < t < b$ and q is continuous on some interval $c < y < d$. We begin by separating variables as done for equation (3.22), dividing by $q(y)$ and integrating to give

$$\int \frac{dy}{q(y)} = \int p(t)dt + c \tag{3.32}$$

where c is a constant of integration. Evaluating the integrals on each side of this equation requires knowing the integrands' antiderivatives. Let us name these antiderivatives

$$Q(y) = \int \frac{dy}{q(y)}, \quad P(t) = \int p(t)dt,$$

by which we mean that $Q(y)$ is a function of y whose derivative with respect to y is $\frac{1}{q(y)}$ and $P(t)$ is a function of t whose derivative with respect to t is $p(t)$. Then (3.32) becomes

$$Q(y) = P(t) + c, \tag{3.33}$$

and this equation describes a family of implicit solutions y of the differential equation (3.31). (The solutions can be made explicit if Q is invertible: $y(t) = Q^{-1}(P(t) + c)$.) To verify that each function $y(t)$ defined by (3.33) is indeed a solution of the differential equation (3.31), we differentiate (3.33) implicitly with respect to t, obtaining $\frac{dQ}{dy}\frac{dy}{dt} = \frac{dP}{dt}$, that is,

$$\frac{1}{q(y(t))}\frac{dy}{dt} = p(t),$$

on any interval on which $q(y(t)) \neq 0$, and this shows that $y(t)$ satisfies the differential equation (3.31). There may be constant solutions of (3.31) which are not included in the family (3.33). A constant solution to the original equation (3.31) corresponds to a solution of the equation $q(y) = 0$.

If we wish to solve an initial value problem and find the particular solution of equation (3.31) which satisfies the initial condition $y(t_0) = y_0$, then instead we should make the indefinite integrals in equation (3.32) definite, integrating forward from the initial condition (t_0, y_0):

$$\int_{y_0}^{y} \frac{du}{q(u)} = \int_{t_0}^{t} p(s)\, ds. \tag{3.34}$$

(We use u and s as integration variables in place of y and t so as not to confuse them with the bounds of integration.) Evaluation of this integral defines y implicitly in terms of t.

Alternatively, to solve the initial value problem we can take the general solution (3.33) and (since the constant of integration c is arbitrary) specify $Q(y)$ to be the indefinite integral of $\frac{1}{q(y)}$ such that $Q(y_0) = 0$, and $P(t)$ to be the indefinite integral of $p(t)$ such that $P(t_0) = 0$. Then, because of the fundamental theorem of calculus, we have

$$Q(y) = \int_{y_0}^{y} \frac{du}{q(u)}, \quad P(t) = \int_{t_0}^{t} p(s)\, ds,$$

and we may write the solution (3.33) in the form

$$\int_{y_0}^{y} \frac{du}{q(u)} = \int_{t_0}^{t} p(s)\, ds + c.$$

Now substitution of the initial conditions $t = t_0$, $y = y_0$ gives $c = 0$, and the solution of the initial value problem is again given implicitly by (3.34).

We have now solved the initial value problem in the case $q(y_0) \neq 0$. Since $y(t_0) = y_0$ and the function q is continuous at y_0, $q(y(t))$ is continuous, and therefore $q(y(t)) \neq 0$ on some interval containing t_0 (possibly smaller than the original interval $a < t < b$). On this interval, $y(t)$ is the unique solution of the initial value problem.

If $q(y_0) = 0$, the situation is more complicated. We now have the constant function $y = y_0$ as a solution of the initial value problem. However, the condition $q(y_0) = 0$ also makes the integral on the left-hand side of equation (3.34) improper. If it diverges, then equation (3.34) is meaningless. If it converges, then (3.34) gives another solution in addition to the constant solution $y = y_0$. It is possible to show that if $q(y)$ is differentiable, so that the existence and uniqueness theorem stated in Section 3.3 applies, then this improper integral must diverge, and the initial value problem has only the constant solution (which exists regardless of the convergence of the integral). However, if $q(y)$ is not differentiable at y_0, there may be more than one solution: a constant solution and another solution found by separation of variables (see Exercise 29 below).

We shall see in Section 4.1 that the constant solutions of a separable differentiable equation (3.31) play an important role in describing the behavior of all solutions as $t \to \infty$.

EXAMPLE 1.
Find all solutions of the differential equation $y' = -y^2$.

Solution: We divide the equation by y^2, permissible if $y \neq 0$, to give

$$\frac{1}{y^2}\frac{dy}{dt} = -1.$$

We then integrate both sides of this equation with respect to t. In order to integrate the left side of the equation we must make the substitution $y = y(t)$, where $y(t)$ is the (as yet unknown) solution. This substitution gives

$$\int \frac{1}{y^2}\frac{dy}{dt}\,dt = \int \frac{dy}{y^2} = -\frac{1}{y} + c$$

and we obtain

$$-\frac{1}{y} = -t - c \qquad (3.35)$$

or (since $Q(y) = -1/y$ is easily inverted here) $y = \frac{1}{t+c}$, with c a constant of integration. Observe that no matter what value of c is chosen this solution is never equal to zero. We began by dividing the equation by y^2, which was legitimate provided $y \neq 0$. If $y = 0$, we cannot divide, but the constant function $y = 0$ is a solution. We now have all the solutions of the differential equation, namely the family (3.35) together with the zero solution. □

In practice, it may be more convenient to use the more colloquial form of separation of variables mentioned in analysis of the pacemaker model, rather than the formal procedure of Example 1. We can write the differential equation $y' = -y^2$ in the form $\frac{dy}{dt} = -y^2$ and separate variables by dividing the equation through by y^2 and multiplying by dt to give

$$-\frac{dy}{y^2} = dt.$$

Although, as noted earlier, this form of the equation is meaningless without an interpretation of differentials, the integrated form

$$-\int \frac{dy}{y^2} = \int dt$$

is meaningful. Carrying out this integration yields equation (3.35) as before.

EXAMPLE 2.
Solve the initial value problem

$$y' = -y^2, \; y(0) = 1.$$

<u>Solution:</u> The results of Example 1 give the family of solutions $y = \frac{1}{t+c}$ for the differential equation $y' = -y^2$. When we substitute $t = 0$ and $y = 1$, we get $1 = \frac{1}{0+c} = \frac{1}{c}$, which implies that $c = 1$. Thus the solution to the initial value problem is $y = \frac{1}{t+1}$, defined for all $t \geq 0$.

Alternatively, separating variables and integrating as before yields

$$\int \frac{dy}{y^2} = \int -1 \, dt + c,$$

or, in its definite form using the initial condition $(t, y) = (0, 1)$,

$$\int_1^y \frac{du}{u^2} = \int_0^t -1 \, ds.$$

Evaluating both integrals yields

$$-\frac{1}{u}\bigg|_1^y = -s\bigg|_0^t,$$

or $-\frac{1}{y} - (-\frac{1}{1}) = (-t) - (-0)$, which simplifies to $y = \frac{1}{t+1}$. □

EXAMPLE 3.
Solve the initial value problem

$$y' = -y^2, \; y(0) = 0.$$

<u>Solution:</u> The procedure used in Example 1 leads to the family of solutions

$y = \frac{1}{t+c}$ for the differential equation $y' = -y^2$. When we substitute $t = 0$, $y = 0$ and attempt to solve for the constant c, we find that there is no solution. When we divided the differential equation by y^2 we had to assume $y \neq 0$, but the constant function $y \equiv 0$ is also a solution of the differential equation, as may easily be verified. Since this function also satisfies the initial condition, it is the solution of the given initial value problem. □

EXAMPLE 4.
Solve the initial value problem

$$y' = ty^3, \ y(0) = 1.$$

Solution: Division by y^3 (assuming $y \neq 0$) and integration with respect to t gives

$$\int \frac{dy}{y^3} = \int t\,dt.$$

Carrying out the integration, we obtain

$$-\frac{1}{2y^2} = \frac{t^2}{2} + c.$$

The initial condition $y(0) = 1$ gives $c = -\frac{1}{2}$, which gives the solution implicitly by

$$-\frac{1}{y^2} = t^2 - 1.$$

In this example it is easy to find the solution explicitly. We have $y^2 = (1 - t^2)^{-1}$. Since $y(0) = 1$, we must take the positive square root and thus we obtain the solution $y = (1 - t^2)^{-1/2}$ defined on the interval $-1 < t < 1$. □

3.4.3 Solution of logistic equations

We now return to the logistic differential equation

$$y' = ry\left(1 - \frac{y}{K}\right) \tag{3.36}$$

whose solutions were described in Section 3.2. Recall that this equation describes the growth of a population y whose natural per capita birth rate r is reduced by resource limitations in an environment capable of supporting a population of K members. In Section 3.2, we verified that the solutions were given by equation (3.12) but did not show how to obtain these solutions. The differential equation (3.36) is separable, and separation of variables and integration gives

$$\int \frac{K}{y(K-y)}\,dy = \int r\,dt$$

provided $y \neq 0$, $y \neq K$. In order to evaluate the integral on the left-hand side, we use the algebraic relation (which may be obtained by partial fractions)

$$\frac{K}{y(K-y)} = \frac{1}{y} + \frac{1}{K-y}$$

and then rewrite the left-hand side as

$$\int \frac{K\,dy}{y(K-y)} = \int \frac{dy}{y} + \int \frac{dy}{K-y} = \log|y| - \log|K-y| = \log\left|\frac{y}{K-y}\right|.$$

The right-hand side simplifies to $rt+c$, so we can now exponentiate both sides of the equation to obtain

$$\left|\frac{y}{K-y}\right| = e^{rt+c} = e^{rt}\,e^{c}.$$

If we remove the absolute value bars from the left-hand side, we can define a new constant C on the right-hand side, equal to e^{-c} if $0 < y < K$, or to $-e^{-c}$ otherwise, thus rewriting the right-hand side as $\frac{1}{C}e^{rt}$. Finally, we solve for y, and obtain the family of solutions

$$y = \frac{K}{1 + Ce^{-rt}} \tag{3.37}$$

as in Section 3.2. This family contains all solutions of (3.36) except for the constant solution $y = 0$. A qualitative observation is that if $r > 0$ every positive solution approaches the limit K as $t \to \infty$. (The only other non-negative solution is the constant solution $y \equiv 0$.) The derivation of the family of solutions is the same if $r < 0$, and in this case every solution which begins less than K approaches the limit 0 as $t \to \infty$. (For $r < 0$ all solutions which begin above the constant solution $y \equiv K$ increase without bound.)

EXAMPLE 5.

In Example 1 of Section 3.2 we saw that the population of the United States can be described reasonably well by a logistic growth curve with intrinsic (maximum) growth rate $r = -0.03$/yr and carrying capacity $K = 265$ (million people). However, in the year 2000 the U.S. population was estimated at 281.4 million. Adapt the model to account for the observed population exceeding the apparent carrying capacity.

Solution: One possible explanation is an increased carrying capacity due to improved agricultural techniques (or the converting of forest land into farmland). In practice, such changes are continuous and ongoing, but making K a continuously varying function of time keeps the model from being separable. A simple way to incorporate an increase in K is to choose a single year in which to consider the carrying capacity to have risen. Since, from the data given in Example 1 of Section 3.2, 1970 was the first since 1910 in which the real population surpassed the model prediction, let us choose 1970 as our new starting point. This will give us an initial condition of $y_0 = 202$. To determine the increased carrying capacity, we may use estimates that have been made of 470 million acres of arable cultivated land in the U.S. and the average need per person of 1.2 acres to calculate $K = 470/1.2 \approx 392$ (million people). We substitute into the solution (3.37) to obtain

$$y(t) = \frac{392}{1 + Ce^{-0.03t}};$$

if we let the year 1970 correspond to $t = 0$, we get $202 = 392/(1 + C)$, which gives a value of $C = 0.94$. Using these values, we predict a population in the year 2000 of

$$y(30) = \frac{392}{1 + 0.94e^{-0.03(30)}} = 283.6 \,(\text{million people}),$$

which is within less than 1% of the actual value. □

3.4.4 Discrete-time metered population models

In Section 2.4, we described some models for fish species which have an annual reproductive cycle. The number of new adults recruited each year is the number of fish born the preceding year (a function of the adult population size the preceding year) multiplied by the fraction of newborns who survive until the next year. Thus the model formulated is discrete, but has a recruitment function which is the result of a continuous process. We are now ready to derive these recruitment functions by solving the separable differential equations that describe the survivorship of newborn fish.

Species of fish which suffer cannibalism in the egg, larval and/or early juvenile stages have a per capita mortality rate which is proportional to the size of the adult population (which serves as predators). We shall also assume that the initial number of newborn fish (larvae, or eggs that hatch) is directly proportional to the size of the adult population. If we denote the adult population at the beginning ($t = 0$) of the kth year by y_k, and the number of surviving newborn/juvenile fish at time t in the kth year by $u(t)$, then we can write these two assumptions mathematically as

$$u'(t) = -(\hat{b}y_k)\,u(t), \quad u(0) = ry_k, \tag{3.38}$$

where $\hat{b}$ and r are constants which measure, respectively, the per-adult consumption rate of young and the average number of young produced per adult. This system can be solved using separation of variables (in fact, it is linear in u):

$$\int \frac{du}{u} = -\int (\hat{b}y_k)\,dt$$

$$\log u(t) - \log u(0) = -(\hat{b}y_k)t$$

$$u(t) = u(0)e^{-(\hat{b}y_k)t}.$$

Substituting the initial condition and a final time value of T corresponding to one year, we have

$$u(T) = ry_k e^{-(\hat{b}y_k)T}.$$

Defining a new constant $b = \hat{b}T$, the average number of young consumed per adult in one year, and renaming r as a, we have, finally the recruitment function

$$g(y_k) = ay_k\, e^{-by_k} \tag{3.39}$$

which leads to the Ricker model obtained in Section 2.4. (Beverton and Holt also showed that if instead we assume that mortality depends upon size of the newborn fish, and that the time required to grow to the critical size beyond which predation drops off is proportional to the adult population, we also get a Ricker model. See Exercise 21 below for details.)

The other common stock-recruitment model assumes instead that newborn mortality depends not upon the density of the adult population, but upon the density of newborns themselves, due to such factors as competition for food among the juveniles (often referred to as "nursery competition"). In this case, one assumes a per capita death rate of newborn fish which is a linear function of the current newborn population size. This leads to a more complicated but still separable differential equation with the same initial condition as for the Ricker model:

$$u' = -(\mu_0 + \mu_1 u)u, \quad u(0) = ry_k.$$

Separating variables and integrating, we have

$$\int \frac{du}{u(\mu_0 + \mu_1 u)} = -\int dt.$$

To continue, we need the result, obtained by the method of partial fractions, that

$$\frac{1}{u(\mu_0 + \mu_1 u)} = \frac{1}{\mu_0}\left(\frac{1}{u} - \frac{1}{u + \frac{\mu_0}{\mu_1}}\right).$$

Substituting this expression and making the integrals definite, we have

$$\int_{u(0)}^{u(T)} \frac{1}{\mu_0}\left(\frac{1}{v} - \frac{1}{v + \frac{\mu_0}{\mu_1}}\right) dv = -\int_0^T dt,$$

$$\frac{1}{\mu_0}\left[\log\left(\frac{u(T)}{u(T) + \frac{\mu_0}{\mu_1}}\right) - \log\left(\frac{u(0)}{u(0) + \frac{\mu_0}{\mu_1}}\right)\right] = -T.$$

Solving for $u(T)$, we obtain (after some work)

$$u(T) = \frac{e^{-\mu_0 T}u(0)}{\frac{\mu_1}{\mu_0}(1 - e^{-\mu_0 T})u(0) + 1}.$$

We now substitute the initial condition $u(0) = ry_k$ and define new constants

$a = re^{-\mu_0 T}$ and $b = r\frac{\mu_1}{\mu_0}(1 - e^{-\mu_0 T})$, so that $u(T)$ gives the Beverton-Holt recruitment function

$$g(y_k) = \frac{ay_k}{1 + by_k} \tag{3.40}$$

obtained in Section 2.4.

In practice, models are sometimes used to determine basic demographic rates by fitting observed data to the given functions, and using estimates for model parameters (a and b in the models above) to calculate the underlying rates (such as the mortality parameters in the models above). For example, Schmidt et al.[18] studied the coral reef damselfish *Dascyllus trimaculatus*, the three-spot dascyllus, in lagoons of Moorea, French Polynesia, and found its stock-recruitment data fit the Beverton-Holt function fairly well (see Exercise 22 below for a caveat), with estimated parameter values of $a = 0.696$ and $b = 0.0711$ per (sub-adult per $0.1m^2$ anemone). They measured survivorship over 6 months rather than a year, so we take $t = 1/2$ yr. If we assume an average number of fertilized eggs per adult of $r = 10,000$ (observed values range from a few thousand to 50,000 or more), then we can back-calculate the corresponding mortality rates as follows:

$$\mu_0 = -\frac{\log(a/r)}{T} = -\frac{\log(0.696/10,000)}{1/2} = 19.1 \text{ yr}^{-1},$$

$$\mu_1 = \frac{b\mu_0}{r(1 - e^{-\mu_0 T})} = 0.000136 \text{ yr}^{-1}\text{pop}^{-1}$$

where pop is the population unit given above.

As another example, we consider the study by Dumas and Prouzet[19] of Atlantic salmon in the Nivelle River in southwestern France. They fit their data to a Ricker model and estimated values of $a = 8.135 10^{-2}$ and $b = 3.84 10^{-6}$ per adult for survivorship of parr (salmon which have not yet left freshwater) over a season (let us again assume $T = 1/2$ yr). Here the estimation is easier: the reproductive ratio $r = a$ and the mortality rate is $\hat{b} = b/T = 7.68 10^{-6}$ per adult-yr.

Finally, we observe that some studies adapt these common stock-recruitment models to incorporate additional variables which they find important in explaining variations in survivorship. Jung and Houde[20] studied the stock-recruitment relationship in bay anchovies in Chesapeake Bay and found the data fit best to a modified Ricker model

$$g(y_k) = ay_k e^{-(b_1 y_k + b_2 \Delta L)},$$

[18] Russell J. Schmidt, Sally J. Holbrook, and Craig W. Osenberg, Quantifying the effects of multiple processes on local abundance: a cohort approach for open populations, *Ecology Letters* 2: 294–303, 1999.

[19] J. Dumas and P. Prouzet, Variability of demographic parameters and population dynamics of Atlantic salmon *Salmo salar* L. in a south-west French river, *ICES Journal of Marine Science* 60(2): 356–370, April 2003.

[20] Sukgeun Jung and Edward D. Houde, Recruitment and spawning-stock biomass distribution of bay anchovy (*Anchoa mitchilli*) in Chesapeake Bay, *Fishery Bulletin* 102(1): 63–77, 2004.

where ΔL is the mean latitude difference (weighted by biomass) from the mouth of the bay (37° N), a variable which accounted for differences in the amount of dissolved oxygen in the waters of the bay, a factor in juvenile anchovy development. This model corresponds to a per capita mortality rate of the form $-(\hat{b}_1 y_k + \hat{b}_2 \Delta L)$, where $\hat{b}_2$ describes the differential mortality rate per degree of latitude.

3.4.5 Allometry

Let $x(t)$ and $y(t)$ be the sizes of two different organs or parts of an individual at time t. There is considerable empirical evidence to suggest that the relative growth rates $\frac{dx}{dt}/x$ and $\frac{dy}{dt}/y$ are proportional. This means that there is a constant k, depending on the nature of the organs, such that

$$\frac{1}{y}\frac{dy}{dt} = k\frac{1}{x}\frac{dx}{dt}. \tag{3.41}$$

The relation (3.41) is called the *allometric law*, and the identification of such differential growth ratios is called *allometry* (the scientist Julian Huxley referred to it as *heterogony*). The single equation (3.41) does not provide enough information to determine x and y as functions of t, but we can eliminate t and obtain a relation between x and y. If we consider y as a function of x, then according to the chain rule of calculus,

$$\frac{dy}{dx} = \frac{dy}{dt} \Big/ \frac{dx}{dt}. \tag{3.42}$$

Combining (3.41) and (3.42) we have

$$\frac{dy}{dx} = k\frac{y}{x},$$

which can be solved by separation of variables $\left(\int dy/y = k \int dx/x\right)$ to give

$$\log y = k \log x + a, \tag{3.43}$$

and finally

$$y = e^{k \log x} e^a = x^k e^a = cx^k. \tag{3.44}$$

In order to determine the constant c, we need the values of x and y at some starting time.

In experiments, one might measure both x and y at various times and then use these measurements to plot $\log y$ against $\log x$, that is, to plot the experimental data on logarithmically scaled graph paper. According to the relation (3.43), the graph should be a straight line with slope k and y-intercept $a = \log c$. Because of experimental error, the points may not line up perfectly, but it should be possible to draw a line fitting the data well. One can then measure the slope and y-intercept of the line to determine the constants k and c in the relation (3.44).

EXAMPLE 6.

Thompson's classic work *On Growth and Form*[21] reports the following data from Huxley for the weights of the claw and body of the fiddler crab *Uca pugnax*:

Mass of claw (g)	5	9	14	25	38	53	59	78	105	135	165	196
Mass of body without claw (g)	58	80	109	156	200	238	270	300	355	420	470	536

Use these data to determine the constants c and k in the allometric law $y = cx^k$ describing the relation between claw mass x and body mass y.

Solution: Transforming the data, we have:

$\log x$	1.61	2.20	2.64	3.22	3.64	3.97	4.08	4.36	4.65	4.91	5.11	5.28
$\log y$	4.06	4.38	4.69	5.05	5.30	5.47	5.60	5.70	5.87	6.04	6.15	6.28

Linear least-squares regression for $\log y = k \log x + a$ gives $k = 0.602$, $a = 3.09$, and $c = e^a = 22.0$. Thus we have the relation $y = 22x^{0.602}$ which allows prediction of body mass given claw mass, or vice versa. Figures 3.15 and 3.16 show the fit of this curve to Thompson's data. □

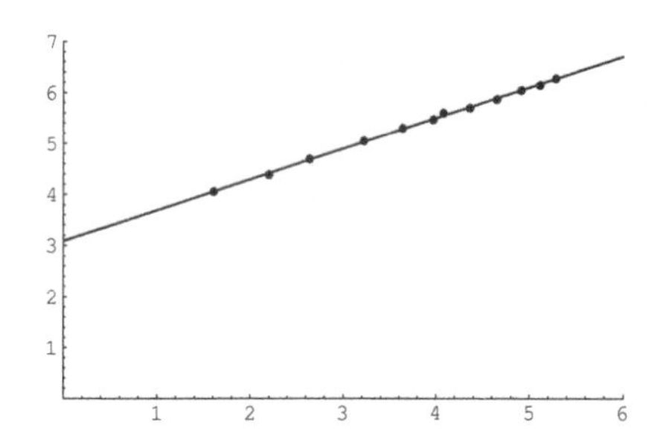

FIGURE 3.15: Log-log plot of Huxley's allometric data, and the line of best fit.

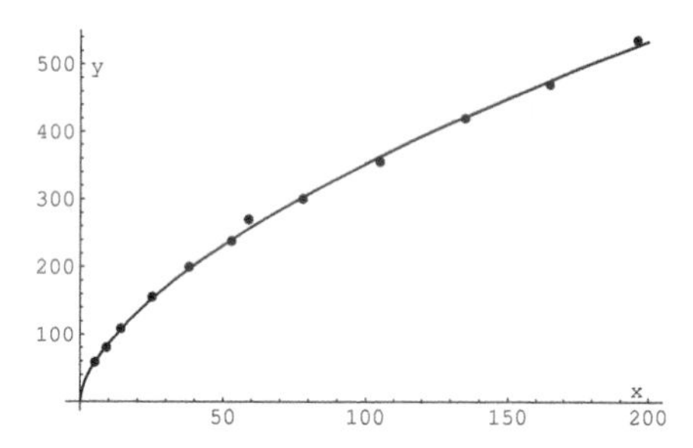

FIGURE 3.16: Linear plot of Huxley's allometric data, and the corresponding curve.

A caveat: Thompson warned that allometric laws hold only during periods of rapid growth, and claimed that simpler relationships are likely to hold at other times (see Exercise 24 for Thompson's discussion of this with regard to the fiddler crab). It is certainly true that all models have limitations; this one is similar to those we observed in earlier sections for linear and logistic models.

[21] D'Arcy Wentworth Thompson, *On growth and form*, Vol. I, 2nd ed., Cambridge: Cambridge University Press, 1942, pp. 205–212.

Exercises

In Exercises 1–8 below, find all solutions of the given differential equation. Do not overlook constant solutions.

1. $y' = t^2 y$
2. $y' = 7y^3 \cos t$
3. $y' = \frac{t}{y}$
4. $y' = ty^2$
5. $y' = -2ty$
6. $y' = \frac{y^2}{1-t}$
7. $y' = t^2 y - 4t^2$
8. $y' = \frac{y+1}{t}$

In Exercises 9–16 below, find the solution of the given differential equation which satisfies the given initial condition.

9. $y' = t^2 y$, $y(5) = 1$
10. $y' = 7y^3 \cos t, y(1) = 0$
11. $y' = -2ty, y(0) = y_0$
12. $y' = \frac{t}{y}$, $y(0) = 1$
13. $y' = ty^2$, $y(0) = -2$
14. $y' = \frac{y^2}{1-t}, y(-1) = \frac{1}{\log 2}$
15. $y' = t^2 y - 4t^2, y(0) = 0$
16. $y' = \frac{y+1}{t}$, $y(4) = \frac{1}{4}$

17. We can rewrite the pacemaker model (3.22) to emphasize how the pacemaker signal y follows the control u: $y' = -a(y-u)$. Since we are supposing the control signal u to be piecewise constant, then $(y-u)' = y'$, and we can make a substitution of variables, as is sometimes done in evaluating integrals in calculus courses, to rewrite this equation in terms of the difference $z = y - u$: $(y-u)' = -a(y-u)$ becomes

$$z' = -az. \tag{3.45}$$

   (a) Interpret equation (3.45) biologically.

   (b) Solve the differential equation $z' = -az$ for $z(t)$, and then back-substitute to return the solution to terms of y and u. Verify explicitly that the solution obtained in this way is equivalent to that obtained in this section.

18. In this problem we consider the effects of irregularities in the threshold voltages for the pacemaker model (3.22) studied in this section.

   (a) Suppose that for some reason the control signal becomes less sensitive to the pacemaker voltage, in such a way that y_H begins to rise toward u_H. For example, with $u_H = +40$mV, suppose y_H rises from +30 mV to +35 mV. What important effect does this have on $y(t)$?

(b) Given the nature of the solution $y(t)$, what will happen as y_H approaches u_H? (Hint: Consider equation (3.29).)

(c) If the control signal becomes sufficiently desensitized, it may become necessary to introduce an external pacemaker which drives the control based explicitly on timing rather than on values of y. That is, if the neuron does not fire within a certain amount of time after last resetting, the artificial pacemaker forces the signal to switch. This has the effect of controlling T_1 and T_2 directly.

If an external pacemaker of this type is applied to a heart with the parameter values given in this section, and the maximum period T_1 allowed by the external pacemaker before causing the control to switch off is 800 msec, then how high can the threshold y_H rise before the external pacemaker activates?

19. In the pacemaker model of this section, what happens to the firing times T_1 and T_2 when $u_H - y_H = y_L - u_L$?

20. In the logistic model, suppose that we have a population whose reproductive rate undergoes seasonal fluctuations—many species of animal reproduce only at certain times of year. This has the effect of making the growth rate r a function of time. If we let $t = 0$ correspond to January and suppose that most reproduction occurs in June or July, then we might model the growth rate as

$$r(t) = r_0 t + r_1 \sin 2\pi t,$$

with $r_0 \geq r_1 > 0$. Substitute this expression into the logistic equation (3.36) and apply the method of separation of variables to solve it. What does this model predict about the long-term behavior of this population? (You may want to consider a graph.)

21. Beverton and Holt showed[22] another way to derive the Ricker recruitment function. If we assume that newborn fish suffer higher mortality while very small, and that the time t_c required to grow to the critical size below which mortality drops is directly proportional to the adult population (since a greater adult population means more food resources are unavailable for the growing young), then we can model newborn survivorship with a per capita mortality rate of μ_1 for $0 \leq t < t_c$ and μ_2 for $t_c \leq t \leq T$ (where T represents one year).

(a) Write the linear differential equation describing newborn survivorship during the period $0 \leq t < t_c$. Solve the equation, using the initial condition $u(0) = r y_k$ and following the method in the text. Find the value of $u(t_c)$.

[22] Raymond J.H. Beverton and Sidney J. Holt, *On the dynamics of exploited fish populations.* Fishery Investigations Series 2 (19). London: Great Britain Ministry of Agriculture.

(b) Write the linear differential equation describing newborn survivorship during the period $t_c \leq t \leq T$. Solve the equation, using the initial condition for $u(t_c)$ given in part (a) (and t_c as a lower bound for integration). Find the value of $u(T)$.

(c) Define new constants a and b so that $a = re^{-\mu_2 T}$ and $(\mu_1 - \mu_2)t_c = by_k$, where the latter equation reflects the assumption that t_c is directly proportional to y_k. Show that $u(T)$ gives the Ricker recruitment function $g(y_k)$ as in (3.39).

22. In their study of the damselfish, Schmidt et al. (1999) initially used the Maynard Smith and Slotkin model (equation 1.27, p. 27), a generalization of the Beverton-Holt model, and estimated the exponent n which measures the nature of the intraspecies competition at 1.14 ($n = 1$ corresponds to strictly contest competition and the Beverton-Holt model). Using the parameter estimates $a = 0.696$, $b = 0.0711$ per (sub-adult per $0.1m^2$ anemone) and $r = 10,000$ eggs per adult given in the text, determine the relative (percentage) error incurred in calculating the per capita mortality parameters μ_0 and μ_1.

23. Jung and Houde (2004) estimated parameter values for their modified Ricker model of $a = 365$, $b_1 = 0.19$ per metric ton, and $b_2 = 1.35$ per degree. They measured survivorship over an approximately 6-month span.

(a) Use these values to calculate the differential mortality rates $\hat{b}_1$ and $\hat{b}_2$.

(b) Solve the corresponding differential equation using the per capita mortality rate given in the text. Substitute the parameter values obtained in (a) to estimate survivorship over a full year.

24. Thompson's response to the heterogony identified by Huxley and others through what we called in this section the allometric law was to caution that such differential growth ratios hold only during unconstrained growth. In particular, he claimed that the fiddler crab's growth ratio was linear rather than geometric past a certain point. Below are the rest of Huxley's data for the fiddler crab.

Mass of claw (g)	243	319	418	461	537	594
Mass of body without claw (g)	618	743	872	983	1080	1166
Mass of claw (g)	617	670	699	773	1009	1380
Mass of body without claw (g)	1212	1299	1363	1449	1808	2233

(a) Determine the constants in the allometric law that best fit these additional data.

(b) Now try instead a simple linear fit, of the form $y = mx + b$.

(c) Which of the two models above (allometric or linear) appears to be a better fit to these data?

(d) Repeat the analysis in parts (a)–(c) above, using the entire data set (24 points). Which model is the better fit for the entire data set? How would a model defined piecewise (allometric for part of the data set, linear for the rest) compare to the two one-piece models?

25. As in Exercise 24, determine, evaluate and compare the allometric and linear models for the following data on the stag beetle *Lucanus cervus* from Huxley, republished by Thompson (p. 208).

Length of mandible (mm)	6.0	7.8	9.0	10.0	11.2	11.9	12.8	14.4
Total length (mm)	31.0	38.8	40.5	42.6	45.0	46.9	49.2	53.6

26. As in Exercise 24, determine, evaluate and compare the allometric and linear models for the following data on the reindeer beetle *Cyclommatus tarandus* from Huxley, republished by Thompson (p. 209).

Length of mandible (mm)	3.9	10.7	14.1	19.9	24.0	30.7	34.5
Total length (mm)	20.4	33.1	38.4	47.3	54.2	66.1	74.0

27. Suppose a population satisfies a logistic model with $r = 0.4$, $K = 100$, $y(0) = 5$. Find the population size when $t = 10$.

28. Suppose a population satisfies a logistic model with $r = 0.3$, $K = 50$, $y(0) = 2$. Find the population size when $t = 25$.

29. Find the limit as $t \to \infty$ of the solution of each of the following initial value problems:

(a) $y' = -y + 1$, $y(0) = 0$

(b) $y' = -y + 1$, $y(0) = 100$

(c) $y' = y^2$, $y(0) = -1$

(d) $y' = -y^2$, $y(0) = 1$.

30*. Find two solutions of the initial value problem $y' = 3y^{2/3}$, $y(0) = 0$.

31*. Find all differentiable functions $f(t)$ such that $[f(t)]^2 = \int_0^t f(s)\,ds$ for all $t \geq 0$.

32*. Find all continuous (not necessarily differentiable) functions $f(t)$ such that $[f(t)]^2 = \int_0^t f(s)\,ds$ for all $t \geq 0$.

3.5 Mixing processes, and first-order linear models with constant coefficients

The autonomous first-order linear differential equation

$$\frac{dy}{dt} = -ay + b, \tag{3.46}$$

where a and b are given constants, serves as the basis for many models of biological systems in two senses: first, it can serve as a simplified model for many processes that occur in biological systems; and, second, the solution of linear equations is the foundation for the qualitative analysis of nonlinear systems, as will be seen in Section 4.1.In this section we shall consider (3.46) as a model for mixing processes in which chemicals, cells, organisms, and even heat distribute themselves evenly within a system. The overall quantity y of a given substance (or population) present within a biological system changes as the substance enters the system at a rate b (here taken to be constant; it is, at least, independent of y), and leaves the system at a rate ay proportional to the quantity of the substance already present.[23] Here we assume that the concentration of the substance within the system is relatively uniform, or at least that the concentration leaving the system is close to the average concentration throughout the system. See Figure 3.17 for an illustration. This assumption of a uniform, or homogeneous, distribution within the given system has given the name "mixing processes" to such phenomena. The assumption is often fairly accurate in describing the mixing of simple, nonreactive chemicals in a vat, but may be a gross oversimplification in describing the movement of more complex populations. It is nevertheless useful as a first step in developing and understanding more complex models.

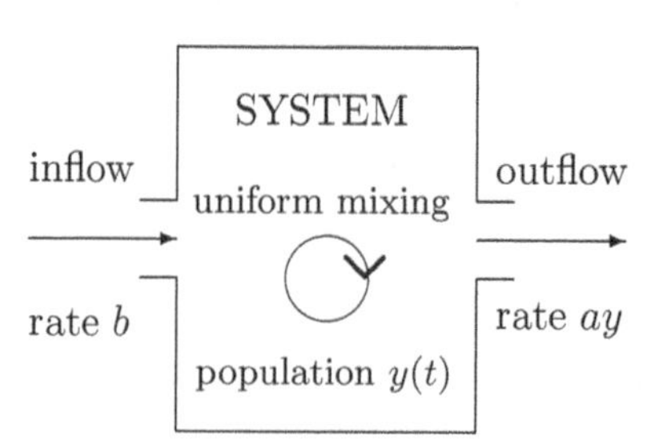

FIGURE 3.17: A schematic representation of a simple mixing process.

Before we consider some common applications of this model to biological systems, we will first show how to solve equation (3.46) and related initial value problems. There are several techniques for solving first-order linear differential

[23]Here we assume that $a > 0$, although all our calculations will be equally valid for $a \leq 0$.

equations[24]; here we will introduce one which will also be of use when we consider the more general first-order equation in Section 3.6, in which the coefficients a and b may be functions of the independent variable t (that is, they may change over time).

The technique of an *integrating factor* involves using the product rule for derivatives in reverse, much like integration by parts. We begin by writing (3.46) as

$$\frac{dy}{dt} + ay = b$$

and then look for a way to rewrite the left-hand side as an exact derivative: that is, the two-term derivative of a product of functions. If we multiply through by a function $g(t)$, then the left-hand side becomes $gy' + agy$. The derivative of the product (gy) is $gy' + g'y$, so we have a match if $g' = ag$. Our integrating factor g should therefore be $e^{\int a\,dt}$ (making $g' = ag$), which for constant a reduces to e^{at}. Now multiplying the original equation by g yields

$$e^{at}\frac{dy}{dt} + ae^{at}y = be^{at},$$

which by the above line of reasoning we can rewrite as

$$\frac{d}{dt}\left(e^{at}y\right) = be^{at}. \tag{3.47}$$

From this point we can apply either an indefinite or definite integral. In the first case, we get

$$e^{at}y = \int be^{at}\,dt;$$

for a and b constant, the right-hand side simplifies to $\frac{b}{a}e^{at}$ plus an arbitrary constant of integration C, which gives us the solution

$$y = \frac{b}{a} + Ce^{-at}. \tag{3.48}$$

If we have the initial condition $y(0) = y_0$, then the definite integral of (3.47) becomes

$$\int_0^t \frac{d}{d\tau}\left(e^{a\tau}y(\tau)\right)\,d\tau = \int_0^t be^{a\tau}\,d\tau$$

(using τ as the variable of integration), which simplifies to

$$e^{at}y(t) - e^0 y(0) = \frac{b}{a}\left(e^{at} - e^0\right);$$

that is,

$$y(t) = y_0 e^{-at} + \frac{b}{a}\left(1 - e^{-at}\right), \tag{3.49}$$

[24]Including separation of variables, cf. the cardiac pacemaker model of section 3.4.

a time-weighted average of the initial value y_0 and the final (asymptotic) value b/a. (In terms of (3.48), then, $C = y_0 - \frac{b}{a}$.)

In the discussions of related models that follow, we shall take as given the solution (3.48) to the differential equation (3.46), and the solution (3.49) to the corresponding initial value problem. While we wish to point out the breadth and variety of examples, our main goal is to encourage some understanding of the modeling process in areas of interest to the reader.

3.5.1 Chemostats

The *chemostat* is a laboratory device proposed in 1950 by Novick and Szilard[25] and Monod[26] to reproduce classic continuous linear diffusion in cultures of bacteria, and has since become a regular fixture in many biology laboratories. It consists of a large container, the chemostat reactor, to which are connected inflow and outflow tubes (connected, respectively, to a nutrient supply tank and an outflow collector), as well as other instruments which help maintain constant conditions inside the reactor. The underlying assumption governing chemostat populations is that growth is limited by one key nutrient (the rest being present in abundant supply). The chemostat is a physical model of a natural biological system in much the same way as a differential equation such as (3.46) is a mathematical model of a biological system. In addition to its use as a constant supply of bacterial cultures in the lab, it is used to approximate and study the workings of systems from populations of cells to entire ecosystems, especially where field and *in vivo* studies are impossible. Figure 3.18 shows a typical setup.

Many chemostat models are complex, incorporating multiple populations or substances within the reactor (for example, nutrient supply and the organisms that consume them), multiple spatial layers within the reactor, etc., but here we shall consider the simplest possible model, in which we track a single substance or population within a system. For example, Carman and Woodburn[27] used a chemostat to study the impact of low levels of antibiotics used in food-producing animals on human intestinal flora: in particular, the extent to which low levels of the common antibiotic ciprofloxacin may induce resistance in intestinal bacteria. Such drugs are commonly used in food animals, but residues may remain in meat ingested by consumers in levels low enough to produce resistance (to the same or related drugs) in gut bacteria. Carman and Woodburn used a chemostat containing 500 mL of culture, with a nutrient inflow of 35 mL/hr. If we assume a constant balancing outflow of

[25] Aaron Novick and Leo Szilard, Description of the chemostat, *Science* 112: 715–716, 1950.

[26] Jacques Monod, La technique de culture continue: théorie et applications, *Ann. Inst. Pasteur* 79: 390–410, 1950.

[27] Robert J. Carman and Mary Alice Woodburn, Effects of low levels of ciprofloxacin on a chemostat model of the human colonic microflora, *Regulatory Toxicology and Pharmacology* 33: 276–284, 2001.

Photo courtesy Thomas Chrzanowski

Photo courtesy Alan M. Hughes

FIGURE 3.18: Left: A chemostat, perhaps the most classical physical model of a linear mixing process. In the two chemostats pictured, the containers in front are the chemostat reactors, the large bottles above are the nutrient reservoirs, and the metal boxes circulate water around the reactors to maintain temperature. The outflow collectors are not in view. Right: Loch Linnhe in western Scotland, the fjordic lake studied by Ross et al. (2001), in which the nutrient dynamics behave like a chemostat.

35 mL/hr and an initial concentration of 0 g of ciprofloxacin in the chemostat, how long will it take for the concentration of cipro in the chemostat to reach 0.1 μg/mL for each of the inflow concentrations used by Carman and Woodhouse: 0.43, 4.3, and 43 μg/mL? What will the concentration be in each case after an hour?

In order to answer these questions, we let $y(t)$ be the total amount of cipro (in μg) in the chemostat t hours after the inflow is turned on. The inflow b is then $0.43\,\mu g/mL \times 35\,mL/hr = 15.05\,\mu g/hr$ in the first case (multiply by 10 or 100 for the larger inflows), and the outflow a, whose units are per-time, is $35\,mL/hr \div 500\,mL = 0.07/hr$, i.e., 7% of the chemostat's volume is exchanged each hour. From (3.49) with $y_0 = 0$, we then have $y(t) = 215\left(1 - e^{-0.07t}\right)$. After one hour, the amount of cipro in the chemostat should be

$$y(1) = 215\left(1 - e^{-0.07}\right) \approx 14.5\,\mu g,$$

making the concentration about $14.5\,\mu g/500\,mL = 0.029, \mu g/mL$ (multiply this by 10 or 100 for the higher inflow rates). In order for the concentration to reach 0.1 μg/mL, we need a total amount of $0.1\,\mu g/mL \times 500\,mL = 50\,\mu g$, so we set $y(t)$ to 50 and solve for t:

$$50/215 = 1 - e^{-0.07t}, \text{ so } e^{-0.07t} \approx 0.767 \text{ and } t \approx -\log(0.767)/0.07 \approx 3.78\,hr.$$

For the higher inflow rates, we use 2150 or 21500 in place of 215 and obtain time values of 0.336 hr and 0.0333 hr, respectively (about 20 min and 2 min). We may observe that the primary adjustment necessary in applying (3.49) here is the conversion between concentration and total amount; this is a common issue with simple nonbiological mixing problems as well.

Jannasch wrote[28] that in practice, the number of biological systems that can be reasonably modeled by a chemostat is extremely limited, because of the assumptions of time-independence and constancy of conditions that are rarely true even in controlled continuous cultures. Nevertheless, the chemostat model continues to be used, in mathematics and in laboratories, as a starting point for understanding the dynamics of well-mixed systems with known inflows and outflows. Perhaps the largest recent modeling application of this metaphor involves an entire ecosystem: a fjordic lake, whose high levels of freshwater runoff (inflow) and tidal exchange (inflow and outflow) give it characteristics not unlike the chemostat. (Most lakes turn over so little of their volumes per day that they are more usually modeled as closed systems.) Ross et al.[29] studied the food chain in Loch Linnhe, on the western coast of Scotland (see Figure 3.18), which was found to have an average of about 40% of its surface-layer volume exchanged daily via the tides, and about 3.5% of its surface-layer volume per day through freshwater runoff. Their model not only counted separately the carnivores, zooplankton, phytoplankton, and nutrients in the lake, but also differentiated levels of the lake by depth (fjordic lakes are typically elongated, deep for their width, and open to the ocean). The surface layer (about 10 m deep), which includes just under a quarter of the lake's total volume, accounts for most of the daily exchange. Ross et al. found that the inorganic nutrient dynamics did indeed resemble those of a chemostat, but that the metaphor was less accurate for the more complex dynamics of plankton and carnivores in the lake. The main difference between sea loch and chemostat dynamics lay in the occasional irregularities in the former due to sporadic mixing among the different layers in the lake, which can cause brief net outward nutrient flows when nutrients settled in the bottom are stirred into the levels of higher exchange nearer the surface.

If we interpret simplistically the exchange rates given above for Loch Linnhe (and assume a constant total volume for the lake), the coefficient $a = -0.435/day$ accounts for outflow in the surface layer (compare with 0.07/hr or 1.68/day in the Carman and Woodburn chemostat). This coefficient can be used in tracking any of various populations in the lake's surface layer, such as nutrients like carbon or nitrogen, or chemicals collected in runoff. For instance, suppose a given chemical accumulates from runoff, entering the loch at a rate of $b = 1$ kg/day. The limiting value this model predicts is then $b/a \approx 2.3$ kg of the substance in the loch. Although seasonal disturbances will prevent the true accumulation from approaching this steady state monotonically, a quick calculation shows that within a week, there will be

$$y(7) = 2.3\left(1 - e^{-0.435\times 7}\right) \approx 2.2$$

[28] Holger W. Jannasch, Steady state and the chemostat in ecology, *Limnology and Oceanography* 19(4): 716–720, 1974.

[29] Alex H. Ross, William S. C. Gurney, Michael R. Heath, Steven J. Hay, and Eric W. Henderson, A strategic simulation model of a fjord ecosystem, *Limnology and Oceanography* 38(1): 128–153, 1993.

kg of it in the lake, which is within 5% of its steady-state value.

3.5.2 Drug dosage

When a dosage of a drug is administered to a patient, the concentration of the drug in the blood increases. Over time, the concentration decreases as the drug is eliminated from the body. The question we would like to model is how the drug concentration changes with repeated doses of the drug. This area of study is called *pharmacokinetics*.

Experimental evidence indicates that the rate of elimination is proportional to the concentration of drug in the bloodstream. Thus we assume that, if $C(t)$ is a function representing the concentration of drug in the blood at time t, its derivative is given by

$$C'(t) = -qC(t).$$

Here q is a positive constant, called the elimination constant of the drug. We assume also that when a drug is administered, it is absorbed completely immediately. This is certainly at best an approximation to the truth, probably a better approximation for a drug injected directly into the bloodstream than for a drug taken by mouth. Let us assume that the drug is administered regularly at fixed time intervals of length T with a dose capable of raising the concentration in the blood by A. Then if we begin with an initial dose A at time $t = 0$, $C(t)$ satisfies the initial value problem $C'(t) = -qC(t)$, $C(0) = A$ until the second dose. Thus for $0 \leq t \leq T$, we have $C(t) = Ae^{-qt}$, and just before the second dose at time T, $C(T^-) = Ae^{-qT}$.

We let C_i be the concentration at the beginning of the ith interval and R_i the residual concentration at the end of the ith interval; we have shown that $C_1 = A$, $R_1 = Ae^{-qT}$. After the second dose, at time T, the concentration jumps from R_1 to $C_2 = R_1 + A = A(1 + e^{-qT})$. Now we have a new initial value problem, $C'(t) = -qC(t)$, $C(T) = C_2$, for $T < t \leq 2T$, whose solution we can show to be $C(t) = C_2e^{-q(t-T)}$. Just before the third dose, then, at time $2T$, we have $R_2 = C(2T^-) = C_2e^{-qT}$.

More generally, the residual concentration R_i at the end of each interval between doses is the concentration C_i at the beginning of the interval multiplied by e^{-qT}, and the level C_{i+1} at the beginning of the next interval (just after the next dose is given) is this residual concentration plus A. Thus for the second interval we have

$$C_2 = A(1 + e^{-qT}), \quad R_2 = A(1 + e^{-qT})e^{-qT} = A(e^{-qT} + e^{-2qT}).$$

Next we see that for the third interval

$$C_3 = A + A(e^{-qT} + e^{-2qT}) = A(1 + e^{-qT} + e^{-2qT}), R_3 = A(e^{-qT} + e^{-2qT} + e^{-3qT}).$$

We may give a simpler expression for each of these by using the formula for the sum of a finite geometric series,

$$1 + r^2 + r^3 + \ldots + r^n = \frac{1 - r^{n+1}}{1 - r},$$

to obtain

$$C_3 = A\frac{1-e^{-3qT}}{1-e^{-qT}}, \quad R_3 = Ae^{-qT}\frac{1-e^{-3qT}}{1-e^{-qT}}.$$

From this we may conjecture (and prove by induction) that for every positive integer n,

$$C_n = A\frac{1-e^{-nqT}}{1-e^{-qT}}, \quad R_n = Ae^{-qT}\frac{1-e^{-nqT}}{1-e^{-qT}}.$$

As $n \to \infty$, $e^{-nqT} \to 0$ because $e^{-qT} < 1$. Thus C_n and R_n approach limits C and R, respectively, as $n \to \infty$, where

$$C = \frac{A}{1-e^{-qT}}, \quad R = \frac{Ae^{-qT}}{1-e^{-qT}}. \tag{3.50}$$

It is easy to see that both the initial and residual concentration increase with each dose but never exceed the limit values. For example, if a drug with an elimination constant of 0.1 hours^{-1} is administered every 8 hours in a dosage of 1 mg/L, the limiting residual concentration is $\frac{e^{-0.8}}{1-e^{-0.8}} = 0.816$ mg/L.

If the time interval T between doses is short, then the ratio of residual concentration to dose,

$$\frac{R}{A} = \frac{e^{-qT}}{1-e^{-qT}} = \frac{1}{e^{qT}-1},$$

is large (for small T, $e^{qT} - 1$ is near 0), as each new dose builds upon the previous one before it has had time to dissipate. On the other hand, if the time interval between doses is long, each R_n is close to zero, and each C_n is close to A; each individual dose dissipates almost completely before the next dose is given.

EXAMPLE 1.

Harms et al.[30] studied the pharmacokinetics of oxytetracycline, an antibiotic also used as a marker to label bone for age studies, in the loggerhead sea turtle *Caretta caretta* (see Figure 3.19). They administered a single dose of the drug and monitored its gradual elimination from the turtles' blood plasma. They reported results in terms of the elimination half-life, which is the amount of time required to eliminate half of the drug from the blood. For a single dose of 25 mg/kg administered intramuscularly to turtles with an average mass of 8 kg, the mean elimination half-life was 61.9 hours.

What is the minimum [intramuscular] dosage required in order to keep the level of oxytetracycline in an 8 kg loggerhead's blood plasma above 100 mg, if doses are given once every day?[31] Once every three days? Once a week?

[30] Craig A. Harms, Mark G. Papich, M. Andrew Stamper, Patricia M. Ross, Mauricio X. Rodriguez, and Aleta A. Holm, Pharmacokinetics of oxytetracycline in loggerhead sea turtles (*Caretta caretta*) after single intravenous and intramuscular injections, *Journal of Zoo and Wildlife Medicine* 35(4): 477-488, 2004.

[31] Concentrations are typically reported in units of μg/mL; here we simplify by considering the total amount.

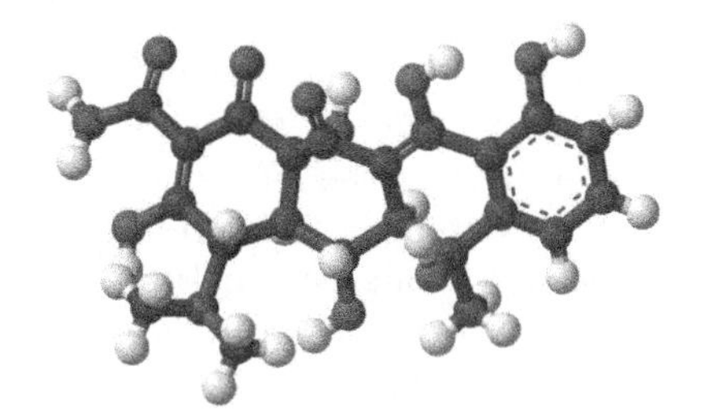

FIGURE 3.19: Left: A loggerhead sea turtle fitted with a tracking device. Right: The molecular structure of oxytetracycline, so named for the oxygen (dark atoms along top and bottom) bonded to four (tetra) hydrocarbon rings.

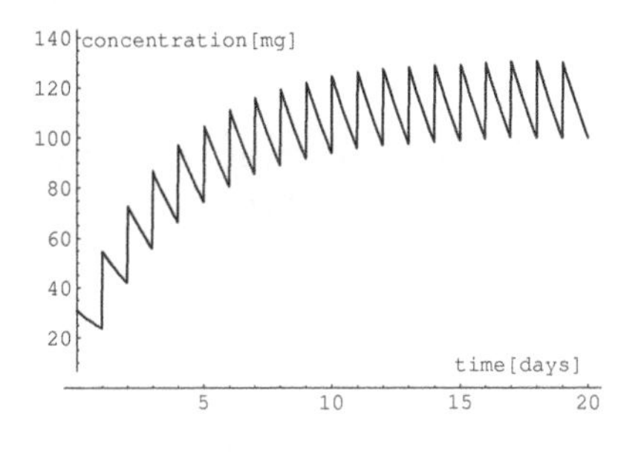

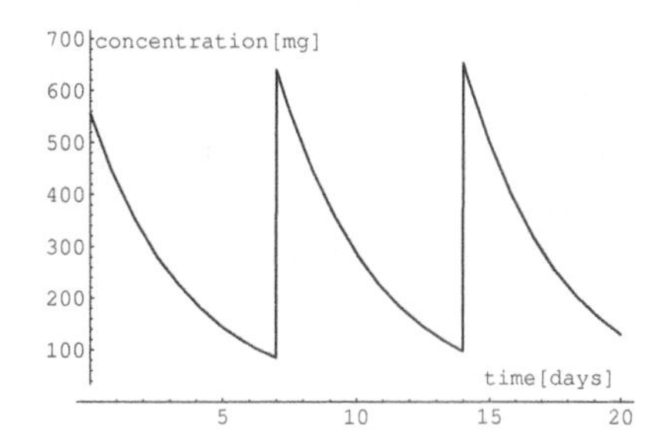

FIGURE 3.20: Left: Projected drug concentration with daily doses of 31 mg. Right: Projected drug concentration with weekly doses of 556 mg.

Solution: The equation $\frac{1}{2} = e^{-qt_{1/2}}$ for the elimination half-life $t_{1/2}$ can be rewritten for the elimination constant q:

$$q = -\log\left(\frac{1}{2}\right)/t_{1/2}.$$

A half-life of 61.9 hours therefore corresponds to a value $q = 0.0112/\text{hr}$. We are interested in the eventual residual concentration R. From (3.50), for $T = 24$ hr we have $R = 3.24A$, for $T = 72$ hr, $R = 0.807A$, and for $T = 168$ hr $R = 0.180A$. This means that daily doses A must be at least $100/3.24 \approx 30.9$ mg, while doses given once every three days must be at least 124 mg, and weekly doses must be at least 556 mg. (For an 8 kg turtle, these figures correspond to concentrations of 3.86, 15.5, and 69.5 mg/kg.) Observe that when dose frequency is reduced by a factor of 7 (from once per day to once per week), the dosage size needed increases by a factor of 18 (see Figure 3.20). □

3.5.3 Newton's law of cooling

When a body at some temperature is placed in surroundings at a different temperature, heat flows from the body to the surroundings if the body is hotter than the surroundings, and from the surroundings to the body if the surroundings are hotter than the body. *Newton's law of cooling*, which is based on balance equations for the transfer of heat, states that the rate at which body temperature changes is proportional to the temperature difference between the body and its surroundings. Let us assume that the surroundings are at a temperature T^*, called the *ambient temperature*, and that the body is small enough compared to the surroundings that the heat exchange has a negligible effect on the temperature of the surroundings. We assume also that all of the surroundings are at the same temperature; that is, that heat is circulated throughout the surroundings. According to Newton's law of cooling, the temperature T of the body at time t satisfies a differential equation of the form

$$T' = -k(T - T^*) = -kT + kT^*.$$

Here, k is a constant of proportionality which must be positive in order to make $T' > 0$ when $T < T^*$ and $T' < 0$ when $T > T^*$. Then T satisfies an initial value problem

$$T' = -kT + kT^*, T(0) = T_0, \tag{3.51}$$

where T_0 is the temperature of the body when it is placed into the surroundings. Since (3.51) is the same as (3.46) with a and b replaced by k and kT^* respectively, its solution (from (3.49)) is

$$T = T_0 e^{-kt} + T^*(1 - e^{-kt}), \quad \text{or} \quad T(t) = T^* + (T_0 - T^*)e^{-kt}. \tag{3.52}$$

From this we see that the temperature of the body will approach the ambient temperature as $t \to \infty$. In practical applications, the constant of proportionality k cannot be measured directly but must be calculated from temperature readings at more than one time.

EXAMPLE 2.

In an experiment published in 1937, James Hardy tried to determine a constant of proportionality for heat loss from the human body.[32] Immediately there are many complicating factors which must be acknowledged, including wind, body position (which affects the amount of surface area exposed), and substances such as water or fabric (clothes) which affect the transfer of heat from the skin to the surrounding air. The body's own defenses, such as the chill or shivering reflex, which attempts to create heat through expending [kinetic] energy, also interrupt the heat flow described by Newton. Hardy therefore limited his primary investigations to the transfer of heat from dry

[32] James D. Hardy, The physical laws of heat loss from the human body, *PNAS* 23: 631–637, 1937.

skin directly to still air, in the limited range of 23°C to 28°C, with his human subjects lying perfectly still and straight. Hardy also wrote the heat exchange rate coefficient as being directly proportional to surface area, $k = KA$, where K is a constant rate per unit area and A is the surface area across which heat is being exchanged.

Hardy used two human subjects, as well as an elliptical iron cylinder for comparison purposes (the cylinder had dimensions roughly those of an adult human torso, and afforded more detailed measurements). He reported heat exchange rate coefficients of 5.2 cal/°C hr m^2 for the human subjects and 6.75 cal/°C hr m^2 for the cylinder. Hardy gave this latter figure as being equivalent to 1.64 cal/°C hr overall (from which we may infer an effective area of about 0.24 m^2, though Hardy's figure is twice this), and gave the cylinder's mass as 24.98 kg. Using the common mass and surface area estimates of 68 kg and $1.8m^2$ for an adult human male, determine the two coefficients k for Newton's law of cooling with the human subjects and the cylinder, and the amount of time required for each to cool from 28°C to 23°C in a 20°C environment. The specific heats of iron and the human body are 829 cal/kg°C and 112 cal/kg°C, respectively.

Solution: Hardy wrote in terms of heat exchange; we wish to write in terms of cooling, or temperature change, so we need to use the specific heat of each object to convert Hardy's data (in calories) into purely temperature-based terms. For the iron cylinder, we have

$$k = 1.64\frac{cal}{°C\,hr} \div 112\frac{cal}{kg\,°C} \div 24.98kg = 0.000586/hr.$$

For the human subjects, we multiply 5.2 cal/°C hr m^2 by the estimated surface area of $1.8m^2$ and then proceed similarly:

$$k = 5.2\frac{cal}{°C\,hr\,m^2} \times 1.8m^2 \div 829\frac{cal}{kg\,°C} \div 68kg = 0.000166/hr.$$

Substituting into the second form of (3.52), we have $23 = 20 + (28 - 20)e^{-kt}$, or $t = -\log(3/8)/k$, which gives us 1674 hours and 5909 hours, respectively, for the iron cylinder and the human body. □

EXAMPLE 3.

Newton's law of cooling may also be applied forensically, to estimate the time of death based on body temperature measurements after death. This application is especially common at the beginning and end of hunting seasons, to determine whether game animals such as deer were killed within the legal season. Kienzler et al.[33] conducted a study on the cooling rate for white-tailed deer.[34] They found a linear cooling rate constant of $k = 0.087656$/hr,

[33] J.M. Kienzler, P.F. Dahm, W.A. Fuller, A.F. Ritter, Temperature-based estimation for the time of death in white-tailed deer, *Biometrics* 40(3): 849–854, 1984.

[34] Their model incorporated the effects of weight and ambient temperature as well as

with a standard deviation of 0.004912/hr. The normal body temperature for white-tailed deer is 38.8°C.

If a deer carcass found in 0°C weather has a thigh temperature of 15°C, how long is it likely to have been dead? Give a mean estimate as well as a 95% confidence interval (2 standard deviations above and below the mean), assuming variation in k only.

Solution: The second form of (3.52) gives us $15°C = 0°C + (38.8°C - 0°C)e^{-0.087656t}$, from which $t = \log(15/38.8)/(-0.087656)hr \approx 10.8\,hr$. Using the endpoints of the 95% CI [0.077832,0.09748] for k yields t values of 12.2 and 9.75 hours, respectively. □

EXAMPLE 4.

We can also sometimes use data to generate predictions directly without calculating k explicitly. If a loaf of freshly baked bread cools from 100°C to 60°C in the first 10 minutes it is on a cooling rack, when the surroundings are at 20°C, what will the loaf's temperature be 30 minutes after it was placed on the cooling rack?

Solution: From the second form of (3.52) with $T_0 = 100°C$, $T^* = 20°C$, we have $T(10) = 20 + (100 - 20)e^{-10k} = 60$, and $T(30) = 20 + (100 - 20)e^{-30k}$. From the first of these, $e^{-10k} = 40/80 = 1/2$, so that $e^{-30k} = (e^{-10k})^3 = 1/8$. Thus $T(30) = 20 + 80(1/8) = 30°C$. □

3.5.4 Migration

Another process which can be described in terms of mixing is the migration of populations, between or among groups in general, and into and out of a single group in the context of one-equation models. From cells to organisms and households, migration adds an important dimension to the dynamics of populations that are not entirely closed. Cell migration is a necessary mechanism for such processes as cell differentiation and the healing of wounds, and the migration of tumor cells is metastasis, the defining characteristic of malignant cancers. Many animals such as insects and birds undertake periodic seasonal migrations, and transnational immigration of humans has become a major political issue in the early twenty-first century.

Although some behavioral and social scientists would distinguish between seasonal or temporary movements and those movements of populations intended to be permanent, we will include in our definition of migration any movement which can be captured on the time scale of interest to us. If we are dealing with migration over the course of years, seasonal, month-to-month movements into and out of the zone under study will only manifest in our model to the extent that they accumulate. On a time scale of hours, however, every daily activity has the potential to affect the model. When a population

---

time since death; see Exercise 15. A linear regression of their data assuming proportional dependence on time only would yield a different value for k than the one used here.

TABLE 3.3: United States population and per capita demographic rates for selected years.

Year	Population (millions)	Births (/yr)	Deaths (/yr)	Immigrations (/yr)	Emigrations (/yr)
1950	152	0.0241	0.0096	0.000681	0.000185
1960	180	0.0237	0.0095	0.001397	0.000236
1970	204	0.0184	0.0095	0.001628	0.000441
1980	226	0.0159	0.0088	0.001988	0.000520
1990	250	0.0167	0.0086	0.002935	0.000640
2000	281.4	0.0144	0.0085	0.003232	0.000831

gains or loses a significant proportion of its size through connections with other populations, it is important to incorporate them into a model. These movements can be classified as either *immigration* (movement into the population from outside) or *emigration* (departure from within the population). Many demographic studies tally only net migration, but as immigration and emigration may have different causes, they must often be modeled separately.

As we are limiting ourselves in this chapter and the next to the study of single populations, we will give here only a single application of this idea, and consider the exchange of members among multiple populations in later chapters. For this example we return to the modeling of the United States population first undertaken in Example 1 of Section 3.2. Table 3.3 lists per capita birth, death, immigration, and emigration rates for the United States for every tenth year beginning in 1950, as well as the corresponding population sizes. (To get the total births, etc. for the given year, we would multiply the per capita rates by the population size.) A population model designed to incorporate migration will include a term for each of these. The simplest approach might be to write an equation containing a linear term for each of these processes:

$$\frac{dx}{dt} = bx - dx + \iota x - \epsilon x, \tag{3.53}$$

where x is the population size and b, d, ι, and ϵ are the four per capita rates that drive it.

The assumption underlying (3.53), however, is that each of these processes occurs at a rate proportional to the size of the population. This may be reasonable for processes that involve the population itself, including births, deaths, and emigrations (ignoring for the moment the logistic refinement suggested in Section 3.2). Immigration, however, may be driven by forces outside the population under study. For example, the number of people who immigrate to the United States in a given year may depend more on the populations of other countries, or on political factors, than on the present population size of the United States. If so, then the term describing immigration should not be a function of x. One option would be to introduce a second variable to describe the factor which drives the immigration rate, but as we wish to keep our model

TABLE 3.4: United States population predictions (in millions) and percent errors for models (3.54) and (3.53), alongside the actual values.

	Updated parameters				1950 parameters				
Year	Const. immig.	% error	Prop'l immig.	% error	Const. immig.	% error	Prop'l immig.	% error	True value
1950	152	—	152	—	152	—	152	—	152
1960	176.5	−1.94%	176.6	−1.89%	176.5	−1.94%	176.6	−1.89%	180
1970	209.7	2.78%	209.9	2.89%	204.8	0.38%	205.2	0.57%	204
1980	225.5	−0.23%	225.7	−0.15%	237.4	5.05%	238.4	5.47%	226
1990	246.0	−1.59%	246.2	−1.51%	275.1	10.03%	276.9	10.77%	250
2000	277.0	−1.57%	277.4	−1.43%	318.5	13.19%	321.7	14.33%	281.4

to a single variable, the only alternative is to make this rate a constant (or, conceivably, a function of time alone: see Exercise 16 below for an exploration of this possibility). This assumption yields an alternative model,

$$\frac{dx}{dt} = bx - dx + I - \epsilon x, \tag{3.54}$$

where I is the total immigration rate (in, say, people per year).

One way to decide which assumption about the factors driving immigration gives a better description of population growth is to compare the predictions of both models with actual data. Both (3.53) and (3.54) are of the form (3.46), so we may give their solutions, from (3.49), as

$$x(t) = x(0)e^{(b-d+\iota-\epsilon)t} \text{ and } x(t) = x(0)e^{(b-d-\epsilon)t} + \frac{I}{b-d-\epsilon}\left(e^{(b-d-\epsilon)t} - 1\right),$$

respectively. Since both models make the simplifying assumption that the associated per capita rates are constant over time, as well as the assumption that the underlying forces are not constrained by resource limitations (hence the linear terms, rather than, say, logistic), the model should only be used for limited ranges of time. Table 3.4 compares the predictions made by these two models in two ways: first using the data for each given year to estimate only the next decade's population, and then using the true values for the next decade to generate the next prediction; and then using the initial (1950) data to project as far as 2000. One can see that the proportional immigration model (3.53) always makes higher predictions, which is not surprising since the immigration rate in (3.53) increases continuously with the population, rather than remaining fixed as in (3.54). Thus the proportional immigration model is slightly more accurate than the constant migration model when they undershoot, and less accurate when they overshoot. In general, however, the difference in predictions is less than one percent, which implies that the difference between these two assumptions on immigration rates is less important for predicting population size than other factors.

As can be seen in Table 3.3, the per capita birth, death, and migration rates are not really constant over time. In the following section we shall consider the more general linear first-order differential equation, in which coefficients may be functions of time.

Exercises

1. At time $t = 0$, a tank contains 3 lb. of salt dissolved in 100 gal. of water. A mixture containing $\frac{1}{2}$ lb. of salt per gallon is pumped in at a rate of 5 gal./min., and the mixture is pumped out at the same rate. Find the weight of salt in the tank at time t. How much salt is there after one hour? Does the weight of salt in the tank at time t have a limit as $t \to \infty$? [Note Exercises 1–3 involve nonzero initial populations.]

2. A tank contains 100 liters of water and 10 kg. of salt, thoroughly mixed. Pure water is added at a rate of 5 liters/minute and the mixture is poured off at the same rate. How much salt is left in the tank after 1 hour, assuming instantaneous and complete mixing?

3. A tank contains 30 lb. of a pollutant dissolved in 200 gallons of water. Fresh water is poured into the tank at a rate of 10 gal./min. and the solution is pumped out at the same rate, with immediate and complete mixing. Find the concentration of pollutant as a function of time, and the time required for the concentration to drop to 1 lb. per 100 gallons.

4. Water containing 1 pound of salt per gallon is poured into a 50 gallon tank of water at a rate of 2 gallons per minute and the mixture runs out at the same rate. After 20 minutes, this process is stopped, and fresh water is pumped into the tank at a rate of 2 gallons per minute, with the mixture running out at the same rate. Find the amount of salt in the tank after 10 more minutes.

5. At time $t = 0$ cigarette smoke containing 4% carbon monoxide is introduced into a previously carbon monoxide-free room with volume 1800 cubic feet, at a rate of 0.6 cubic feet per minute. The mixture leaves the room at the same rate. After how long will the carbon monoxide in the room reach a concentration of 10^{-4} (which is considered harmful)? [Make the simplifying assumption that the gas has constant density.]

6. Use either the technique of separation of variables or an integrating factor to show that the solution to the initial value problem $C'(t) = -qC(t)$, $C(T) = C_2$, for $T < t \leq 2T$ (between the first and second drug doses), is $C(t) = C_2 e^{-q(t-T)}$.

7. Suppose that a dose of d milligrams of a drug is injected into the bloodstream. Assume that the drug is eliminated from the blood at a rate proportional to the amount of the drug present in the blood. Assume in addition that half of the drug dose has been eliminated after 1 hour. Find the time at which the amount of drug in the bloodstream is 20% of the original dose.

8. A drug with an elimination constant of 0.1 hours^{-1} is administered every 8 hours in a dosage of 1 mg/L. Find the residual concentration at the end of each of the first five intervals.

9. The same drug as in Exercise 8 is administered and it is known that the maximum safe concentration is 0.5 mg/L (the concentration must never be allowed to exceed 0.5 mg/L).

   Find the shortest time interval T between doses which will be safe.

10. Solve the differential equation (3.51) by using the substitution $u = T - T^*$ to eliminate the constant term, and then back-substituting in the resulting purely exponential solution.

11. If a body cools from 100° C to 60° C in surroundings at 20°C in 10 minutes, how long will it take for the body to cool to 25°C?

12. Water is heated to the boiling point (100°C) and is then removed from the heat and placed in a room with a constant temperature of 20°C. After 3 minutes the water temperature is 90°C.

    (a) Find the water temperature after 6 minutes.

    (b) How long will it take for the water temperature to reach 75°C?

13. A container of ice cream with temperature 0°F is put in a room with temperature 75°F. After 15 minutes it is observed that the temperature has risen to 10°F. How long will it take for the ice cream to reach a temperature of 32°F?

14. An iron bar at 400°F is moved to a room with temperature 80°F. After 5 minutes the temperature of the bar is 120°F. When will the temperature of the bar reach 90°F?

15. The cooling model used by Kienzler et al. (1984) for white-tailed deer was actually more complex than (3.52): their exponent was more than a simple linear function of time. In place of $-kt$, they used $\beta_1 t + \beta_2 t^2 + \beta_3 W t + \beta_4 T^* t + \beta_5 Z(t)$, where W is weight and $Z(t) = \min((t-3)^2 - 9, -9)$. They estimated $\beta_1 = -0.087656/\text{hr}$, $\beta_2 = -0.001174/\text{hr}^2$, $\beta_3 = 0.000515/\text{hr kg}$, $\beta_4 = -0.001971/\text{hr °C}$, and $\beta_5 = -0.006168/\text{hr}^2$. Use their model to repeat the calculation of Example 3 for a 68 kg deer. What difference does it make in the estimate?

16. Suppose that the immigration rate for the United States is not constant, as in (3.54), but a linear function of time, $I(t) = \iota_0 + \iota_1 t$. Solve the resulting differential equation.

17*. Using the data in Table 3.3, compute the total immigration per year, and use linear regression to determine the best-fit coefficients ι_0 and ι_1 for the model in Exercise 16. Compare the resulting predictions against those given in Table 3.4 and evaluate the results.

3.6 First-order models with time dependence

The models presented so far for biological systems have involved parameters which represent various aspects of the environment within which the system operates: both internal characteristics such as natural birth and death rates and external factors such as resource limitations, inflow and outflow. For simplicity we have assumed these parameters, which appear as coefficients of various terms in the models, to be constant. However, there are many situations in which either this assumption makes model predictions wildly inaccurate, or the variations in these coefficients are precisely what we want to study. In mixing processes such as chemostat systems or migrating populations, inflow and outflow may not be balanced, or may vary periodically. An environment's resource capacity may increase or decrease over time. In such cases, we must adapt our model by relaxing this simplifying assumption, and writing one or more coefficients as functions of time (or, in some cases, of population size).

In this section we will look at some ways to solve first-order differential equations with variable coefficients. Sometimes we can use one of the techniques we have already seen, with no more than minor adaptations, but at other times it will be more convenient to use an extension of these methods that shows us more directly the structure associated with the general first-order linear differential equation. We will begin by developing and illustrating these two kinds of solution methods, and then look at some applications to chemostat, migration, fishery management, and population growth models with variable coefficients.

3.6.1 Superposition

In order to understand the solution methods to be presented in this section, it is helpful to understand how solutions to a linear ODE can be written in terms of the solutions to a special, simplified version of the equation. The idea is that each solution is the sum of two functions, one of which is a solution to this simpler equation. This idea is called the *superposition principle.* Although it applies, properly speaking, only to linear differential equations, for reasons shown below, we will also use some helpful substitutions to allow us to apply it to some nonlinear models such as the logistic equation.

The first-order linear differential equation seen in the previous section can be written in its most general form as

$$\frac{dy}{dt} = -a(t)y + b(t), \tag{3.55}$$

where the coefficients $a(t)$ and $b(t)$ are functions of t. (Recall that a linear differential equation $y' = f(t, y)$ is one in which the function $f(t, y)$ is linear in

y.) In the special case that a and b are constants, equation (3.55) is separable, and we found in the previous section that its solution has the form

$$y = \frac{b}{a} + ce^{-at} \tag{3.56}$$

for some constant c. This solution is a sum of two terms, the first a specific solution $y = \frac{b}{a}$ of $y' + ay = b$ and the second a family of solutions $y = ce^{-at}$ of the corresponding *homogeneous differential equation* $y' + ay = 0$. It turns out that an analogous result holds for the more general linear ODE (3.55) — that is, it turns out that solutions to (3.55) can be written as a sum of two terms, as done in equation (3.56), even if a and b are not constants. First let us state this result more precisely; then we will see why it is true.

We will call a differential equation such as (3.55) *homogeneous* if the function $b(t)$ on the right side is the identically zero function, so that (3.55) takes the form

$$\frac{dy}{dt} = -a(t)\, y. \tag{3.57}$$

We will say that the differential equation (3.55) is *non-homogeneous* (or *heterogeneous*) if $b(t)$ is *not* the identically zero function. For a non-homogeneous differential equation (3.55), the differential equation (3.57) is said to be the corresponding homogeneous differential equation. A homogeneous equation (3.57) is separable, and by separation of variables has a family of solutions of the form

$$y = c\, y_H(t) = c\, e^{-\int a(t)\, dt}. \tag{3.58}$$

Here, $y_H(t)$ is a solution of the homogeneous differential equation (3.57), and every solution y of (3.57) is a constant multiple of $y_H(t)$ (because this is a first-order equation[35]). The relation between the solutions of the general equation (3.55) and the homogeneous equation (3.57) is called the *superposition principle.*

THEOREM (SUPERPOSITION PRINCIPLE): Every solution of the non-homogeneous differential equation (3.55) is the sum of a particular solution of (3.55) and some solution of the corresponding homogeneous differential equation (3.57).

This means that once we find a single solution of the non-homogeneous differential equation (3.55) (perhaps a constant solution, if there is one), we can obtain all the other solutions of (3.55) by adding the set of all solutions of the homogeneous equation (3.57), as given by (3.58).

[35] As we will see in later chapters, the solution space of an nth-order linear ODE is n-dimensional; that is, every solution is a combination of constant multiples of n different functions.

To establish the superposition principle, we can show that any two solutions $y_1(t)$ and $y_2(t)$ of (3.57) differ by some constant multiple of y_H. We first observe that

$$y_1'(t) = -a(t)y_1(t) + b(t), \; y_2'(t) = -a(t)y_2(t) + b(t),$$

and then calculate

$$\begin{aligned} \frac{d}{dt}(y_2(t) - y_1(t)) = y_2'(t) - y_1'(t) &= [-a(t)y_2(t) + b(t)] - [-a(t)y_1(t) + b(t)] \\ &= -a(t)[y_2(t) - y_1(t)]. \end{aligned}$$

Thus $y_2 - y_1$ is a solution of the homogeneous equation (3.57), for any two solutions y_1 and y_2 of the general equation (3.55). Once we know a solution y_1, then, we can add the collected solutions to (3.57) to get the collected solutions to (3.55). None will be left out. For this reason, finding solutions y_H to the homogeneous equation is often a first step in solving the general equation.

3.6.2 Integrating factors

The solution y_H to the homogeneous equation (3.57) can now be seen as the motivation behind the technique of integrating factors introduced in the previous section for the case of constant coefficients. In Section 3.5 we suggested looking at the term $y' + ay$ as the derivative of a product of two functions (one of them y). From equation (3.58) we can see that $y_H' = -ay_H$, so that $a = -y_H'/y_H$, which transforms (3.55) into

$$y' - y\,\frac{y_H'}{y_H} = b(t).$$

This suggests the integrating factor $1/y_H$, by which we multiply to get

$$y'\,\frac{1}{y_H} + y\left(-\frac{1}{y_H^2}\,y_H'\right)\left(y\,\frac{1}{y_H}\right)' = \frac{b(t)}{y_H}.$$

Integration will then allow us to solve for y.

EXAMPLE 1.
Find all solutions of the differential equation

$$y' + 2y = 2\,e^t.$$

Solution: We begin with the corresponding homogeneous differential equation $y' + 2y = 0$, for which we find (by separation of variables, for instance) the family of solutions $y_H = ke^{-2t}$ (we will arbitrarily choose $k = 1$ in what follows). This makes our integrating factor $1/y_H = e^{2t}$, which makes the equation

$$y'e^{2t} + y\,2e^{2t} = \left(ye^{2t}\right)' = 2\,e^{3t}.$$

Integrating, we have

$$ye^{2t} = \frac{2}{3}e^{3t} + c$$

for arbitrary constant of integration c. Thus we have the family of solutions

$$y = \frac{2}{3}e^{t} + ce^{-2t}. \ \square$$

EXAMPLE 2.

Find the solution of the initial value problem

$$y' + 2y = 2e^{t}, \ y(0) = 4.$$

Solution: In Example 1 we found all solutions to the given differential equation, so now we need only determine for which value of c the initial condition is satisfied. Substituting $t = 0$, $y = 4$ into the solution above, we have $4 = \frac{2}{3} + c$, from which $c = \frac{10}{3}$. Thus the solution of the initial value problem is $y = \frac{2}{3}e^{t} + \frac{10}{3}e^{-2t}$. $\square$

EXAMPLE 3.

Find all solutions of the differential equation

$$y' + \frac{1}{1+t}y = 1.$$

Solution: The corresponding homogeneous differential equation is

$$y' = -\frac{1}{1+t}y,$$

and separation of variables gives the solution

$$\int \frac{dy}{y} = -\int \frac{dt}{1+t},$$

$$\log|y| = -\log|1+t| + k.$$

Exponentiation leads to the family of solutions

$$y_H = ce^{-\log|1+t|} = \frac{c}{|1+t|}.$$

If we take $t > -1$ we can drop the absolute value sign. This y_H suggests the integrating factor $(t+1)$, which yields

$$(t+1)y' + y = ((t+1)y)' = (t+1).$$

Integrating, we obtain

$$(t+1)y = \frac{1}{2}(t+1)^2 + c, \ y = \frac{t+1}{2} + \frac{c}{t+1}.$$

(Integrating $t+1$ to $t^2/2 + t + c$ merely yields a different value for c.) $\square$

3.6.3 Substitution and integration

We see that in applying an integrating factor, we end up first solving for the quantity (y/y_H). Giving this quantity a name (typically u) and making the substitution formal produces another approach to solving nonhomogeneous equations. In this approach, we define $u(t)$ by

$$y(t) = y_H(t)\, u(t) \tag{3.59}$$

and (once we know y_H) use this relation to transform the equation for y into an equation for u which may be easier to solve outright. This kind of substitution based on guessing a partial form for the solution was one of the chief techniques developed in the days before computers were widely available to solve differential equations numerically (or symbolically), when outright solution of a differential equation might be the only convenient way to interpret the predictions of the corresponding model.

The change of variable defined in (3.59) actually reduces the solution of the differential equation to an integration. Differentiating (3.59), we find

$$\begin{aligned} y'(t) &= y_H(t)u'(t) + y_H'(t)u(t) = y_H(t)u'(t) + [-a(t)y_H(t)]u(t) \\ &= y_H(t)u'(t) - a(t)y(t), \end{aligned}$$

so that (comparing to (3.55)) $y_H(t)u'(t) = b(t)$, or

$$u'(t) = \frac{b(t)}{y_H(t)}. \tag{3.60}$$

We can now find $u(t)$ by integration, and then find $y(t) = y_H(t)\, u(t)$.

It is possible to give an explicit formula for $y(t)$,

$$y(t) = y_H(t) \int \frac{b(t)}{y_H(t)}\, dt,$$

and an even more complicated formula may be obtained by using the formula (3.58) for $y_H(t)$. However, we advise against attempting to memorize such a formula, and recommend instead using the substitution (3.59) and obtaining the solution by integration.

Note that, from (3.58) and by continuity of the exponential function, the homogeneous solution $y_H(t)$ is continuous and different from zero on any interval on which the function $a(t)$ is continuous. The solution given above, and hence every solution of (3.55), is defined on every interval on which the coefficients $a(t)$ and $b(t)$ are continuous, in contrast to separable differential equations, for which there may be values of t where the functions in the differential equation are smooth but solutions do not exist.

EXAMPLE 4.

Solve the initial value problem

$$ty' + 2y = t^3, \;\; y(1) = 0.$$

Solution: We first put the differential equation in standard form (where the coefficient of y' is 1) by dividing both sides of the differential equation by t to give the initial value problem

$$y' + \frac{2}{t}y = t^2, \ y(1) = 0.$$

Next, we need to solve the corresponding homogeneous differential equation

$$y' + \frac{2}{t}y = 0.$$

Separation of variables gives the solution

$$\int \frac{dy}{y} = -2 \int \frac{dt}{t},$$

$$\log|y| = -2\log|t| + k,$$

$$y = \pm e^{-2\log t} e^k = \frac{c}{t^2}.$$

With $y_H = 1/t^2$, we can make the substitution $y = u/t^2$, from which

$$y' = \frac{u'}{t^2} - \frac{2u}{t^3}.$$

Substituting for y' and y in the original equation gives us

$$\left(\frac{u'}{t^2} - \frac{2u}{t^3}\right) + \frac{2}{t}\left(\frac{u}{t^2}\right) = t^2, \text{ or } \frac{u'}{t^2} = t^2,$$

from which $u' = t^4$ and thus $u(t) = t^5/5 + c$. (Alternatively, we can substitute directly into (3.60) to find u'.) Now we substitute y_H and u into (3.59) to find the general solution

$$y(t) = \frac{1}{t^2}\left(\frac{t^5}{5} + c\right) = \frac{t^3}{5} + \frac{c}{t^2}.$$

Substituting $t = 1$, $y = 0$ gives that $\frac{1}{5} + c = 0$, so $c = -\frac{1}{5}$, and the solution to the initial value problem is $y = \frac{t^3}{5} - \frac{1}{5t^2}$. □

EXAMPLE 5.
Solve the initial value problem

$$y' + 2ty = 2t, \ y(0) = y_0.$$

Solution: The corresponding homogeneous equation $y' + 2ty = 0$ has solutions given by

$$\int \frac{dy}{y} = -2 \int t\,dt$$

so that $\log|y| = -t^2 + k$, and $y = ce^{-t^2}$. Thus we may use $y_H = e^{-t^2}$ and (3.59) or (3.60) to derive that $u' = 2te^{t^2}$, from which $u = e^{t^2} + c$. Then

$$y = y_H u = e^{-t^2}\left(e^{t^2} + c\right) = 1 + ce^{-t^2}.$$

Imposition of the initial condition $y(0) = y_0$ gives $y_0 = c + 1$, or $c = y_0 - 1$, and this gives

$$y = 1 + (y_0 - 1)e^{-t^2}. \ \square$$

EXAMPLE 6.
Solve the initial value problem

$$y' + 2ty = \sin t, \ y(0) = y_0.$$

Solution: The corresponding homogeneous equation is the same as in Example 5, so we can use $y_H = e^{-t^2}$ and (3.60) to derive that $u' = e^{t^2}\sin t$. Thus

$$u(t) = \int e^{t^2} \sin t \, dt + c,$$

and, writing the solution as a definite integral,

$$y = y_H u = e^{-t^2}\int_0^t e^{s^2} \sin s \, ds + ce^{-t^2}.$$

Imposition of the initial condition shows $c = y_0$ and gives, finally,

$$y = e^{-t^2}\int_0^t e^{s^2} \sin s \, ds \; + \; y_0 e^{-t^2}. \tag{3.61}$$

Although the integral in (3.61) cannot be evaluated in terms of elementary functions, it gives a function whose value for $t = 0$ is zero and whose derivative is $e^{t^2}\sin t$. Using this information and the product rule for differentiation we may verify that the function defined by (3.61) is actually a solution of the initial value problem. (See Exercise 15 below.) □

3.6.4 Mixing processes with variable coefficients

We have already looked at some mixing problems which led to linear differential equations whose coefficient functions $a(t)$ and $b(t)$ were constants. These differential equations were not only linear but also separable, and we treated them in Section 3.5. In these problems, the total volume of fluid remained constant because the rates of flow in and out of the system were the same. If the rates of flow into and out of the system are not constant, we will obtain linear differential equations which are *not* separable. Following are two examples of mixing processes discussed in the previous section in which the simplifying assumption of constant inflow and outflow has been removed.

EXAMPLE 7.
A culture tank contains 10 liters of water. A solution containing 1 kg/liter of nutrient is added at a rate of 0.5 liters/hour, and the mixture is drained off at a rate of 0.3 liters/hour. Find the concentration of the nutrient solution after 24 hours.

Solution: Let $s(t)$ be the weight of nutrient in the tank at time t. Since the volume of solution in the tank at time t is $(10+0.2t)$ liters, the concentration of nutrient at time t is $s(t)/(10+0.2t)$. Thus the weight of nutrient being removed at time t is $0.3s(t)/(10+0.2t)$ kg/hr. The weight of nutrient being added is 0.5 kg/hr. Thus the rate of change of the nutrient weight is given by

$$\frac{ds}{dt} = 0.5 - \frac{0.3s}{10+0.2t},$$

which we write in standard form as

$$s' + \frac{3}{100+2t}s = 0.5.$$

To find the appropriate integrating factor, we solve the corresponding homogeneous differential equation by separation of variables,

$$\int \frac{ds}{s} = -\int \frac{3\,dt}{100+2t},$$

which leads to $\log s = -\frac{3}{2}\log(100+2t) + k_1$. Exponentiating, we have

$$s = e^{-\frac{3}{2}\log(100+2t)}e^{k_1} = (100+2t)^{-\frac{3}{2}}k_2.$$

Since we need only one solution to obtain an integrating factor, we may omit the constant of integration and write $s_H = (100+2t)^{-\frac{3}{2}}$. Then the integrating factor is $(100+2t)^{\frac{3}{2}}$, multiplication by which gives

$$(100+2t)^{\frac{3}{2}}s' + (100+2t)^{\frac{1}{2}}s = [(100+2t)^{\frac{3}{2}}s]' = 0.5(100+2t)^{\frac{3}{2}}.$$

Integration gives

$$(100+2t)^{\frac{3}{2}}s = \int 0.5(100+2t)^{\frac{3}{2}}\,dt = 0.1(100+2t)^{\frac{5}{2}} + c.$$

Substitution of the initial condition $s(0) = 0$ gives $0 = 0.1(100)^{\frac{5}{2}} + c$, $c = -0.1(100)^{\frac{5}{2}} = -10^4$, leading to the solution

$$(100+2t)^{\frac{3}{2}}s = 0.1(100+2t)^{\frac{5}{2}} - 10^4,$$

$$s = 0.1(100+2t) - 10^4(100+2t)^{-\frac{3}{2}}.$$

When $t = 24$, $s = 0.1(148) - 10^4(148)^{-\frac{3}{2}} \approx 9.25$ kg. Also, when $t = 24$, the volume of water in the tank is 14.8 liters, and the concentration of nutrient is $\frac{9.25}{14.8} \approx 0.625$ kg/liter. The concentration at time t, when the volume of water is $(10 + 0.2t)$ liters, is

$$\frac{10 + 0.2t - 10^4(100 + 2t)^{-\frac{3}{2}}}{10 + 0.2t} = 1 - \left(\frac{10}{10 + 0.2t}\right)^{-\frac{5}{2}} \text{ kg/liter. } \square$$

EXAMPLE 8.

If we examine the per capita demographic rates for the United States given in Table 3.3, we can see clear trends of change in each rate: decreasing birth and death rates and increasing immigration and emigration rates. Graphs of these rates show that the changes can be approximately described as linear in time. Assume rates of this form and the hypothesis discussed in Section 3.5 that immigration is independent of the U.S. population size, to derive the linear model $x' = r(t)x + I(t)$, where $r(t) = b - d - \epsilon$. Compare the resulting population predictions with those of the constant-coefficient model of Section 3.5. Does this refinement affect the model's predictions significantly?

Solution: Let us define t as time in years since 1950, as was done in the previous section. To make the demographic coefficients linear in time, we write $r(t) = r_0 + r_1 t$ and $I(t) = I_0 + I_1 t$, and substitute into the original equation to obtain

$$x'(t) = (r_0 + r_1 t)x + (I_0 + I_1 t), \tag{3.62}$$

for which the corresponding homogeneous equation $x'(t) = (r_0 + r_1 t)x$ can be solved by separation of variables:

$$\int \frac{dx}{x} = \int (r_0 + r_1 t)dt$$
$$\log|x| = r_0 t + r_1 \frac{t^2}{2} + k$$
$$x_H = ce^{r_0 t + r_1 t^2/2}$$

and we again take $e^k = c = 1$ for simplicity. Then from (3.60) we have

$$u'(t) = (I_0 + I_1 t)e^{-(r_0 t + r_1 t^2/2)},$$

from which

$$u(t) = x_0 + \int_0^t (I_0 + I_1 s)e^{-(r_0 s + r_1 s^2/2)} ds,$$

since $u(0) = x(0)/x_H(0) = x_0/1$. We can simplify part of the integral via the substitution $w = r_0 s + r_1 s^2/2$: then

$$I_0 + I_1 s = \left(I_0 - I_1 \frac{r_0}{r_1}\right) + \frac{I_1}{r_1}(r_0 + r_1 s) = \left(I_0 - I_1 \frac{r_0}{r_1}\right) + \frac{I_1}{r_1}\frac{dw}{ds},$$

and

$$\begin{aligned}\int \frac{I_1}{r_1}(r_0 + r_1 s)e^{-(r_0 s + r_1 s^2/2)}ds &= \frac{I_1}{r_1}\int e^{-w}dw = -\frac{I_1}{r_1}e^{-w} + c \\ &= -\frac{I_1}{r_1}e^{-(r_0 s + r_1 s^2/2)} + c.\end{aligned}$$

Thus

$$u(t) = x_0 + \left(I_0 - I_1\frac{r_0}{r_1}\right)\int_0^t e^{-(r_0 s + r_1 s^2/2)}ds - \frac{I_1}{r_1}\left(e^{-(r_0 t + r_1 t^2/2)} - 1\right),$$

and

$$\begin{aligned}x(t) = x_H u = \left(x_0 + \frac{I_1}{r_1}\right)e^{r_0 t + r_1 t^2/2} - \frac{I_1}{r_1} \\ + \left(I_0 - I_1\frac{r_0}{r_1}\right)e^{r_0 t + r_1 t^2/2}\int_0^t e^{-(r_0 s + r_1 s^2/2)}ds.\end{aligned}$$

As in Example 6, the unresolved integral cannot be evaluated in terms of elementary functions, but values can be obtained through numerical integration for any given value of t, once we have values for the demographic coefficients.

To determine coefficient values for the demographic rates, we use simple linear regression on the values in Table 3.3. This yields $b_0 = 0.0240095/\text{yr}$, $b_1 = -0.000205714/\text{yr}^2$, $d_0 = 0.00971905/\text{yr}$, $d_1 = -0.0000254286/\text{yr}^2$, $\epsilon_0 = 0.000152571/\text{yr}$, $\epsilon_1 = 0.0000129171/\text{yr}^2$ (so that $r_0 = 0.0141379/\text{yr}$, $r_1 = -0.000193203/\text{yr}^2$), $I_0 = 63728.6$ people/yr, $I_1 = 15982.9$ people/yr^2. We compare the resulting solution with that obtained in Section 3.5 for fixed coefficients, taking for the coefficient values here the averages of those in Table 3.3, which yield $r = 0.0093/\text{yr}$ and $I = 463,300$ people/yr. Figure 3.21 compares the two solutions with the actual population values; one can see that the model with coefficients linear in time makes predictions with about half the error of the model predictions using constant coefficients for most of the period given. Only toward the end of the period does a sudden jump in census population throw the variable-coefficient model's estimate off. Overall, the two estimates are within 3% of each other during this period. □

3.6.5 Bernouilli equation

Earlier in this section we observed the use of a substitution (change of variables) to transform a differential equation into another equation which may be easier to solve. The methods outlined in this section apply only to linear differential equations, but there are classes of nonlinear ODEs which can be transformed into linear ODEs — and hence solved — via a substitution. One such class of equation is the *Bernouilli equation* (named after Jacob Bernouilli (1654–1705)), of the form

$$y' + a(t)y = b(t)y^n, \tag{3.63}$$

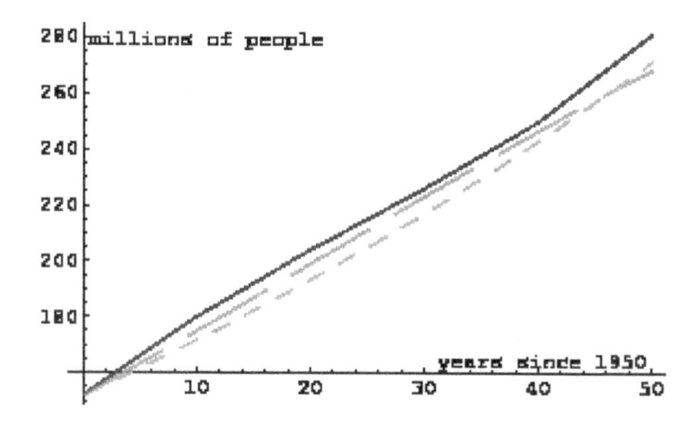

FIGURE 3.21: Migration model predictions for the U.S. population using fixed coefficients (short dashes) and coefficients linear in time (long dashes), compared to census values (solid line).

where n is a constant which is neither 0 nor 1. (If $n = 0$ or $n = 1$ the differential equation (3.63) is linear.) The change of dependent variable $z = y^{1-n}$ in (3.63) gives

$$y = z^{1/(1-n)},\ y' = \frac{1}{1-n} z^{n/(1-n)} z',$$

so that

$$\frac{1}{1-n} z^{n/(1-n)} z' + a(t) z^{1/(1-n)} = b(t) z^{n/(1-n)},$$

$$z' + (1-n)a(t)z = (1-n)b(t). \tag{3.64}$$

Since (3.64) is linear it may be solved by finding the appropriate integrating factor and integrating. Note, however, that the identically zero solution of (3.63) may not be obtained by this method, because we have divided through by $z^{n/(1-n)}$.

EXAMPLE 9.
Solve the initial value problem

$$y' - y = -ty^2,\ y(0) = -\frac{1}{2}.$$

Solution: The given differential equation is a Bernouilli equation with $n = 2$. Therefore we let $z = y^{-1}$, or $y = \frac{1}{z}$, $y' = -\frac{z'}{z^2}$, and the differential equation becomes

$$-\frac{1}{z^2} z' - \frac{1}{z} = -\frac{t}{z^2},$$

$$z' + z = t.$$

The appropriate integrating factor is e^t, and multiplying by it yields

$$z'e^t + ze^t = (ze^t)' = te^t,$$

$$ze^t = \int te^t \, dt = (t-1)e^t + c.$$

Then we have

$$y = \frac{1}{z} = \frac{1}{(t-1) + ce^{-t}}.$$

Substitution of the initial condition $y(0) = \frac{1}{2}$ gives $\frac{1}{2} = \frac{1}{c-1}$, or $c = 3$. Thus the solution of the given initial value problem is $y = [(t-1) + 3e^{-t}]^{-1}$. □

Bernouilli equations may arise in many different areas, as an artifact of the mathematics involved. For example, Benchekroun and Van Long studied a problem in international fishing by casting it in terms of game theory; the resulting equation for the optimal reactive strategy took the form of a Bernouilli equation[36] with exponent $n = 2$. In the situation they modeled, migratory fish that travel along a coastline belonging to two nations, such as the Canadian salmon which travel along the Alaskan coastline on their way back to their Canadian breeding grounds, are subject to catching by groups from both countries, with the first player in an interceptor role and the second in a reactive role. Overfishing will drive the population to extinction. The optimal reaction of the second player in cases where the first player deviates from his optimal catch can be described by a differential equation which the authors transform into a Bernouilli equation.

However, by far the most common Bernouilli equation to arise in the study of biological systems is the logistic equation, in which again $n = 2$. We have already solved the logistic equation with constant coefficients using separation of variables, but when the coefficients — the intrinsic growth rate r and the carrying capacity K — vary over time the equation is no longer separable. In practice, these coefficients do vary with time: on a short time scale, periodic (daily or annual) fluctuations affect system resources and reproduction, while on a long time scale gradual changes in both the environment and the organisms which live in it can accumulate and become appreciable.

We have seen in Example 9 that the change of variables $y = \frac{1}{z}$ transforms a Bernouilli equation with $n = 2$ into a first-order linear differential equation. The result of this substitution is the same whether r and K are constants or functions of t. Thus we can use this substitution to transform the logistic differential equation with variable coefficients

$$y' = r(t)y\left(1 - \frac{y}{K(t)}\right) \tag{3.65}$$

[36] Hassan Benchekroun and Ngo Van Long, Transboundary fishery: a differential game model, *Economica* 69: 207–221, 2002. See their Appendix B for details.

into the linear first-order differential equation

$$z' = -r(t)z + \frac{r(t)}{K(t)}. \tag{3.66}$$

In order to find an integrating factor, we find a solution of the corresponding homogeneous equation $z' = -r(t)z$, namely $z_H(t) = e^{-\int r(t)dt}$, and this indicates an integrating factor $1/z_H(t) = e^{\int r(t)dt}$. Multiplication of (3.66) by this integrating factor gives

$$z' e^{\int r(t)dt} + zr(t)e^{\int r(t)dt} = \left[z\, e^{\int r(t)dt}\right]' = \frac{r(t)\, e^{\int r(t)dt}}{K(t)},$$

and integration gives

$$z\, e^{\int r(t)dt} = \int \frac{r(t)\, e^{\int r(t)dt}}{K(t)}\, dt + c, \tag{3.67}$$

from which we can obtain an explicit formula for z and then an explicit formula for the original unknown function $y = 1/z$.

If we assume the carrying capacity K to be constant in order to study the effects of a variable intrinsic growth rate, (3.67) gives

$$z\, e^{\int r(t)dt} = \frac{1}{K}\int r(t)\, e^{\int r(t)dt}\, dt + c = \frac{e^{\int r(t)dt}}{K} + c,$$

and we obtain

$$z = \frac{1}{K} + c\, e^{-\int r(t)dt},$$

$$y = \frac{1}{\frac{1}{K} + c\, e^{-\int r(t)dt}} = \frac{K}{1 + cKe^{-\int r(t)dt}}. \tag{3.68}$$

(Compare with the solution (3.37) for the case where r is constant, too.) We see that if the infinite integral $\int^\infty r(t)dt$ diverges, $\int^\infty r(t)\, dt \to \infty$, then $\lim_{t\to\infty} y(t) = K$. This will be the case whenever the growth rate $r(t)$ does not die down to zero fast enough: if, for example, r is constant, or periodic (seasonal), say $r(t) = r_0 + a\cos\omega t$ with $r_0 > 0$. When $r(t)$ is integrable as $t \to \infty$, however, $\int^\infty r(t)\, dt \to L$, then $\lim_{t\to\infty} y(t) < K$. Thus (3.68) gives the quantitative behavior of the population when the growth rate varies, and if in addition $r(t)$ dies down to zero fast enough (for whatever biological reason) the population will settle at a lower level.

If instead r is constant but the carrying capacity K is a function of t, (3.67) gives

$$ze^{rt} = \int \frac{re^{rt}}{K(t)}\, dt + c.$$

In terms of definite integrals we may write

$$z = e^{-rt}\int_0^t \frac{r\, e^{rs}}{K(s)}\, ds + z(0)\, e^{-rt}. \tag{3.69}$$

In order to find $\lim_{t\to\infty} z(t)$, we may use L'Hôpital's rule for the indeterminate form

$$\lim_{t\to\infty} \frac{\int_0^t \frac{r\,e^{rs}}{K(s)}\,ds}{e^{rt}} = \lim_{t\to\infty} \frac{\frac{r\,e^{rt}}{K(t)}}{r\,e^{rt}} = \lim_{t\to\infty} \frac{1}{K(t)}.$$

Thus if $K(t)$ approaches a limit as $t \to \infty$, then $z(t) \to \frac{1}{\lim_{t\to\infty} K(t)}$, and thus $y(t)$ and $K(t)$ have the same limit. If the carrying capacity does not approach a limit (for instance, if it is seasonal), then the population will not do so, either (since its "target level" is moving). However, the solution can be shown (see Exercise 15 below) to be asymptotically bounded within the range of values taken on by $K(t)$.

EXAMPLE 10.

In Section 3.2 we discussed the use of the logistic model with constant coefficients in accounting for Australian sheep populations, noting that the model described the initial period of rapid growth well but failed to account for the sustained significant (10% or more) fluctuations seen after the population reached its carrying capacity. One hypothesis has been that the pasture land required some time to recover from the overgrazing it experienced. Assume that the pastures undergo cycles of overgrazing followed by recovery. To what extent can this assumption account for the variations in the Tasmanian sheep population seen in Figure 3.7?

Solution: From the figure, if the period 1846–1856 represents the period of initial overgrazing (causing growth greater than expected for the amount of system resources), then 4 cycles of fluctuation can be observed during the approximately 68 years which follow, with amplitude approximately 170 (thousand sheep). This can be modeled most simply by adding a sine wave of amplitude 170 and period $17 (= 68 \div 4)$ to the estimated carrying capacity of 1670 (thousand sheep) given in the figure caption, making

$$K(t) = 1670 + 170 \sin \frac{2\pi}{17}(t - 1856),$$

where t is the year. If we assume the sheep's intrinsic growth rate still constant at $r = 0.13125/\text{yr}$, then we have the second special case described above, which leads to (3.69). Taking our starting point as $t_0 = 1857$, for which the original solution $y(t) = 1670/[1 + \exp(240.81 - 0.13125\,t)]$ has the value 1585, so that $z_0 = 1/1585$, we can then rewrite (3.69) as

$$z(t) = z_0 e^{-r(t-t_0)} + e^{-rt} \int_{t_0}^{t} \frac{re^{rs}}{K(s)}\,ds.$$

Finally, recall $y(t) = 1/z(t)$.

Although the integral in the expression for z cannot be evaluated in terms of elementary functions, it can be evaluated numerically for any given value of t. We can therefore plot a graph of y and compare it with the given data; see

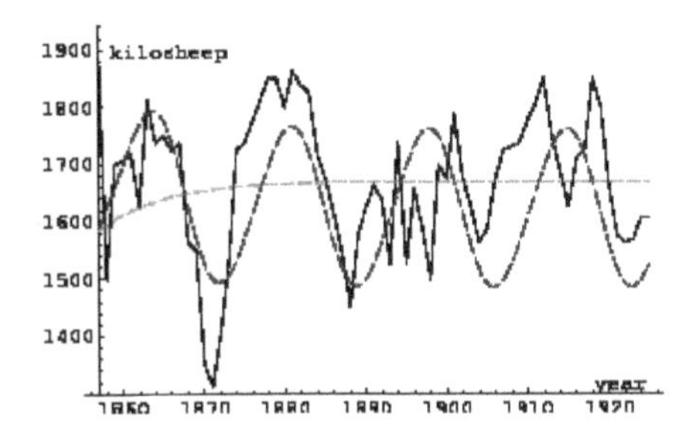

FIGURE 3.22: A logistic model with sinusoidally varying carrying capacity approximates observed fluctuations in the Tasmanian sheep population.

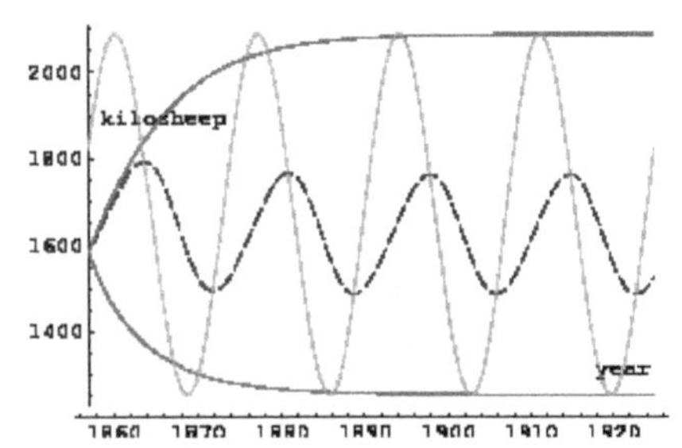

FIGURE 3.23: A population (dashed curve) governed by a logistic model with periodic carrying capacity (system resources) oscillates out of phase with the carrying capacity (solid curve). Also shown are simple logistic bounds (see Exercise 15).

Figure 3.22. We see that the population size does not approach the variable carrying capacity asymptotically; rather, it increases for that part of the cycle when $y(t) < K(t)$, and decreases as long as $y(t) > K(t)$ (see Figure 3.23). We also see that the fluctuations appear to have a negative effect overall on the population, in that the population predicted using a variable carrying capacity is less than the average carrying capacity more than half of the time (that is, it is below the line more often than above it). In general, we can see that this model is more consistent with observed population values than the original, though we should note that this is still only a rough approximation at best. □

Exercises

In Exercises 1–12, solve the given initial value problem.

1. $y' + 2y = 2$, $y(0) = \frac{1}{2}$
2. $y' + 2y = e^t, y(0) = 1$
3. $y' + 3y = e^{-2t}, y(0) = 5$
4. $y' + y = t$, $y(1) = -2$
5. $y' + \frac{1}{t}y = 1$, $y(1) = 1$
6. $y' - \frac{1}{t}y = 1$, $y(1) = 1$
7. $ty' + 2y = t^2$, $y(-2) = 5$
8. $t^2y' + 2ty = 1, y(1) = -1$
9. $ty' + y = e^t$, $y(1) = 1$
10. $ty' - y = t^2, y(1) = 1$
11. $y' = -2t(y - t^2)$, $y(0) = 0$
12. $(t+1)(y' + y) = e^{-t}$, $y(0) = 0$

13. Find the solution of $y' + \frac{y}{t} = \frac{y^3}{t^2}$ passing through the point (6,3).
14. Find the solution of $y' - y = ty^2$ passing through the point $(1, -1)$.

15. Find the solution of the differential equation $ty' = 2t^2y + y\log y$ through the point (1,1). [Hint: Make the change of dependent variable $v = \log y$.]

16*. Discuss the behavior of solutions of the initial value problem $y' = \lambda y$, $y(0) = y_0$ as $t \to \infty$ for each of the cases $\lambda > 0$, $\lambda = 0$, and $\lambda < 0$. (We will see in Section 4.1 that this is a key issue in the qualitative analysis of differential equations.)

17*. Solve and then discuss the behavior of solutions as $t \to \infty$ for each of the following differential equations satisfying the generic initial condition $y(t_0) = y_0$, where t_0 and y_0 are given constants: (a) $y' = -2y + e^{-t}$ (b) $y' = -2y + e^t$ (c) $y' = -2y + 1$ (d) $y' = -2y + \frac{1}{1+t^2}$ [Hint: Use L'Hôpital's rule to evaluate the limit.]

18*. Find the solution of the initial value problem $y' + y = g(t)$, $y(0) = 0$, where $g(t) = 1$ if $0 \leq t < 2$ and $g(t) = 0$ if $t \geq 2$. This equation models a system with an external influence (i.e., independent of the population y) which shuts off after a given time.

19*. Solve the integral equation $y(t) = \int_0^t y(s)\,ds + t + a$. [Hint: Differentiate the equation and note that $y(0) = a$.]

20*. (a) Show (by constructing it) that the differential equation $y' + y = f(t)$ has a unique solution which is bounded for $-\infty < t < \infty$ if $f(t)$ is continuous and $|f(t)| \leq M$ for $-\infty < t < \infty$. [Hint: Consider the solution y_A with $y_A(-A) = 0$, where $A > 0$, and let $A \to \infty$, $y(t) = \lim_{A\to\infty} y_A(t)$.] (b) If the function $f(t)$ is periodic with period 2π (i.e., $f(t + 2\pi) = f(t)$ for $-\infty < t < \infty$), show that the solution obtained in part (a) is periodic with period 2π. [Hint: Show by a suitable change of the variable of integration that the solution obtained in part (a) satisfies $y(t + 2\pi) = y(t)$ for every t.]

21*. Let a be a positive constant and let $\lim_{t\to 0+} f(t) = b$. By choosing an appropriate initial condition, show that the differential equation

$$ty' + ay = f(t)$$

has a unique solution that is bounded as $t \to 0+$, and find the limit of this solution as $t \to 0+$.

22. Verify that the function defined by equation (3.61) is the solution of Example 6.

23. A tank contains 10 liters of water to which is added a nutrient solution containing 0.3 kg of nutrient per liter. This nutrient solution is poured in at the rate of 3 liters/min, is thoroughly mixed, and then the mixture is drained off at the rate of 2 liters/min. How much nutrient is in the tank after 5 minutes?

24. A 500 gallon runoff containment tank contains 200 gallons of a solution containing 15 lb. of pollutant, when heavy rains flood the system. Fresh water pours into the tank at a rate of 12 gallons per minute, and the mixture is pumped out for processing at a rate of 6 gallons per minute. Find the amount of pollutant in the tank when the tank overflows.

25. A catfish pond has 1000 L of water containing 2 kg of dissolved salt. Flooding causes seawater to enter the pond at 60 L/min. (Seawater has an average salt content of 36 g/L.) At the same time, a drainage system turns on and pumps the water out at a rate of 100 L/min. (a) Find the amount of salt in the tank at any time t. (b) Find the concentration of salt (in grams per liter) as a function of time. Chervinski found[37] that young catfish can survive salinity levels up to 1/4 that of seawater. How long does the pond's caretaker have to rescue the catfish? (c) Sketch the graphs of the amount of salt and the concentration of salt against time, and determine the absolute maximum of each quantity.

26. A 400 liter tank contains a mixture of water and chlorine with 0.05 grams of chlorine per liter. In order to reduce the concentration of chlorine in the tank, fresh water is pumped into the tank at a rate of 4 L/sec. After thorough stirring, the mixture is pumped out at a rate of 10 L/sec. Find the amount of chlorine in the tank as a function of t.

27. A very crude model of a population with migration is given by

$$y' + ay = M(t), \; y(0) = y_0,$$

where a is the proportional rate of decrease without migration and $M(t)$ is the rate at which members are added to the population. Here, positive values of $M(t)$ correspond to immigration, and negative values of $M(t)$ correspond to emigration. Find the population size as a function of t if $M(t)$ is a positive constant M_0.

28. Find the population size y modeled by $y' + ay = M_0 + A \sin \omega t, \; y(0) = y_0$ where the sine term models fluctuations in migration. [Observe that if $|M_0| < |A|$, then the migration rate takes on both positive and negative values, indicating that there are periods of net immigration and periods of net emigration.]

29. Solve the logistic equation with constant coefficients $y' = ry(1 - y/K)$ via the Bernouilli substitution.

30. Suppose that a population governed by the logistic model (3.65) has a carrying capacity bounded by constants $K_{min} \leq K(t) \leq K_{max}$ for all time (and a constant intrinsic growth rate r). Show that the resulting population size $y(t)$ is bounded by the two solutions of the form

[37] J. Chervinski, Salinity tolerance of young catfish, Clarias lazera (Burchell), *Journal of Fish Biology* 25(2): 147, August 1984.

(3.13) to the logistic equation with constant carrying capacities K_{min} and K_{max}, respectively (as illustrated in Figure 3.23), and thus asymptotically bounded by the constants K_{min} and K_{max}. [Hint: Use the above inequality to derive bounds on the integrand in (3.69), and use the resulting bounds on z to derive bounds for $y = 1/z$.]

Miscellaneous exercises

In Exercises 1–4, verify that the given family of functions is a solution of the given differential equation for every value of the constant c.

1. $y = ce^{2t} + \dfrac{\sin t - 2\cos t}{5}$, $y' - 2y = \cos t$
2. $y = ce^{t^2/2}$, $y' - ty = 0$
3. $y = c\dfrac{e^{-1/t}}{t}$, $t^2 y' = y(1-t)$
4. $y = ct - t\log t$, $y' = -1 + \frac{y}{t}$

In Exercises 5–8, draw a direction field for the given differential equation on the given rectangle.

5. $y' - 2y = \cos t$; $0 \le t \le 5$, $-2 \le y \le 2$
6. $y' - ty = 0$; $-2 \le t \le 2$, $-2 \le y \le 2$
7. $t^2 y' = y(1-t)$; $\frac{1}{2} \le t \le 4$, $-1 \le y \le 1$
8. $y' = -1 + \frac{y}{t}$; $\frac{1}{2} \le t \le 4$, $0 \le y \le 4$

For each of the differential equations in Exercises 9–12, construct the direction field and use it to sketch the solution passing through the origin.

9. $y' = t + y$
10. $y' = \dfrac{t}{y - t}$
11. $y' = (y+1)(t+1)$
12. $y' = y^3 - 1$

In Exercises 13–26, find the solution of the given initial value problem.

13. $y' = t^2 y^2$, $y(1) = 1$
14. $y' = ty\log y$, $y(0) = e$
15. $y' = te^y e^{t^2}$, $y(1) = 0$

16. $y' = (1 + y^2)(1 + t^2)$, $y(0) = 0$

17. $yy' = t$, $y(1) = 0$

18. $y' = \tan y$, $y(\frac{\pi}{4}) = 1$

19. $y' = \frac{y}{t} + t^2$, $y(1) = 1$

20. $y' + ty = t^2$, $y(0) = 0$

21. $ty' - y = ty$, $y(1) = 4$

22. $y' - y = e^t$, $y(0) = 1$

23. $y' = ty + y^{\frac{1}{2}}$, $y(-1) = 1$

24. $ty' + y = t^2y^2$, $y(0) = -1$

25. $(1 + t^2)y' + 2ty = \dfrac{1}{1 + t^2}$, $y(0) = 0$

26. $(1 + t^2)y' = 1 + ty$, $y(1) = 0$

27. A population of organisms governed by the law of simple population growth has a birth rate of 0.35 per member per week. How long does it take for the population size to triple?

28. A population governed by the law of simple population growth has a birth rate of 0.20 per member per week. If the initial population size is 100 members, what is the population size after 6 weeks?

29. The half-life of radium is 1620 years. How long does it take for one quarter of a given amount of radium to disintegrate?

30. The half-life of plutonium is 50 years. How long does it take for one tenth of a given amount of plutonium to disintegrate?

31. A population grows according to the logistic law with an initial population size of 100 members and a limiting population size of 500 members. The population size reaches 250 in 1 week. What is the population size after 2 weeks?

32. A population grows according to the logistic law with an initial population size of 100 members, a population size of 200 members after 1 week, and a limiting population size of 500 members. Find the *rate* at which the population size is growing after 1 week.

33. A tank contains 1 gallon of a salt solution consisting of 1 lb. salt per gallon. A solution containing 2 lb. salt per gallon runs into the tank at a rate of 5 gallons per minute and the mixture runs out of the tank at the same rate. Find the weight of salt in the tank as a function of time.

34. To a tank containing 100 lb. salt dissolved in 100 gallons of water is added a solution containing 2 lb. salt per gallon at a rate of 2 gallons per minute. The mixture runs out at a rate of 4 gallons per minute. Find the weight of salt and the concentration of the mixture after 10 minutes.

35. A body with initial temperature 100°C cools in air at 20°C, taking 5 minutes to cool to a temperature of 80°C. How long does it take to reach a temperature of 60°C?

36. A hot body is cooled in air at 20°C, reaching temperatures of 45°C 10 minutes after it starts to cool and 40°C 20 minutes after it starts to cool. What was its original temperature when cooling began?

37*. Explain why the initial value problem $y' = \sqrt{1-y^2}$, $y(0) = 2$ has no solution.

38*. Show that the function $y(t)$ defined as $y = 0$ for $0 \le t < 1$ and as $y = t^3$ for $1 \le t \le 2$ is differentiable for $0 \le t \le 2$ and solves the initial value problem $y' = 3y^{2/3}$, $y(0) = 0$.

39. The differential equation $y' = \frac{ry}{y+A} - y$ may be thought of as a limiting case of the Verhulst difference equation $y_{k+1} = \frac{ry_k}{y_k+A}$ (think of $y_{k+1} - y_k$ as an approximation to the derivative y'). Compare the behaviors of the solution of the difference equation and the solution of the differential equation, with $A = 1$ and various values of r. For what values of r do the two equations behave similarly?

40. (a) Formulate a differential equation which is a limiting case of the Beverton-Holt difference equation

$$y_{k+1} = \frac{ay_k}{1+by_k}.$$

(b) For what values of the parameters a and b do the solutions of the difference equation and the differential equation derived in part (a) behave similarly?

41*. As has been mentioned in the text, use of a continuous variable to represent a discrete population is an approximation, the error involved in which we are usually willing to overlook if it is small relative to the population size; that is, if a model predicts a population of 761.23 birds at a given time, we can use the general size rounded to the nearest integer as an indication that the population will include something close to 761 birds. However, if we keep in mind the fact that deterministic models such as differential equations represent an average over all possible outcomes, we can interpret non-integer (or indeed all) results in these terms. Write probabilistic interpretations of the results in the first two simple examples of Section 3.1.

Chapter 4

Nonlinear Differential Equations

In the previous chapter, we developed techniques for solving first-order differential equations which are either linear or separable. However, in general the models we develop to describe biological systems will not be linear—indeed, we shall shortly see several types of biological systems which are driven by distinctly nonlinear interactions. Most nonlinear differential equations cannot be solved outright; we can, however, develop tools to analyze the long-term behavior of these models, as was done for difference equations in Chapter 2, and often the long-term behavior of the biological system is what really interests us. (If we are interested in short-term behavior of the system, we typically use computers to approximate the solution numerically.[1])

In this chapter, we shall first develop these qualitative analysis tools for single first-order differential equations, taking care to use notation which will help us extend them to systems of differential equations in the next chapter. Following this, we shall study particular types of nonlinear interactions such as harvesting (last seen in Section 2.5) and contact processes. One of the fundamental results in biological modeling is the description of *threshold quantities* which control sharp changes in the nature of a system—for example, the difference between survival and extinction—and we shall also take time to examine common biological thresholds under the mathematical lens of bifurcations.

4.1 Qualitative analysis tools

As an example of a simple nonlinear model for a biological system, let us consider the model developed by Beekman, Sumpter, and Ratnieks[2] to describe the foraging of Pharaoh ants, *Monomorium pharaonis*. Like many

[1]Before computers became readily available, researchers often expended great effort to solve differential equations outright; this fact accounts for the many solution techniques that have historically been taught in courses on ordinary differential equations.

[2]Madeleine Beekman, David J. T. Sumpter, and Francis L. W. Ratnieks, Phase transition between disordered and ordered foraging in Pharaoh's ants, *PNAS* 98(17): 9703–9706, 14 August 2001; and David J. T. Sumpter and Madeleine Beekman, From nonlinearity to optimality: pheromone foraging by ants, *Animal Behaviour* 66: 273–280, 2003.

ants, these foragers lay pheromone trails connecting any food sources they find back to their nests, in order to help nestmates find the food (see Figure 4.1). These pheromone trails are volatile, lasting only about 10 minutes unless ants still on the trail lay down more. Beekman, Sumpter, and Ratnieks studied the effects of colony size, distance from food source to nest, and strength of the pheromone trail on the number of ants using the trail. Their model hypothesized, first, that ants begin to forage at a given food site either by finding it independently, at a rate α inversely proportional to the food site's distance from the nest, or by being drawn to the pheromone trail, at a rate dependent both on the number of ants laying the trail down and on the quality of the food source, measured by a parameter β. If we let $x(t)$ represent the number of ants foraging on the trail at time t, and denote the colony size by n, then the rate at which an ant begins using the trail is given by $\alpha + \beta x$, the number of ants *not* using the trail is $n-x$, and thus the total rate at which ants join the trail is $(\alpha+\beta x)(n-x)$. Their model also hypothesized that ants lose or stop using the pheromone trail at a rate dependent, in part, on the strength of the trail. More specifically, they hypothesized that the rate at which an individual ant loses the trail is inversely proportional to the number of ants x renewing the trail, in such a way that the total rate at which ants leave the trail never passes a maximum rate s. (Note the distinction here between the rate for a single ant, or per capita rate, and the total rate.) This hypothesis gives a trail loss rate of $\frac{sx}{x+K}$, where K is the number of ants on the trail at which the total trail loss rate reaches $s/2$. Since the total rate never surpasses s, this function is said to saturate; we shall see other examples of biological processes that saturate in the sections ahead.

The model that results from the above hypotheses is

$$\frac{dx}{dt} = (\alpha + \beta x)(n - x) - \frac{sx}{x + K}. \tag{4.1}$$

Note that although equation (4.1) is separable, the technique of separation of variables does not lead to a closed-form expression for $x(t)$ because of the algebraic complexity of the terms. We can nevertheless learn a lot about the behavior this set of rules produces, results which Beekman, Sumpter, and Ratnieks confirmed experimentally and interpreted to reveal a remarkable shift in the efficiency of such behavior as colony size grows. In order to do so, we now develop the tools we will need.

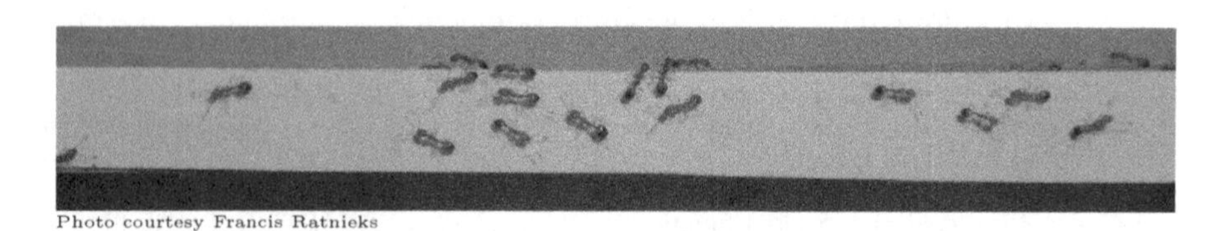

Photo courtesy Francis Ratnieks

FIGURE 4.1: Foraging Pharaoh ants following a pheromone trail.

4.1.1 Possible end behaviors

We might begin our consideration of long-term behavior by asking what can happen to a biological system in the end. A quick review of the continuous-time systems we studied in Chapter 3 reveals two types of behavior observed so far: the quantity or population in question either (1) settles down to a constant level, called an equilibrium, or (2) in the case of our simplest models, grows unbounded. Recalling the period doubling and chaos exhibited in Chapter 2 by nonlinear discrete-time models describing a population under scramble competition (such as the discrete logistic equation), we might ask whether it is possible for continuous-time models $\frac{dy}{dt} = f(y)$ of a single population to predict sustained oscillations and other complicated behaviors. It turns out that (under some very simple hypotheses, such as $f(y)$ being a continuous function) they cannot: they can only do one of the two things listed above. We can see why this is true if we think for a moment about the nature of the model. Since the rate of change $\frac{dy}{dt}$ of y depends only on y, then for a given y-value, say y_1, every time the solution reaches that value, its rate of change will be the same, $f(y_1)$. More specifically, if $f(y_1) > 0$, then the solution will always increase from y_1; if $f(y_1) < 0$, then the solution will always decrease from y_1. This means that a solution can never return to a point it has already passed: if $f(y_1) > 0$, then once the solution has passed above y_1, it can never pass below it, because any solution that gets close enough to y_1 is forced upward. In other words, every solution is *monotone*: either always increasing or always decreasing (or constant).

Note that this argument holds only for one-dimensional, continuous-time models: solutions to discrete-time models make discrete jumps, enabling a solution to pass from above y_1 to below y_1 without being affected by $f(y_1)$, and solutions to multi-dimensional systems can pass below (or even return to) a given point without first passing through it.

Before continuing, we should state this result a little more precisely. We restrict our study here to *autonomous* differential equations, which do not depend explicitly on the independent variable t, of the general form

$$\frac{dy}{dt} = f(y). \tag{4.2}$$

(Autonomous differential equations are always separable, but solution by separation of variables may be impractical if the integral $\int \frac{dy}{f(y)}$ is difficult or impossible to evaluate, as with the ant model (4.1) above.) We will also assume that the function $f(y)$ is smooth enough that the existence and uniqueness theorem of Section 3.3 holds, and there is a unique solution to (4.2) for each given initial condition. For simplicity, we shall consider only non-negative values of t, and we will think of solutions as determined by their initial values for $t = 0$. Under these constraints, we can make use of certain properties of autonomous differential equations, beginning with the role of fixed points of the system.

PROPERTY 1: EQUILIBRIA. The constant solutions of the differential equation $y' = f(y)$ are precisely the zeroes of the function $f(y)$. That is, for a given number y^*, $y = y^*$ is a constant solution of (4.2) if and only if $f(y^*) = 0$.

This is true because the derivative of a constant function is the zero function, so if $y = y^*$, then $\frac{dy}{dt} = 0$, and this function obeys the differential equation $\frac{dy}{dt} = f(y)$ if and only if $f(y^*) = 0$. Such a solution is called an *equilibrium* or *fixed point* of the differential equation. Qualitative analysis of differential equations begins with the study of the equations' equilibria.

Because any solution which includes an equilibrium remains at that equilibrium for all time, by virtue of the informal argument raised earlier, the equilibria of a single differential equation separate the state space (the set of all possible values of $y(t)$) into disjoint intervals: no solution can cross an equilibrium (or else two solutions pass through that point, violating uniqueness), so any solution must stay within a single interval. Furthermore, since $f(y)$ is continuous, and is zero only at equilibria, $f(y)$ cannot change sign within an interval (between equilibria): it will either be uniformly positive, in which case all solutions in that interval will increase monotonically, or uniformly negative, in which case all solutions in that interval will decrease monotonically. If we consider the graph of $y(t)$ over time, these intervals become horizontal bands on the graph.

PROPERTY 2: MONOTONICITY. The graph of every solution curve of the differential equation $y' = f(y)$ remains in the same band and is either monotone increasing ($y'(t) > 0$) or monotone decreasing ($y'(t) < 0$) for all $t \geq 0$, depending on whether $f(y) > 0$ or $f(y) < 0$, respectively, in the band.

It may be easiest to see these properties at work via an example already familiar to us.

EXAMPLE 1.
Describe the bands for the logistic differential equation

$$\frac{dy}{dt} = ry\left(1 - \frac{y}{K}\right) \tag{4.3}$$

with $r, K > 0$, and find which solutions are increasing, and which decreasing.

<u>Solution:</u> The bands are bounded by equilibria, which are the roots of $f(y) = ry\left(1 - \frac{y}{K}\right)$, namely $y = 0$ and $y = K$. There are therefore three bands to

consider (Figure 4.2). If $y > K$ (Band 1), then $\left(1 - \frac{y}{K}\right) < 0$, and the function $f(y) = ry\left(1 - \frac{y}{K}\right)$ is negative. Therefore, solutions $y(t)$ of (4.3) with $y(0) > K$ are decreasing for all t. If $0 < y < K$ (Band 2), the function $f(y)$ is positive, and therefore solutions $y(t)$ with $0 < y(0) < K$ are increasing for all t. If $y < 0$ (Band 3), then $f(y)$ is negative, and therefore solutions $y(t)$ with $y(0) < 0$ are decreasing for all t. (Compare Figure 4.2 with the direction field in Figure 3.11, where $r = K = 1$.) □

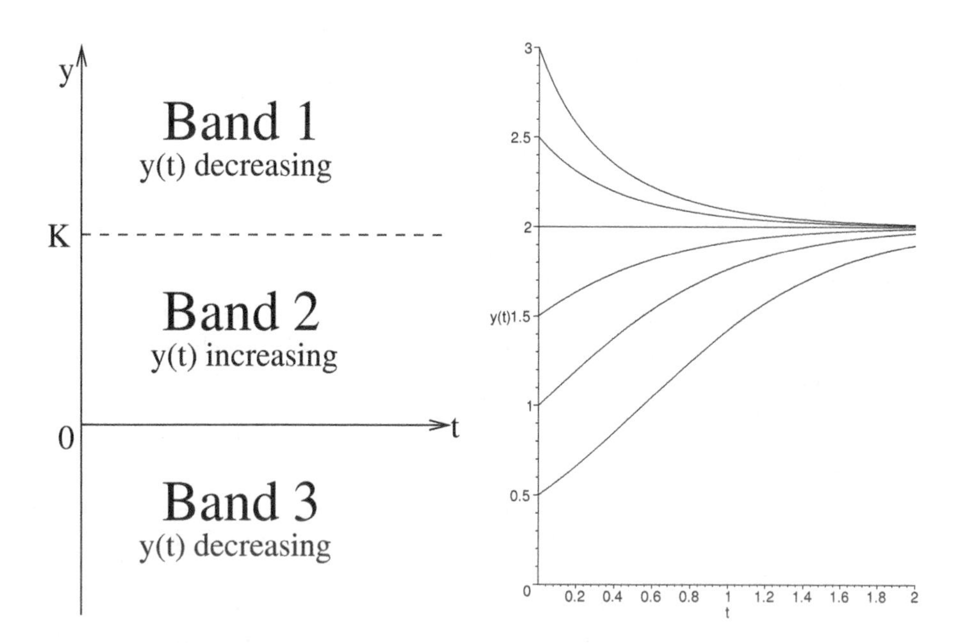

FIGURE 4.2: The $t - y$ plane divided into bands by the equilibria of the logistic equation.

FIGURE 4.3: Some solutions of the logistic equation (4.3) with $r = 2$, $K = 2$.

Note that if $y(0)$ is above the largest equilibrium of (4.2), then the interval or band containing the solution is unbounded, and if $f(y) > 0$ in this band, then the solution $y(t)$ may be (positively) unbounded and does not necessarily exist for all $t \geq 0$. Likewise, if $y(0)$ is below the smallest equilibrium of (4.2) and $f(y) < 0$ in the band containing the solution, then the solution may be unbounded (negatively) and may fail to exist for all $t \geq 0$. In the example above of the logistic equation, solutions with $y(0) < 0$ (Band 3) grow negatively without bound, but nevertheless exist for all $t \geq 0$.

In practice, it is often easy to check for unbounded growth in solutions of a single differential equation, by seeing whether the growth rate $f(y)$ is bounded away from zero for large y. A solution may become unbounded positively (i.e., $y(t) \to +\infty$) if it is monotone increasing when y is large. Models of most biological systems account for limitations on system resources, so that

$f(y) < 0$ for very large y, and in such a case every solution remains bounded, because solutions which become large and positive must be decreasing. If, as is also frequently the case in applications, $y = 0$ is an equilibrium and only non-negative solutions are of interest, then solutions cannot cross the line $y = 0$ or become negatively unbounded (i.e., $y(t) \to -\infty$). In many applications, y stands for a quantity such as the number of members of a population, the mass of a radioactive substance, the quantity of money in an account, or the height of a particle above ground which cannot become negative. In such a situation, only non-negative solutions are significant, and if $y = 0$ is not an equilibrium but a solution reaches the value zero for some finite t, we will consider the population system to have collapsed and the population to be zero for all larger t. If this is the case, we need not be concerned with the possibility of solutions becoming negatively unbounded even if $y = 0$ is not an equilibrium.

In the above case of the logistic equation (4.3), we can see that every non-negative solution remains bounded, by observing that $f(y) < 0$ (and thus $dy/dt < 0$) when $y > K$ (Band 1).

We can now combine the properties of boundedness and monotonicity to argue that any solution of a differential equation (4.2) which does not grow unbounded must instead approach a limit. From calculus, any function which is either bounded above and monotone increasing, or bounded below and monotone decreasing, must approach a limit. (Consider the analogy of a car stuck in forward gear approaching a wall along a narrow lane. At some point—the wall, if not sooner—it must approach a stopping point.) However, it can be shown that if a differentiable, monotone function—a solution of (4.2)—tends to a limit as $t \to \infty$, then its derivative meanwhile must tend to zero. That is, if $y(t) \to L$, then $\frac{dy}{dt} \to \frac{d}{dt}L = 0$. At the same time, by continuity of f, $f(y(t)) \to f(L)$; since $\frac{dy}{dt} = f(y(t))$, this suggests that $f(L) = 0$: that is, that L is a root of $f(y)$. By Property 1, this means that L must be an equilibrium of the system. Thus every bounded solution of (4.2) approaches an equilibrium.

PROPERTY 3: BOUNDED SOLUTIONS. Every solution of the differential equation $\frac{dy}{dt} = f(y)$ which remains bounded for $0 \leq t < \infty$ approaches an equilibrium of the equation as $t \to \infty$.

In the case of the logistic equation (4.3), this means that since all non-negative solutions are bounded, all non-negative solutions must approach an equilibrium, either $y = 0$ or $y = K$. Note, however, that different solutions to the same equation may approach different equilibria. Here it follows from Property 3 and the arguments in Example 1 that (as seen in Section 3.2) all non-negative solutions approach $y = K$, except for the constant solution

$y(t) = 0$, which is already at the zero equilibrium. Figure 4.3 illustrates the behavior with $r = 2$, $K = 2$.

Note also that these last two properties are carefully worded to apply only to single differential equations, and not to systems of several differential equations. We shall see in Chapter 5 that solutions to systems of two or more differential equations need not be monotone, and consequently their bounded solutions need not approach equilibria.

Before continuing on, we invite the reader to consider how these properties apply to the other equation familiar to us from Chapter 3: the first-order linear differential equation $y' = -ay + b$. We will discuss the results in an example later in this section.

4.1.2 Stability

The question of whether or not solutions approach a given equilibrium motivates us to study more formally the concept of *stability*, which has to do with whether solutions approach (and remain near) a given equilibrium. In applications, the initial condition usually comes from observations and is subject to experimental error. For a model to be a plausible predictor of what will actually occur, it is important that a small change in the initial value not produce a large change in the solution. For example, the solution of the logistic differential equation with initial value zero remains zero for all values of t, but every solution with positive initial value, no matter how small, approaches the limit K as $t \to \infty$. Because of its extreme sensitivity to changes in the initial value, we do not ascribe practical significance to the equilibrium zero as a limiting value for solutions of the logistic equation.

An equilibrium y^* of (4.2) is said to be *stable* if every solution with initial value close enough to y^* remains close to it for all time. The equilibrium is said to be *locally asymptotically stable* when every solution with initial value close enough to y^* actually approaches it as $t \to \infty$. The equilibrium is said to be *globally asymptotically stable* when every solution approaches it, regardless of initial condition. If there are solutions which start arbitrarily close to an equilibrium but move away from it, then the equilibrium is said to be *unstable*. For the logistic differential equation (4.3) the equilibrium $y = 0$ is unstable, and we can see from examining Bands 1 and 2 in Example 1 that the equilibrium $y = K$ is locally asymptotically stable; in fact, it is globally asymptotically stable in the positive ($y > 0$) state space. (In asserting the global nature of the stability of $y = K$, we discount the initial condition $y = 0$: in applications, unstable equilibria have no significance, because they can be observed only if the initial condition is "just right.")

Note that the process for determining global asymptotic stability, which requires knowledge of other equilibria, the state space, and boundedness of solutions, is distinct from that for determining local asymptotic stability, which requires only information about what happens very close to one equilibrium point. Therefore, the standard qualitative analysis process we shall now develop involves three steps: (1) identifying all equilibria, (2) performing a local stability analysis on each equilibrium, and (3) assembling the pictures of local behavior into a single (global) portrait of how solutions may behave in general. From Property 1 we already have a methodical way to find equilibria of a differential equation (4.2): they are the zeroes of the function $f(y)$.

There is also a methodical way to see whether or not a given equilibrium is locally asymptotically stable or not, beyond testing the sign of $f(y)$ just above and below the equilibrium (this latter method also does not extend gracefully to systems of more than one equation, where there are many directions in which one can be close to an equilibrium). As was the case with discrete-time models, this way involves the derivative (slope) of the growth function $f(y)$ at the equilibrium.

Let us look at the piece of the graph of $f(y)$ near an equilibrium y^* of (4.2). By Property 1, $f(y^*) = 0$; normally this means that the graph of $f(y)$ crosses the y-axis from one side to the other at y^* (in a few unusual cases, f might turn around and go back the way it came, as in Figure 4.4(c)). We can determine the direction of the crossing from the sign of the derivative at the equilibrium, $f'(y^*)$. If $f'(y^*) < 0$, then $f(y)$ has a negative slope there, so that, close to y^*, $f(y)$ is positive if $y < y^*$ and negative if $y > y^*$ (Figure 4.4(a)). In this case solutions above the equilibrium decrease toward the equilibrium, while solutions below the equilibrium increase toward the equilibrium. This shows that an equilibrium y^* with $f'(y^*) < 0$ is locally asymptotically stable.

By a similar argument, if $f'(y^*) > 0$, solutions above the equilibrium increase and solutions below the equilibrium decrease, with both moving away from the equilibrium (Figure 4.4(b)). In this case we see that the equilibrium is unstable. The only case in which this approach fails to determine an equilibrium's stability is when $f'(y^*) = 0$, one example of which is illustrated in Figure 4.4(c). In such a case (which occurs relatively rarely in practice), both stability and instability are possible, and we must investigate the sign of $f(y)$ above and below the equilibrium. If $f(y) < 0$ above the equilibrium and $f(y) > 0$ below the equilibrium, then the equilibrium is locally asymptotically stable.

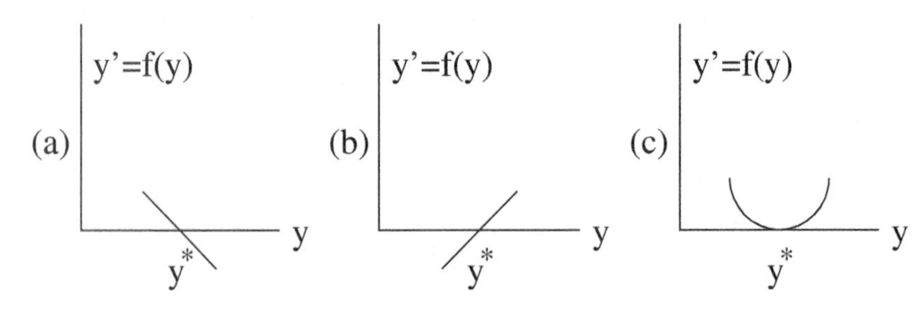

FIGURE 4.4: Equilibria y^* with (a) $f'(y^*) < 0$, (b) $f'(y^*) > 0$, (c) $f'(y^*) = 0$.

We can state this test more formally as follows:

PROPERTY 4: STABILITY. An equilibrium y^* of $y' = f(y)$ with $f'(y^*) < 0$ is (locally) asymptotically stable; an equilibrium y^* with $f'(y^*) > 0$ is unstable.

We now apply this test on a few examples.

EXAMPLE 2.

For each equilibrium of the logistic differential equation (4.3), determine whether it is locally asymptotically stable or unstable.

Solution: We have $f(y) = ry\left(1 - \frac{y}{K}\right)$, $f'(y) = r - \frac{2ry}{K} = r\left(1 - \frac{2y}{K}\right)$. Since $f'(K) = -r < 0$, the equilibrium $y = K$ is asymptotically stable, and since $f'(0) = r > 0$, the equilibrium $y = 0$ is unstable. □

EXAMPLE 3.

Show that the equilibrium $y = 0$ of the differential equation $y' = -y^3$ is locally asymptotically stable.

Solution: $f'(0) = -3(0)^2 = 0$, so we cannot apply Property 4 here. However, for y small and negative we see that $-y^3 > 0$, while for y small and positive $-y^3 < 0$. Thus all solutions approach the limiting value zero, and the equilibrium $y = 0$ is locally asymptotically stable. The behavior of solutions is shown in Figure 4.5. □

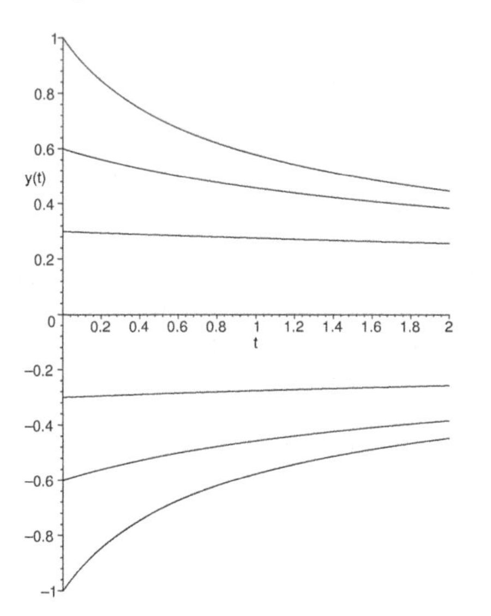

FIGURE 4.5: Some solutions of the equation $y' = -y^3$ near 0.

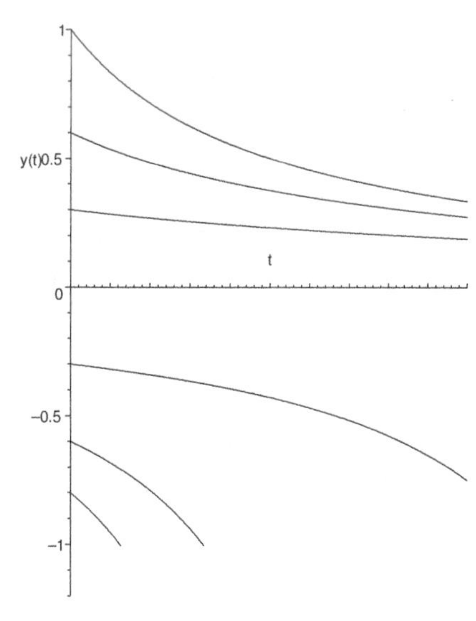

FIGURE 4.6: Some solutions of the equation $y' = -y^2$ near 0.

EXAMPLE 4.
Show that the equilibrium $y = 0$ of the differential equation $y' = -y^2$ is unstable.

Solution: The function $-y^2$ is negative for all y except $y = 0$. Thus solutions of the differential equation $y' = -y^2$ are decreasing. Since solutions with negative initial value move away from $y = 0$, the equilibrium $y = 0$ is unstable. The behavior of solutions is shown in Figure 4.6. Note that for each of the differential equations in Examples 3 and 4 we have $f(0) = 0$, $f'(0) = 0$; no conclusion about stability can be drawn from $f'(0) = 0$ without further examination. □

Property 4 is the principal result of this section. Before stating it formally as a theorem, we shall extend it to consider the special case when the per capita growth rate is written explicitly in the differential equation.

In many applications, the population's overall growth rate is written so as to make the per capita growth rate explicit: that is, the function $f(y)$ has the form $f(y) = yg(y)$, which has the effect of guaranteeing that $y = 0$ is an equilibrium. Also, any nonzero equilibrium y^* must have $g(y^*) = 0$. We calculate $f'(y) = g(y) + yg'(y)$ and note that at the equilibrium $y = 0$, $f'(0) = g(0)$, while at any nonzero equilibrium y^*, $f'(y^*) = y^* g'(y^*)$ (since $g(y^*) = 0$). Thus the equilibrium $y = 0$ is locally asymptotically stable if $g(0) < 0$ and unstable if $g(0) > 0$; a nonzero equilibrium $y^* > 0$ is locally asymptotically stable if $g'(y^*) < 0$ and unstable if $g'(y^*) > 0$.

With the logistic equation (4.3), for instance, the per capita growth rate is $g(y) = r(1 - y/K)$. Thus $g(0) = r$, so that the equilibrium $y = 0$ is unstable, while $g'(y) = -r/K$ for all y, so that the equilibrium $y = K$ is asymptotically stable.

EQUILIBRIUM STABILITY THEOREM: An equilibrium y^* of $y' = f(y)$ with $f'(y^*) < 0$ is asymptotically stable; an equilibrium y^* with $f'(y^*) > 0$ is unstable. The equilibrium $y = 0$ of $y' = yg(y)$ is asymptotically stable if $g(0) < 0$ and unstable if $g(0) > 0$, while a nonzero equilibrium y^* is asymptotically stable if $y^*g'(y^*) < 0$ and unstable if $y^*g'(y^*) > 0$.

We note that this theorem does not cover the case $f'(y^*) = 0$. As we have seen in Examples 3 and 4 above, a situation with $f(y^*) = f'(y^*) = 0$ must be examined individually.

Let (y^*) be an equilibrium of a differential equation (4.2), that is, a solution of the equation $f(y) = 0$. We assume that the equilibrium is *isolated*, that is, that there is an interval around y^* that does not contain any other equilibrium. We shift the origin to the equilibrium by letting $y = y^* + u$, and then use Taylor's theorem to approximate $f(y^* + u)$. Our approximation is

$$f(y^* + u) = f(y^*) + f'(y^*)u + h_1 = f'(y^*)u + h \tag{4.4}$$

where h is a function that is "quadratic" in u in the sense that it is negligible relative to the linear term in (4.4) when u is small (i.e., close to the equilibrium).

The linearization of the system (4.2) at the equilibrium y^* is defined to be the linear equation with constant coefficients obtained by omitting the higher order term h in (4.2).

$$u' = f'(x^*)u. \tag{4.5}$$

If $f'(x^*) < 0$, all solutions of (4.5) approach zero as $t \to \infty$, while if $f'(x^*) > 0$, all solutions of (4.5) grow unbounded as $t \to \infty$. There is a theorem stating that if $f'(x^*) \neq 0$ all solutions of (4.2) starting sufficiently close to x^* behave like the corresponding solution of (4.5). In other words, if $f'(x^*) < 0$, the equilibrium x^* is asymptotically stable and if $f'(x^*) > 0$ the equilibrium x^* is unstable. We have been able to obtain this result (in fact a somewhat stronger result) by the arguments described in this section. However, the arguments used here do not extend to systems of differential equations and in Chapter 5 we will see how to use the idea of the linearization to obtain a criterion for determining the asymptotic stability of an equilibrium of a system.

The final step in the qualitative analysis of a differential equation involves moving from a local to a global perspective, assembling all the information about behaviors near each equilibrium to form a picture of how all possible solutions of the equation behave. As noted earlier in this section, the equilibria of an autonomous differential equation (4.2) separate the state space into distinct intervals, within each of which solutions are either all increasing or all decreasing. After determining the local stability of each equilibrium, one must determine whether the model allows any unbounded solutions. If $f(y)$ is negative for values of y above the largest equilibrium, then no solutions become positively unbounded. If $f(y)$ is positive for values of y above the largest equilibrium, then this equilibrium is unstable, and solutions with initial value above this equilibrium become unbounded. If $f(y)$ is negative for values of y below the smallest equilibrium, then this equilibrium is likewise unstable, and solutions with initial value below this equilibrium become negatively unbounded.

If, within the state space (the set of y values of interest), all solutions approach the same locally asymptotically stable equilibrium, then that equilibrium is said to be *globally asymptotically stable.* If, instead, some solutions grow unbounded, or there are multiple locally stable equilibria, then there is no globally asymptotically stable equilibrium, and the end behavior of the model depends upon initial conditions. In the latter case, accurate measurement of initial conditions is important to the extent that it places solutions on the correct side of the threshold values which separate one end behavior from another, and these threshold values are key to understanding the biological system being studied.

4.1.3 Phase portraits

As is the case with discrete-time models, it sometimes helps to have a graphical approach to complement our algebraic analysis tools. Continuous-time models' behavior can be described pictorially by graphing the state space and marking important points and boundaries, including stability and direction of "flow" of solutions, in what are called *phase portraits.* Since the state space of a single differential equation is one-dimensional, we refer to such a diagram as the equation's *phase line*; however, in later chapters we will also consider *phase planes* for two-dimensional systems. (Phase portraits for systems with more than two dimensions are difficult to produce in their entirety, for obvious reasons.)

A one-dimensional phase portrait consists of the phase line (which may or may not be infinite in both directions, depending on how the state space is defined) with the equilibria marked on it, and an arrow on each interval between equilibria, indicating the direction of change on that interval. To see the relationship between the phase line and the time-derivative dy/dt, it may help to begin by graphing the function $f(y)$ and examining the y-axis. An example is given in Figure 4.7. Note first that the graph of $f(y)$ crosses the axis at each equilibrium. In between equilibria, the graph of $f(y)$ is either always above the axis, corresponding to an increasing $y(t)$, or always below the axis, corresponding to a decreasing $y(t)$. We therefore draw arrows on the y-axis, to the right where the graph is above the axis, and to the left where the graph is below the axis. From the directions of the arrows to either side of an equilibrium we can tell whether the equilibrium is locally asymptotically stable or unstable; we mark stable equilibria with a solid dot and unstable equilibria with a hollow dot. After enough experience with this approach, we no longer need the graph of $f(y)$ superimposed, and can simply draw the phase line itself. (Note that we can also construct a phase line portrait by identifying the equilibria analytically, and then substituting a representative point from each interval to determine the sign of $f(y)$ on that interval.)

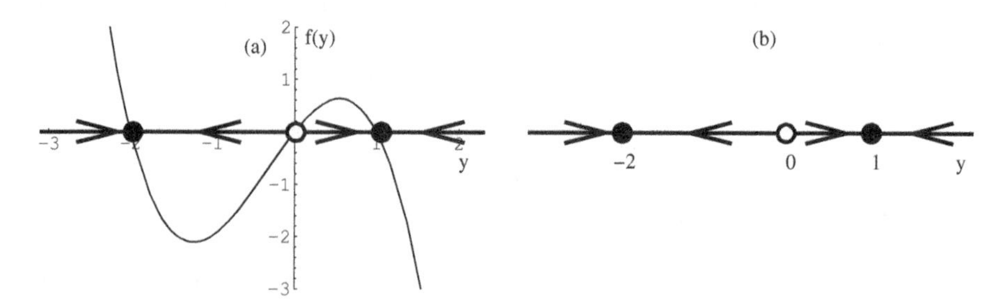

FIGURE 4.7: A phase line (a) superimposed on the corresponding graph of $f(y)$, and (b) by itself.

To read a phase line portrait, we think of the phase line as the state space within which the solution curve $y(t)$ moves, thinking of t as a parameter.

Thus the solution is described by motion along the line, in the direction given by the arrows. The graph of Figure 4.7 describes a situation in which there are asymptotically stable equilibria at $y = -1$ and $y = 2$, and an unstable equilibrium at $y = 0$.

We illustrate this approach with two examples, the first the familiar linear differential equation studied in Chapter 3.

EXAMPLE 5.

Describe the asymptotic behavior of solutions of the differential equation $\frac{dy}{dt} = -ay + b$, analytically and with a phase line.

Solution: Here $f(y) = -ay + b$, $f'(y) = -a$ for all y, and the only equilibrium is $y = \frac{b}{a}$. If $a > 0$, then any equilibrium is asymptotically stable (since $f'(y) = -a < 0$), and there are no unbounded solutions (since $f(y) < 0$ if y is large and positive, $f(y) > 0$ if y is large and negative). This means that every solution is bounded and approaches the limit $\frac{b}{a}$.

If $a < 0$, however, the equilibrium is unstable, as now $f'(y) = -a > 0$ for all y. Further, since $f(y) > 0$ above the equilibrium and $f(y) < 0$ below the equilibrium, every solution grows unbounded, either positively or negatively. Figure 4.8 shows the two corresponding phase lines. □

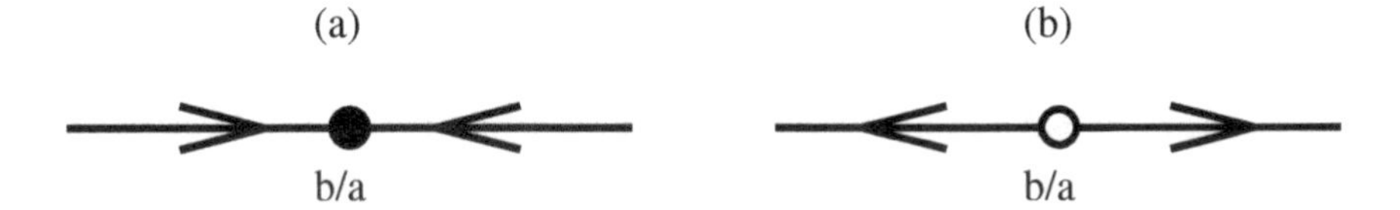

FIGURE 4.8: Phase lines for $\frac{dy}{dt} = -ay + b$ with (a) $a > 0$, (b) $a < 0$.

For our second example we consider a model which has been proposed as an alternative to the logistic model. Recall that the logistic model was originally conceived to account for the effect of resource limitations on population growth, by making the per capita growth rate a decreasing function of population size, positive for small y but negative for large y. Another way to do so is with a per capita birth rate re^{-y} and a constant per capita death rate d, which yields the differential equation

$$\frac{dy}{dt} = y\left(re^{-y} - d\right). \tag{4.6}$$

Since the precise form of the logistic model was derived empirically, and we want model predictions that are robust in the sense of being tied more to the qualitative assumptions than to the specific function(s) used, we can now apply the analysis tools derived in this section to see whether this alternative model predicts the same kind of population growth as the logistic model.

EXAMPLE 6.
Describe the asymptotic behavior of solutions with $y(0) \geq 0$ of the differential equation (4.6), with $r, d > 0$.

Solution: The equilibria of (4.6) are the solutions of $y(re^{-y} - d) = 0$. Thus there are two equilibria, namely $y = 0$ and the solution y^* of $re^{-y} = d$, which is $y^* = \log \frac{r}{d}$. If $r < d$, then $y^* < 0$, and only the equilibrium $y = 0$ is of interest. In this case, $f(y) < 0$ for $y > 0$, and solutions with $y(0) > 0$ decrease to 0 (Figure 4.9). That is, the zero equilibrium is globally asymptotically stable in the non-negative state space.

If instead $r > d$, we define $K = \log \frac{r}{d} > 0$, so that the positive equilibrium is $y = K$, and $e^{-K} = d/r$. We may now rewrite the differential equation (4.6) as $y' = ry\left(e^{-y} - e^{-K}\right)$. For the function $f(y) = ry\left(e^{-y} - e^{-K}\right)$, we have $f'(y) = r\left(e^{-y} - e^{-K}\right) - rye^{-y}$ and $f'(0) = r(1 - e^{-K}) > 0$, implying that the equilibrium $y = 0$ is unstable. The equilibrium $y = K$ is asymptotically stable since $f'(K) = -rKe^{-K} < 0$. All positive solutions are bounded, because $f(y) < 0$ for $y > K$. Thus every solution with $y(0) > 0$ tends to the limit K (which is thus globally asymptotically stable in the non-negative state space), while the solution with initial value zero is the zero function and has limit zero (Figure 4.10).

Alternatively, using the per capita growth rate form of the equilibrium stability theorem, we have a per capita growth rate of $g(y) = re^{-y} - d$, so that $g'(y) = -re^{-y}$. Then $g(0) = r - d$, making the equilibrium $y = 0$ stable ($g(0) < 0$) precisely when $r < d$, while $g'(K) = -re^{-K} = -d < 0$, making the equilibrium $y = K$ stable ($Kg'(K) < 0$) precisely when $r > d$ ($K > 0$). □

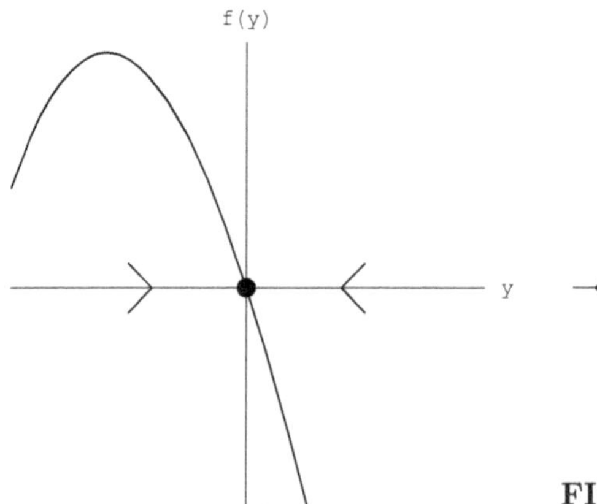

FIGURE 4.9: A phase line and graph of $f(y)$ for (4.6), with $r < d$.

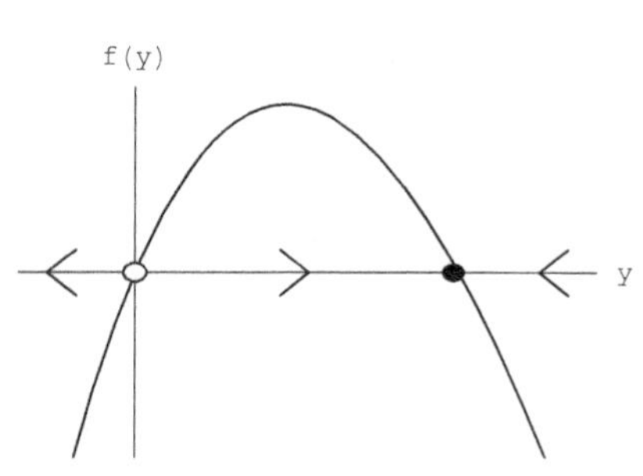

FIGURE 4.10: A phase line and graph of $f(y)$ for (4.6), with $r > d$.

Example 6 shows that solutions of the differential equation (4.6) behave qualitatively in the same manner as solutions of the logistic equation. Some other differential equations which have been proposed as population models and which exhibit the same behavior are suggested in Exercises 9–12 at the end of this section. It turns out that one can show that every population model of the form (4.2) for which the per capita growth rate is a decreasing function of the population size y, and which is positive for $0 < y < K$ and negative for $y > K$, has the property that every solution with $y(0) > 0$ approaches the limit K as $t \to \infty$ (see Exercise 13 below).

4.1.4 Foraging ants and phase transitions

We now return to the model of ant foraging developed by Beekman et al., who wished to investigate the notion that in order for insect collectives to function effectively, the population must be larger than some critical size. In the case of Pharaoh ants, the question was whether the effective use of pheromone trails to guide fellow ants from the same colony to a given food source is dependent upon colony size. Their model, given in equation (4.1), provides a means to determine whether the underlying assumptions imply a dependence on colony size in terms of the number of ants eventually following the pheromone trail.

To apply the qualitative analysis techniques developed in this section, we first write the equilibrium condition $dx/dt = 0$, or

$$f(x) = (\alpha + \beta x)(n - x) - \frac{sx}{x + K} = 0.$$

We see immediately that $x = 0$ is not an equilibrium of this system (since $f(0) = \alpha n > 0$: if no ants are initially on a trail to the given food source, some will nonetheless find the food source and (re)establish the trail). Likewise the number of ants on the trail cannot increase without bound; as we would expect, it is in particular bounded by the colony size n (and note $f(n) < 0$, and $f(x) < 0$ for $x > n$). We can multiply the equation through by $x + K$ to make it polynomial, obtaining the cubic equation

$$\beta x^3 + (\alpha + \beta(K - n))x^2 + (s + \alpha(K - n) - \beta K n)x + (-\alpha K n) = 0 \quad (4.7)$$

(cf. equation 1 in Beekman et al.). Since $f(0) > 0$ and $f(n) < 0$, by continuity of f there must be at least one equilibrium x^* in between; we see from the cubic equilibrium condition that there is in fact either one or three. There is a cubic formula which can be used to find any equilibria, but the resulting expressions have many terms and may be difficult to interpret.

Since the expressions for the equilibria are complicated, the first derivative test given in the equilibrium stability theorem may be difficult to apply in general. Instead, we shall use the phase line, together with the fact that the one or three equilibria are simple zeroes of f, to determine their stability. If there is only one equilibrium, then the phase line is split into two intervals, on the lower of which f is increasing, and on the upper of which f is decreasing

(Figure 4.11(a)). The unique equilibrium is therefore locally asymptotically stable. Since there are no unbounded solutions, the equilibrium is also globally asymptotically stable.

If, however, there are three equilibria, the phase line is comprised of four intervals, with the sign of f alternating from one to the next (+, −, +, −; Figure 4.11(b)). In this case, the first and third equilibria (E_1 and E_3) are both locally asymptotically stable, while the second (E_2) is unstable. Here there is no globally asymptotically stable end state: all solutions with initial value $x(0) < E_2$ therefore approach E_1, while solutions with initial value above E_2 approach E_3. The unstable equilibrium E_2 acts as a critical threshold number of ants using the trail, above which a high proportion of ants uses the trail, and below which the ants' foraging behavior remains disorganized.

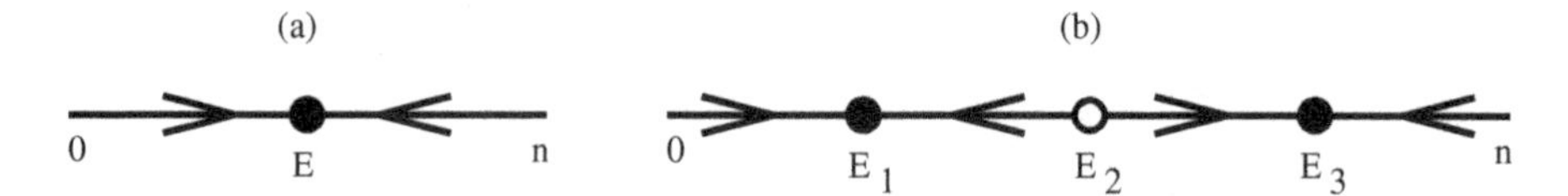

FIGURE 4.11: Phase lines for (4.1) with (a) one equilibrium, (b) three equilibria.

Since the model's behavior differs significantly depending on whether there is one or three equilibria, we can use a quantity called the cubic discriminant Δ to determine which is the case (this can be found in references on solving cubic equations). If $\Delta < 0$, there are three equilibria; if $\Delta > 0$, there is one. For (4.7) the discriminant is

$$\begin{aligned}\Delta =& 4\beta[s+\alpha K-(\alpha+\beta K)n]^3+27\alpha^2\beta^2K^2n^2\\ &-[(\alpha+\beta K)-\beta n]^2[s+\alpha K-(\alpha+\beta K)n]^2-4\alpha Kn[(\alpha+\beta K)-\beta n]^3\\ &+18\alpha\beta Kn[(\alpha+\beta K)-\beta n][s+\alpha K-(\alpha+\beta K)n].\end{aligned}$$

Although this quantity is complicated, we can investigate two specific assertions made in Beekman et al.: first, that the more complicated behavior (with three equilibria) occurs when food sources are difficult for ants to locate without a pheromone trail (i.e., when α is small), and second, that pheromone-based foraging is more efficient in larger colonies. If we set $\alpha = 0$, then $\Delta = 4\beta[s-\beta Kn]^2[s-\beta(K+n)^2/4]$, so that $\Delta < 0$ whenever $n > \sqrt{4s/\beta}-K$. Beekman et al. measured the actual value of α as 0.0052/min. and (in Sumpter and Beekman) used values for the other parameters of β=0.00125/ant-min. (an average of two values used), $s = 1$ ant/min., and $K = 10$ ants. This gives a threshold of about $\sqrt{4s/\beta}-K = \sqrt{4/0.00125}-10 = 57$ ants when $\alpha = 0$, and a threshold of about 628 ants for the measured value of α. The threshold colony size of 57 ants for $\alpha = 0$ is so small that we can verify that the more complicated behavior occurs when food sources are sufficiently difficult to find, and the threshold colony size of 628 ants for the measured value of α agrees with Beekman et al.'s experimental results, which found a marked improvement in foraging organization for colony sizes above 600 ants. We can also observe that if we expand the expression for Δ as a polynomial in n, the lead term is $-\beta^2(\alpha - \beta K)^2 n^4$, implying that for sufficiently large colonies $\Delta < 0$ and the more complicated behavior occurs.

We can also address the hypothesis of more efficient foraging (i.e., a higher proportion of ants using the trail) in larger colonies by dividing the equilibrium condition (4.7) by n^3 in order to rewrite it in terms of the equilibrium proportion $y = x/n$ of ants using the trail:

$$\beta y^3 + \left(\frac{\alpha + \beta K}{n} - \beta\right) y^2 + \left(\frac{s + \alpha K}{n^2} - \frac{\alpha + \beta K}{n}\right) y - \frac{\alpha K}{n^2} = 0.$$

If we let $n \to \infty$, so that $1/n \to 0$, then this equation simplifies to $\beta y^3 - \beta y^2 = 0$, the solutions of which are $y = 0$ (twice) and $y = 1$. Thus as colony size increases, the upper equilibrium approaches 1 (all ants are on the trail), while the middle (unstable threshold) equilibrium approaches 0, making the upper equilibrium nearly globally asymptotically stable (since very few initial conditions will be less than E_2).

Therefore in general this model predicts the existence of a critical threshold (E_2) in pheromone trail usage when food sources are difficult enough to find, and when a colony is large enough (roughly 700 ants or more); in addition, as colony size increases, the initial proportion of the colony's ants required to establish the trail (in order for it to become well used) decreases, and the proportion of the colony's ants who will then use the trail increases toward 1.

The transition from one to three equilibria marks a major change in the global behavior of this model, and we will return to this equation later in this chapter, when we shall study the topic of bifurcations more generally, both mathematically and in terms of their biological consequences.

Exercises

For each of the differential equations in Exercises 1–8, draw the phase line, find all equilibria and describe the behavior of solutions as $t \to \infty$.

1. $y' = y$
2. $y' = -y$
3. $y' = y^3$
4. $y' = \sin y$
5. $y' = y(1-y)(2-y)$
6. $y' = -y(1-y)(2-y)$
7. $y' = y^2(y+1)$
8. $y' = -c(y-1)^2$

For each of the differential equations in Exercises 9–12, describe the behavior of solutions with $y(0) > 0$.

9. $y' = ry \log \frac{K}{y}$

10. $y' = ry\left(-1 + \frac{K}{y}\right)$

11. $y' = \frac{ry(K-y)}{K+Ay}$

12. $y' = ry\left[1 - \left(\frac{y}{K}\right)^{\theta}\right]$

13. * Consider a differential equation $y' = g(y)$ with $g(y) = yh(y)$, where $h(y)$ is a decreasing function of y, so that $h'(y) < 0$, and $h(y) > 0$ for $0 < y < K$, but $h(y) < 0$ for $y > K$. Show that the only equilibria are $y = 0$ and $y = K$, and that the equilibrium $y = 0$ is unstable while the equilibrium $y = K$ is asymptotically stable.

14. * (a) Show that if $y(t)$ is a solution of an autonomous differential equation $y' = g(y)$, then $y''(t) = g'\{y(t)\}\, g\{y(t)\}$. (b) Deduce from part (a) that if the solution $y(t)$ has an inflection point, this inflection point must occur at a point (t, y) with $g'(y) = 0$.

15. In this section, we analyzed the long-term behavior of Beekman et al.'s model for ants following a pheromone trail to a single food source. In what sense is this behavior long-term, i.e., what is the time scale implicit in this discussion? How does this time scale relate to the assumption that the colony size is constant?

16. How would the ant foraging model (4.1) change if

    (a) the ants did not use pheromone trails?

    (b) ants using the trail never left it?

    (c) ants using the trail left it at a constant per capita rate σ?

    In each case, indicate how the resulting model's behavior differs from that of Beekman et al.

17. How would the interpretation of the rate at which ants leave the trail be affected if it were modeled by

$$s\left(\frac{x}{x+K}\right)^{p},$$

    for some number $p > 1$?

18. In practice, why is it advantageous for some ants using the trail to stop using it (before the food source is exhausted)?

4.2 Harvesting

In Section 2.5 we studied harvesting, or population management, in discrete-time models, where both reproduction and harvesting occur at specific discrete times. If the harvesting events are frequent enough, they can be considered to be nearly continuous and ongoing, in the same way that we consider ongoing reproduction and mortality for many biological systems. In this section we will look at the management of populations from a continuous-time perspective, including not only such activities as hunting, fishing and logging, but also production of bacteria in laboratory settings and even the dispersal of human populations.

In our previous study of population harvesting we considered primarily *constant-yield harvesting*, a practice in which members of a population are removed at a constant rate. Constant-yield harvesting models[3] are appropriate for systems in which there is a set quota or deliberate, controlled management of the population. In other situations, however, such as the notorious bycatch of dolphins in the tuna industry in the late 20th century that eventually led to changes in fishing practices, harvesting may be either an incidental byproduct of an activity not directly designed to control that population, or a process with a duration or intensity set by factors other than the population size: for instance, occasional fishing in a small lake or pond, where fishermen go a given number of times, and the number of fish caught depends upon the availability (population density) of fish in the pond. These situations are better modeled by *constant-effort harvesting*, in which members of the population are removed at a rate proportional to the population size.[4] In this section we shall consider biological systems of both types.

4.2.1 Constant-yield harvesting

Population harvesting affects populations of every size, from microscopic organisms to blue whales, the largest animals on Earth. Constant-yield harvesting occurs when a fixed quota is set for removing members of the population, as often occurs in hunting. In general, if a population's natural growth without harvesting is described by the model $y' = f(y)$, then under constant-yield harvesting it is modeled by the differential equation

$$y' = f(y) - H, \tag{4.8}$$

[3]A primary reference is Fred Brauer and David A. Sánchez, Constant rate population harvesting: equilibrium and stability, *Theor. Pop. Biol.* 8(1): 12–30, August 1975.

[4]A primary reference is M.B. Schaefer, Some aspects of the dynamics of populations important to the management of commercial marine fisheries, *Bull. Inter-Amer. Trop. Tuna Comm.* I: 25–56, 1954.

where $H > 0$ is the constant *harvest rate* (yield per time) at which members are removed from the population. (Note that the units of H here are population per time, whereas in discrete-time harvesting models it had simply population units.) We assume that the rate of removal is uniform (not seasonal as in many types of hunting).

Although with a few reasonable assumptions we can prove results describing the behavior of any constant-yield harvesting model (4.8), for illustrative purposes we will assume that the populations under study in this section obey a logistic law, so that $f(y) = ry(1 - y/K)$. Then the population size satisfies a differential equation of the form

$$y' = ry\left(1 - \frac{y}{K}\right) - H, \tag{4.9}$$

where r, K, and H are positive constants. While this equation can be solved explicitly by separation of variables, the integration must be handled by examining three different cases depending on the values of the constants (see Exercise 1). It is simpler (and probably more informative) to carry out a qualitative analysis.

The equilibria of (4.9) are the solutions of $ry - \frac{r}{K}y^2 - H = 0$, or

$$y^2 - Ky + \frac{HK}{r} = 0. \tag{4.10}$$

These are given by

$$y = \frac{1}{2}\left[K \pm \sqrt{K^2 - \frac{4HK}{r}}\right]. \tag{4.11}$$

We must distinguish three cases:

(i) $K^2 - \frac{4HK}{r} > 0$, i.e., $H < \frac{rK}{4}$,

(ii) $K^2 - \frac{4HK}{r} = 0$, i.e., $H = \frac{rK}{4}$,

(iii) $K^2 - \frac{4HK}{r} < 0$, i.e., $H > \frac{rK}{4}$.

The number and nature of the equilibria are different in the three cases.

In case (i) there are two distinct equilibria given by (4.11), and it is clear that one, which we shall call y_1, is smaller than $\frac{K}{2}$ while the other, which we shall call y_2, is larger than $\frac{K}{2}$. If we let $f(y) = ry - \frac{r}{K}y^2 - H$, then $f'(y) = r - \frac{2r}{K}y = \frac{2r}{K}\left(\frac{K}{2} - y\right)$. Thus $f'(y) > 0$ if $y < \frac{K}{2}$, and $f'(y) < 0$ if $y > \frac{K}{2}$. We now see, using the equilibrium stability theorem, that any equilibrium y^* of (4.9) with $y^* > K/2$ is asymptotically stable, while any equilibrium y^* with $y^* < K/2$ is unstable. More specifically, in case (i), we see that y_1 is unstable while y_2 is asymptotically stable.

To consider the global picture for case (i), we note that since $f(y) < 0$ for $0 < y < y_1$, any solution $y(t)$ with $0 < y(0) < y_1$ is monotone decreasing and reaches zero in finite time. When this happens, we consider the population to have been wiped out and the model to have collapsed. If instead $y(0) > y_1$, the solution $y(t)$ approaches the limit y_2 as $t \to \infty$. This is illustrated in Figure 4.12 with $r = 2$, $K = 2$, $H = \frac{5}{9}$.

In case (ii) there is a single equilibrium $\frac{K}{2}$ which is a double root of (4.10), and $f'(\frac{K}{2}) = 0$. The equilibrium stability theorem does not apply here, but if we rewrite $f(y)$ as

$$f(y) = ry - \frac{r}{K}y^2 - \frac{rK}{4} = -\left(\frac{r}{K}y^2 - ry + \frac{rK}{4}\right) = -\frac{r}{K}\left(y - \frac{K}{2}\right)^2,$$

we see that $f(y) < 0$ if $y \neq \frac{K}{2}$. Thus every solution is monotone decreasing (except the equilibrium solution), and solutions starting above $\frac{K}{2}$ approach the equilibrium $\frac{K}{2}$, while solutions starting below $\frac{K}{2}$ reach zero in finite time. This behavior is illustrated in Figure 4.13 with $r = 2$, $K = 2$, $H = 1$.

In case (iii) the roots of (4.10) are complex, so (4.9) has no equilibria. Since $f(y)$ has no real zeroes and $f(0) = -H < 0$, $f(y) < 0$ for all y. Thus every solution is monotone decreasing and reaches zero in finite time; the population is wiped out no matter what the initial population size is, as shown in Figure 4.14 with $r = 2$, $K = 2$, $H = 1.5$.

Phase portraits for all three cases are shown in Figure 4.15.

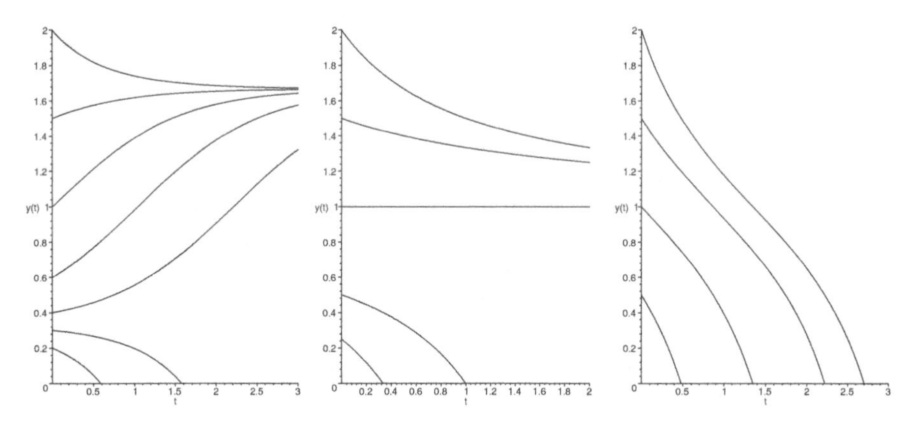

FIGURE 4.12: Solutions to (4.9), with $H < rK/4$.

FIGURE 4.13: Solutions to (4.9), with $H = rK/4$.

FIGURE 4.14: Solutions to (4.9), with $H > rK/4$.

If we now compare the three cases, we can understand the overall effect of constant-yield harvesting on a population with logistic growth. Recall that with no harvesting ($H = 0$) the population approaches the carrying capacity K regardless of its initial size. As H increases from 0 to $\frac{rK}{4}$, the limiting population size y_2 decreases monotonically from K to $K/2$, while at the same time the minimum population size y_1 required for survival increases from 0 toward $K/2$. That is, low harvest rates affect the population in two ways:

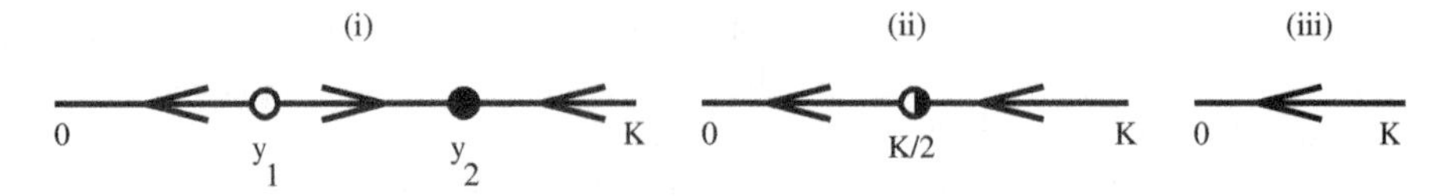

FIGURE 4.15: Phase lines for (4.9) in the three cases given in the text. Note the "semi-stable" equilibrium in case (ii).

by lowering its long-term size and by requiring a minimum size for survival. As the harvest rate increases, the long-term size and minimum size approach each other, narrowing the window for survival. When the harvest rate reaches the critical value $H_c = \frac{rK}{4}$, these two values meet at $K/2$, half the original carrying capacity.

However, for harvest rates beyond the critical value ($H > \frac{rK}{4}$), the population is wiped out in a finite amount of time. Thus the limiting population size decreases continuously (from K to $\frac{K}{2}$) as H increases from 0 to $rK/4$, and then drops to zero as H passes through the critical value $rK/4$. Such a discontinuity in the limiting behavior of a system is called a (mathematical) *catastrophe*. The corresponding biological catastrophe is the extinction of the population.

The implications of this model for the management of populations under a quota system are entailed in the critical harvest rate $rK/4$. In cases such as fishing and logging where harvesting is tied to industry, the economic goal of maximizing the harvest rate must be tempered by an awareness of the fragility created by harvesting. In this case, the harvest rate must be held sufficiently far below the critical rate to ensure that occasional or seasonal fluctuations in reproduction or mortality do not cause the population to drop below the minimum survival size.

As suggested at the beginning of this section, the logistic growth model is representative of a wide range of population growth functions. It turns out that for any general population model of the form $y' = f(y)$ whose per capita growth rate is a monotone decreasing function of population size, and positive for small y but negative for large y,[5] the response to constant-yield harvesting is similar. (The per capita growth rate in the logistic model, $r(1 - y/K)$, fits this description.) For such models, with initial conditions close to the natural carrying capacity, we can show (see Exercise 3d for a special case) that the limiting population size is positive when H is less than a critical harvest rate H_c, but drops to zero as H passes H_c. We observed similar behavior for discrete models in Section 2.5, but for discrete models the situation is complicated by the possibility that all equilibria may be unstable.

Note that for the logistic model, the critical harvesting rate $H_c = rK/4$ is the maximum value of the function $f(y) = ry(1 - y/K)$. This is not a

[5]Note that these conditions are similar to our definition of scramble competition for discrete-time models in Chapter 2.

Photo courtesy Wisconsin Dept. of Natural Resources

FIGURE 4.16: A white-tailed buck.

Wisconsin Deer Population, 1960-2003

Prehunt Population
Posthunt Population
Posthunt Goal

Number of Deer

2,000,000
1,800,000
1,600,000
1,400,000
1,200,000
1,000,000
800,000
600,000
400,000
200,000
0

1960 1965 1970 1975 1980 1985 1990 1995 2000

Year

Graph courtesy Wisconsin Dept. of Natural Resources

FIGURE 4.17: Wisconsin white-tailed deer population.

coincidence; in general $H_c = \max_y f(y)$, as can be seen graphically (see again Exercise 3d).

We will now consider some case studies to which our constant-yield harvesting model can be applied.

Example 1.
Wisconsin is one of many states in the U.S. where human expansion has removed most of the predators of the deer native to the area, with the result that the deer population grows unchecked, until deer frequently appear on roads and in yards, causing accidents and damage to gardens and shrubs. As a result, the Wisconsin Department of Natural Resources (WDNR) issues hunting permits for a fixed number of deer to be removed each hunting season, in order to manage the burgeoning population.

The Wisconsin Department of Natural Resources (WDNR) first established population goals for white-tailed deer (Figure 4.16) in 1962. As deer range expanded and hunting interest increased, the goal has grown, to the point that in the early 21st century it stands at about 700,000. As can be seen in Figure 4.17, the deer population has continued to grow steadily nonetheless; beginning in the mid-1990s, winter populations have been about 50% over goal. In the 2006–2007 season, just over 500,000 deer were hunted, but by the fall of 2007, the population had reached about 1.7 million deer, in keeping with the reproductive rate of about 0.5/yr which can be estimated from the graph.

(a) What is the minimum equilibrium population value we can infer from the continued growth shown in the graph together with these values for H and r?

(b) If hunting half a million deer per year eventually keeps the population around 3 million deer, how many deer would there be with no hunting? (This question requires many unrealistic simplifying assumptions, such as keeping the deer within a fixed habitat.)

(c) Using this value of K, what are the critical harvesting rate and resulting equilibrium level?

(d) Using this value of K, what harvesting rate H would meet the WDNR goal of 700,000 deer?

Solution:

(a) From the graph, we can see that the population keeps growing even with 1/2 million deer harvested per year. Therefore, harvesting is presently below the critical rate, that is, $H < rK/4$. Solving for K, we have $K > 4H/r = 4(0.5)/0.5 = 4$ million deer as a minimal carrying capacity. From (4.11), the stable equilibrium must be greater than $K/2 = 2$ million deer.

(b) Solving the equilibrium condition (4.11) for K, we get (after some algebra)

$$K = \frac{y^{*2}}{y^* - \frac{H}{r}}.$$

Substituting $r = 0.5/\text{yr}$, $H = 0.5\text{M deer/yr}$, $y^* = 3\text{M}$ deer yields a value of $K = 4.5\text{M}$ deer.

(c) $H_c = rK/4 = 0.5625\text{M deer/yr}$, and the equilibrium there is $K/2 = 2.25\text{M}$ deer.

(d) As mentioned in the answer to (a), the constant-yield harvesting model (4.9) cannot give a stable equilibrium value lower than $K/2$. Of course, our estimates for r and y^* are very rough, but in general the true carrying capacity must be less than twice the goal, in order for the goal to be reachable through constant-yield harvesting. Since present pre-hunt populations are more than twice the present population goal, the goal can only be reached with variable yield harvesting: that is, harvesting that depends upon the population size. WDNR continues to adjust their harvest quotas annually; we will consider one type of population-dependent harvesting later in this section. □

The great whales once numbered in the millions, but whaling emerged as a major industry beginning in the eighteenth century, hunting whales of several types for their oil and other products, and by the late twentieth century many species had been reduced to just a few thousand members each (see Table 4.1[6]). As a result, in 1986 the International Whaling Commission (IWC) implemented a moratorium on commercial whale hunting, and began a careful study of the different species' ability to recover, while setting guidelines for

[6] Data from G. A. Knox, The key role of krill in the ecosystem of the southern ocean with special reference to the convention on the conservation of Antarctic marine living resources, *Ocean Management* 9: 113–156, 1984.

TABLE 4.1: Population data (in thousands) for five species of whales prior to the 1986 IWC moratorium, from Knox, 1984.

Species \ Year	1900	1910	1920	1930	1940	1950	1960	1970	1980
Fin	400	380	360	320	275	220	140	90	80
Blue	200	185	150	95	30	20	10	7	3
Humpback	100	95	70	50	35	20	10	5	3
Sei	95	90	87	83	80	76	70	50	20
Minke	25	25	25	30	34	38	65	96	100

future whale hunting quotas. At present, no commercial whaling is allowed (aboriginal groups such as those in Alaska continue their traditional subsistence whale hunting, but under fixed quotas); this ban is to remain in effect until species have recovered to at least 54% of their pre-whaling populations, which will probably take at least 50–100 years. Norway and Japan continue to harvest whales, Norway by declaring its own quotas for commercial whaling in exception of the IWC moratorium, and Japan by using a loophole in the regulation which allows limited whale hunting for scientific research purposes. Both countries hunt minke whales, now one of the most numerous species, Norway harvesting an average of 577 per year from 1997 to 2006 and Japan about 300 per year.

Biologists worldwide are hoping the damage done can be reversed. Complex chain reactions are already occurring as a result of the overharvesting of whales: Over-harvesting of great whales (1946–1979), a major food source for killer whales, forced killer whales to hunt smaller marine mammals: harbor seals (1970s), fur seals (mid-70s to mid-80s), sea lions (1980s–1990s), and finally sea otters (1990s–present). Sea urchins, a major food source for sea otters, then grew unchecked, decimating the forests of kelp eaten by the sea urchins.

EXAMPLE 2.

The International Whaling Commission decided in 1965 to protect blue whales, whose population decreased to fewer than 2000 before the moratorium was enacted, in spite of an estimated carrying capacity of 200,000 whales and intrinsic growth rate of 0.05/yr, as a result of annual catches averaging 13,000 whales per year over the period 1926–1940. (a) Assuming that blue whales obey a logistic law, estimate the maximum number of whales that can be caught per year without wiping out the population. (b) The IWC estimated a population of 2,300 blue whales (excluding pygmy blues) in the southern hemisphere in 1997–1998, and a growth rate of 0.082/yr. If we make the simplistic assumption that the southern hemisphere's carrying capacity for blue whales is half that for the entire planet, how long will it take before the population reaches the IWC's goal of 54% of carrying capacity using their figures?

Solution: (a) The critical harvest rate is $rK/4 = (0.05)200,000/4 = 2500$ whales/yr, less than one fifth the historical catch rate cited.

(b) With no harvesting, we assume the population obeys a simple logistic law, for which equation (2.13) gives the solution as a function of time. Solving this equation for time yields

$$t_f = \frac{1}{r} \ln \left[\frac{K - y_0}{y_0} \left(\frac{K}{y(t_f)} - 1 \right) \right].$$

Substituting $r = 0.082/\text{yr}$, $K = 100,000$ whales, $y_0 = 2300$ whales, and $y(t_f) = 0.54K$ gives $t_f \approx 48$ years. Under this prediction, southern hemisphere blue whales could reach IWC goal levels by 2046. □

EXAMPLE 3.

Another species of whale hunted to near-extinction in the early twentieth century is the right whale,whose post-moratorium fate in the two hemispheres has been quite different. Southern right whales off the coast of Argentina numbered about 7500 at the end of the twentieth century, according to the IWC, and reproduction rates of approximately 0.07/yr have been cited. They numbered about 100,000 a century earlier. However, the North Atlantic right whale is now extinct in the eastern Atlantic, and numbered about 300 in the western Atlantic as of the end of the twentieth century. In the first fifteen years of the IWC moratorium, their mortality rate increased more than fivefold, although they too numbered about 100,000 a century before. This population has been observed to be declining at a rate of 0.024/yr.[7] (a) If some southern right whales were to be transported somehow to bolster the northern population, what removal rate could the present southern population sustain and still make progress toward carrying capacity? (b) If we suppose the northern right whales to be in exponential decline at the given rate (a logistic law makes no sense in this case) and add whales from the southern hemisphere at the rate identified in (a), what will the expected equilibrium value be?

Solution: (a) In order for removal not to threaten the southern population, the present population level must be above the lower equilibrium y_1 in (4.11). If we set $y_1 = 7500$ whales, $r = 0.07/\text{yr}$, and $K = 100,000$ whales and solve for H, we get $H = 485$ whales/yr.

(b) Exponential decline with importation at a constant rate yields the mixing equation $y' = -ay + b$ studied in Section 3.5, the equilibrium value for which is b/a. Taking $b = 485$ whales/yr and $a = 0.024/\text{yr}$ yields $b/a = 20,200$ whales, a large enough population that one might hope its decline could be reversed. □

The plan suggested in the above example is overly simplistic for many reasons but illustrates the strategy of importing members of related populations to save endangered populations that has been used successfully with some

[7] Hal Caswell, Masami Fujiwara, and Solange Brault, Declining survival probability threatens the north Atlantic right whale, *Proceedings of the National Academy of Sciences USA* 96(6): 3308–3313, March 16, 1999.

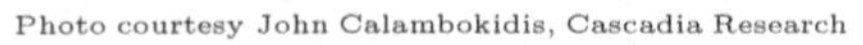

FIGURE 4.18: An aerial view of two blue whales.

Photo courtesy NOAA

FIGURE 4.19: A right whale.

other species, especially birds. The failure of some species, like the northern right whale, to recover following the cessation of decades of overharvesting is not yet understood, and remains the subject of intense study. One possibility is a lack of sufficient diversity (genetic and otherwise) in small populations, but there are many other, behavioral aspects of these populations' lives that have been completely disrupted and are not well-known. Some dolphin species are suffering a similar problem, as we shall see later in this section. A few other whale populations are considered in the exercises at the end of this section.

Logging is another major industry, producing both wood and paper for uses around the world. This type of harvesting has great impact on the world's forests. For example, clear-cutting—a technique in which all the trees in a given area are cut down, leaving bare ground (see Figure 4.20)—has destroyed many primary (old-growth) forest habitats. In population terms, clear-cutting has the effect of reducing the area of wooded land and thus the carrying capacity of the system. However, in some places lumber and paper companies replant trees on the lands where they cut, so that, with many such areas in a given region, each in a different stage of regrowth, the total wooded area remains constant, even though the harvesting is ongoing. In the United States, conservation efforts are holding the amount of forested land roughly constant. About one third of the U.S., or 303 million hectares,[8] is forested (including Alaska). Of this, just over one third (104.2 million hectares) is classified as primary forest, the most biodiverse form of forest. In general, the world's temperate forests (including those of the U.S. and Canada) are being roughly conserved. However, the world's tropical forests, almost all of them in developing countries, are disappearing at a rate of 8% per decade, while global wood-consumption is growing at 26% per decade.

Logging yields depend on the intensity of forest management. Intensive management (used only in Europe as of the end of the twentieth century)

[8]A hectare is 10,000 m^2.

Photo by Steve Hillebrand, courtesy U.S. FWS

FIGURE 4.20: This photo of a clear-cut hill in Oregon was taken in a study of habitat loss for the northern spotted owl.

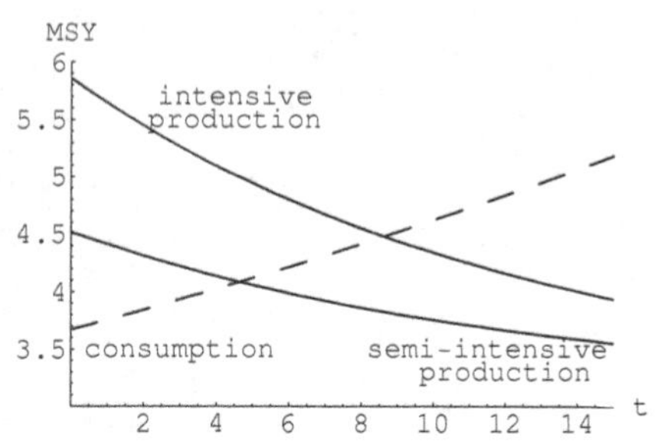

FIGURE 4.21: Graphs of estimated global wood production vs. consumption, in $10^9 m^3/yr$ vs. years since 1984, in Example 4.

TABLE 4.2: Estimated maximum sustainable stemwood yield for the world's forests as of 1984, assuming intensive management and, for tropical forests, a 75-year rotation cycle.

Forest type	Area ($10^6 km^2$)	Productivity ($m^3/km^2/yr$)	MSY ($10^9 m^3/yr$)
Cold temperate closed	10.56	180	1.91
Cold temperate open	3.09	27	0.09
Moderate temperate closed	3.88	280	1.09
Moderate temperate open	1.73	42	0.08
Tropical closed	11.9	207	2.46
Tropical open	7.3	31	0.23
Total	38.46	152	5.86

involves gradual complete replacement of all trees in a given region on a fixed cycle under high-quality supervision, and implies the eventual replacement of nearly all old growth. Non-intensive management may preserve some old growth but generally gives sustainable yields about half that of intensive management. Yields per area also depend on the density of trees: closed forests have dense woods, with trees as close together as natural growth allows, while open forests are sparser. Optimal harvest-cycles are 3–4 decades for wood pulp and firewood, and 9–12 decades for sawtimber. Shorter harvest cycles reduce sustainable wood-yield. Table 4.2 gives one published estimate for maximum sustainable yields of the world's forests as of 1984. The *maximum sustainable yield* (MSY) is the maximum harvest rate which can be sustained without driving the population in question to extinction. In the table, temperate forests are divided into two classes, as those in harsher climates (Canada and the former USSR) have historically had lower yields.

EXAMPLE 4.

If we make the wildly unrealistic assumption that all the world's forests began intensive management in 1984, and take the conservation rates and consumption growth rates given in the text, along with an estimated 1984 global wood consumption rate of 3.67×10^9 m^3/yr, in what year would simple exponential models predict consumption to outpace production? What if we assume intensive management of only temperate forests?

Solution: An exponential model for production under the intensive maximum sustainable yield levels in Table 4.2 would give

$$MSY(t) = 3.17 + 2.69(0.92)^t$$

where t is measured in years since 1984 and MSY in $10^9 m^3/yr$; the first (constant) component comes from temperate woods and the second (exponential) from tropical woods, shrinking at the rate of 8% per year. Consumption, meanwhile, can be calculated in the same units as $3.67(1.26)^{t/10}$ (growth at the rate of 26% per decade). Although we cannot set these two expressions equal and solve for t in closed form, by graphing both quantities as functions of time we can easily find the point at which they intersect, which occurs for $t = 8.63$, that is, in 1992 or 1993.

If we instead assume intensive management of only temperate forests, production is $MSY(t) = 3.17 + 1.35(0.92)^t$. In this case, consumption outpaces production at $t = 4.65$, i.e., in 1988 or 1989. Both production curves are graphed vs. consumption in Figure 4.21. Of course, without intensive management consumption is already beyond production at $t = 0$. □

EXAMPLE 5.

Brazil is home to one of the largest tropical forests in the world. 57%—or about 477.7 million hectares (Mha)—of Brazil was forested as of 2005. Between 1990 and 2005, Brazil lost 8.1% of its forest cover, or around 42.33 million hectares. This land was largely clear-cut for agricultural or other non-logging reasons, so that the habitat is effectively lost. Suppose, however, that in 2005 developed countries began providing subsidies to protect and restore the remaining forest land. If individuals, meanwhile, continued clear-cutting at the same rate (by 1980, over 72% of Brazil's forest clearing was due to ranching), what effective inherent growth rate r would be necessary in order for replanting to save the forests?

Solution: The average harvesting rate is $H = 42.33$Mha/15yr= 2.822 Mha/yr. In order for the constant-yield harvesting model to have an equilibrium, we must have $H < rK/4$, from which $r > 4H/K = 4(2.822)/477.7 = 0.02363$/yr. At this rate, however, the equilibrium forest area is only half its 2005 size. This solution also ignores issues of old (primary) vs. new growth. □

EXAMPLE 6.
A less controversial example of harvesting is the use of laboratory bacteria cultures to provide a source of bacteria for use in experiments. If a given culture of bacteria doubles in 12 hours in a nutrient dish capable of sustaining 10^6 bacteria, what is the maximum sustained rate at which bacteria can be removed from the dish? How would the answer change if we assume simple exponential growth rather than logistic?

Solution: The inherent reproduction rate r can be written either as $(\ln 2)/12hr = 0.058/hr$ or as $(\ln 2)/0.5day = 1.39/day$, using the methods of Section 3.1. The MSY under the logistic constant-yield harvesting model is $rK/4 = (1.39)10^6/4 = 347,000$ bacteria/day. □

An exponential constant-yield harvesting model would have

$$y' = ry - H,$$

whose only equilibrium is H/r. However, stability analysis of this equilibrium—with $f(y) = ry - H$, $f'(y) = r > 0$—shows that it is unstable. This means that, for a given harvest rate H, if the initial condition is greater than H/r, the population will grow unbounded (just as in Malthus's unharvested exponential growth model), while if the initial condition is less than H/r, the harvesting will cause the population to crash (reach zero in finite time). In either case, the model becomes unrealistic at some point in time, just as the Malthus model does. Also, there is no maximum harvesting rate, if one is willing to wait long enough before beginning harvesting. If we define harvesting to begin at time zero, then we must have $H < ry(0)$. In an environment with limited resources, of course, the initial condition is limited by the environment's carrying capacity, but the exponential model cannot reflect this limitation. This criterion is identical to that for the logistic harvesting model if we take $y(0) = K/4$, at which point (in the logistic model) growth has already been reduced to $\frac{3}{4}r$. These ambiguities are why we do not use the above exponential harvesting model.

4.2.2 Constant-effort harvesting

In *constant-effort harvesting*, members are removed from a population by a process dependent upon both effort E (a rate or frequency with units of 1/time) and the population size y, under the assumption that the larger the population, the more often members of it will be removed through this process. For example, if a fisherman goes fishing on a private pond a certain number of times per year, and each trip lasts the same length of time, the number of fish he catches each year may be taken as dependent upon the pond's population, rather than upon his effort, which we assume to be fixed. Alternatively, if a certain type of animal eats a farmer's crop, the farmer will attempt to kill or remove the animal when he sees it in his fields; he does not set out with the idea of finding a certain number of animals, but the more of them there are in his fields, the more often he will find them. The key distinction from constant-yield harvesting is that here the process is not tied to a harvesting quota; instead, the frequency of encounters is assumed to be proportional to the population size (an idea central to the notion of *contact processes*, which we shall investigate in the next section). For a population governed by the model $y' = f(y)$, constant-effort harvesting produces the equation $y' = f(y) - Ey$.

Once again we shall consider the specific case of a population governed by a logistic law. In this case the model becomes

$$y' = ry\left(1 - \frac{y}{K}\right) - Ey. \tag{4.12}$$

Clearly $y = 0$ is one equilibrium of this equation; the other we calculate to be

$$y_H^* = K\left(1 - \frac{E}{r}\right). \tag{4.13}$$

Note that this second equilibrium is only biologically meaningful (i.e., positive) if $E < r$, that is, if harvesting events occur less frequently than reproduction. If instead $E > r$, then the zero equilibrium is the only equilibrium.

To determine local asymptotic stability of the equilibria, we compute the derivative of the right-hand side of (4.12) and use the criterion

$$r\left(1 - \frac{2y^*}{K}\right) - E < 0,$$

which simplifies to

$$y^* > \frac{1}{2}K\left(1 - \frac{E}{r}\right) = \frac{1}{2}y_H^*.$$

If $0 \leq E < r$, then this criterion is clearly true for the positive equilibrium y_H^* and false for the zero equilibrium. If, however, $E > r$, then $y_H^* < 0$, so that the stability criterion is true for the zero equilibrium (and false for the negative equilibrium y_H^*). In both cases, the local asymptotic stability extends to global asymptotic stability in the non-negative state space, since there are no unbounded solutions ($f(y)$ is negative for large y). Figure 4.22 gives phase line diagrams for both cases.

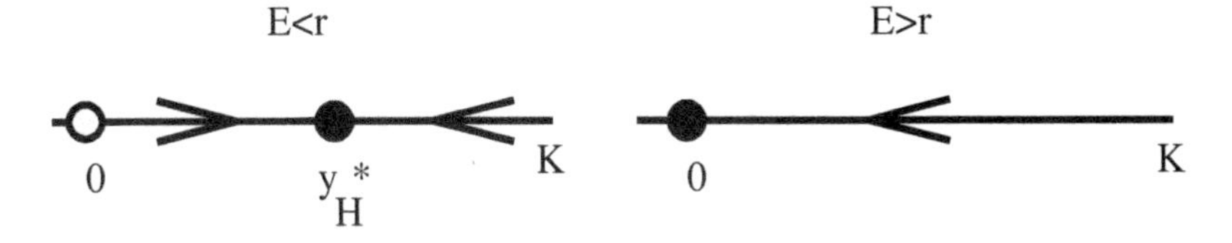

FIGURE 4.22: Phase lines for (4.12) in the two cases given in the text.

Constant-effort harvesting therefore reduces the size of a logistic population by a proportion E/r of the carrying capacity, unless $E \geq r$, in which case it gradually drives the population extinct. The extinction in the latter case is gradual rather than sudden (as with constant-yield harvesting) because as the population becomes scarcer, its members become harder to find with the same level of effort. (One might argue that when the population becomes very small, its members may become too difficult to find to guarantee a constant yield. Exercise 15a at the end of this section suggests a way to address this scenario.)

We may wish to compare the maximum sustainable yield (MSY) for constant-effort harvesting with that obtained for constant-yield harvesting ($rK/4$ for a population with logistic growth). The maximum sustainable effort is a little below r, but recall that the actual harvest rate is Ey rather than E. We therefore find the maximum of Ey_H^* over all possible efforts:

$$MSY = \max_E \, Ey_H^* = \max_E \, E\,K\left(1 - \frac{E}{r}\right) = \max_E \, K\left(E - \frac{E^2}{r}\right). \tag{4.14}$$

To find the maximum of this last expression with respect to E, we set its derivative to zero:

$$K\left(1 - \frac{2E}{r}\right) = 0 \Leftrightarrow E = \frac{r}{2}.$$

Since the MSY is a downward-facing quadratic in E (to see this, just graph it vs. E), we know that this unique critical point is indeed a maximum. If we substitute $E = r/2$ into (4.14), we obtain

$$MSY = K\left(\frac{r}{2} - \frac{(r/2)^2}{r}\right) = K\left(\frac{r}{2} - \frac{r}{4}\right) = \frac{rK}{4}.$$

The reader who is surprised to find the MSY for constant-effort harvesting to be the same as that for constant-yield harvesting should note that, given a constant effort E, the population y will eventually settle down to the corresponding equilibrium y_H^*, so that the product Ey_H^* which represents the yield is asymptotically constant as well. The difference in practice is that constant-yield harvesting is often managed to obtain the MSY, while constant-effort harvesting typically describes scenarios in which the effort is fixed by other considerations than the MSY.

We now consider two applications of constant-effort harvesting to ecology.

A striking example of incidental harvesting that became notorious in the late twentieth century was the bycatch of dolphins in the commercial tuna fishing industry. In the late 1950s, the method of purse-seine fishing was developed to catch yellowfin tuna (*Thunnus albacares*) associated with schools of dolphin in the eastern tropical Pacific Ocean. Although this method was intended to allow dolphins to escape unharmed while trapping the tuna, over the next four decades this approach resulted in the deaths of over six million dolphins, the highest known for any marine harvest. In comparison, the total

Photo courtesy NOAA

FIGURE 4.23: Northeastern offshore spotted dolphins, *Stenella attenuata*.

Photo courtesy NOAA

FIGURE 4.24: Eastern spinner dolphins, *Stenella longirostris orientalis*.

number of whales of all species killed commercially during the entire twentieth century is only two million. The two primary species of dolphins affected are the offshore pantropical spotted dolphin, *Stenella attenuata* (Figure 4.23), and the eastern spinner dolphin, *Stenella longirostris orientalis* (Figure 4.24). An informative and very readable reference on this issue is the report by Tim Gerrodette,[9] from which Figures 4.25–4.28 are taken.

In purse-seine fishing, a net is set around schools of tuna, often (about half the time) associated with schools of dolphins (the reasons the two species swim together are not well understood).[10] Once the net is set around the dolphins and tuna (Figure 4.25), the net is closed, and a "backdown procedure" initiated, in which one end of the net is held slightly underwater, so that the dolphins may escape close to the surface (Figure 4.26). Most of the time, the dolphins do escape (meaning that any given dolphin may be chased, netted and freed many times during its lifetime), but sometimes things go wrong, and the scale and frequency of these operations led over years to the millions of dolphin deaths cited above, reducing the northeastern spotted and eastern spinner dolphin stocks to 20% and 30%, respectively, of their population sizes when the fishery began, as illustrated by the graphs in Figures 4.27 and 4.28. Early 21st-century population estimates for these spotted and spinner dolphin stocks are 640,000 and 450,000, respectively.

What is still not understood about this situation is the failure of these two populations to recover following the halting of this massive harvesting. In

[9]Tim Gerrodette, The tuna-dolphin issue, in W.F. Perrin, B. Würsig and J.G.M. Thewissen, eds. *Encyclopedia of Marine Mammals*. Academic Press, San Diego, 2002. pp. 1269–1273. Updated 2005 at `http://swfsc.noaa.gov/textblock.aspx?Division=PRD&ParentMenuId=228&id=1408`.

[10]One study suggested that the association may be related to reducing predation risk in zones where low oxygen content in the water compresses the tuna habitat into a shallow [oxygen-rich] region near the surface: M.D. Scott, S.J. Chivers, R.J. Olson, P.C. Fiedler, K. Holland, Pelagic predator associations: tuna and dolphins in the eastern tropical Pacific Ocean, *Marine Ecology Progress Series* 458: 283–302, 2012.

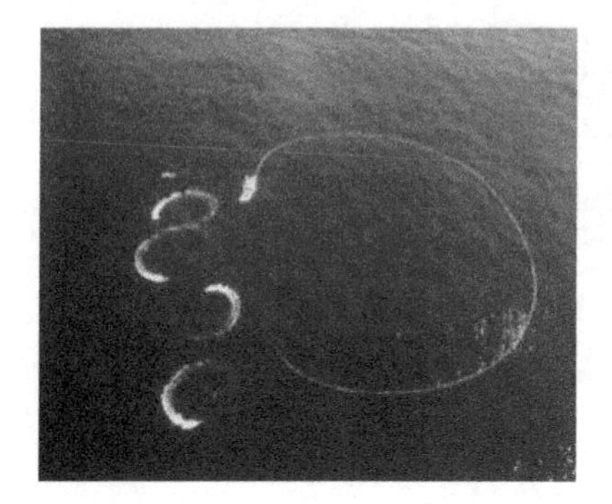

Photo courtesy NOAA SWFSC

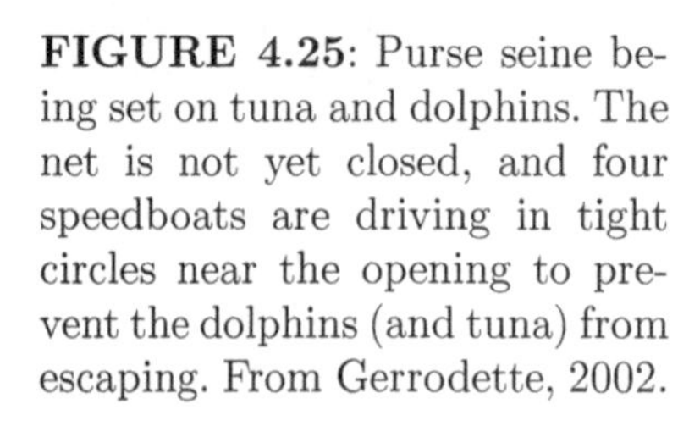

FIGURE 4.25: Purse seine being set on tuna and dolphins. The net is not yet closed, and four speedboats are driving in tight circles near the opening to prevent the dolphins (and tuna) from escaping. From Gerrodette, 2002.

Photo courtesy NOAA SWFSC

FIGURE 4.26: Backdown procedure in progress. As the tuna vessel moves backward, the net is drawn into a long channel. The corkline at the far end is pulled under water slightly, and the dolphins escape. Speedboats along the corkline help keep the net open. From Gerrodette, 2002.

the 1970s the United States passed laws that required American fishing ships to reduce their dolphin bycatch to "insignificant levels approaching zero." As the fleets scaled down, however, tuna fleets from Latin American countries grew, and so it was only through an international effort, including pressure from U.S. consumers, who continued to buy most of the world's commercially processed tuna, that by the end of the century the annual dolphin mortality had been reduced from onetime highs in the hundreds of thousands to under 3000. As can be seen in Figure 4.27, the two affected dolphin populations stopped their steep decline in the mid-1970s, but even into the 21st century, with relatively low mortality rates, the two populations persist at the reduced levels to which purse-seine harvesting brought them. Median recovery rates of 1.7%/yr and 1.4%/yr have been calculated for northeastern offshore spotted and eastern spinner dolphins, respectively.[11]

In studying the present populations, Gerrodette and colleagues have sug-

[11]Two studies with preliminary data suggesting a stronger recovery in the early twenty-first century are: Tim Gerrodette, George Watters, Wayne Perryman, Lisa Ballance, Estimates of 2006 dolphin abundance in the Eastern Tropical Pacific, with revised estimated from 1986-2003, NOAA Technical Memorandum NMFS-SWFSC-422, April 2008, available at `https://swfsc.noaa.gov/publications/TM/SWFSC/NOAA-TM-NMFS-SWFSC-422.pdf`; and International Dolphin Conservation Program Scientific Advisory Board, 7th meeting, Updated estimates of N_{MIN} and stock mortality limits, Document SAB-07-05, La Jolla, California, 30 October 2009, available at `http://www.iattc.org/PDFFiles2/SAB-07-05-Nmin-and-Stock-Mortality-Limits.pdf`.

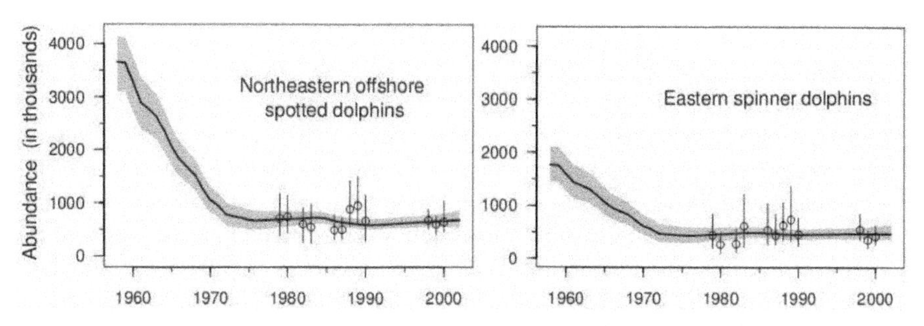

FIGURE 4.27: Estimated population trajectories of northeastern offshore spotted dolphins and eastern spinner dolphins in the eastern tropical Pacific Ocean. The populations declined due to high numbers of dolphins killed in the tuna fishery from 1960–75.

gested several hypotheses to explain the observed failure to recover to earlier levels:[12]

- Dolphin bycatch (including orphaned calves) may be higher than reported.
- The catch-and-release experience, which a given dolphin may experience 2 to 50 times per year, causes stress, increased predation, and separation of mothers and calves. Behavior changes and reduced reproduction have been observed.
- Dolphin habitat is not constant (although no major long-term changes have been measured).
- Many other unmeasured factors may be responsible, such as competitive displacement by other species, disruptions to the population structure, and the ongoing large-scale removal of tuna associated with the dolphins.

The first two hypotheses may be grouped as fishery effects, and the last two as ecosystem effects. Those causes which directly affect individual dolphins may be considered incidental harvesting (whether literal or figurative: that is, whether the individual dies or merely isolates himself from the population), while those which affect the population as a whole through environmental or ecosystemic changes may be considered changes to the baseline parameters (for a logistic model, r and K).

[12] T. Gerrodette and J. Forcada, Non-recovery of two spotted and spinner dolphin populations in the eastern tropical Pacific Ocean, *Marine Ecology Progress Series* 291: 1–21, 2005.

P.R. Wade, G.W. Watters, T. Gerrodette, and S.R. Reilly, Depletion of spotted and spinner dolphins in the eastern tropical Pacific: modeling hypotheses for their lack of recovery, *Marine Ecology Progress Series* 343: 1–14, 2007.

EXAMPLE 7.
Assume the two dolphin species are effectively at equilibrium. (a) If the observed depression of the two dolphin species is to be explained by *de facto* constant-effort harvesting, what is the effective frequency of these harvesting events? (b) If instead harvesting "effort" is limited to the 3000 dolphins per year cited in the discussion above, and the population depressions are to be explained by changes to the baseline parameter r, what reproductive rate would allow such limited harvesting to hold the population so far below carrying capacity? (c) If harvesting effort is set as in (b) and the population equilibrium is to be explained instead by changes to K, what is the resulting carrying capacity from which the present populations are being depressed?

Solution: For the northeastern offshore spotted dolphins we have an estimated present population of 640,000, original (pre-fishing) population of 3.2 million, and $r = 0.017/\text{yr}$. For the eastern spinner dolphins we have an estimated present population of 450,000, original (pre-fishing) population of 1.5 million, and $r = 0.014/\text{yr}$.

(a) If we assume constant-effort harvesting, then the original population is the carrying capacity K and the present population is the equilibrium $y_H^* = K(1 - E/r)$, from which $E = r(1 - y_H^*/K)$. For spotted dolphins this gives $E = 0.017(1 - 0.64/3.2) = 0.0136/\text{yr}$. For spinner dolphins this gives $E = 0.014(1 - 0.45/1.5) = 0.0098/\text{yr}$.

(b) We calculate effort from the given figures for harvesting, Ey^*, and the present equilibrium, y^*. First, we arbitrarily divide the 3000 harvesting deaths between the two species in proportion to their population sizes:

$$3000 \times \frac{640,000}{640,000 + 450,000} = 1761$$

deaths/yr for the spotted dolphins, and the remaining 1239 deaths/yr for the spinner dolphins. Then we divide: $E = 1761/640,000 = 0.00275/\text{yr}$ for the spotted dolphins, and for the spinner dolphins $E = 1239/450,000 = 0.00275/\text{yr}$. We now solve the equilibrium condition (4.13) for r:

$$r = \frac{E}{1 - \frac{y^*}{K}} = \frac{0.00275}{1 - 0.2} = 0.00344/\text{yr}$$

for the spotted dolphins, and similarly we find $r = 0.00393/\text{yr}$ for the spinner dolphins. Both of these values are exceptionally low: that is, the reproductive rates would have to be depressed by an order of magnitude from the usual range of cetacean reproductive rates (a common estimate is 0.04/yr) in order for changes in r to explain the depression. This would include both a reduced birth rate and high calf mortality.

(c) We use the $E = 0.00275/\text{yr}$ value calculated in (b) and the r and equilibrium (y^*) values from the literature to calculate K: for the spotted dolphins

$$K = \frac{y^*}{1 - \frac{E}{r}} = \frac{640,000}{1 - \frac{0.00275}{0.017}} \approx 760,000,$$

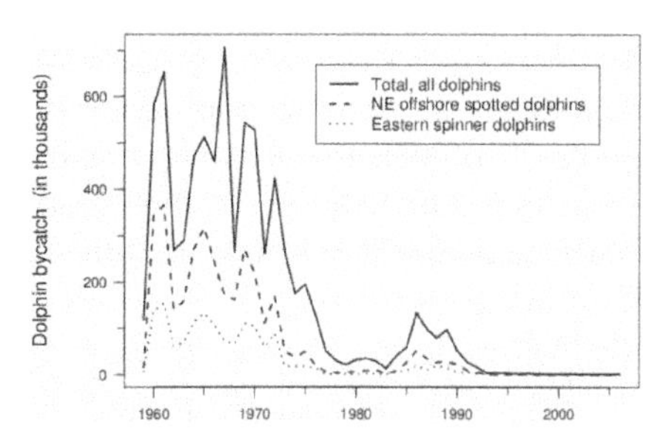

Reprinted from *Encyclopedia of Marine Mammals*, 2nd edition, Tim Gerrodette, The tuna-dolphin issue, pp. 1192–1195, ©2009, with permission from Elsevier.

FIGURE 4.28: Estimated annual number of dolphins killed in the eastern tropical Pacific purse-seine tuna fishery, total for all dolphins and separately for the two dolphin stocks with the highest number killed.

Photo by John and Karen Hollingsworth, courtesy U.S. Fish and Wildlife Service.

FIGURE 4.29: Mississippi sandhill cranes in the National Wildlife Refuge created specifically to preserve them. Once reduced to a population of only 30 birds, this subspecies has now more than doubled that amount.

and for the spinner dolphins we likewise compute $K = 560,000$ as the environmentally reduced carrying capacity. □

The sandhill crane (*Grus canadensis*, Figure 4.29) is the most populous of the world's cranes, with over 500,000 distributed across North America, from the tip of Siberia to Cuba. There are six recognized subspecies, three of them migratory and three nonmigratory subspecies restricted to very small habitats (and populations) in Mississippi, Florida and Cuba. Most migratory sandhill cranes pass through the Platte River in the northern U.S. where they form breeding pairs that are typically lifelong.

Sandhill cranes have lost much of their original wetlands habitats to human development, and this continues to be the main threat to their existence (especially loss of roosting grounds on the Platte). However, overhunting also poses a major threat to some populations in the central U.S. and Canada, along with collisions with vehicles, utility lines, and fences. Some farmers hunt the cranes because of the damage they do to crops, in areas where their original habitat has been drained for agricultural purposes. The rate at which these sandhill cranes encounter farmers (and their crops), and other manmade artifacts, is proportional to the population size, rather than occurring at a fixed rate as with quota-managed populations. For this reason we can model these encounters as constant-effort harvesting (although see Exercise 11 for an application of constant-yield harvesting).

EXAMPLE 8.
It has been estimated[13] that of the 420,000 mid-continental sandhill cranes, between 25,000 and 32,000 cranes are lost each year to hunting or collisions. If we take the current population as an equilibrium level (this population is in fact believed to be stable, although the present population size may not be a true equilibrium), what is the average length of time before a crane has such a fatal encounter, and what is the true carrying capacity K if we assume the cranes' intrinsic growth rate to be 0.098/yr?

Solution: If the current population is at equilibrium ($y^* = 420000$), then the harvest rate is Ey^*. If we take an average of the two extremes given, this gives $Ey^* = 28500$ cranes/yr, from which $E = 28500/420000 = 0.0679$/yr. The average time before an encounter happens is the reciprocal of this figure, or 14.74 years. To find the carrying capacity, we solve the equilibrium condition (4.13) for K, to get

$$K = \frac{y_H^*}{1 - \frac{E}{r}} = \frac{420000}{1 - \frac{0.0679}{0.098}} = 1.37 \text{ million cranes. } \square$$

4.2.3 Migration and dispersal as harvesting

Although we do not typically think of harvesting or population management as pertaining to humans, the processes of human migration and dispersal, typically driven by resource considerations, affect local patterns of growth in the same way as harvesting. Here harvesting corresponds to emigration, and we extend our notion to include immigration as negative harvesting. In Section 3.5 we considered migration through the lens of mixing processes; here we look at migration and dispersal through the lens of harvesting. There we considered migration terms that were either constant or linear (in the population size); these correspond to the two harvesting models discussed in this section. In Section 2.6 we observed for discrete-time models that constant dispersal is equivalent to constant-yield harvesting, and that it may stabilize or destabilize survival (positive) equilibria. For continuous-time models we do not observe the same destabilizing effect but the correspondence of constant dispersal or migration to constant-yield harvesting persists, as does the correspondence of linear dispersal or migration to constant-effort harvesting. Following are two illustrations of how we may turn harvesting terms positive to reflect immigration.

---

[13] Curt D. Meine and George W. Archibald (Eds). 1996. The cranes: Status survey and conservation action plan. IUCN, Gland, Switzerland, and Cambridge, U.K. 294pp. Northern Prairie Wildlife Research Center Online. `http://www.npwrc.usgs.gov/resource/birds/cranes/index.htm` (Version 02MAR98).

EXAMPLE 9.

In Example 1, Section 3.2 we fitted the population of the United States to the curve

$$y = \frac{265}{1 + 69\, e^{-0.03t}}, \tag{4.15}$$

with population size measured in millions and time measured in years from 1790. Suppose the population is governed by a logistic differential equation for which this is the solution but that in addition there was an immigration of one million people per year beginning in 1790. What would be the limiting population size under these conditions?

Solution: For the solution $y(t)$ given in (4.15), $\lim_{t\to\infty} y(t) = 265$, and thus in the logistic model $K = 265$. From the form of the solution of the logistic equation obtained in Section 3.3, comparison with (4.15) shows that $r = 0.03/\text{yr}$. Constant immigration of M per year would change the logistic model to $y' = ry\left(1 - \frac{y}{K}\right) + M$, with equilibria given by $ry - \frac{r}{K}y^2 + M = 0$, whose only positive solution is

$$y = \frac{1}{2}\left[K + \sqrt{K^2 + \frac{4KM}{r}}\right] = \frac{K}{2}\left[1 + \sqrt{1 + \frac{4M}{rK}}\right].$$

With $r = 0.03$, $K = 265$, $M = 1$, this gives the equilibrium (limiting) population 295 million. □

EXAMPLE 10.

An island with resources to support 1000 people becomes so attractive that other people begin to hear about it from the inhabitants and come. On average, each inhabitant's messages convince inspire someone to immigrate there once every fifteen years. If, as in the previous example, the normal reproduction rate is $r = 0.03/\text{yr}$, what will the equilibrium size of the population be?

Solution: This situation corresponds to constant-effort harvesting, where the harvesting (immigration) is proportional to the size of the population. Here the immigration rate is Iy, where $I = 1/(15\text{yr}) = 0.067/yr$, and the equilibrium size, adapting (4.13), is

$$y^* = K\left(1 + \frac{I}{r}\right) = 1000\left(1 + \frac{0.067}{0.03}\right) \approx 3222.$$

Once every fifteen years doesn't sound like much, but in the end, despite the impossibility of further natural growth (because of resource limitations), people are arriving so fast ($Iy^* \approx 215$ people/yr) that the immigration dominates the system dynamics. □

4.2.4 Conclusions

In Section 2.5 we found that for discrete-time constant-yield harvesting of logistic (and other scramble-competition) populations there are both maximum ($H_{max} = \frac{(r-1)^2}{4r}K$) and minimum ($H_{min} = \frac{(r-3)(r+1)}{4r}K$) harvest rates that allow stability since scramble populations can be naturally unstable. In continuous-time harvesting we have only a maximum ($H_{max} = rK/4$, quite close to its discrete counterpart) since continuous-time scramble populations do not have that inherent instability. A similar comparison can be made for constant-effort harvesting: for example, discrete-time models have zero equilibria stable for $r-1 < E < r+1$ (see, e.g., Exercises 9 and 10 in Section 2.5), while the continuous-time model (4.12) has a zero equilibrium stable for $E > r$.

We can also distinguish the results of constant-yield or quota harvesting, where overharvesting leads to catastrophe, from those of constant-effort or proportional harvesting, where overharvesting leads to a gradual decline. Recall, however, that the maximum sustainable harvesting rate turns out to be the same for both models.

Exercises

1. Solve the logistic constant-yield harvesting equation (4.9) using separation of variables and the technique of partial fraction expansion. Note that the quadratic expression in y involved in dy/dt factors differently (or not at all) in each of the three cases discussed in the text.

2. * Solve the logistic constant-effort harvesting equation (4.12) using separation of variables and the technique of partial fraction expansion.

3. * Consider a general population model $y' = f(y)$ for which the population growth rate $f(y)$ is differentiable, has $f(0) = 0$, rises to a unique maximum f_{max}, and decreases thereafter, eventually becoming negative. Under constant-yield harvesting, this population size is governed by a differential equation $y' = f(y) - H$.

   (a) Show graphically that if $0 \leq H < f_{max}$, there are two intersections of the curve $z = f(y)$ and the horizontal line $z = H$, corresponding to equilibria of the differential equation.

   (b) Show that the larger of the two equilibria is asymptotically stable and the smaller of the two equilibria is unstable.

   (c) Show that if $H = f_{max}$ there is only one equilibrium.

   (d) Show that if $H > f_{max}$ there are no equilibria and every solution reaches zero in finite time.

4. Over 40,000 deer-vehicle collisions were reported in 2007, when the deer population rose as high as 1.7 million deer. From this data (and the

value $r = 0.5/\text{yr}$ obtained in this section) we can estimate a per capita encounter rate $E = 0.04M/1.7M = 0.023/\text{yr}$ (for a harvest rate Ey).

(a) If we now consider a deer population at an initial equilibrium of about 3 million, and disregard other harvesting, by how much does this "constant-effort harvesting" reduce the deer population?

(b) Write a model that incorporates both constant-yield (hunting) harvesting and constant-effort (deer-vehicle collisions) harvesting.

(c) Using the same parameters given above, with $H = 0.5$ million deer/yr and $K = 4.5$ million deer as in Example 1 to generate an equilibrium population of 3 million deer under hunting alone, find the impact of deer-vehicle collisions. How does it compare with the simpler estimate obtained in part (a)?

5. It has been estimated that 240,000 humpback whales lived in the North Atlantic prior to commercial whaling, but according to Knox there were only 3,000 by 1980 (see Table 4.1). One international study estimated the 1986 population at 5,500 and the 1993 population at 10,600. Reproduction rate estimates for humpback whales have varied widely across different populations: The IWC estimated an annual rate of population increase of 3.1% in the Gulf of Maine for the period 1979–1993 (note the 1986 moratorium marks the halfway point of this interval) but reported rates of increase of about 7% for the eastern North Pacific during the period 1990–2002, and rates of increase of 12.4% in East Australia (1981–1996), and of 10.9% in West Australia (1977–1991). Use the population estimates for 1986 and 1993, and the given carrying capacity, to estimate r for the North Atlantic population, as well as the critical harvest rate.

6. It has been estimated that 265,000 minke whales lived in the North Atlantic prior to commercial whaling. The IWC estimated 174,000 minke whales in the North Atlantic (Central and Northeastern stocks) from 1996 to 2001. Norway harvested an average of 577 minke whales per year commercially from these stocks between 1997 and 2006. (a) If we take the population estimate as an equilibrium under whaling (an admittedly unrealistic assumption), what is the corresponding reproduction rate r? (b) With the given harvesting rate, what is the expected equilibrium value if the true value of r is $0.03/\text{yr}$? $0.06/\text{yr}$? $0.09/\text{yr}$?

7. The IWC estimated that there were about 10,500 bowhead whales in the Bering-Chukchi-Beaufort Seas stock in 2001, with an observed annual net growth rate of 3.2% during the period 1978-2001. It also estimated that there were 1,230 bowhead whales in a separate stock off western Greenland in 2006. A century earlier, the world bowhead whale population is estimated to have been 120,000. Compare the maximum sustainable yield (harvest rate) of this population with the average annual drop in population during the twentieth century.

8. By the early twentieth century, the population of gray whales in the North Pacific had been reduced to about 2,000, but by the end of the century it had rebounded to about 22,000 (most of them in the Eastern North Pacific, off the coasts of Canada and the U.S.—only 121 were estimated to remain in the Western North Pacific by 2007). Assuming a carrying capacity of 24,000 and the reported (1967–1996) growth rate of 2.5%/yr, estimate the maximum sustainable yield, and compare it with the average annual catch of 174 whales/yr reported during the last 20 years before the IWC moratorium (during which the IWC reported a net growth rate of 3%/yr in this population). What effect does the model predict for a harvest rate of 174 whales/yr?

9. A 2007 study used DNA analysis to estimate a pre-whaling gray whale population of between 76,000 and 118,000. What effect would an annual catch of 174 whales (see previous exercise) have on this population if the carrying capacity were 100,000? (Again use $r = 0.025$/yr.)

10. * Assume that blue whales obey a logistic law in the absence of whaling, with carrying capacity of 200,000 whales and intrinsic growth rate of 0.05/yr, and that the average annual catch of 13,000 whales (1926–1940) held in fact from 1925 to 1965, when the population had fallen to about 2000. Use either numerical integration on a computer, or the case (iii) solution to the logistic constant-yield harvesting equation obtained in Exercise 1, to estimate the blue whale population in 1925.

11. The sandhill crane (*Grus canadensis*) is hunted by farmers because of the damage it causes to crops. Miller and Botkin found[14] that the sandhill crane population that winters in New Mexico, if not harvested, would follow a logistic growth law with carrying capacity 194,600 and intrinsic growth rate 0.098/yr. Find the critical harvest rate which would wipe out the sandhill crane population, and the limiting population size if 3000 sandhill cranes per year are shot by irate farmers.

12. Suppose that a given population of sandhill cranes follows a logistic growth law with carrying capacity 194,600 and intrinsic growth rate 0.098/yr. If a single crane, on average, dies from encounters with man's world roughly 0.04 times per year, by how much is the equilibrium population reduced?

13. Each breeding pair of sandhill cranes needs its own territory of 20–100 acres of marshland. Of the 19,300 acres of land in the Mississippi Sandhill Crane National Wildlife Refuge, about 12,500 acres can be used by cranes. (a) What is the carrying capacity here, if about half of a crane population consists of breeding pairs? (b) As of September 1994, there

---

[14] R.S. Miller and D.B. Botkin, Endangered species: models and predictions, *American Scientist* 62: 172–181, 1974.

were 120 cranes in the refuge. About how fast should the population be growing at that point?

14. Suppose a small breeding population of 10 chipmunks is displaced into a neighborhood capable of feeding 100 of them, but with an unfriendly cat whom they encounter only very occasionally, about 5 times a year. Assume the chipmunks' reproduction satisfies a logistic model with $r = 0.3/\text{yr}$, and that the feline encounters can be considered constant-effort harvesting. How many chipmunks will there be after ten years? At equilibrium? (Note: This is, of course, a facetious example, as at such small sizes, stochastic effects dominate.)

15. In reality, the harvesting rate may depend both upon population size and upon external (e.g., economic) considerations. Consider a scenario in which harvesting activities are driven by a desired (constant-yield) harvest rate H, but when harvesting begins to drive the population size down, harvesting becomes more difficult, and is instead dependent upon the (constant) effort expended to harvest them. In this case the effective harvest rate becomes

$$R(y) = \begin{cases} Ey, & y \leq H/E; \\ H, & y > H/E. \end{cases}$$

    (a) Explain the system's behavior if the population's natural carrying capacity K is less than the threshold value H/E.

    (*b*)*. Explain the system's behavior if $K > H/E$.

16. * Suppose a population of a species that follows a logistic growth pattern is being used as a breeding reserve to produce individuals that can be released back into the animal's native habitat. Given r and K, what is the maximum rate at which individuals in this population can be released back into the wild, if every winter a proportion p of those in the reserve die from cold or hunger?

17. Show that the positive equilibrium is stable, and draw the phase line, for: (a) the constant-yield model with positive immigration M (as in Example 9), (b) the constant-effort model with positive immigration Iy (as in Example 10).

4.3 Mass-action models

One of the key influences on the growth of any population—whether of cells or complete individuals–is encounters with other populations. So-called *contact processes* are at the heart of the dynamics of many population models. Contact processes appear in models as nonlinear terms that describe the rates at which one population interacts in a certain way (for good or ill) with another population. The simplest way to describe such a rate is to assume that it is directly proportional to the size of each population—that is, the more members either population has, the more frequently encounters between the two groups will happen. Under such an assumption, the contact rate has the form $\alpha\, x(t)\, y(t)$, where $x(t)$ and $y(t)$ are the two populations in question, and α is some constant of proportionality that gives the baseline rate at which an encounter happens. This form is called *bilinear*, because it is linear in each of the two variables.

The use of a bilinear contact rate has its origin in the study of chemical reactions. In elementary chemical reactions with two reactants, call them A and B, and one product C, the rate at which the reactants combine to form the product has been determined to have the form k_+AB, where k_+ is a rate constant and $A(t)$ and $B(t)$ give the quantities (or concentrations) of the respective reactants at time t. This description of the reactants "encountering" each other is known as the *law of mass action*, first identified in 1864 by Guldberg and Waage. For this reason, similar bilinear contact rates are said to have *mass-action incidence*, even in contexts outside chemistry (such as population biology).

4.3.1 A simple chemical reaction

For an understanding of what the law of mass action produces in terms of describing this simple reaction, let us complete the model by supposing that the reaction is reversible: that is, that the product C spontaneously decomposes back into the reactants A and B at a rate k_-C proportional to the amount of C. This yields the chemical reaction

$$A + B \underset{k_-}{\overset{k_+}{\rightleftharpoons}} C,$$

and the differential equations

$$\begin{aligned} \frac{dA}{dt} &= -k_+AB + k_-C, \\ \frac{dB}{dt} &= -k_+AB + k_-C, \\ \frac{dC}{dt} &= k_+AB - k_-C. \end{aligned} \tag{4.16}$$

Since $\frac{dA}{dt} + \frac{dC}{dt} = 0$, we can deduce that $A+C$ must be a constant, call it K_A, so that $A(t) + C(t) = A(0) + C(0) = K_A$, and we can write A in terms of C:

$$A(t) = K_A - C(t).$$

Likewise, we have that $B(t) + C(t) = B(0) + C(0) = K_B$. Therefore, we can just keep track of C, rewriting (4.16) as

$$\frac{dC}{dt} = k_+(K_A - C)(K_B - C) - k_-C. \tag{4.17}$$

We can now find the eventual (equilibrium) level of each chemical predicted by the model. The equation (4.17) has two equilibria, which we can write

$$C^*_\pm = \frac{1}{2}\left\{(K_A + K_B + K_d) \pm \sqrt{(K_A + K_B + K_d)^2 - 4K_AK_B}\right\}, \tag{4.18}$$

where the ratio $K_d = \frac{k_-}{k_+}$ of the reverse reaction rate to the forward reaction rate is called the *dissociation constant.* (Note that the units for K_d are in moles (amount), the same as K_A and K_B, because the units for k_- and k_+ are different.) A careful inspection of the expression inside the radical will show that it is always positive, but smaller than $(K_A + K_B + K_d)$. Therefore both equilibria are always positive. To determine their local asymptotic stability, we calculate the derivative of $dC/dt = f(C)$:

$$f'(C) = k_+(2C - K_A - K_B - K_d).$$

We find that $f'(C) < 0$ if and only if $C < \frac{1}{2}(K_A + K_B + K_d)$. Since this is true for C^*_- and false for C^*_+, the former must be locally asymptotically stable, and the latter unstable (see the phase portrait in Figure 4.30).

FIGURE 4.30: A phase portrait for equation (4.17).

This appears to suggest that there is a region in phase space for which the amount of chemical C grows without bound; however, this is illusory, as the equilibria $C^*_\pm$ are defined in terms of the (presumably known) initial condition $C(0)$, and we can show that $C(0)$ is well below the threshold level C^*_+:

$$\frac{C^*_- + C^*_+}{2} = \frac{1}{2}(K_A + K_B + K_d) = C(0) + \frac{1}{2}(A(0) + B(0) + K_d) > C(0),$$

that is, $C(0)$ is below the midway point between C^*_- and C^*_+. Therefore C^*_- is globally asymptotically stable, and the level of chemical C will approach C^*_- (from which we can also calculate the equilibrium levels of A and B; cf. Exercise 6).

EXAMPLE 1.
The dissociation reaction for an acid with chemical formula HA (where A represents the ion that bonds with hydrogen to form the acid) is typically written as

$$H^+ + A^- \underset{k_-}{\overset{k_+}{\rightleftharpoons}} HA,$$

even though in practice this dissociation occurs in water, so that it is more properly

$$H_3O^+ + A^- \underset{k_-}{\overset{k_+}{\rightleftharpoons}} HA + H_2O.$$

The water is commonly omitted since it does not affect the reaction (except to bond with the dissociated hydronium ions H^+). In this context the dissociation constant for acids is often denoted K_a rather than K_d.

Strong acids (distinguished by having $K_a > 1$), such as hydrochloric acid (HCl) and sulfuric acid (H_2SO_4), dissociate almost completely in water, leaving little of the original acid left, while weak acids ($K_a < 1$) dissociate very little. If the dissociation constant for sulfuric acid is 1000, find how much sulfuric acid remains of 0.50 M (moles) when it is left to dissociate in water.

Solution: Here we begin with no reactants, $A(0) = B(0) = 0$, and an initial condition of $C(0) = C_0 = 0.50$ M. From (4.18) we have

$$\begin{aligned} C_-^* &= \frac{1}{2}\left[(2C_0 + K_a) - \sqrt{(2C_0 + K_a)^2 - 4C_0^2}\right] \\ &= \frac{1}{2}\left[1001 - \sqrt{1001^2 - 1}\right] \approx 0.00025M. \ \square \end{aligned}$$

EXAMPLE 2.
Acetic acid, CH_3COOH, is a weak acid, with a dissociation constant of 1.7×10^{-5} M. How much acetic acid would remain if 0.50 M were left to dissociate in water?

Solution: With the same initial conditions as in the previous example, we have

$$\begin{aligned} C_-^* &= \frac{1}{2}\left[(2C_0 + K_a) - \sqrt{(2C_0 + K_a)^2 - 4C_0^2}\right] \\ &= \frac{1}{2}\left[1.000017 - \sqrt{1.000017^2 - 1}\right] = 0.4999915M \approx 0.50M. \end{aligned}$$

In this case one can easily observe that almost none of the acid dissociates, whereas in the previous example almost all of it did. $\square$

The law of mass action has formed the basis for describing contact processes in many contexts within population biology, perhaps most notably predator-prey relationships and the spread of infectious diseases. We shall delay our own study of predator-prey systems until Chapter 6, in which we

Photo courtesy of Hinochika / Shutterstock.com

FIGURE 4.31: On a street near Sannomiya JR Station in Kobe, Japan, people wear face masks during the 2009 outbreak of swine flu. Disease control strategies center around reducing the rate of infectious contacts between susceptible and infected individuals.

examine systems of interacting populations. However, as we saw in Section 3.2, a simple two-compartment epidemic model can be represented with a single equation in the case when the total population remains constant. We shall now take advantage of this fact to use such a model to consider how mass action incidence and its alternatives affect predictions about whether an outbreak of infectious disease can persist.

4.3.2 The spread of infectious diseases, revisited

The driving process for the spread of any infectious disease is contacts between infectious individuals and susceptible individuals (see Figure 4.31). From its inception, mathematical epidemiology has used some version of the law of mass action to describe the rate at which these contacts occur. Compartmental models such as those we shall consider assume a certain amount of homogeneity within each compartment, so that we can describe contact rates in terms of averages for each compartment and ignore individual differences. In Section 3.2 we presented the differential equation

$$\frac{dI}{dt} = \beta SI - \frac{1}{\tau}I = \beta(N-I)I - \frac{1}{\tau}I, \tag{4.19}$$

where the total population N (assumed constant) is composed of infectives I and susceptibles S, as a simple model for the spread of an infectious disease. The fact that $S = N - I$, so that as the number of infectives increases, the number of susceptibles available to be infected decreases, introduces into the mass-action term βSI the self-limiting term that makes (4.19) fit the logistic form. In this way we were able to use the solution to the logistic equation in order to derive the threshold quantity that determines whether an outbreak will persist or die out. We will now reconsider this model from a slightly more general framework of contact processes that will help us understand precisely what the law of mass action means as a description of contact rates between two populations.

Let us suppose, as before, that we have a population of N individuals classified as either infectives I or susceptibles S. Also as before, infected individuals recover (and become susceptible again) after an average τ units of time. Now let us assume that the rate at which individuals contact other members of the population is a function $c(N)$ of the population size, with the property that $c(N)$ is a nondecreasing function of N. That is, the larger the population, the more (or possibly the same) potentially infectious contacts one makes each day. (In other words, having more people around should not mean fewer contacts.) We will say more about the specific form of this function shortly.

If we now also assume that everyone in this population mixes homogeneously, so that a person's infection status (infected or not) has no effect on the [infection] classes of people one contacts, then the proportion of one's contacts that are made with susceptible individuals is just S/N, and the proportion of contacts made with infected individuals is I/N. (This assumption is known as *proportional mixing* or *random mixing*.) Thus, each day an average infected individual contacts $c(N)$ individuals, $c(N)\,S/N$ of which are with susceptibles. Then a total of $(c(N)\,S/N)I$ infectious contacts are made each day (or other unit of time).[15]

We can now write this more general model as

$$\frac{dI}{dt} = c(N)\frac{SI}{N} - \frac{1}{\tau}I = c(N)\frac{(N-I)I}{N} - \frac{1}{\tau}I. \tag{4.20}$$

Here the important distinction is that the per capita contact rate is not a constant β, but a function $c(N)$ of the population size. In fact, for populations of small or moderate size, this per capita contact rate increases roughly linearly with population size, say $c(N) = \beta N$, in which case equation (4.20) simplifies to the mass-action model (4.19) which we have seen before. However, for very large populations the contact rate *saturates*, approaching a constant maximum, say $c(N) = \alpha$, at which point it is simply not possible to come into [potentially infectious] contact with more people in a single day. This assumption, which generates a rate of new infections of $\alpha SI/N$, is called *standard incidence*. Standard incidence is not appropriate for small populations, or populations in danger of being driven to extinction, because it suggests that even

[15] An alternative, equivalent way to derive the new infections term is to start with a single susceptible's daily contacts $c(N)$, multiply by the proportion of infectious contacts I/N and the number of susceptibles S.

as $N \to 0$ the contact rate remains steady. In reality, the contact rate should be described as increasing more or less linearly with population size at first, and then leveling off, possibly but not necessarily smoothly, as the population size increases.

We can now take a qualitative approach to analyzing this more general model. By inspection, one equilibrium of (4.20) is $I = 0$, called the *disease-free equilibrium.* Factoring I out of the equilibrium condition, we are left with

$$c(N)\,\frac{N-I}{N} - \frac{1}{\tau} = 0,$$

from which we find the *endemic equilibrium*

$$I^* = N\left(1 - \frac{1}{c(N)\tau}\right).$$

To determine each equilibrium's local asymptotic stability, we differentiate the function $f(I) = dI/dt$, calculating:

$$f'(I) = c(N)\left(1 - 2\frac{I^*}{N}\right) - \frac{1}{\tau}.$$

Thus (after some algebra)

$$f'(0) = c(N)\left(1 - \frac{1}{\beta(N)\tau}\right), \quad f'(I^*) = -c(N)\left(1 - \frac{1}{\beta(N)\tau}\right).$$

Regardless of the form of $c(N)$, therefore, the persistence of the infection is determined (as before) by its *basic reproductive number*, usually denoted R_0, and given in this case by $R_0 = c(N)\tau$. (Note this quantity is dimensionless). If $R_0 < 1$, then $f'(0) < 0$, making the zero equilibrium locally asymptotically stable, while the endemic equilibrium I^* is not only unstable but outside the state space altogether, for its value in this case is negative. On the other hand, if $R_0 > 1$, then $f'(0) > 0$, so that the zero equilibrium is unstable, while $I^* > 0$ and $f'(I^*) < 0$, so that the endemic equilibrium is locally asymptotically stable, and the infection will persist in the population. Note, finally, that in either case, the unique locally asymptotically stable equilibrium is in fact globally asymptotically stable.

We can interpret the expression for R_0 biologically by recalling from the original discussion in Section 3.2 that an "average" infective makes $c(N)$ potentially infective contacts in unit time, and remains infective an average of τ units of time. Therefore $R_0 = c(N)\tau$ represents the average number of secondary infections caused per infected individual before recovery (hence the term *basic reproductive number*). If each infection is able to replace itself and more ($R_0 > 1$), then we should expect the disease to persist, while if each infection cannot, on average, replace itself before the individual recovers ($R_0 < 1$), then we would say the disease is doing a poor job of reproducing itself and

Left: photo by C.M. Kribs. Right: photo by Daniel Bowen/Public Transport Users Assoc., Melbourne, Australia

FIGURE 4.32: Left, Main Street in Hope, Arkansas, USA (population 10,000); right, a commuter train in Melbourne, Victoria, Australia (population 3.74 million). The number of [potentially infectious] contacts one makes in a day depends very heavily upon population density, but beyond a certain point it saturates, and can increase no further.

should die out. This kind of insight illustrates the reason why such mathematical models are useful. For this model, one might be able to make this same argument without going through all the details of a qualitative analysis, but not so for the more complicated models which arise in studying the problems of interest to us. *Mathematical* results give rise to *biological* insights, and that is our motivation.

So what difference does the form of the contact rate $c(N)$ make? In this case, we have assumed the population size N to be constant, in order to be able to study the system using a single differential equation. However, in many more complicated systems, the population size may change during an outbreak (especially for endemic situations). For populations whose size is on the order of hundreds, or a few thousands, the mass-action assumption that the contact rate is directly proportional to population size is reasonable: If you live in a small town and the population doubles overnight, then you probably will encounter double the usual number of people throughout the course of your day: busier sidewalks and stores, fuller buses, etc.

However, at some point one's ability to contact others in a way that might transmit infection begins to saturate as the population grows. If you live in a major metropolitan area like New York or Tokyo and the population doubles overnight, you will probably not come into contact with twice as many people on your daily routine, because the sidewalks, subways, and stores are already all filled to capacity (compare the photos in Figure 4.32). In such a situation, where the contact rate has saturated, we might instead assign $c(N) = a$, a constant value independent of changes in population size (as long as the population remains at saturation levels). As discussed above, this assumption produces the standard incidence model

$$\frac{dI}{dt} = \alpha \frac{SI}{N} - \frac{1}{\tau} I,$$

with a resulting constant $R_0 = \alpha\tau$, as opposed to that generated by mass-action incidence as in Section 3.2: $R_0 = \beta N \tau$, which is proportional to the population size (and in particular may cross the critical value of 1 as the population size changes during an outbreak). The conclusion to draw from this discussion is that for low population densities the basic reproductive number may change if the population changes, whereas for high population densities it is not sensitive to fluctuations in population size.

In a scenario where the population size may change drastically during the course of an outbreak—for example, if the mortality rate from infection is quite high, as with the Ebola virus—it may be necessary to incorporate both the unsaturated, mass-action phase for low population densities and the saturated, standard incidence phase for high population densities. In this case, the function $c(N)$ would rise almost linearly near the origin and then taper off, either gradually or with a ramp-like cutoff. Figure 4.33 illustrates two different possible contact rate functions, both of which saturate at a rate α for large populations: a smooth, gradual saturation c_{sm} and a ramp-like function c_{sw} that switches sharply between linear and constant phases at some threshold density A. In both cases, the overall form $R_0 = c(N)\tau$ of the basic reproductive number remains the same, but its dependence on population size changes, again in ways that may prevent a disease from driving a population extinct, because as the population size drops, so does R_0. (We should note that although the expression we derived above for R_0 came from a model in which the total population is constant, the overall dependence of R_0 on $c(N)$ remains true for models in which the population size may vary, as we shall see in later chapters.)

EXAMPLE 3.

Suppose that the infectious contact rate for the common cold saturates sharply (like c_{sw} in Figure 4.33) at a level of $\alpha = 1.6$/day for population sizes above 1 million, and increases linearly for smaller populations. (a) If the average infectious period for the cold is 1 day, what is the critical population size above which an outbreak of the cold will persist? (b) For large populations, at what endemic level will the infection establish itself? (c) Despite model predictions, outbreaks are observed in small populations, and outbreaks come and go in large populations, rather than remain in a constant endemic state. What limitations on the model are involved in these apparent contradictions?

Solution: (a) For large populations, we have the saturated $R_0 = \alpha\tau = (1.6/day)(1 day) = 1.6 > 1$, so that the model predicts persistence of the infection in populations of over 1 million. For smaller populations, we have $c(N) = (1.6/day)(N/10^6)$ and $R_0 = (1.6/day)(N/10^6)(1 day) = 1.6N/10^6$, so that $R_0 > 1$ when $N > 10^6/1.6 = 625,000$. That is, the model predicts that the infection will persist in populations of 625,000 people or more.

(b) The endemic equilibrium for populations of over 1 million has

$$\frac{I^*}{N} = \left(1 - \frac{1}{R_0}\right) = \left(1 - \frac{1}{1.6}\right) = 0.375.$$

That is, the model predicts that, eventually, 37.5% of the population will be infected at any given time.

(c) The occurrence of outbreaks in small populations does not contradict the model's predictions, which simply state that when $R_0 < 1$, eventually the outbreak will die out. It is possible, however, that an outbreak might grow within a small subset of the population which makes [potentially infectious] contacts much more often than the general population—for instance, children enrolled in school; this is a limitation of the model, which assumes that everyone in the population has roughly the same average contact rate. A more detailed model could subdivide the population by contact rate, into high-risk and low-risk groups.

The periodic disappearance and reappearance of outbreaks of the common cold in large populations is most often attributed to seasonal factors which are also not considered in this model, the most evident being weather. □

As we have seen before, re-examining the assumptions underlying a given model may help us understand better why the model makes the predictions that it does. We will here consider an alternative to one of the assumptions underlying this simple epidemic model, and reconsider the others in the exercises at the end of this section.

EXAMPLE 4.

In formulating the epidemic model (4.20), we assumed that the outbreak is occurring on such a short time scale that demographic changes such as births and deaths are negligible. How can the model be adapted to account for such changes without contradicting the assumption of a constant total population size?

Solution: As discussed in our initial attempts to model population growth, established populations run up against the carrying capacity of their environmental resources. For an established population, it is therefore reasonable to assume that, on a time scale of a few years or less, the total birth rate roughly evens out the total death rate, so that the population size remains constant even though it is constantly gaining and losing members. We can incorporate this assumption via a constant per capita mortality rate μ, which can be estimated by noting that $1/\mu$ is then the average lifetime in the system. If we assume that the disease is not transmitted vertically, so that all newborns (or new recruits) enter the susceptible class, then our model becomes

$$S'(t) = \mu N + \frac{1}{\tau} I - \beta(N) SI/N - \mu S,$$
$$I'(t) = c(N) SI/N - \frac{1}{\tau} I - \mu I.$$

Adding the two equations yields $dN/dt = dS/dt + dI/dt = \mu N - \mu S - \mu I = 0$, so N remains constant, and we can again write simply the equation for I, substituting $S = N - I$.

We leave the analysis of this model as an exercise but note that the basic qualitative results remain unchanged from the original model (4.20), and the quantitative results change only to incorporate μ into R_0. □

4.3.3 Contact rate saturation and the "Pay It Forward" model

We shall now apply the above discussion of how contact rates saturate as populations grow large to a new context. Many other phenomena besides the transmission of infectious diseases are driven by human interactions, and can be modeled by descriptions of contact processes. Studies of numerous sociological phenomena driven by peer-pressure influence have extended the metaphor of mass-action contact rates by applying epidemic modeling techniques to eating disorders, crime and gang activity, drug and substance abuse, grassroots political movements, and even learning environments. Although these behaviors are far removed from the chemical reactions for which the law of mass action was developed, such descriptions of the rate at which individuals change their behavior can lead to powerful insights about why phenomena like these can be so robust, and difficult to eradicate once established.

One study[16] considered the effects of contact rate saturation on the growth of a grassroots movement which depends upon individuals' actions in order to motivate others to participate. In this case, the phenomenon in question was the philosophy of proactive charitable acts described in Catherine Ryan Hyde's novel *Pay It Forward.*[17] The growth of this movement occurs when participating individuals "pay it forward" (do a significant favor for one person in return for one done to them by someone else) to non-participating individuals, inspiring those who benefit to "pay it forward" themselves. Potential recruits' inclination to join the movement upon receiving such a favor may be conditioned on their familiarity with the movement from secondary sources like the news and anecdotes from friends. In this way, the effective contact rate becomes an increasing function of the size of the movement (the larger the movement, the more likely a person is to have heard of it, and thus be predisposed to join upon receiving a favor).

If we let $N(t)$ denote the number of people in the movement at time t and P be the (constant) size of the overall population, then we can describe the growth of the phenomenon as follows:

$$N' = f(N) = c(N)N(P - N)/P - \mu N, \tag{4.21}$$

where $c(N)$ is the per capita contact rate (making $c(N)(P - N)/P$ the per capita recruitment rate) and μ is the per capita rate at which individuals die or drop out of the movement. We shall consider two ways to model the saturation

[16] C.M. Kribs-Zaleta, To switch or taper off: the dynamics of saturation, *Math. Biosci.* 192(2): 137–152, Dec. 2004.

[17] Catherine Ryan Hyde, *Pay it forward*, Simon and Schuster, New York, 1999.

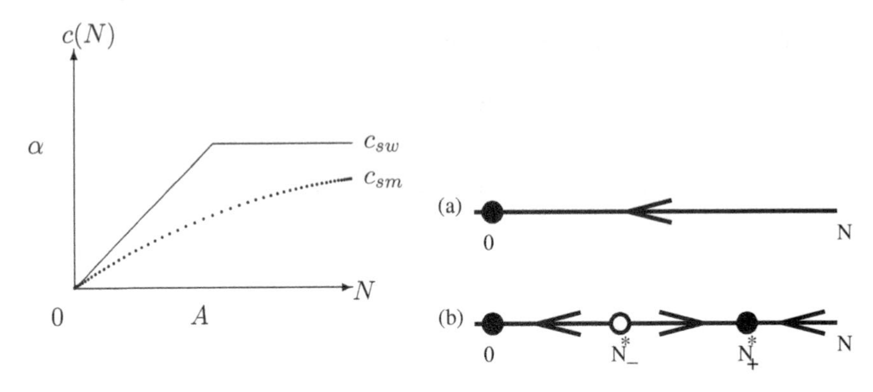

FIGURE 4.33: Two different ways to model contact rate saturation for large populations.

FIGURE 4.34: Phase portraits for the "Pay It Forward" model (4.21).

that occurs as the movement reaches a critical size (at which point the movement is so well known that further growth does not make non-participants any more familiar with it): smooth and sharp saturation corresponding to those graphed in Figure 4.33. If we keep α as the maximum per capita contact rate and A as the critical population level, then we can define the two contact rates as

$$c_{sw} = \alpha \min(N/A, 1), \quad c_{sm} = \alpha N/(N + A).$$

A qualitative analysis of the model (4.21) using smooth saturation c_{sm} reveals that, in addition to the zero equilibrium, a pair of positive equilibria

$$\frac{N^*_\pm}{P} = \frac{1}{2}\left((1-m) + \sqrt{(1-m)^2 - 4ma}\right)$$

exists if and only if $m \leq 1 + 2a - 2\sqrt{a(a+1)}$, where $m = \mu/\alpha$ and $a = A/P$. Stability analysis using the derivative

$$f'(N) = \frac{\alpha}{P}\frac{(N+A)(2PN - 3N^2) - N^2(P-N)}{(N+A)^2} - \mu$$

shows that the zero equilibrium is locally asymptotically stable ($f'(0) = -\mu < 0$), and when the positive equilibria exist, some algebra (which we omit here) can show that the lower one N^*_- is always unstable and the upper one N^*_+ is always stable. This leaves us with a situation where, if individuals retire from the movement faster than new ones can ever be recruited ($m > 1$), or if the critical size at which the movement is well known to the public is large enough ($a > (1-m)^2/4m$), the movement is sure to die out (the zero equilibrium is globally asymptotically stable, Figure 4.34(a)). If, on the other hand, the maximum recruitment rate exceeds the retirement rate and that maximum occurs at a small enough critical size, then survival of the movement depends upon having a large enough initial core ($N(0) > N_-^*$, see Figure 4.34(b)). In this latter case, we have two locally asymptotically stable equilibria, 0 and N_+^*; for this reason the equation is said to be *bistable.*

The analysis of this same model using the switching function c_{sw} to describe the contact rate is a little more complicated, but not more difficult. It yields qualitatively similar results to those above (an indication that these results are *robust*), and we leave the details as an exercise (see Exercise 12).

We may wonder how the phase portrait in Figure 4.34(a) can become the one in Figure 4.34(b) simply by shifts in some parameters, or more generally what consequences this change from a globally stable extinction equilibrium to a bistable state implies for the future of the movement. These questions can best be answered through the lens of bifurcations, first mentioned in Section 2.6 as changes in the qualitative behavior of a dynamical system. In the next section, we will develop techniques to study bifurcations, and study or revisit several situations (including this one) where bifurcations identify the important changes in a system's dynamics.

Exercises

1. In a chemical reaction a substance S_1 with initial concentration K is transformed into a substance S_2. Let $y(t)$ be the concentration of S_2 at time t, so that $K - y(t)$ is the concentration of S_1 at time t. If the reaction is autocatalytic (meaning that the reaction is stimulated by S_2), then $\frac{dy}{dt}$ is proportional to y and $K - y$. Thus

$$\frac{dy}{dt} = \alpha y(K - y)$$

for some constant α. If the reaction is started at time $t = 0$ by introducing an initial concentration A of S_2, find the concentration of S_2 as a function of t.

2. In a second-order chemical reaction, a molecule of a substance S_1 and a molecule of a substance S_2 interact to produce a molecule of a new substance S_3. Suppose the substances S_1 and S_2 have initial concentrations a and b, respectively, and let $y(t)$ be the concentration of S_3 at time t. Then the concentrations of S_1 and S_2 at time t are $a - y(t)$ and $b - y(t)$, respectively. The rate at which the reaction occurs is described by the differential equation

$$\frac{dy}{dt} = \alpha(a - y)(b - y),$$

where α is a positive constant (cf. (4.17)). Find the concentration y as a function of t if $y(0) = 0$.

3. Find the concentration $y(t)$ in Exercise 2 if $a = b$.

4. Find the amount of each of the following acids that remains at equilibrium after 0.50 M of it is left to dissociate in water:

   (a) boric acid, $B(OH)_3$ ($K_a = 5.8 \times 10^{-10}$ M)

   (b) iodic acid, HIO_3 ($K_a = 0.17$ M)

   (c) trichloroacetic acid, Cl_3COO_2H ($K_a = 0.22$ M)

5. Give the units for the reaction rate constants k_+ and k_- in (4.17).

6. Show that the expression for C^*_- in (4.18) is equivalent to the formula used by chemists,
$$K_d = \frac{A^* B^*}{C^*}.$$

7. Sketch a complete phase portrait for the epidemic model (4.20).

8. Complete the substitution suggested in Example 4 and write the differential equation for $I(t)$ describing the epidemic model with demographic renewal. Then complete the model's qualitative analysis.

9. In formulating the epidemic model (4.20), we assumed that the infectious contact rate was a function of population size, $\beta(N)$. This is a reasonable assumption in most cases, but outbreaks of disease can themselves change the nature of the daily contact structure, either because of the severity of their symptoms (people may be scared into wearing breathing masks, as in the 2003 SARS epidemic, or staying home altogether) or because of the size of the outbreak (healthcare facilities may become overburdened and unable to cope, as in the 1918 Spanish flu pandemic). How would the model and its analysis change if the contact rate were instead dependent on $I(t)$, and what form would you suggest for the function $\beta(I)$?

10. In formulating the epidemic model (4.20), we assumed that no deaths occurred due to the disease. If instead we assume that infected individuals die at a rate of δ/day, how does the model change? *Hint: Consider this assumption's impact on the system overall.*

11. Some researchers[18] have suggested a contact rate with exponents to describe the incidence of new infections, something of the form $\beta S^m I^n$, where at least one of $m, n > 1$. In each case this unusual form was motivated by a particular characteristic of the scenario being modeled.

    (a) Formulate and analyze such a model for the case $m = 1$, $n = 2$. How does the behavior of this model differ from (4.20) with the usual mass-action incidence?

---

[18] e.g., Pauline van den Driessche and James Watmough, A simple SIS epidemic model with a backward bifurcation, *J. Math. Biol.* 40(6): 525–540, 2000, Example 4.1.

(b) Formulate and analyze such a model for the case $m = 2$, $n = 1$. How does the behavior of this model differ from (4.20) with the usual mass-action incidence, and from the model in part (a)?

12. * The analysis of switching models, such as the "Pay It Forward" model (4.21) using $r_{sw}(N)$, has three parts: first, analyze the part of the model that falls below the switching point; second, analyze the part above the switch; and third, address the continuity of these results at the switch point.

    (a) Complete a qualitative analysis of (4.21) with $r(N) = \beta_0 N/A$, and note under which conditions the equilibrium values N^* have $N^* < A$.

    (b) Complete a qualitative analysis of (4.21) with $r(N) = \beta_0$, and note under which conditions the equilibrium values N^* have $N^* > A$.

    (c) Make a composite phase portrait for (4.21) with $r_{sw}(N)$ by pasting together the part of the portrait for part (a) below the switch point A and the part of the portrait for part (b) above A. Show that the switching model is bistable when $a \leq 1 - m$ or $1/2 < a < 1/(4m)$ (where a and m are as defined in this section).

13. Predator-prey systems, like the other systems studied in this section, are driven by a contact process: encounters between predator and prey. This process is also commonly modeled with a mass-action term kxy, where k is an encounter rate, x is the number of predators, and y is the number of prey.

    (a) If the prey exhibit logistic growth in the absence of predators (with, say, growth rate r and carrying capacity K), what differential equation describes their growth dy/dt in the presence of predators?

    (b) Suppose the prey are the predators' only food source, and that the predator birth rate is a mass-action term (with a different rate constant, say c, reflecting the efficiency with which predation is "converted" into reproduction), while the predators' per capita natural mortality rate is a constant μ. Write the differential equation describing the predators' growth rate dx/dt.

    (c) Now suppose that in one predator-prey system, the prey are so plentiful that we can consider their population size a constant Y, no matter how great the predation (for instance, because any depletion is quickly compensated for by migration from nearby habitats). Analyze the behavior of the predator model in part (b) under this assumption.

    (d) Suppose that in a different predator-prey system, the predators have many other food sources in addition to the prey, so that their population is a constant X. Analyze the behavior of the prey model in part (a) under this assumption.

4.4 Parameter changes, thresholds, and bifurcations

We have now seen several examples of models for biological systems in which the nature of a population's long-term behavior depends upon the values of one or more parameters in the model: beginning with the discrete logistic [difference] equation, which we saw in Section 2.6 exhibits period doubling for some values of the intrinsic growth rate r, and earlier in the present chapter with models of foraging ants, harvesting, infectious diseases, and collective behaviors. In all of these models, small changes in a parameter may cause a system to cross a threshold beyond which the model's behavior changes qualitatively and sometimes abruptly. Even the simplest of all differential equations we have studied, that attributed to Malthus, has solutions which behave differently (grow unbounded or dwindle away to zero) depending upon the sign of the coefficient r in $dx/dt = rx$. We will use the term *bifurcation* to refer to such qualitative changes, which involve the appearance, disappearance, or change in stability of the equation's equilibria.

Although a bifurcation gives a mathematical description of a change in a model's behavior, it usually also has important biological consequences, most commonly the survival or extinction of the population under study. There are two principal reasons why we should be interested in knowing the behavior a given model predicts for a wide range of parameter values. In some cases, values of model parameters may change over time, as the corresponding biological system reacts to external factors; in others, model parameters may remain constant but be difficult or impossible to measure accurately and precisely. *Bifurcation diagram* (graphs) and *threshold quantities* (the combinations of model parameters that mark changes in a system's behavior) are mathematical tools that help us interpret bifurcations in the context of the systems in which they occur. In this section we shall develop some familiarity with these tools by revisiting the models mentioned above to look at the underlying bifurcations and get an idea for the kinds of systemic changes they herald. We shall not try to give an exhaustive list of the possibilities, but the examples we already have in hand will illustrate the most common types of bifurcation that can occur in a single differential equation.

We begin by recalling the simple epidemic model presented in Sections 3.2 (equation (2.19)) and 4.3 (equation (4.19)). We have previously rewritten it to fit the form of a logistic equation, but now we will instead use rescaling to reduce the number of parameters to 1:

$$\frac{dI}{dt} = \beta(N - I)I - \frac{1}{\tau}I = \beta\left[\frac{\beta N\tau - 1}{\beta\tau}I - I^2\right]. \tag{4.22}$$

The technique of rescaling, first mentioned in Chapter 1, involves defining new variables, and often new corresponding parameters, relative to certain benchmark quantities: populations relative to a given size, time relative to a given period or rate. It also involves using the Chain Rule from calculus to write a new differential equation: $\frac{dz}{dx} = \frac{dz}{dy}\frac{dy}{dx}$. In this case, we need rescale only the time variable, to eliminate the β; we define $s = \beta t$, rename $y = I$, and define the bifurcation parameter $p = \frac{\beta N\tau - 1}{\beta\tau} = \frac{R_0 - 1}{\beta\tau}$; then, using the Chain Rule,

$$\frac{dy}{ds} = \frac{dI}{dt}\frac{dt}{ds} = \beta[py - y^2]\,\frac{1}{\beta},$$

so that the epidemic model (4.22) becomes

$$y' = py - y^2. \tag{4.23}$$

In this simplified form, we can better see the effects of the bifurcation parameter p.

Applying qualitative analysis techniques, we find two equilibria, $y^* = 0$ and $y^* = p$. We have $f(y) = py - y^2$, $f'(y) = p - 2y$, so that $f'(0) = p$ and $f'(p) = -p$. Thus if $p < 0$, the equilibrium $y = 0$ is asymptotically stable and the equilibrium $y = p$ is unstable, while if $p > 0$, the equilibrium $y = p$ is asymptotically stable and the equilibrium $y = 0$ is unstable. (It is left as an exercise to draw the two corresponding phase portraits and verify that there are no non-negative unbounded solutions, so that local stability extends to global.)

We can represent this information graphically in a *bifurcation diagram* by graphing the equilibrium values y^* as functions of the bifurcation parameter p. We draw solid lines or curves to represent stable equilibria and broken or dashed lines to represent unstable equilibria. The bifurcation diagram for equation (4.23) is shown in Figure 4.35. We see that as p increases through zero, the two equilibria exchange stability; this type of exchange is called a *transcritical bifurcation*. We can also now identify the *bifurcation point* via the coordinates $p = 0$ and $y = 0$, since the two equilibria coincide at $y = 0$ when $p = 0$.

We can also return to the original equation (4.22) and describe the bifurcation in terms of I^*, the equilibrium number of infectives, and R_0, the infection's basic reproductive number (which has more biological meaning than p). The endemic equilibrium $y^* = p$ becomes $I^* = (R_0 - 1)/\beta\tau$, and its stability criterion $p > 0$ becomes $R_0 > 1$. Since we consider only a nonnegative state space $I \geq 0$ and by definition our epidemiological parameter $R_0 = \beta N\tau \geq 0$, we include in the bifurcation diagram (Figure 4.36) only the positive quadrant. In general, this bifurcation at $R_0 = 1$, $I^* = 0$, which is so central to the analysis of even the most complex epidemic models, is normally transcritical.

The "Pay It Forward" model studied in Section 4.3 exhibits a different type of bifurcation. Equation (4.21) with smooth saturation is

$$\frac{dN}{dt} = \frac{\alpha N}{N + A}\,N\,\frac{P - N}{P} - \mu N. \tag{4.24}$$

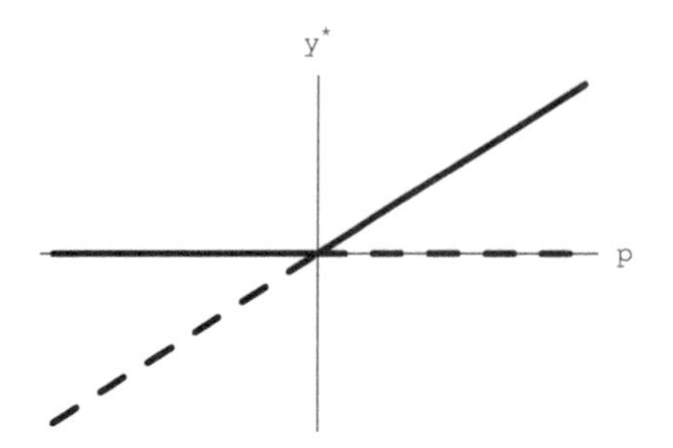

FIGURE 4.35: Bifurcation diagram for the transcritical bifurcation in (4.23).

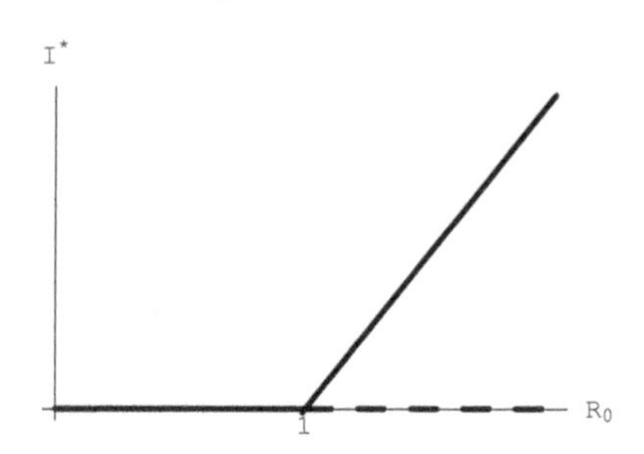

FIGURE 4.36: Bifurcation diagram for the transcritical bifurcation in (4.22).

One equilibrium is $N^* = 0$; if we define $a = A/P$ and $b = \alpha/\mu$ (in terms of the analysis in Section 4.3, $b = 1/m$), there are two additional equilibria

$$\frac{N^*_{\pm}}{P} = \frac{b}{2}\left[(b-1) \pm \sqrt{(b-1)^2 - 4ab}\right]$$

if and only if $b > b_c = 1 + 2a + 2\sqrt{a(a+1)}$. If we fix the parameter a and use b as our bifurcation parameter, we can use the equilibrium stability information determined in Section 4.3 to sketch the bifurcation diagram shown in Figure 4.37. Note that here, rather than an exchange of stability between two existing equilibria, we have a bifurcation point $b = b_c$, $N^*/P = b(b-1)/2$ on either side of which two equilibria (one stable and one unstable) appear and disappear (respectively). This type of bifurcation is called a *saddle-node* bifurcation, because the two equilibria thus created are commonly referred to in higher dimensions as a saddle point (the unstable one) and a stable node. (We shall see the reasons for these names when we study systems of equations beginning in the next chapter).

Formal bifurcation analysis often involves shifting a bifurcation point to the origin to simplify the equation. In its simplified form, the saddle-node bifurcation is given by the equation

$$y' = p - y^2. \tag{4.25}$$

This has no equilibria if $p < 0$, and the two equilibria $y = \pm\sqrt{p}$ if $p > 0$. Since $f(y) = p - y^2$, $f'(y) = -2y$, and we have $f'(\pm\sqrt{p}) = \mp 2\sqrt{p}$. Thus the positive equilibrium is asymptotically stable and the negative equilibrium is unstable. Figure 4.38 shows the bifurcation diagram for (4.25). The term bifurcation literally means "splitting," and it is from situations such as this that the phenomenon takes its name.

We may also observe that bifurcations are the reason why in some cases we have needed to draw more than one phase portrait for a given differential equation. In fact, phase portraits can be seen as vertical "slices" of a

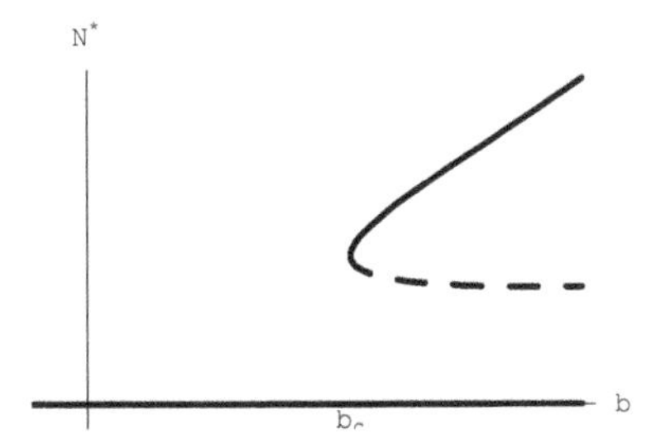

FIGURE 4.37: Bifurcation diagram for the saddle-node bifurcation in (4.24).

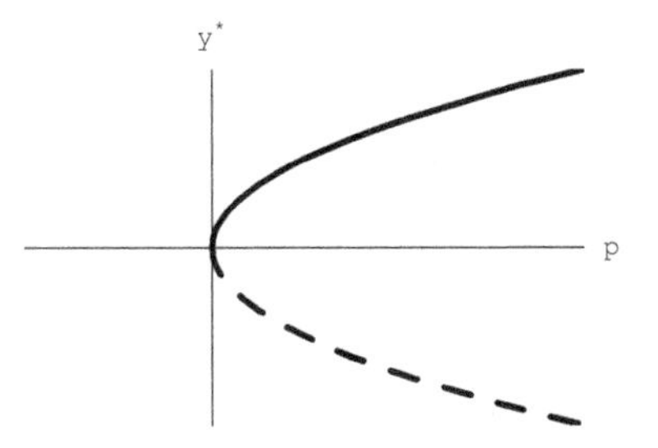

FIGURE 4.38: Bifurcation diagram for the saddle-node bifurcation in (4.25).

bifurcation diagram, on which the bifurcation parameter is constant. Drawing vertical lines to either side of the bifurcation in Figure 4.37, for example, and marking the equilibria on those lines with circles (solid for stable and hollow for unstable) produces the two phase portraits given for the "Pay It Forward" model in Figure 4.34 in Section 4.3.

A third common type of bifurcation, which combines elements of both of the previous two types, is illustrated by the following simplified model used by Bernard et al.[19] in their study of the dynamics of white blood cell production. They identified three processes that affect the population of hematopoietic stem cells (see Figure 4.40) in the resting phase: differentiation into mature white blood cells, assumed to occur at a per-cell rate F (here assumed constant); re-entry into the proliferative phase, at a per-cell rate driven by negative feedback to limit the number of proliferating cells; and entry from the proliferative phase by the newly divided cells. The negative feedback to avoid over-proliferation was modeled by Bernard et al. using a Hill function, of the form

$$K(S) = \frac{k\theta^s}{\theta^s + S^s}, \tag{4.26}$$

where k is the maximum rate of re-entry into the proliferative phase (when $S \approx 0$) and θ is the half-saturation constant [cell density]. Here the exponent s measures the sharpness of the saturation; for $s = 1$ we have a Verhulst function like the one used in the smooth-saturation "Pay It Forward" model. Finally, the proliferative phase is assumed to last precisely τ units of time, during which proliferating cells may experience apoptosis (cell death), at a per-cell rate of γ. Therefore, if $S(t-\tau)K(S(t-\tau))$ cells re-enter the proliferative phase at time $t-\tau$, then at time t (τ units later) they will return to the quiescent (resting) phase, multiplied in number by two (because of proliferation) and reduced by a proportion $e^{-\gamma\tau}$ due to apoptosis. (To understand this

[19]Samuel Bernard, Jacques Bélair, and Michael C. Mackey, Bifurcations in a white-blood-cell production model, *Comptes Rendus Biologies* 327: 201–210, 2004.

last factor, consider a fixed initial population P of cells in the proliferative phase undergoing apoptosis: we have $P' = -\gamma P$, so that $P(t) = P(0)e^{-\gamma t}$.) All together, these assumptions yield the model

$$\frac{dS}{dt} = -F\,S(t) - S(t)\,K(S(t)) + 2e^{-\gamma\tau}\,S(t-\tau)\,K(S(t-\tau)). \tag{4.27}$$

Because of the fixed time delay involved, equation (4.27) is a delay differential equation, rather than an ordinary differential equation. As such, its analysis is beyond the scope of this course, but we can nevertheless identify its equilibria, since at an equilibrium $S(t)$ and $S(t-\tau)$ are the same. We can see immediately that $S^* = 0$ is one equilibrium; factoring this out of the equilibrium condition, we are left with the equation $-F - K(S^*) + 2e^{-\gamma\tau}K(S^*) = 0$, which (using (4.26)) we can solve to get

$$S^* = \theta\left(\frac{k(2e^{-\gamma\tau}-1)}{F} - 1\right)^{1/s}.$$

If the exponent s is odd, then this expression gives a single (possibly negative) value of S^* regardless of parameter values. In this case, were we to choose a single parameter, say the maximum reactivation rate k, as a bifurcation parameter, we would find a transcritical bifurcation at $k = F/(2e^{-\gamma\tau} - 1)$ where the nonzero equilibrium crosses the zero equilibrium and exchanges stability with it. (We leave sketching the bifurcation diagram as an exercise for the reader.) However, if the exponent s is even, then the expression for S^* gives either 2 equilibria (the positive and negative roots), if $k > F/(2e^{-\gamma\tau}-1)$, or none if $k < F/(2e^{-\gamma\tau} - 1)$ (since negative numbers have no real sth roots for s even). Bernard et al. show that the zero equilibrium is stable when $k < F/(2e^{-\gamma\tau} - 1)$, and that near the bifurcation point the nonzero equilibria are stable, leading to the bifurcation diagram shown in Figure 4.39. In reality we assume that k must exceed this critical value, since the stem cell density remains positive, constantly producing new blood cells.

As the figure suggests, this type of behavior is called a *pitchfork bifurcation.* In its simplest form, a pitchfork bifurcation has the equation

$$y' = py - y^3. \tag{4.28}$$

This equation has equilibria $y^* = 0$ and $y^* = \pm\sqrt{p}$ (if $p > 0$). To determine stability, we calculate $f'(y) = p - 3y^2$. Since $f'(0) = p$, the zero equilibrium is asymptotically stable if $p < 0$ and unstable if $p > 0$. Since $f'(\pm\sqrt{p}) = -2p$, the two nonzero equilibria which exist if $p > 0$ are both locally asymptotically stable. As p increases through $p = 0$, a single, globally asymptotically stable equilibrium loses its stability and gives rise to two locally asymptotically stable equilibria. The corresponding bifurcation diagram showing the three equilibria as functions of p is given in Figure 4.41. Here we have both the appearance/disappearance of equilibria associated with saddle-node bifurcations and the stability exchange seen in transcritical bifurcations.

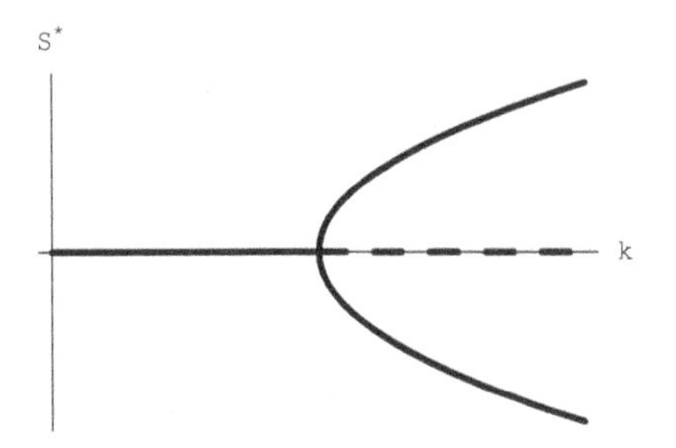

FIGURE 4.39: Bifurcation diagram for the pitchfork bifurcation in (4.27) with $s = 2$, using F as bifurcation parameter.

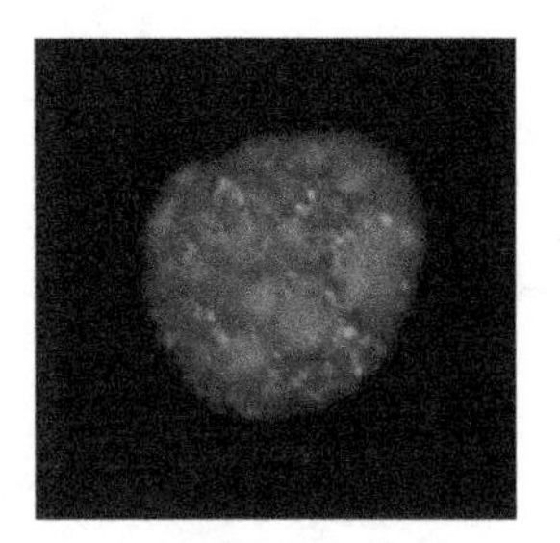

Photo courtesy Patricia Ernst

FIGURE 4.40: Hematopoietic stem cells (HSCs) develop into many different types of blood cells. These HSCs may proliferate (subdivide) many times before differentiating into mature white blood cells.

Having now seen the three most common types of bifurcation that a single differential equation may exhibit, we can now revisit some other models that involve bifurcations, to see what type of bifurcation each model involves.

EXAMPLE 1.
Classify any bifurcations exhibited by the constant-yield harvesting model (4.9) of Section 4.2.

Solution: Since H is an externally controlled parameter (as opposed to r and K, which are innate), it is the most sensible choice for bifurcation parameter. As discussed in Section 4.2, the model's only equilibria, given by equation (4.11), exist if and only if $H < rK/4$. At $H = rK/4$ the equilibria coalesce into a single point, and for $H > rK/4$ there are no equilibria. From this description, from their stabilities as determined earlier, and from the corresponding bifurcation diagram graphing the equilibria as functions of H (see Exercise 8), we can identify the bifurcation at $H = rK/4$, $y^* = K/2$ as a saddle-node. □

EXAMPLE 2.
Classify any bifurcations exhibited by the constant-effort harvesting model (4.12) of Section 4.2.

Solution: Here the effort E is the most reasonable choice of bifurcation parameter. As discussed in Section 4.2, the model's two equilibria cross at $E = r$, $y^* = 0$, with each one stable on a different side of the bifurcation point. From this description and the bifurcation diagram (see Exercise 9), the bifurcation can be identified as transcritical. □

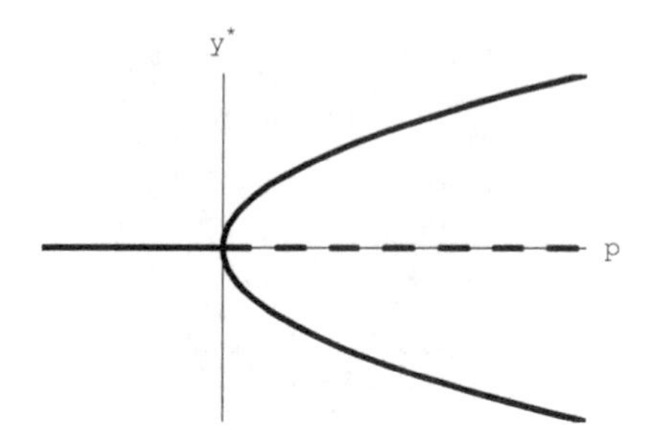

FIGURE 4.41: Bifurcation diagram for the pitchfork bifurcation in (4.28).

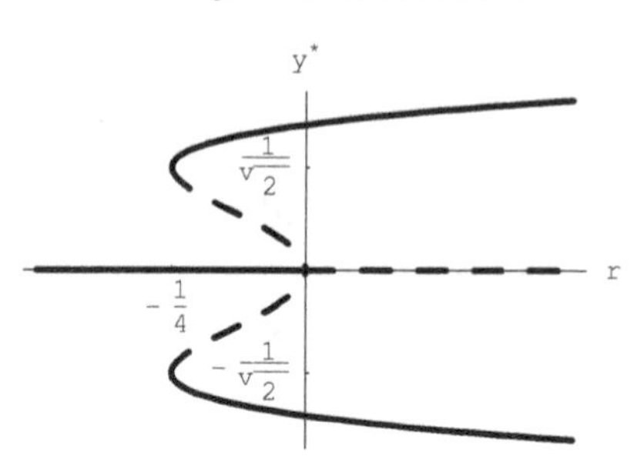

FIGURE 4.42: Bifurcation diagram for Example 3.

EXAMPLE 3.
Describe the behavior of solutions of the differential equation $y' = ry + y^3 - y^5$.

Solution: Equilibria are $y = 0$ and the solutions of $r + y^2 - y^4 = 0$, which is quadratic in y^2; solutions to the latter equation are given by

$$y^2 = \frac{1 \pm \sqrt{1+4r}}{2}$$

if $r \geq -\frac{1}{4}$. To determine stability, we substitute into $f'(y) = r + 3y^2 - 5y^4$ and find that $f'(0) = r$, so that the zero equilibrium is locally asymptotically stable if and only if $r < 0$, while

$$f'\left(\frac{1}{2}(1 \pm \sqrt{1+4r})\right) = -\sqrt{1+4r}\left(\sqrt{1+4r} \pm 1\right).$$

Thus for the two equilibria given by $y^{*2} = (1 + \sqrt{1+4r})/2$ which exist for $r \geq -1/4$ we have $f'(y^*) = -\sqrt{1+4r}(\sqrt{1+4r}+1)$, implying stability, while for the two given by $y^{*2} = (1 - \sqrt{1+4r})/2$ which exist for $-1/4 \leq r \leq 0$ we have $f'(y^*) = -\sqrt{1+4r}(\sqrt{1+4r}-1) < 0$, implying instability. We plot the bifurcation curve (Figure 4.42), which appears to show a pitchfork bifurcation at $r = 0$, $y^* = 0$ and two saddle-node bifurcations at $r = -\frac{1}{4}$, $y^* = \pm 1/\sqrt{2}$. □

If we consider what would happen to a solution of the differential equation in Example 3 as the parameter r changes through values in the range $[-1/4, 0]$, we will see that the bifurcations cause sudden jumps from one equilibrium value to another. If $r < -1/4$ initially, for instance, the solution y will approach the zero equilibrium, but if r increases past 0, the zero equilibrium becomes unstable, and the two outer equilibria become stable. If we suppose y to represent a population, so that $y(t) \geq 0$, then the solution $y(t)$ will jump toward the upper equilibrium, approaching it asymptotically. If later r decreases again, the solution will remain close to the upper equilibrium until the latter vanishes at $r = -1/4$, at which point the solution will jump back down toward zero. This discontinuous dependence of equilibrium population size on a parameter is known as *hysteresis*, a phenomenon with important enough biological consequences to warrant a brief discussion of its own.

4.4.1 Hysteresis

Hysteresis can be described mathematically by having two (or more) bifurcations at each of which a stable equilibrium either vanishes or becomes unstable, without a direct "hand-off" of stability to an intersecting equilibrium. These multiple discontinuities in the limiting value for solutions give the system what is thought of as "memory": in the interval between the discontinuities, there are multiple locally asymptotically stable equilibria, and the one which a given solution approaches depends upon which of the two regions beyond this interval the bifurcation parameter has most recently occupied. In the case of Example 3 above, if $-1/4 < r < 0$, then the solution $y(t)$ approaches the upper equilibrium if, prior to entering the interval $[-1/4, 0]$, r was most recently greater than 0, and approaches the zero equilibrium if r was most recently less than $-1/4$. In this way a directed loop is formed.

Hysteresis loops have significant biological consequences as well. First, a small change in environmental parameters may cause a sudden and very large change in population size. Second, once the change occurs, it may be difficult to undo: returning to the previous equilibrium requires a large change in the environmental parameter(s), far beyond the value(s) held before the change. For instance, in Example 3, if r increases past 0, in order to return the population to the zero equilibrium one must get r below $-1/4$. These consequences will be easier to understand in context, and so in the remainder of this section we will look at the effects of hysteresis through the lens of two biological systems which exhibit it.

The first such system is the ant foraging model (4.1) with which we began this chapter,

$$\frac{dx}{dt} = (\alpha + \beta x)(n - x) - \frac{sx}{x + K}.$$

In Section 4.1 we saw that this equation may have either one or three equilibria. The appearance or disappearance of two equilibria signals a bifurcation. Although the equilibrium condition (4.7) is cubic in the number x of ants foraging at a given source, it is linear in the colony size n, and we can easily solve for n in terms of x:

$$n = \left(1 + \frac{s}{(\alpha + \beta x)(x + K)}\right) x,$$

which allows us to plot a bifurcation diagram of x^* versus n. More generally, when studying the relationship between a bifurcation parameter and some property of a model (such as equilibria), we may often derive expressions which are complicated in the state variable (here, x) but simple, even linear, in the bifurcation parameter. Therefore, although we would really like to write an

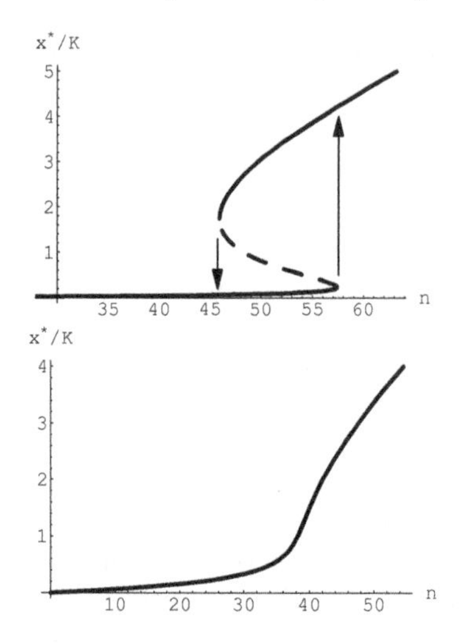

FIGURE 4.43: Bifurcation diagrams for the ant foraging model (4.1) with and without hysteresis.

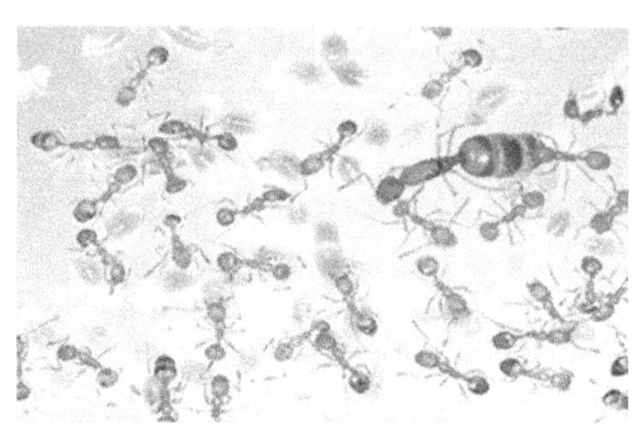

Photo courtesy Duncan Jackson

FIGURE 4.44: Foraging Pharaoh ants, *Monomorium pharaonis*, are considered a major nuisance in many places such as hospitals. Their ability to send out new colonies makes them difficult to eradicate.

expression for the state variable as a function of the parameter, we may instead do the reverse, writing the parameter as a function of the state variable, in order to graph it. This approach involves the notion of *inverse functions*, and rather than graphing the independent variable on the horizontal axis and the function on the vertical, we graph the variable (x, which we wish to interpret as a function of the bifurcation parameter) on the vertical, and the inverse function (which gives the parameter value) on the horizontal, thus giving us the graph we wanted in the beginning. This approach may also be referred to as defining the equilibrium value *implicitly* in terms of the bifurcation parameter.

Using the parameter estimates given by Beekman et al. and used in Section 4.1, we see that the behavior varies with α, the rate at which ants find the food source independently. The two possibilities are given in Figure 4.43: either there are no bifurcations, and the stable equilibrium is always unique, or else there are two saddle-node bifurcations, and a hysteresis loop exists in the region where three equilibria exist. In the latter case, a growth in colony size may result in a sudden leap in Pharaoh ants' (Figure 4.44) collective foraging ability at a single site, whereas a drop in colony size below the lower bifurcation point will suddenly wipe out the ants' ability to make use of their fellows' pheromone trails. Either of these sudden changes will be difficult to reverse, in that a much larger change in population size is necessary to undo the change than was necessary to cause it in the first place.

A caveat is in order here: as can be seen from the two cases in Figure 4.43, n is not the only bifurcation parameter. In fact, through rescaling we can show (see Exercise 12 below) that our model has three fundamental bifurcation parameters. It is important to be aware of how many model parameters really affect a system's bifurcation behavior. In all the other equations analyzed in this section, we have derived explicit conditions for bifurcations; for instance, each of the harvesting models from Section 4.2 only has one real bifurcation parameter, since for any given parameter for which we sketch a bifurcation diagram for those models, the graph remains qualitatively the same for all allowable values of the other model parameters. This is not true, however, for Bernard et al.'s white blood cell production model, nor is it true for the ant foraging model.

Note that we have now classified all the bifurcations observed in models presented in earlier sections and chapters, save one: the discrete logistic model. Single differential equations cannot undergo period doubling as single difference equations can, because state variables change continuously rather than making discrete jumps, and therefore cannot jump past an equilibrium value. We will, however, see an equivalent behavior in systems of differential equations in later chapters.

4.4.2 The spruce budworm

The spruce budworm *Choristoneura fumiferana* (see Figure 4.45) is an insect which attacks spruce and fir trees in the forests of eastern North America. Most years its population level is relatively low, but in some years (historically roughly every forty years) it exhibits outbreaks in which the population may increase by a factor of 1000. During an outbreak, budworms may eat enough new needles in an evergreen forest to kill 80% of the trees in the forest and effectively destroy the forest (see Figure 4.46). The destruction of the forest also eliminates the budworms' food supply, causing a collapse of the budworm population, after which the forest can begin a slow recovery. A full model of this behavior would include both the spruce budworm and the forest, and these operate on very different time scales (weeks vs. decades). Here, we shall describe a model for the budworm population (the fast variable) but rather than also modelling the forest (the slow variable) we shall allow some of the parameters of the model to vary to describe the change in the forest in response to the budworm dynamics. This technique is often used to simplify the analysis of models for systems with processes or components that act on vastly different time scales.

This insect has been extensively studied, notably by Ludwig, Jones and Holling.[20] Consistent with this prior work, we let y denote the budworm population density (in larvae/acre) and assume that its natural growth is logistic, with capacity K a function of the forest's size (in particular, the average leaf surface area per tree), and maximum growth rate r intrinsic to budworm biology.

[20] D. Ludwig, D.D. Jones and C.S. Holling, Qualitative analysis of insect outbreak systems: the spruce budworm and forest, *Journal of Animal Ecology* 47(1): 315–332, February 1978.

FIGURE 4.45: Mature spruce budworm larva.

FIGURE 4.46: Spruce budworm induced whole tree mortality in a natural conifer stand.

In addition, the spruce budworm is subject to predation by birds and parasites which saturates at a level H for high budworm densities. We assume that the predators are opportunistic, choosing the most abundant prey, and therefore preying little upon spruce budworms until they exceed a threshold density A. This assumption leads us to use what is called a Holling type III function to describe the predation: in particular, a second-order Hill function, like that used in the blood cell production model presented earlier in this section. The reason for this choice is that Hill functions stay close to zero before the population passes the threshold A, and rise quickly to the maximum thereafter (see Figure 4.47). Ludwig et al.'s research suggested that the saturation threshold A is, like K, proportional to the average leaf surface area per tree.

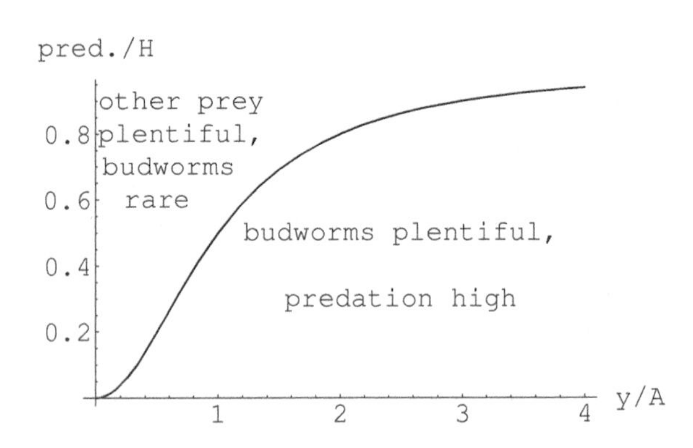

FIGURE 4.47: A second-order Hill function illustrates Holling type III predation, with opportunistic predators preying only on the most populous prey.

These assumptions lead to the following model for spruce budworm growth:

$$y' = ry\left(1 - \frac{y}{K}\right) - H\,\frac{y^2}{y^2 + A^2}. \tag{4.29}$$

The model involves the four parameters r, K, H, and A. As discussed in Chapter 1, it is often convenient to reduce the number of parameters by rescaling the model: rescaling both population and time relative to certain benchmark values will reduce the number of parameters by two (and with only two left, we can more easily graph the different cases that arise). There are several options for the rescaling benchmarks. Ludwig et al. used the half-saturation constant A and the per-budworm-density predation rate H/A (see Exercise 13); here we shall instead use the carrying capacity K as our population benchmark and [the reciprocal of] the budworms' natural growth rate r as our time benchmark, as did Robert May in his review of the work of Ludwig et al.[21] The qualitative results will be the same, and we encourage the interested reader to compare our analysis below with that of Ludwig et al.

We therefore define the rescaled variables $x = y/K$ (budworm larva density relative to the forest's carrying capacity for them) and, if we call the original time variable τ, the new time scale $t = r\tau$, measured in budworm "generations" ($t = \tau/\frac{1}{r}$, where $\frac{1}{r}$ is the average time for a budworm to reproduce). We will also find it convenient to define the rescaled parameters $a = A/K$, which gives the threshold density for predation relative to the forest's carrying capacity, and $h = H/H_c = H/\frac{rK}{4}$, the predation rate expressed as a proportion of the critical harvesting rate H_c (identified in Section 4.2). Invoking the Chain Rule twice,

$$\frac{dx}{dt} = \frac{dy}{d\tau}\,\frac{dx}{dy}\,\frac{d\tau}{dt},$$

with $dx/dy = 1/K$, $d\tau/dt = 1/r$, we obtain the rescaled model

$$\frac{dx}{dt} = x(1 - x) - \frac{h}{4}\,\frac{x^2}{x^2 + a^2}. \tag{4.30}$$

Looking for equilibria, we find that either $x^* = 0$ or $(1-x^*) - \frac{h}{4}\frac{x^*}{x^{*2}+a^2} = 0$. In the latter case we can multiply through by $x^{*2} + a^2$ to make the equation polynomial:

$$G(x^*) \equiv (1 - x^*)(x^{*2} + a^2) - \frac{h}{4}\,x^* = 0.$$

The function G is cubic with $G(0) = a^2 > 0$ and $G(1) = -\frac{h}{4} < 0$, so it has either 1 or 3 roots in (0,1). To determine local asymptotic stability, we differentiate $f(x)$, the right-hand side of equation (4.30), and find that

$$f'(x) = 1 - 2x - \frac{h}{4}\,\frac{2a^2x}{(x^{*2} + a^2)^2}.$$

[21] Robert M. May, Thresholds and breakpoints in ecosystems with a multiplicity of stable states, *Nature* 269: 471–477, 06 October 1977.

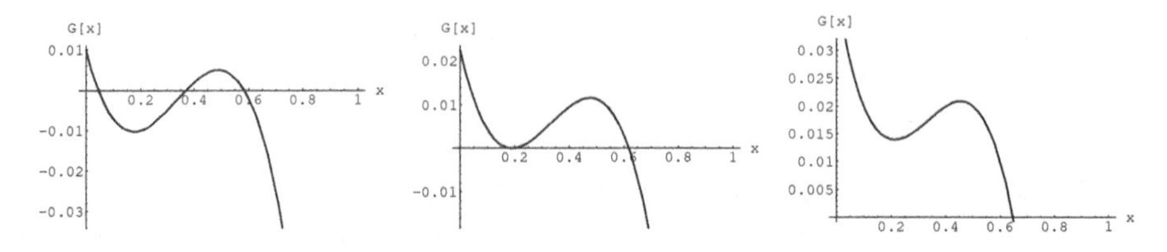

FIGURE 4.48: As model parameters a and h change, the polynomial $G(x)$ goes from having three roots in (0,1) to having only one; the change occurs at a bifurcation point where G has a double root (center graph).

We calculate $f'(0) = 1 > 0$, so the equilibrium $x^* = 0$ is unstable. This means that the graph of $f(x)$ crosses 0 from below at $x = 0$, so the next time it crosses 0, it must do so from above, making $f'(x^*) < 0$ for the first positive equilibrium, and alternating in this way thereafter. Thus the first positive equilibrium is locally asymptotically stable, and, if there are three positive equilibria, the second and third will be unstable and stable, respectively.

Returning to the question of bifurcations, we can identify the combinations of the parameters a and h where bifurcations occur by noting that, at a bifurcation point, there must be a double root of $G(x)$: that is, G goes from crossing the x-axis once on the interval (0,1) to crossing it thrice (or vice versa) via a situation in which it just touches the x-axis tangentially at a single point (see Figure 4.48). At such points, not only is $G(x) = 0$, but also $G'(x) = 0$. Although these equations are cubic and quadratic, respectively, in x, they are linear in both a^2 and h, so (again using an implicit approach) we can solve them to obtain a and h parametrically, in terms of x:

$$G'(x) = 2x - 3x^2 - a^2 - \frac{h}{4} = 0 \Leftrightarrow a^2 = 2x - 3x^2 - \frac{h}{4}.$$

We substitute this expression for a^2 into the equation $G(x) = 0$:

$$(1 - x)\left(x^2 + \left[2x - 3x^2 - \frac{h}{4}\right]\right) - \frac{h}{4}x = 0.$$

Solving for h, we get $h = 8x(1 - x)^2$, from which $a^2 = x^2(1 - 2x)$. Since by definition $a > 0$, we take the positive square root: $a = x\sqrt{1 - 2x}$. This last expression gives us the necessary bounds on the parameter x: $0 \leq x \leq 1/2$; that is, different combinations of a and h produce bifurcations at x values anywhere from 0 to 1/2, and letting x vary from 0 to 1/2 will allow us to trace out on the a–h plane precisely the combinations that produce those bifurcations. (We can refer to this approach as reverse parametrization.) Modern scientific computing software such as Mathematica, Maple and MATLAB can create such parametric plots, and the result is given in Figure 4.49.

Thus for most values of a and h there are only two equilibria (the unstable zero equilibrium and a globally stable positive one), but for small a

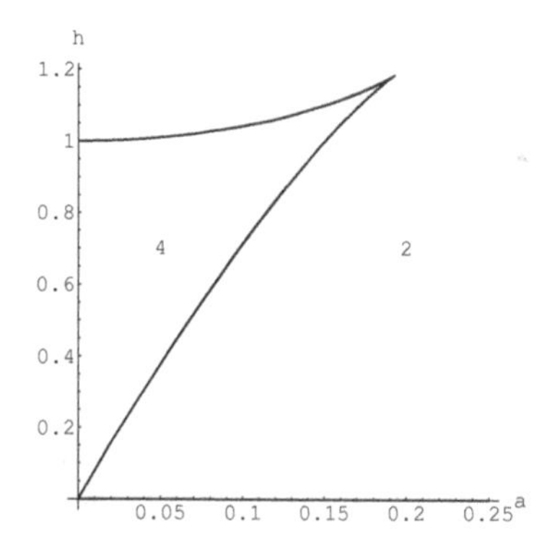

FIGURE 4.49: The parametric curve $a = x\sqrt{1-2x}$, $h = 8x(1-x)^2$ $(0 \le x \le \frac{1}{2})$ marks bifurcations dividing the a–h plane into regions where (4.30) has either two or four equilibria.

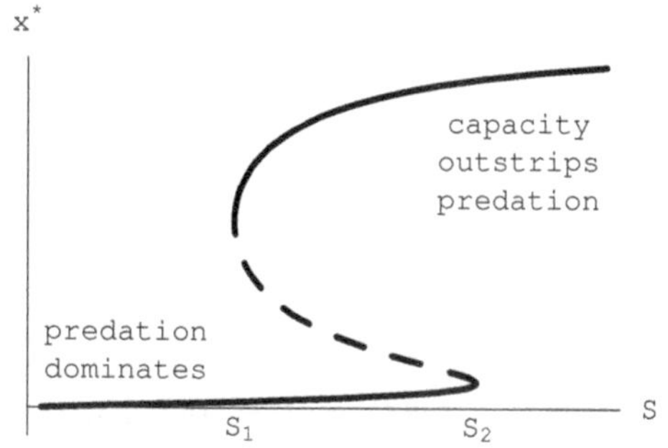

FIGURE 4.50: A bifurcation diagram for (4.30) showing equilibrium budworm density as a function of forest size S.

and small enough h—that is, when predation saturates ("turns on") for even relatively low budworm densities, and is too weak to keep up with budworm reproduction—there are four (the unstable zero equilibrium, and two stable positive equilibria separated by an unstable one).

We now return to the context in which we developed the model: the budworm population developing on a fast timescale, and the forest on a slow timescale. In order to see what effect the forest's growth and decay has on the budworm population, we need to establish what relationship exists between the two. Of the model's four original parameters, the carrying capacity K and predation saturation threshold A were both found to be directly proportional to the state of the forest (in particular, to the leaf surface area density, measured in branches per acre). If we consider our rescaled parameters $a = A/K$ and $h = H/\frac{rK}{4}$, we find a to be independent of forest development as the two linear factors—say $A = \alpha S$ and $K = \kappa S$ for foliage density S—cancel each other out, while $h = 4H/r\kappa S$ varies inversely with foliage density. Since the equilibrium condition $G(x) = 0$ is linear in h, it is a simple matter to solve it for h in terms of x, and then to substitute into the expression $S = \frac{4H}{r\kappa h}$, to get the equilibrium budworm density x^* implicitly as a function of S:

$$S = \frac{H}{r\kappa} \frac{x}{(1-x)(x^2+a^2)}.$$

Figure 4.50 shows the resulting bifurcation diagram for a representative value of a in the interval $(0, 1/3\sqrt{3})$ where the bifurcations occur (see Exercise 14 for the identification of these limiting values).

TABLE 4.3: Empirical parameter estimates for (4.30), from Ludwig et al. (1978).

Parameter	Value	Units
r	1.52	/yr
α	1.11	larvae/branch
κ	355	larvae/branch
H	43 200	larvae/acre/yr

Looking at the bifurcation diagram, we see that in young forests (or those largely degraded by infestation), with S small, the budworm density is kept relatively low, first because the carrying capacity $K = \kappa S$ is low, and second because the predation threshold $A = \alpha S$ is also low, so that predators choose the budworms as primary prey. Here A is low because there are as yet so few [healthy] branches that it does not take predators long to find budworms. In this way, predation keeps the budworm population in check. As the forest grows, both the carrying capacity and the predation threshold increase, leading to a gradual increase in budworm density. When S reaches the upper bifurcation point, marked S_2 in Figure 4.50, the food supply has become so great that predation can no longer contain the budworm reproduction, and the budworm density rises exponentially quickly toward the upper equilibrium, where budworm growth is limited only by the carrying capacity of the food supply. At this upper equilibrium, the budworm population begins to degrade the forest, and S begins to decrease. If the deforestation caused by the budworms causes it to decrease below the lower bifurcation point, marked S_1 in Figure 4.50, then the food supply—and cover for the budworms—disappear, leaving the budworms again dominated by predation. The upper equilibrium disappears, and the population crashes back down to the lower equilibrium, where they remain while the forest begins to rebuild, thus completing the hysteresis loop. Since the forest growth takes decades, these cycles will also, explaining the roughly 40-year budworm irruptions that were long observed.

The remaining question, given that hysteresis only occurs in our model for a certain range of parameter values, is whether in fact observations yield parameter estimates within this range. Ludwig et al. report two sets of parameter estimates, one based upon general principles and the other based upon empirical observations. In most cases the two estimates are fairly close. We reproduce in Table 4.3 the relevant parameters from their Table 1. From them, we can calculate $a = \alpha/\kappa = 1.11/355 \approx 0.00313$, which is well within the range $a < 1/3\sqrt{3} \approx 0.192$ where hysteresis occurs.

The phenomenon of hysteresis has been observed in the outbreak and subsequent crash of many different insect populations. There is, however, another possible behavior for the system which has been observed in practice. This involves human intervention to keep the insect population down by spraying, in order to avoid the collapse of the forest (of course, this is not incorporated in our model). In this case, the insect population remains at a high equilibrium level, held in check only by ongoing intervention. Such behavior is called "perpetual outbreak."

Exercises

In Exercises 1–7, describe the behavior of solutions of the given differential equation as the parameter r is varied.

1. $y' = r + y^2$
2. $y' = ry - ye^y$
3. $y' = ry + y^2$
4. $y' = r - y + \frac{y^2}{y^2+1}$
5. $y' = y + ry^3$
6. $y' = ry(1-y) - \frac{y}{y+1}$
7. $y' = ry + 1 + y^2$
8. Draw a bifurcation diagram for the constant-yield harvesting model (4.9) of Section 4.2 using H as a bifurcation parameter, and show how it includes the three phase portraits given in Figure 4.15.
9. Draw a bifurcation diagram for the constant-effort harvesting model (4.12) of Section 4.2 using E as a bifurcation parameter, and show how it includes the two phase portraits given in Figure 4.22.
10. Use the bifurcation diagrams for transcritical, saddle-node, and pitchfork bifurcations to sketch all [qualitatively distinct] phase line portraits for each model, superimposed on the bifurcation diagrams.
11. Use Figure 4.43 to sketch all [qualitatively distinct] phase line portraits for equation (4.1), superimposed on the bifurcation diagrams.
12. Rescale both the colony size x and the time variable t in the foraging ant model (4.1) to reduce the number of model parameters to three.
13. (a) Rescale the spruce budworm model (4.29) using the benchmarks Ludwig et al. originally used, defining the new variables $u = y/A$ and $s = \tau H/A$ and the new parameters $Q = K/A$ and $R = rA/H$. Which of these parameters is more reasonable to use as a bifurcation parameter?

    (b) A technique that we used frequently in Chapter 2 to analyze equilibria for discrete-time models was to consider equilibria as the intersections of the graphs of two functions $f(x)$ and $g(x)$, in cases where the equilibrium condition can be written as $f(x) - g(x) = 0$ or $f(x) = g(x)$. Use this technique to identify and analyze the stability of the equilibria for the rescaled budworm model you derived in part (a).

    (c) Interpret the results in (b) in terms of forest growth through the bifurcation parameter chosen in (a).

14. We can identify the range of values of the parameter a for which the rescaled budworm model (4.30) undergoes two bifurcations as h (or S) varies, by using the fact that the parametric curve $a = x\sqrt{1-2x}$, $h = 8x(1-x)^2$ $(0 \leq x \leq 1/2)$ identified in the text "turns around" at the cusp point. That is, as x increases from 0, the curve moves up and to the right from the origin (that is, $da/dx > 0$ and $dh/dx > 0$). At the cusp point, however, the curve begins to move down and to the left. Therefore, at the cusp point $da/dx = dh/dx = 0$. Use this fact to find the values of a and h at which this occurs.

4.5 Numerical analysis of differential equations

We began this chapter by observing that most differential equation models for biological systems are nonlinear and cannot be solved outright. Section 4.1 presented some qualitative tools for analyzing the long-term behavior of such models. In this section we shall conside briefly how numerical (quantitative) analysis is used to study the short-term behavior of such models—what will the spruce budworm density be in this forest a month from now? How long will it take for an epidemic outbreak to drop below a given number of cases? (Quantitative analysis can also be used to develop intuition about the long-term behavior of systems in cases where we cannot complete a qualitative analysis.) Numerical analysis, which uses computers to approximate solutions of dynamical systems such as differential equations, has a long and detailed history, and today many different methods (some quite specialized) are used to analyze mathematical models numerically. Even a full introduction to numerical analysis would fill an entire book, so in this section we shall simply try to give the reader an understanding of the general approach behind numerical methods, the relationship between the true solution to a differential equation and the approximation to that solution produced by numerical methods, and the trade-offs involved in quantitative modeling.

There is a great deal of computer software available to generate numerical approximations of solutions to differential equations. Some of this software is part of a larger computing system, such as Mathematica, MATLAB or Maple; other programs are smaller and more narrowly aimed. Most useful programs can generate both a table of values for a solution and a graphic portrayal of solution curves. We encourage the reader to become familiar with some computer software for approximating solutions of differential equations, and to experiment with it, but not to use it to the exclusion of all other approaches. In particular, numerical approximations are usually not a substitute for the information provided by the qualitative approach of Section 4.1. One reason for this is that numerical approximations require as inputs specific values for all model parameters, and consequently the results only give information about the system's behavior for those particular parameter values; the system's behavior for other parameter values may be quite different, even if the differences in values are slight. Another reason is that all approximations involve errors, as we will discuss in more detail below.

The basic problem we would like to solve is to estimate the value of the population $y(t)$ at some particular moment in time T, given an initial population size y_0 and a description (a differential equation) of how that population changes over time. In mathematical terms, we want to solve the initial value problem

$$y' = f(t, y), \quad y(t_0) = y_0. \tag{4.31}$$

Numerical analysis works by converting this continuous-time problem into a discrete-time problem (i.e., a difference equation $y_{k+1} = g(k, y_k)$), which can be solved one time-step at a time, just as we did when we first encountered nonlinear difference equations in Section 2.1. In order to accomplish this conversion, we first subdivide the time interval of interest, $t_0 \leq t \leq T$, into N subintervals of length h, by defining $t_1 = t_0 + h$, $t_2 = t_1 + h = t_0 + 2h$, ..., $t_N = T = t_0 + Nh$. The number of subintervals N and the length of each subinterval h are related by $T - t_0 = Nh$; that is, once the time interval is known, choosing the time-step h is equivalent to choosing the number of time-steps $N = (T - t_0)/h$. The fundamental idea is to proceed to $y(T)$ by first approximating $y(t_1)$, then $y(t_2)$, then $y(t_3)$, etc. Each calculation is based on an approximate value for $y'(t)$ at that step, obtained using the value of the solution at one or more points (but not involving any limiting process). In other words, at each step we try to make the approximation y_k as close as possible to the real solution at that point, $y(t_k)$, so that the final value y_N is as close as possible to the real solution $y(t_N) = y(T)$. A numerical method must then have a way to use the function $f(t, y)$ from the original differential equation to derive a function $g(k, y)$ that will produce faithful estimates.

4.5.1 Approximation error

Any numerical approximation method is subject to two sources of error:

(i) *Discretization error*, caused by the approximations used in formulating the method (converting the differential equation into a difference equation).

(ii) *Round-off error*, caused by the limitation of the computer or calculator in the number of digits it can retain in its calculations.

The errors from previous steps may tend to propagate and build up. In general, discretization error can be reduced by choosing time-steps small enough that the discrete approximation cannot diverge far from the original system's solution. (More specifically, if we can make the approximation as close as we like to the true solution by choosing a small enough time-step, then the numerical method is said to *converge*, a prerequisite for any useful method.)

However, a smaller time-step also means more calculations to reach the final time T, which in turn increases the cumulative round-off error.

In practice, therefore, the choice of step size h is a challenge: If the step size is too large, the discretization error at each stage may be large, and these errors may accumulate catastrophically. On the other hand, if the step size is too small the computation may require a long time, and the large number of steps may produce large accumulated round-off errors. One common procedure is to vary the step size, selecting an optimal h value at each stage to minimize the total error of both kinds, but one may instead simply repeat the calculation with successively smaller values of h until further reduction of h makes no significant change in the approximate value of $y(T)$ for the chosen time T. In practice, computer software often carries out the numerical approximation automatically without asking the user to provide the step size and without indicating what method is being used.

We should also be concerned that the qualitative behavior of the original (continuous-time) system and its (discrete-time) approximation agree: for instance, that any equilibrium of one is also an equilibrium of the other, and that each equilibrium is stable for the same range of parameter values in both models. Since we have seen that even with a simple logistic model can behave very differently in continuous and discrete time, we must verify each of these properties for any given numerical method.

4.5.2 Euler's method

In this section we will describe three common numerical methods. In each case we will only use a constant step size h, although in practice many sophisticated methods adjust the step size during the approximation process. The first, and simplest, of these methods is known as *Euler's method*, originated by the great mathematician Leonhard Euler (1707–1783).

The idea behind this method is to extrapolate at each step along a line tangent to the true solution: in effect, to linearize from one step to the next. If we imagine a direction field (see Section 3.3) imposed on a graph, we begin at the initial condition and jump forward h time units in the direction signaled by the direction field at that point. Then, from this new point, we again jump ahead following the direction field. Although the direction field varies continuously, the idea is that if we make small enough jumps we should not go too far afield. While the Euler method is much less accurate than the approximation methods used in standard computational software, it gives us a great deal of insight into the underlying principles of numerical approximation of solutions to differential equations.

Mathematically, Euler's method is based on the difference quotient

$$y'(t) \approx \frac{y(t+h) - y(t)}{h}$$

(true for small h). We can rewrite this approximation as

$$y(t+h) \approx y(t) + h\,y'(t).$$

Euler's method therefore replaces the differential equation (6.18) with the difference equation

$$y(t+h) = y(t) + h\,f(t,y), \tag{4.32}$$

or in discrete terms ($t_k = t_0 + kh$)

$$y_{k+1} = y_k + h\,f(t_k, y_k), \quad k = 0, 1, 2, \ldots, N. \tag{4.33}$$

If we start with $t = t_0$, we can use the difference equation (4.32) to calculate an approximation

$$y_1 = y(t_0) + hf(t_0, y(t_0)) = y_0 + hf(t_0, y_0)$$

to $y(t_1)$. We now take $t = t_1$ and use the approximation y_1 for $y(t_1)$ in (4.32) to obtain an approximation

$$y_2 = y_1 + hf(t_1, y_1)$$

to $y(t_2)$. We repeat this procedure, letting y_k be the approximation to $y(t_k)$ for $k = 1, 2, \ldots, N$ and calculating these approximations recursively using (4.33) until we have our estimate y_N for $y(T) = y(t_N)$. In general, a numerical method is defined by the difference equation it uses to approximate a differential equation; the Euler method is identified by (4.33). The Euler method is usable on a microcomputer or programmable calculator, or even on a simple non-programmable hand calculator if speed and convenience are not vital.

EXAMPLE 1.
Use Euler's method with a step size of 1 to approximate the solution of the logistic equation $y' = ry(1 - y/K)$ on the interval $0 \le t \le 10$ for $r = 1$, $K = 10$, and $y_0 = 1$.

Solution: Following (4.33), we compute

$$\begin{aligned} y_1 &= y_0 + hry_0(1 - y_0/K) = 1 + 1(1)1(1 - 1/10) = 1.9, \\ y_2 &= y_1 + hry_1(1 - y_1/K) = 1.9 + 1(1)1.9(1 - 1.9/10) = 3.439, \\ y_3 &= y_2 + hry_2(1 - y_2/K) = 3.439 + 1(1)3.439(1 - 3.439/10) = 5.6953279, \end{aligned}$$

and so forth, up to $y_{10} \approx 10$. The graphs of the true solution (given by equation (2.13)), the approximation $u(t)$ based on the above values for y_k, and a coarser approximation based on a step size of $h = 1.5$ are shown in Figure 4.51. □

In the example above and in Figure 4.51, we use impractically large step

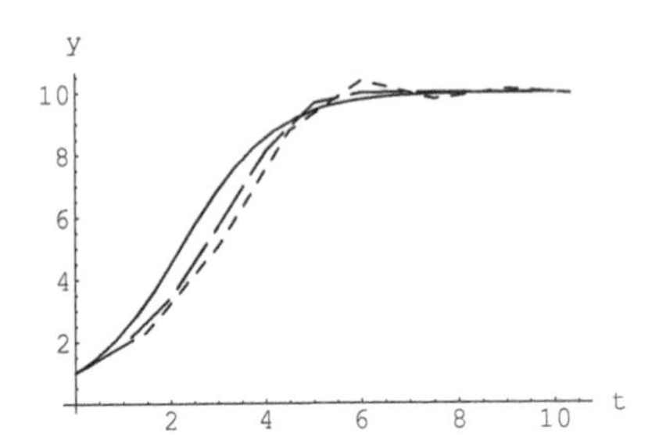

FIGURE 4.51: Comparison of the actual (solid curve) and two approximate solutions ($h = 1$ long dashes, $h = 1.5$ short dashes) to the logistic equation of Example 1.

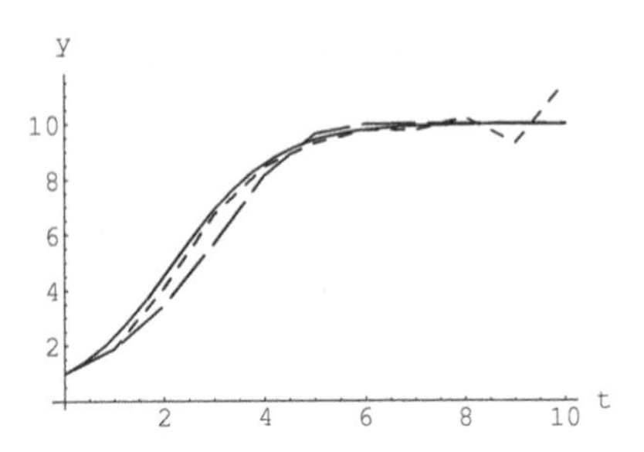

FIGURE 4.52: Comparison of the true solution (solid curve) to the logistic equation of Example 1 and (4.32) with the approximations ($h = 1$) generated by Euler's method (long dashes) and the modified Euler method (short dashes).

sizes to illustrate some basic properties of approximations (in practice we would use step sizes orders of magnitude smaller, perhaps one-millionth the length of the entire interval). Note first of all that the places where the approximations diverge most from the true solution are those where the derivative $y'(t) = f(t, y)$ changes rapidly; because Euler's method essentially assumes that dy/dt is constant on each subinterval, the discretization error is greatest following a subinterval on which the true value of dy/dt changes significantly. For the S-shaped logistic curve, this means underestimating the initial jump in growth and then overshooting when the population levels off near the carrying capacity K. Second, note that the smaller step size ($h = 1$) produces a closer approximation than the larger one ($h = 1.5$).

Properly speaking, the approximations produced by any numerical method are discrete sequences rather than continuous functions, and should be graphed discretely, with dots; however, in order to compare such approximations with the true solution, we commonly graph the approximations by connecting the points y_k to form continuous functions $u(t)$ which can be used to approximate $y(t)$ at any point in time. In this way, the graph of the function $u(t)$ defined on each subinterval by

$$u(t) = y_k + (t - t_k)f(t_k, y_k) \ \ [t_k \leq t \leq t_{k+1}]$$

is made up of straight line segments joining the points (t_k, y_k) and (t_{k+1}, y_{k+1}). Each straight line segment has the same direction as the element of the direction field (Section 3.2) at the point (t_k, y_k). A good approximation is one for which the graph of the function $u(t)$ is close to the graph of the true solution $y(t)$.

We can also use this simple example to look at errors in more detail. Since we used large step sizes in Example 1, the approximation error is due primarily to discretization. For example, we calculated $y_3 = 5.6953279$, while the true solution has $y(3) = \frac{10}{1+9e^{-3}} \approx 6.90568$, an error of about 17.5%. However, if we had rounded the value of $y_2 = 3.439$ to 3.4 in calculating y_3, we would have obtained the estimate $3.4+1(1)3.4(1-3.4/10) = 5.644$ for y_3, which introduces an additional 0.0513279, or 0.74%, round-off error. Furthermore, if we then round this estimate for y_3 to 5.6 in order to calculate y_4, the round-off error will accumulate (note the true value of y_3 rounds to 5.7). Although rounding to two digits is an extreme example, the true value of y_{10} in Example 1 requires over 1000 decimal places to represent completely, which is more than even the most precise software normally uses, so there is round-off error even at high precision.

It is possible to show using Taylor approximations (see Appendix B) that, in general, the discretization error of the Euler method in each step is no greater than Mh^2, where M is some constant whose value depends on the function $f(t, y)$ but not on h. Taking a smaller value of h will decrease the discretization error at each step, but taking a smaller value of h means taking more steps and may increase the round-off error. The best accuracy may require an intermediate value of h. The accumulated discretization error of the Euler method after $N = (T - t_0)/h$ steps is then at most $Mh^2(T - t_0)/h = M(T - t_0)h$, that is, Kh for some constant K. The size of the round-off error (relative to h) depends on the precision used in the calculations.

We can also compare the qualitative behaviors of solutions to the discrete-time model used by a given numerical method with the solutions to the original continuous-time model. For Euler's method, we compare the difference equation (6.10) with the differential equation (4.31). Any equilibrium y^* of the differential equation (4.31) must have $f(t, y^*) = 0$ for all t; substituting this into (6.10) gives $y_{k+1} = y_k$, so y^* must also be a fixed point of the difference equation (6.10), and vice versa. We should be concerned about the possibility that the equilibrium y^* might be asymptotically stable for (4.31) but unstable for (6.10). An equilibrium y^* of (4.31) is asymptotically stable if $f'(y^*) < 0$, while an equilibrium y_∞ of (6.10) is asymptotically stable if $|1 + hf'(y_\infty)| < 1$, or

$$-2 < hf'(y_\infty) < 0.$$

The upper inequality is equivalent to $f'(y_\infty) < 0$, the asymptotic stability condition for (4.31). The other condition is $f'(y_\infty) > -\frac{2}{h}$, and this condition can always be satisfied if h is sufficiently small. Thus there is no instability problem using the Euler method if h is chosen sufficiently small.

EXAMPLE 2.

Use the Euler method to estimate the value of e as $y(1)$, where $y(t)$ is the solution of the initial value problem $y' = y$, $y(0) = 1$, and determine the discretization error.

Solution: The Euler method gives

$$y_0 = 1,\ y_1 = 1 + hy_0 = 1 + h,\ y_2 = y_1 + hy_1 = y_1(1 + h) = (1 + h)^2,\ \ldots$$

If we take $h = 0.1$, so that $N = 10$, then the approximation to e is $y_{10} = (1+h)^{10} = (1.1)^{10} \approx 2.5937$. If we take $h = 0.01$, we would have $y_{k+1} = (1+h)y_k = 1.01\, y_k$, and this would approximate e by $y_{100} = (1.01)^{100} \approx 2.7048$. Similarly, $h = 0.001$ would lead to the approximation $(1.001)^{1000} = 2.7169$. In general, the approximation to e given by Euler's method is $(1+h)^{1/h}$, and we know from calculus that the limit of this expression as $h \to 0$ is e. Here we can see that reducing h from 0.01 to 0.001 changes our approximation of e by about 0.01, so we may be confident that our approximation is correct to the nearest tenth, i.e., $e \approx 2.7$. (The actual value of e is approximately 2.7183, which yields discretization errors of about 0.1246, 0.0135, and 0.0014, or 4.6%, 0.5%, and 0.05%, for $h = 0.1$, 0.01, and 0.001, respectively.) □

4.5.3 Other numerical methods

Although the Euler method can provide accurate approximations for small enough step sizes, in practice more sophisticated methods are used to approximate the solutions of differential equations. They all share the ideas of division into subintervals and the use of difference equations, but make more complicated calculations at each step, in order to generate less error.

As an example, the *modified Euler method* uses a slight change in method to improve results. The Euler method is based on the approximation $y'(t) \approx [y(t+h)-y(t)]/h$, which uses the derivative at the beginning of each subinterval as the derivative over the entire interval. We might suppose that the derivative in the middle of a subinterval might be a better approximation of the derivative over the entire interval, and indeed it can be shown (although we leave the proof to a text on numerical analysis) that such an approach generates a smaller error in general. There are several numerical methods based upon this idea; one is the modified Euler method, which uses the difference quotient

$$y'(t) \approx \frac{y(t+h) - y(t-h)}{2h}.$$

(For another such method, see the midpoint method in Exercise 9.) Here we are actually using the derivative at t as the derivative over the double subinterval $[t-h, t+h]$. To implement this method, we substitute for $y'(t)$ in (4.31):

$$\frac{y(t+h) - y(t-h)}{2h} = f(t, y(t))$$

or

$$y(t+h) = y(t-h) + 2hf(t, y(t)).$$

For a given time t_k, we have $t_k + h = t_{k+1}$ and $t_k - h = t_{k-1}$, so that we can rewrite the difference equation as

$$y_{k+1} = y_{k-1} + 2hf(t_k, y_k) \;\; [k = 1, 2, \ldots, N-1]. \tag{4.34}$$

If the initial value $y_0 = y(t_0)$ and an approximation y_1 for $y(t_1)$ are known, then the approximation y_2 for $y(t_2)$ is given by

$$y_2 = y_0 + 2hf(t_1, y_1).$$

In order to use this method, it is necessary to estimate y_1 by some other means. For example, we could use the Euler method (preferably with a smaller value of h) to approximate $y(t_1)$.

It is possible to show that the discretization error in the modified Euler method at each step is at most $\tilde{M}h^3$, as compared to Mh^2 for the Euler method. The higher power of h in this error bound indicates greater accuracy for the modified Euler method (since h is small). However, the constants in the two bounds are different, and the constant $\tilde{M}$ in the bound for the modified Euler method may be larger than M. Thus the improvement may be less than expected unless h is small enough.

EXAMPLE 3.

Use the modified Euler method with a step size of 1 to approximate the solution of the logistic equation $y' = ry(1 - y/K)$ on the interval $0 \leq t \leq 10$ for $r = 1$, $K = 10$, and $y_0 = 1$. Compare the accuracy to that of Euler's method using the same step size.

Solution: As noted above, before we can apply the modified Euler method we must first obtain an estimate y_1 for $y(t_1)$. To keep things simple, let us use the estimate $y_1 = 1.9$ given by the original Euler method (with $h = 1$) in Example 1 (a smaller h would give a better y_1). Then, following (4.34), we compute

$$\begin{aligned}
y_2 &= y_0 + 2h\,ry_1(1 - y_1/K) = 1 + 2(1)\,1(1.9)(1 - 1.9/10) = 4.078, \\
y_3 &= y_1 + 2h\,ry_2(1 - y_2/K) = 1.9 + 2(1)\,1(4.078)(1 - 4.078/10) = 6.7299832, \\
y_4 &= y_2 + 2h\,ry_3(1 - y_3/K) = 4.078 + 2(1)\,1(6.7299832)(1 - 6.7299832/10) \\
&= 8.479431625543553,
\end{aligned}$$

etc., up to $y_{10} \approx 10$. If we compare the results with those given by the simpler Euler method, we find significant differences immediately: for $y(2) \approx 4.50853$, the modified Euler method gives $y_2 = 4.078$, an error of less than 10%, rather than the Euler method's $y_2 = 3.439$ which has an error of 23.7%. At the next step, for $y(3) \approx 6.90568$, the modified Euler method has an error of 2.5% compared to the Euler method's 17.5% error. However, in this case the equilibrium $y_\infty = K$ of the difference equation (4.34) corresponding to the logistic differential equation is not stable, unlike both the true solution and the Euler method approximation with $h = 1$. Figure 4.52 graphs the approximations $u(t)$ based on the Euler method and modified Euler method along with the true solution $y(t)$ for the given interval; one can see the instability evidenced by the oscillation as the modified Euler method solution approaches K. Therefore, although the modified Euler method gives a much better approximation here during the initial period of growth (say $0 \leq t \leq 5$) than the simple Euler method, it fails to converge to the carrying capacity K and hence provides a markedly worse estimate for the settled period $t > 5$. □

The final numerical method we shall mention in this section is perhaps the most well known for this purpose: the *fourth-order Runge-Kutta method* (also called RK4), which uses the approximation

$$y_{k+1} = y_k + h\,\frac{K_1 + 2K_2 + 2K_3 + K_4}{6},$$

where

$$K_1 = f(t_k, y_k), \quad K_2 = f\left(t_k + \frac{h}{2}, y_k + \frac{h}{2}\,K_1\right),$$

$$K_3 = f\left(t_k + \frac{h}{2}, y_k + \frac{h}{2}\,K_2\right), \quad K_4 = f(t_k + h, y_k + h\,K_3).$$

An interpretation of these terms is that K_1 is the same approximation to $y'(t_k)$ given by the Euler method, K_2 is a first approximation to $y'(t_k + \frac{h}{2})$ (the slope halfway between t_k and t_{k+1}) based on K_1, K_3 is a second approximation to $y'(t_k + \frac{h}{2})$ based on K_2, and K_4 is an approximation to $y'(t_{k+1})$ (the slope at the end of the interval) based on K_3. y_{k+1} then uses a weighted average of these estimates of the slope $y'(t)$ over the time interval $[t_k, t_{k+1}]$.

It is possible to show that the discretization error at each step is no greater than a constant multiple of h^5 for this method, so that the accumulated discretization error is no greater than a constant multiple of h^4. This considerable improvement is evident in the following example.

EXAMPLE 4.
Use the fourth-order Runge-Kutta method with a step size of 1 to approximate the solution of the logistic equation $y' = ry(1-y/K)$ on the interval $0 \le t \le 10$ for $r = 1$, $K = 10$, and $y_0 = 1$. Compare the accuracy to that of Euler's method and the modified Euler method using the same step size.

Solution: Here we must first calculate the four K_i prior to finding each y_k. We therefore proceed as follows: To calculate y_1, we find

$$\begin{aligned}
K_1 &= r y_0(1 - y_0/K) &&= (1)1(1 - 1/10) = 0.9,\\
K_2 &= r\left(y_0 + \tfrac{h}{2}\,K_1\right)\left[1 - \left(y_0 + \tfrac{h}{2}\,K_1\right)/K\right] &&= (1)1.45(1 - 1.45/10) = 1.23975,\\
K_3 &= r\left(y_0 + \tfrac{h}{2}\,K_2\right)\left[1 - \left(y_0 + \tfrac{h}{2}\,K_2\right)/K\right] &&\approx (1)1.62(1 - 1.62/10) \approx 1.35748,\\
K_4 &= r(y_0 + K_3)(1 - (y_0 + K_3)/10) &&\approx (1)1.68(1 - 1.68/10) \approx 1.80171,
\end{aligned}$$

and then finally obtain

$$y_1 = y_0 + h\,\frac{K_1 + 2K_2 + 2K_3 + K_4}{6} = 1 + 1\,\frac{0.9 + 2(1.24) + 2(1.36) + 1.80}{6} \approx 2.31603.$$

If we compare the discretization errors involved in just this first time-step across all three of the methods presented so far, we find errors $[y_1 - y(1)]/y(1)$

of 18.09% for Euler's method (and the modified Euler method) but only 0.16% for RK4. This dramatic (by a factor of 100) improvement in accuracy continues throughout the interval of interest, with the error in later time-steps never even reaching 1/5 of 1% for RK4, whereas the average magnitude of error is 7% for Euler's method and almost 6% for the modified Euler method over the interval [0,10]. Table 4.4 gives the true solution and the approximations for all three methods at each step on this interval, rounded to 4 decimal places. □

TABLE 4.4: Comparison of the true solution to the logistic equation of Examples 1, 2, and 3 with the approximations ($h = 1$) generated by Euler's method, the modified Euler method, and RK4.

t (k)	True solution $y(t)$	Euler method y_k	Modified Euler method y_k	RK4 y_k
0	1	1	1	1
1	2.3197	1.9	1.9	2.32
2	4.5085	3.44	4.0780	4.5007
3	6.9057	5.6953	6.7300	6.8976
4	8.5849	8.1470	8.4794	8.5759
5	9.4283	9.6566	9.3087	9.4187
6	9.7818	9.9882	9.7665	9.7747
7	9.9186	10.0000	9.7649	9.9144
8	9.9699	10.0000	10.2257	9.9678
9	9.9889	10.0000	9.3033	9.9879
10	9.9959	10.0000	11.5220	9.9955

```
In[1]:= r=1; K=10; y0=1;
In[2]:= truesol=NDSolve[{y'[t]==r*y[t]*(1-y[t]/K),y[0]==y0},
{t ,0,10}]
Out[2]= {{y[t] → InterpolatingFunction[{{0.,10.}},<>] [t]}}
In[3]:= truey[t_]=y[t]/.truesol[[1]]
Out[3]= InterpolatingFunction[{{0.,10.}},<>][t]
In[4]:= truey[2]
Out[4]= 4.50854
```

FIGURE 4.53: Sample syntax for solving a differential equation numerically in *Mathematica*.

Common computational software normally uses even more sophisticated methods with variable step sizes h from one sub-interval to the next and even variable order (RK4 is fourth-order because it involves four levels of estimation at each step). Figures 4.53, 4.54 and 4.55 show the syntax used to obtain numerical solutions of differential equations in Mathematica, Maple,

```
> r=1; K=10; y0=1;
> p:= dsolve({D(y)(x) = r*y(x)*(1-y(x)/K), y(0)=y0}, y(x),
type=numeric):
> p(2);
```

$$x = 2,\ y(x) = 4.50854...$$

FIGURE 4.54: Sample syntax for solving a differential equation numerically in *Maple*.

in file LOGISTIC.M:

```
function yout = logistic(t,y)

r = 1;
K = 10;
yout = r*y*(1-y/K);
```

in file MAIN.M:

```
t0 = 0;
tf = 10;
y0 = 1;
[t,y] = ode45('logistic',[t0
tf],y0);
ans = interp1(t,y,2)
```

FIGURE 4.55: Sample syntax for solving a differential equation numerically in *MATLAB*.

and MATLAB, respectively. Note that in each case the precise step size and method are left implicit, although documentation accompanying each software package describes the methods used. Here we solve the logistic differential equation used in the foregoing examples.

We may now consider some examples for which the point is not the mechanics of the approximation method, but rather the insights that it produces.

EXAMPLE 5.
In the budworm model (4.29) of Section 4.4, with the parameter values as given in Table 4.3 (and $A = \alpha S$, $K = \kappa S$ as discussed therein), suppose that an initial budworm density is measured of approximately 30,000 larvae/acre for a given forest.

(a) Find the range of forest densities $[S_1, S_2]$ (cf. Figure 4.50) for which a hysteresis loop exists.

(b) If the forest density is 500 branches/acre, what will the budworm density be in 1 year? 2 years? How long will it take for the budworm density to be within 10% of its equilibrium value?

(c) If the forest density is 5000 branches/acre, what will the budworm density be in 1 year? 2 years? How long will it take for the budworm density to be within 10% of its equilibrium value?

(d) Interpret and explain the differences in the answers to (b) and (c).

Solution: (a) Using the parameter values, we can use a computer or graphing calculator to graph S as a function of the equilibrium rescaled budworm

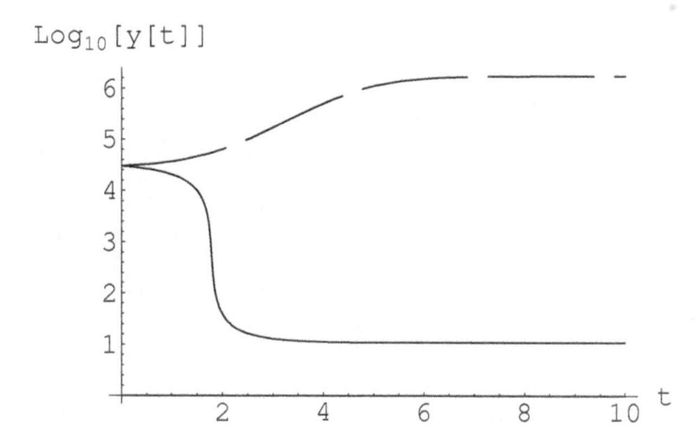

FIGURE 4.56: Solutions to the differential equation (4.29) showing the log budworm densities of Example 5 over time, for forest densities of 500 (solid curve) and 5000 (dashed curve) branches/acre.

density x, as described in Section 4.4, and find the two places where the graph has a horizontal tangent. These prove to be $(x, S) = (0.0031366, 12842)$ and $(0.49998, 320.22)$, so that $S_1 = 320.22$ branches/acre and $S_2 = 12842$ branches/acre.

(b) Using a computer to solve the differential equation numerically, we find that $y(1) \approx 20,644$ larvae/acre, $y(2) \approx 39$ larvae/acre, and it will take about 3.36 years for the population density to come within 10% of its equilibrium value of 10.84 larvae/acre.

(c) Numerical solution with $S = 5000$ branches/acre indicates that $y(1) \approx 36,620$ larvae/acre, $y(2) \approx 63,613$ larvae/acre, and it will take about 6.12 years for the population density to come within 10% of its equilibrium value of 1.75 million larvae/acre.

(d) In the smaller forest ($S = 500$ branches/acre), the budworm density drops by a factor of 1,000 in just 2 or 3 years, while in the larger forest ($S = 5000$ branches/acre), the same initial density leads to a slower growth, by a factor of nearly 100. Both forest densities fall within the hysteresis range, but in one case the initial density given falls below the separatrix (the unstable equilibrium), causing the density to crash, while in the other that same initial density lies above the separatrix, leading the density to skyrocket to a major infestation. In ecological terms, the tenfold increase in forest density allows the given budworms to reproduce beyond the predators' ability to keep them in check, whereas the lower forest density does not provide an abundant enough food supply for reproduction to overcome predation. Figure 4.56 shows the two populations over the first ten years, graphed on a logarithmic scale since the equilibria are orders of magnitude apart. □

Photo courtesy River Alliance of Wisconsin

Photo by Jeff Schmaltz, MODIS/Courtesy NASA

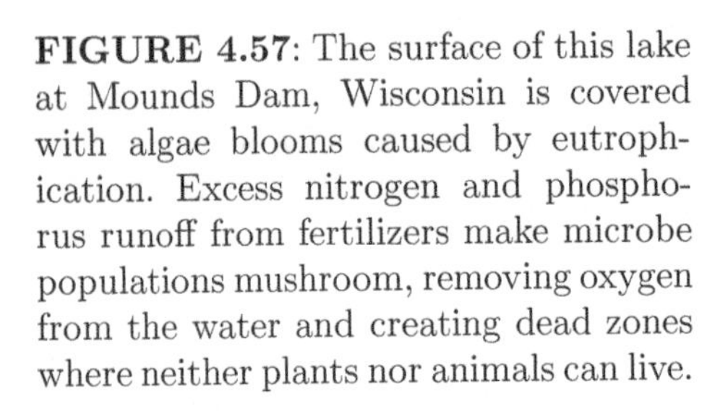

FIGURE 4.57: The surface of this lake at Mounds Dam, Wisconsin is covered with algae blooms caused by eutrophication. Excess nitrogen and phosphorus runoff from fertilizers make microbe populations mushroom, removing oxygen from the water and creating dead zones where neither plants nor animals can live.

FIGURE 4.58: This satellite photo from June 2008 shows phytoplankton blooms from extensive eutrophication in the Sea of Azov (upper right) and where the Danube River empties into the Black Sea (upper left). The pale swirls in the Black Sea reflect sedimentation.

4.5.4 Eutrophication

Aquatic ecosystems are affected strongly by the concentrations of many different substances in the water, which determine the ability of plant and animal species to survive. Some of the substances found in bodies of water serve to feed the organisms in the ecosystem, in particular oxygen and various nutrients, while others may be toxic. The extent to which sunlight is blocked may also affect plants that use photosynthesis. The level of nutrients in the water is characterized on a spectrum between oligotrophic, a state with relatively low levels of nutrients and plant production, and consequently clear water, and eutrophic, a state with high levels of nutrients and plant production, and consequently murky water. Eutrophic lakes actually suffer from their overabundance of nutrients, as phytoplankton (microscopic photosynthetic plants that live near the water's surface) and other microbes use the excess nutrients to reproduce exponentially, creating so-called blooms which deplete the oxygen in the water that larger plants and animals need in order to survive. Left unchecked, the eutrophication process can ultimately wipe out an ecosystem.

Eutrophication of lakes and even larger bodies of water has become a major problem in the past century because of the rise of chemical fertilizers in agriculture and urban lands, which pump high concentrations of nitrogen and phosphorus into the soil, much of which runs off into the local water table, streams, rivers, and lakes. In the twenty-first century, eutrophication has been observed not only in ponds and lakes (see Figure 4.57) but in the world's seas and oceans (see Figure 4.58), including the Sea of Azor and the

Gulf of Mexico. Because the effects of eutrophication can be devastating and in some cases irreversible, it is important to understand how each of the various contributing factors affects eutrophic dynamics.

Carpenter, Ludwig and Brock (1999)[22] studied a model for the amount of phosphorus in a lake over time, as a function of three factors: runoff, natural clearance, and recycling. The primary source of new phosphorus is from runoff, and is assumed to enter the lake at a constant rate (L). The lake's ability to flush the excess (free) phosphorus out is proportional to the concentration of phosphorus in the water. The recycling of phosphorus, however, is a major factor in sustaining eutrophic environments: Living organisms in the water ingest phosphorus and other nutrients, taking them out of the water and binding them into themselves, notably the benthic plants along the lake bottoms. When eutrophication begins, the multiplication of microbes on the surface kills these other organisms, and the dead matter that remains then decomposes, releasing the phosphorus back into the ecosystem. At low phosphorus levels, little of this recycling occurs, because the lake is in an oligotrophic state favorable to larger organisms, but at some point there is a threshold beyond which eutrophication causes recycling of phosphorus from sediment at a high rate. In eutrophic systems, the rate at which phosphorus is recycled in this way may even exceed the rate at which it accumulates due to runoff. In such cases the eutrophication may become irreversible, as even if the runoff is halted altogether, the recycling can sustain the phosphorus concentration at eutrophic levels.

Carpenter et al. therefore used a Holling Type III, or sigmoid (S-shaped), form to describe the recycling rate, as Ludwig et al. did to describe spruce budworm predation as described in Section 4.4:

$$\frac{P^q}{m^q + P^q},$$

for phosphorus level P, where m is the 50% saturation constant (analogous to the constant A in the budworm model) and q is an exponent greater than 1. This is again a Hill function (of order q), which remains very low for $P < m$ and quickly rises to a maximum (of 1) for $P > m$. We can think of m as the threshold level of phosphorus above which microbial blooms kill the other, larger organisms in the ecosystem. We can therefore write the model as

$$\frac{dP}{dt} = L - sP + r\,\frac{P^q}{m^q + P^q}, \tag{4.35}$$

where s is the phosphorus elimination rate and r is the maximum phosphorus recycling rate. The higher-order terms P^q complicate the analysis of this model, which is why we introduce it here, as an appropriate subject for numerical analysis.

[22] S.R. Carpenter, D. Ludwig, W.A. Brock, Management of eutrophication for lakes subject to potentially irreversible change, *Ecological Applications* 9(3): 751–771, Aug. 1999.

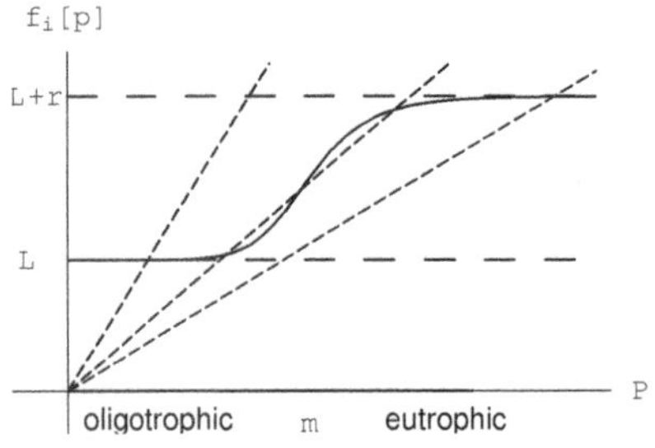

FIGURE 4.59: A graphical equilibrium analysis for (4.35) reveals either one or three equilibria, depending on the slope of $f_2(P)$ (short dashes) and the vertical offset L (long dashes).

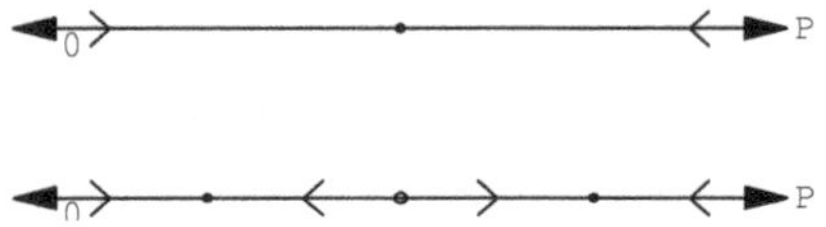

FIGURE 4.60: The two possible phase portraits for (4.35).

If we attempt a qualitative analysis, although we cannot in general write expressions for any equilibria the model may have, we can fairly quickly determine the possible numbers of equilibria and their stability, using graphical methods. If we group the positive (i.e., runoff and recycling) terms in (4.35) together, we can write $dP/dt = f_1(P) - f_2(P)$, where $f_1(P) = L + rP^q/(m^q + P^q) > 0$ and $f_2(P) = sP > 0$. Then equilibria of (4.35) are intersections of the two functions, $f_1(P) = f_2(P)$. We can consider the possible intersections graphically (Figure 4.59). The graph of f_1 is the sigmoid recycling function, shifted vertically by the constant L, while the graph of f_2 is simply a straight line through the origin with slope $s > 0$. As the figure shows, depending upon the slope s of the line and the vertical offset L, there are three possibilities:

- for high slopes s (fast clearance rates), an oligotrophic equilibrium featuring little or no recycling;
- for low slopes s (slow clearance rates), a eutrophic equilibrium with recycling near its maximum;
- for intermediate slopes s and vertical offset L within a given range, there may be three equilibria—one oligotrophic, one eutrophic, and one in between.

We can use phase line analysis to determine the local asymptotic stability of these equilibria, armed with the observations that (for $dP/dt = f(P)$) $f(0) = L > 0$ but $f(P) < 0$ for any $P \geq \frac{L+r}{s}$. Therefore (see Figure 4.60) if there is only one equilibrium, it is globally asymptotically stable, while if there are three, the first and last are locally asymptotically stable, and the middle one is unstable.

Carpenter et al. classified lakes into one of three categories depending upon the values of s and L. If the slope (phosphorus clearance rate) s is

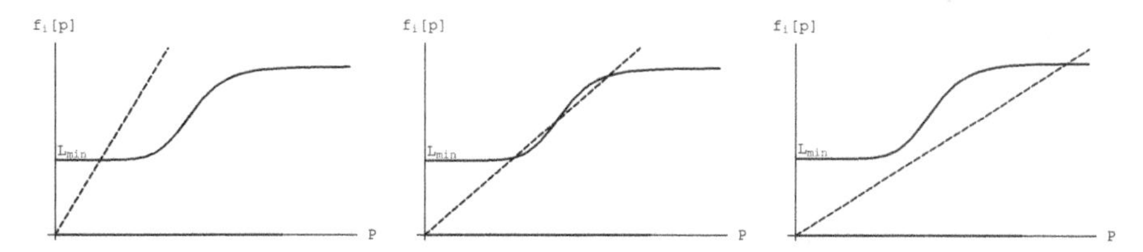

FIGURE 4.61: Graphical descriptions of Carpenter et al.'s three categories of lakes: (a) reversible, (b) hysteretic, (c) irreversible. Hysteretic lakes may also have only an oligotrophic equilibrium when $L = L_{min}$.

higher than the maximum slope (derivative) of the recycling function,[23] then the line can only intersect the sigmoid curve once, and for low enough L the only equilibrium will be oligotrophic (Figure 4.61(a)). Lakes such as these are called *reversible*, since lowering the phosphorus runoff rate L will reverse eutrophication. Reversible lakes tend to be deep and cool (a notable example is Lake Washington, in the U.S. state of Washington).

If the lake's phosphorus clearance rate s is not higher than the recycling rate's maximum slope, then the potential for reversing eutrophication depends upon the extent to which the runoff rate L can be reduced. It is not possible in practice to eliminate all phosphorus runoff, and if the minimum runoff rate that can be achieved, L_{min}, is too high, then there may never be an oligotrophic equilibrium (Figure 4.61(a)). Such lakes are called *irreversible*, since eutrophication is permanent even under the strongest restrictions on runoff. These lakes are often shallow and experience high phosphorus runoff rates (a well-studied example is Shagawa Lake in the U.S. state of Minnesota).

Finally, lakes whose clearance rate s falls below the critical level for reversibility, but whose minimum phosphorus runoff rate L_{min} is low enough to permit an oligotrophic equilibrium (Figure 4.61(b)), are called *hysteretic*, as they undergo hysteresis as a function of runoff control. Such lakes, typically small and shallow with high recycling rates, can have eutrophication permanently reversed through a temporary reduction in runoff rates.

It is possible to carry out further bifurcation analysis (see Exercise 14), but not to find explicit expressions for the equilibrium values, nor conditions under which three equilibria exist (hysteresis), so we turn to numerical analysis for further insights.

Carpenter et al. studied the particular case of Lake Mendota, which is immediately adjacent to the campus of the University of Wisconsin,[24] in the center of the city of Madison. They measured parameter values of $L_{min} = 3800$ kg/yr, $s = 0.817$/yr, $r = 731,000$ kg/yr, $m = 116,000$ kg, $q = 7.88$, and in a related study, Lathrop et al. (1998)[25] published records of the annual

[23] See Exercise 13 for the derivation of this maximum slope.

[24] where both they and the authors of this book worked

[25] Richard C. Lathrop, Stephen R. Carpenter, Craig A. Stow, Patricia A. Soranno, and

phosphorus input into Lake Mendota during the period 1976–1996, which had an average input rate of $L = 34,000$ kg/yr, a range of 15,000–67,000 kg/yr, and a mean phosphorus load of 57,600 kg.

If we use these parameter values in our model (4.35), we find Lake Mendota to be in a hysteretic state, with equilibria at 41,900 kg (oligotrophic), 77,500 kg (unstable), and 936,000 kg (eutrophic). However, during 14 of the 21 years observations were made, the estimated phosphorus load was in the range 44,000–70,000 kg. Therefore, it is likely that the lake is closer to an equilibrium value like the reported mean of 57,600 kg. Indeed, Carpenter et al. estimated as much as a tenfold uncertainty in the value of the maximum recycling rate r (that is, the true value of r might be up to ten times as much as their estimate, or as little as one tenth), and a 100-fold uncertainty in the value of the eutrophication threshold m. There is likewise some uncertainty in L, since Lathrop et al. measured phosphorus runoff coming from the principal streams feeding the lake, but could not measure that coming directly from some urban areas next to the lake shore. Changes in the parameters r and m are not likely to affect the value of the oligotrophic equilibrium much (cf. Figure 4.61(b)), but we can easily estimate the value of L which would correspond to an equilibrium at the observed level: At equilibrium, we have $L = sP - r\,\frac{P^q}{m^q+P^q} = 43,873$ kg/yr, which also generates equilibria at 69,900 kg (unstable) and 948,000 kg (eutrophic). This suggests that the lake is currently in a largely oligotrophic state, although a one-time influx of as little as $69,900 - 57,600 = 12,300$ kg would be enough to cause the lake to experience eutrophication toward the upper equilibrium. Numerical analysis reveals that this change would probably not be detected immediately simply from data: in the first five years, the phosphorus level would only increase by 1000 kg, after seven years it would be up by about 5500 kg (still less than 10%), but after eleven years it would be within 10% of the 948,000 kg eutrophic equilibrium. (We invite the reader to duplicate these results using any standard numerical software and plot the graph of $P(t)$.)

In terms of management, we can also see that aggressive reduction of phosphorus runoff ($L = L_{min}$) could reduce the oligotrophic equilibrium by more than an order of magnitude, down to about 4650 kg. In such a situation, beginning at the present level of 57,600 kg, we can use numerical analysis on our model (4.35) to find that it would then take about 2.81 years to reduce the phosphorus content in the lake to below 10,000 kg.

Numerical analysis allows us to quantify the rates of change in this study in ways that help us recognize it in the data we collect. Carpenter et al. go further to study the economics of managing the phosphorus content in the lake: certain economic benefits that result from having an oligotrophic lake (such as recreational activities and the health benefits associated with water quality) compete with other economic benefits that generate much of the phosphorus

---

John C. Panuska, Phosphorus loading reductions needed to control blue-green algal blooms in Lake Mendota, *Canadian Journal of Fisheries and Aquatic Sciences* 55: 1169–1178, 1998.

runoff (such as agricultural products, farming income, and weed-free lawns). It is difficult to negotiate the tradeoffs between these costs and benefits, but understanding the effects of our collective choices helps provide a framework for making decisions.

Exercises

1. Estimate e by using the modified Euler method to approximate $y(1)$ where $y(t)$ is the solution of the initial value problem $y' = y$, $y(0) = 1$. Compare your results with those of Example 2 using $h = 0.1$.

In Exercises 2–7, use the technology available to you to approximate the solution of the initial value problem for the given interval.

2. $y' = -y$, $y(0) = 1$; $0 \le t \le 5$
3. $y' = \sin y$, $y(0) = 1$; $0 \le t \le 1$
4. $y' = \sin y$, $y(0) = 0$; $0 \le t \le 1$
5. $y' + 2ty = \sin t$, $y(0) = 1$; $0 \le t \le 1$
6. $y' = |y|^{\frac{1}{2}}$, $y(0) = 1$; $0 \le t \le 1$
7. $y' = |y|^{\frac{1}{2}}$, $y(0) = 0$; $0 \le t \le 1$
8. What is the maximum allowable step size h that would ensure that the approximations generated by Euler's method for solutions of the logistic equation with parameters as given in Example 1 approach the (correct) equilibrium?
9. Another relatively simple numerical method which, like the modified Euler method, uses the idea of the derivative at a subinterval's midpoint being a good estimate for the derivative throughout the entire subinterval, is the *midpoint method*, which approximates the differential equation $dy/dt = f(t, y)$ with the difference equation

$$y_{k+1} = y_k + hf\left(t_k + \frac{h}{2}, y_k + \frac{h}{2} f(t_k, y_k)\right).$$

   Compare the accuracy (in terms of discretization error) of this method with those of the three methods explored in this section, by

   (a) using it to solve the logistic equation with $r = 1$, $K = 10$, $y(0) = 1$;

   (b) using it to approximate the value of e as in Exercise 1.

10. Use numerical methods to find how long it would take an ant colony of size 1000 ants governed by equation (4.1) to reach 90% engagement in foraging, if initially only 25 ants find the source and establish the trail. Use Beekman et al.'s parameter values of $\alpha = 0.0052$/min., $\beta = 0.00125$/ant-min., $s = 1$ ant/min., and $K = 10$ ants.
11. Repeat the previous exercise with a colony size of 500 ants, and conjecture a biological explanation for the difference in times.

12. In the foraging ant model (4.1) using the parameter values suggested in Section 4.1 carry out simulations for values of K close to the critical threshold value determined in Section 4.1 to exhibit the change in behavior at the threshold.

13. For the eutrophication model of Carpenter et al., find the maximum slope of the phosphorus inflow rate $f_1(P)$ by first finding the inflection point $\bar{P}$ at which $f_1''(\bar{P}) = 0$, and then calculating $f_1'(\bar{P})$. (Your answer should be r/m times a function of q.) Phosphorus clearance rates s greater than this critical value indicate eutrophication-reversible lakes.

14. * For the eutrophication model of Carpenter et al., use rescaling and the reverse parametrization method applied to the budworm model in Section 4.4 to identify the parameter values for which there exist three equilibria:

    (a) Rescale both the state variable and the time variable to reduce the number of model parameters to two. (Think about meaningful benchmarks by which to rescale each variable.) Write the rescaled differential equation using the new parameters a (replacing L) and b (replacing s).

    (b) Use the fact that at a bifurcation point not only $dp/dt = f(p) = 0$ but also $f'(p) = 0$, to write two equations which are linear in the two new parameters a and b. Solve these equations to find expressions for a and b as functions of p (this is the reverse parametrization).

    (c) Show that (a, b) moves from $(-1, 0)$ to $(q, 0)$ for $0 \leq p < \infty$, with a increasing in p while b hits a (positive) maximum followed by a (negative) minimum.

    (d) Using the known zeroes of $b(p)$, show that the reverse-parametrized curve $(a(p), b(p))$ passes through the positive quadrant if and only if $q > 2$. What is the significance of this result?

15. In their study of eutrophication of Lake Mendota, Carpenter et al. estimated a tenfold uncertainty in their value for the maximum recycling rate r, and a 100-fold uncertainty in their value for the eutrophication threshold m.

    (a) Evaluate the worst-case scenario: that is, imagine their values really are off by respective factors of 10 and 100. What are the corresponding equilibria? What kind of lake would this be?

    (b) Now evaluate the best-case scenario: again, their values for r and m are off by factors of 10 and 100, respectively, but in the other direction. What are the equilibria and lake classification?

    (c) Judging by the equilibrium value in each case, to which extreme is the present situation closer?

Miscellaneous exercises

For each of the differential equations in Exercises 1–4, draw the phase line, find the equilibria, and describe the asymptotic behavior of solutions.

1. $y' = y^2 (y-1)^2$

2. $y' = \frac{y}{1-y}$

3. $y' = ye^{-y}$

4. $y' = y \sin y$

5. (a) Consider the logistic differential equation

$$y' = ry\left(1 - \frac{y}{K}\right), \quad y(0) = A$$

with $K = 100$, $A = 25$ and several different values of r including $r = 1.5$, $r = 2.5$, $r = 3.5$, and $r = 4.5$. Describe the behavior of the solution for each value of r qualitatively.

(b) Consider the discrete logistic equation

$$y_{k+1} = ry_k\left(1 - \frac{y_k}{K}\right), \quad y_0 = A$$

with $K = 100$, $A = 25$ and several different values of r including $r = 1.5$, $r = 2.5$, $r = 3.5$, and $r = 4.5$. Describe the behavior of the solution for each value of r qualitatively.

(c) Consider the alternative discrete version

$$y_{k+1} = ry_k e^{\left(1 - \frac{y_k}{K}\right)}, \quad y_0 = A$$

with $K = 100$, $A = 25$ and several different values of r including $r = 1.5$, $r = 2.5$, $r = 3.5$, and $r = 4.5$. Describe the behavior of the solution for each value of r qualitatively.

Approximate the solutions of the problems in parts (a), (b), (c) numerically for each given value of r. Compare and contrast the behaviors of the three solutions for each r.

6. * Carry out the method of separation of variables for equation (4.1), as follows:

(a) Rewrite the equation $\frac{dx}{dt} = f(x)$ in the form $\int \frac{dx}{f(x)} = \int dt$, and then rewrite the left-hand integrand as a rational polynomial, i.e., write $1/f(x)$ as $A(x)/B(x)$, where $A(x)$ and $B(x)$ are both polynomials.

(b) Use the cubic formula to factor $B(x)$; depending upon parameter values, it will have either one or three real roots. Next, apply the method of partial fraction expansion to $A(x)/B(x)$. In the case where $B(x)$ has three real roots, you can rewrite $A(x)/B(x)$ as the sum of three terms of the form $A_i/(x-r_i)$ (where $i = 1, 2, 3$). In the case where $B(x)$ has only one real root, you can rewrite $A(x)/B(x)$ as the sum $\frac{A_1}{x-r} + \frac{A_2x+A_3}{x^2+B_1x+B_2}$. (You may have to look up the details for both of these techniques.)

(c) Carry out the integration.

(d) Exponentiate both sides of the equation, and use the initial condition $x(0) = x_0$ to rewrite any constant of integration.

7. Explain why no two consecutive equilibria can both be (locally) asymptotically stable. (Consider the behavior of solutions on the interval between them.)

8. Suppose a population of whales grows according to an exponential growth law but with constant-effort harvesting. Then the population size satisfies a differential equation $y' = ry - Ey$. (a) If $r = 0.04/\text{yr}$, what catch per year will maintain the population at a constant size of 8000 whales? (Note the catch is not E but Ey.) (b) If harvesting takes place for ten years with $E = 0.06$, how long will it take after harvesting ceases until the population returns to its original size of 8000 whales?

9. A forest of 10,000 acres suffers from forest fires which destroy an average of 1% of its area annually. If the natural growth rate is $r = 0.05/\text{yr}$, find the equilibrium forest size.

10. * (a) Show that the minimum survival threshold y_1 in the constant-yield harvesting model has $y_1 > H/r$ for all values of K. (*Hint: Work backward.*)

(b) Use calculus to show that $y_1 \to H/r$ as $K \to \infty$: that is,

$$\lim_{K\to\infty} \frac{1}{2}\left[K - \sqrt{K^2 - \frac{4HK}{r}}\right] = \frac{H}{r}.$$

11. In the mid-1990s, the Canadian harp seal population was estimated at 4.8 million seals. In 1997 an estimated 344,000 harp seals were killed. The harp seal's reproductive rate has been estimated at 7%/year. (a) Under a constant-yield scenario, what is the minimum overall carrying capacity, if the observed harvest is sustainable? (b) Use the result of Exercise 10 to evaluate the sustainability of this hunt.

12. In formulating the epidemic model (4.20), we assumed that individuals mix randomly in making contacts with others, so that a proportion I/N

of anyone's contacts are made with infectives, and a proportion S/N are made with susceptibles. If the symptoms of the disease are severe enough, however, it may be more appropriate to assume that infected individuals' contact rates are reduced by some factor σ $(0 < \sigma < 1)$ to $\sigma\beta$ instead of just β, either because they are so ill they must remain in bed, resting, or because they deliberately avoid contact with others whom they might infect. In this case, at what rate would new infections occur? What is the resulting model? *Hint: First determine the new effective population size for contact rate purposes—it's less than N.*

13. * In the previous exercise with less active infectives, determine R_0 and the endemic equilibrium under the assumption of (a) mass-action incidence, (b) standard incidence.

14. In formulating the epidemic model (4.20), we assumed that infected individuals recover from the disease at a given rate $(1/\tau)$. If, instead, infected individuals never recover (as with HIV), (a) what happens to the model? to the expression for R_0? *Hint: Consider what happens to τ under this assumption.* (b) What eventual outcome does the model then predict? Why is this outcome inconsistent with the observed data for unrecoverable infections such as HIV?

15. Describe the behavior of solutions of the differential equation $y' = r - y^2$ as the parameter r is varied.

16. Describe the behavior of solutions of the differential equation $y' = ry - y^2$ as the parameter r is varied.

17. Draw a bifurcation diagram for the two epidemic models in Exercise 11 of Section 4.3.

18. (a) Show graphically that all equations of the form $y' = p - x^n$ exhibit either a saddle-node bifurcation at the origin, if n is even, or no bifurcation at all, if n is odd. (b) Show graphically that all equations of the form $y' = px - x^n$ exhibit either a transctitical bifurcation at the origin, if n is even, or a pitchfork bifurcation at the origin, if n is odd. (c) What changes occur in (a) and (b) above when the $-$ sign in the equation becomes a $+$?

Part II

More Advanced Topics

Chapter 5

Systems of Differential Equations

In Chapter 3 we saw how to model and analyze continuously changing quantities using differential equations. In many applications of interest there may be two or more interacting quantities — populations of two or more species, for instance, or parts of a whole, which depend upon each other. When the amount or size of one quantity depends in part on the amount of another, and vice versa, they are said to be *coupled*, and it is not possible or appropriate to model each one separately. In these cases we write models which consist of *systems* of differential equations. In this chapter, we will find that the quantitative and qualitative approaches we used to analyze individual differential equations in Chapters 3 and 4 extend in a more or less natural way to cover systems of differential equations. Extending them will require some basic multivariable calculus, principally the use of partial derivatives; we provide some support in the worked examples for the unfamiliar reader, but recommend using a calculus text for more complete reference.

We begin below with an important example from population biology, a *predator-prey* system, in which one species preys upon another for its food supply.

5.1 Graphical analysis: The phase plane

One of the landmarks in the development of mathematical ecology is the Lotka-Volterra model for the population sizes of two interacting species. About 1925 the famous Italian mathematician Vito Volterra was asked if it was possible to give a scientific explanation for the large fluctuations in fish populations in the Adriatic Sea. These fluctuations were of great concern to fishermen, both in periods of low population sizes when fish catches were small and there was little income for the fishing industry, and in periods of high population sizes when fish were abundant and the large supply made selling prices low. Volterra constructed a simple model which has become known as the *Lotka-Volterra model* (because A. J. Lotka constructed a similar model about the same time in a different context), based primarily on the hypothesis that fish and sharks were in a predator-prey relationship (see Figure 5.1). Even though it turned out later that this model was unreasonably simplistic and not re-

Photo by Mark Conlin, Southwest Fisheries Science Ctr, NOAA Fisheries Svc

Photo by Stefano Guerrieri

FIGURE 5.1: The most common sharks and fish in the Adriatic Sea, the subject of Volterra's landmark study, are the blue shark, *Prionace glauca* (left), and members of the goby family (pictured at right is one example, *Didogobius schlieweni*).

ally consistent with observations, it gave great insight into the modeling of predator-prey relationships and led to many important developments.

Let $y(t)$ denote the number of fish and $z(t)$ the number of sharks at time t. We make the quite unreasonable simplifying assumption that all fish are of the same kind and that all sharks are identical.[1] We assume that the plankton on which the fish feed is available in unlimited quantities, and thus that the fish population would grow exponentially in the absence of sharks. Then, if there were no sharks, the fish population would satisfy an exponential growth equation

$$y' = \lambda y.$$

The sharks, on the other hand, are assumed to depend on the fish as their food supply. We assume that in the absence of fish the shark population would die out exponentially, and thus that the shark population satisfies an exponential decay equation

$$z' = -\mu z$$

if there are no fish. To describe the interaction between sharks and fish, we assume that the presence of fish produces a linear increase in the per capita shark growth rate, and the presence of sharks produces a linear decrease in the per capita fish growth rate. That is, the rate of encounters between fish y and sharks z is proportional to the size of each group, making it a multiple of yz. Thus we assume per capita growth rates of $\lambda - bz$ for fish and $-\mu + cy$ for sharks, where b and c are constants which describe how sharply the shark population affects the fish reproduction, and vice versa. This gives us a system

[1]This assumption will allow us to write a reasonably simple model and investigate whether the basic idea of interdependence can account for the observed fluctuations.

of two first-order differential equations

$$\begin{aligned} y' &= y(\lambda - bz) \\ z' &= z(-\mu + cy) \end{aligned} \tag{5.1}$$

called the *Lotka-Volterra equations.*

We cannot solve the system (5.1) analytically, but we can use qualitative tools to examine it. In addition to the techniques developed in the previous chapter, which we will apply presently, we can now also eliminate t and reduce the system to a single equation, by using the relation

$$\frac{dz}{dy} = \frac{dz}{dt} \bigg/ \frac{dy}{dt}$$

from calculus (an application of the Chain Rule) to obtain a first-order differential equation for z as a function of y,

$$\frac{dz}{dy} = \frac{z(-\mu + cy)}{y(\lambda - bz)}. \tag{5.2}$$

The differential equation (5.2) has variables separable, and may be solved by the methods of Section 3.4 (which we will now do).

Separation of variables in (5.2) gives

$$\int \frac{-\mu + cy}{y} dy = \int \frac{\lambda - bz}{z} dz,$$

and integration gives

$$-\mu \log y + cy = \lambda \log z - bz + h, \tag{5.3}$$

where h is a constant of integration. We now have a relation between y and z which defines z implicitly as a function of y for each choice of the constant of integration h. (Substituting initial conditions for y and z into (5.3) will allow us to find the appropriate value for h.) If we define a function $V(y, z)$ of two variables by

$$V(y, z) = -\mu \log y + cy - \lambda \log z + bz,$$

the relation (5.3) takes the form

$$V(y, z) = h. \tag{5.4}$$

That is, as y and z change (over time, but we have made that implicit here), they will follow the curve given by (5.4), so that the value of $V(y, z)$ remains constant, at its initial value h.

If we return briefly to (5.1), by setting both equations equal to zero we can find that the system has two equilibria: $y = z = 0$ and $y = \mu/c$, $z = \lambda/b$. $V(y, z)$ is undefined at (0,0) — it approaches ∞ as we approach the origin or either axis — but defined everywhere in the interior of the first quadrant; since we saw above that $V(y,z)$ must remain constant, we can conclude that (y,z) can only approach (0,0) if we begin on one of the axes: in particular, the z-axis. That is, the fish and shark populations will die out entirely only in the uninteresting case where there are no fish, and the sharks die out in the absence of their food supply. (In the remainder of our discussion below we will assume that both fish and sharks are present.)

At the other equilibrium, however, we find that

$$V(y,z) = h_0 \equiv -\mu \log \frac{\mu}{c} + \mu - \lambda \log \frac{\lambda}{b} + \lambda.$$

In fact, we can show that h_0 is the minimum value $V(y,z)$ can ever take on, by examining the first and second derivatives of V, just as with a function of one variable. In particular, if we take the *partial* derivatives of V with respect to y and z, we find[2] that

$$\frac{\partial V}{\partial y} = -\frac{\mu}{y} + c, \quad \frac{\partial V}{\partial z} = -\frac{\lambda}{z} + b, \quad \frac{\partial^2 V}{\partial y^2} = \frac{\mu}{y^2} > 0, \quad \frac{\partial^2 V}{\partial z^2} = \frac{\lambda}{z^2} > 0,$$

so the equilibrium $y = \mu/c$, $z = \lambda/b$ is the only place where both $\frac{\partial V}{\partial y} = 0$ and $\frac{\partial V}{\partial z} = 0$. Furthermore, we see from the second derivatives that the function is concave up in the interior of the first quadrant, so this critical point must be a minimum.[3] (Here we begin to see our first hint of new and interesting behavior: since the value of $V(y,z)$ must remain constant, and $V(\mu/c, \lambda/b)$ is different (less) than the values of V for all the points surrounding it, no trajectories lead toward this equilibrium, either.) Therefore every solution of the system (5.1) is described by the relation (5.4) with some choice of $h \geq h_0$, determined by initial conditions.

Recall now that our purpose in writing this model was to see if the interdependence between the fish and shark populations could account for the fluctuations observed. In order to see that we do have periodic solutions here, we will extend the notion of a *phase line* introduced in Section 4.1. Since we now have a system of two dimensions rather than one, we will speak of the *phase plane*, i.e., the $y - z$ plane. For each value of $h \geq h_0$, the relation (5.4) defines implicitly a curve in this plane. More directly, it gives the relation between the two population sizes from the differential equation (5.2). It does

[2]Note to the reader not accustomed to partial derivatives: the process is just like normal differentiation with respect to the variable in question, while treating all other independent variables as constants. The notation $\partial V/\partial y$ is used above, for instance, instead of dV/dy, in order to show that V is a function of more than one variable. As a short example, consider $A(y,z) = y^2 \sin z + 4$. Then $\partial A/\partial y = 2y \sin z$, while $\partial A/\partial z = y^2 \cos z$. In this case we have *three* different second-order partial derivatives: $\partial^2 A/\partial y^2 = 2\sin z$, $\partial^2 A/\partial z^2 = -y^2 \sin z$, and $\partial^2 A/\partial y \partial z = \partial^2 A/\partial z \partial y = 2y \cos z$.

[3]We can also note that $\frac{\partial^2 V}{\partial y \partial z} = 0$, but we are borrowing just enough from multivariable calculus here to make our point.

not, however, give the populations as functions of the time t. Such a curve is called an *orbit* of the system (5.1).

To show that solutions to (5.1) are periodic, we must show that the curves given by $V(y,z) = h$ are closed. We can, of course, use computer programs such as Maple, Mathematica or MATLAB to generate numerical solutions and plot them, and Figure 5.2 below shows such a plot. An analytical way to show that orbits of (5.1) are closed involves using a polynomial approximation to the logarithms involved in the formula for V, similar to the linearization process introduced in Section 4.1 to analyze stability of equilibria. Since it is often important to be able to show results analytically — here, for example, to distinguish genuine periodic orbits from spirals which grow or decay very slowly — we will now consider this approach.

We begin by making the change of variables

$$y = \frac{\mu}{c} + u, \quad z = \frac{\lambda}{b} + v$$

in (5.4), to center our system (u,v) around the equilibrium $(\frac{\mu}{c}, \frac{\lambda}{b})$. This gives us

$$\begin{aligned} V\left(\frac{\mu}{c} + u, \frac{\lambda}{b} + v\right) &= -\mu \log\left(\frac{\mu}{c} + u\right) + c\left(\frac{\mu}{c} + u\right) \\ &\quad -\lambda \log\left(\frac{\lambda}{b} + v\right) + b\left(\frac{\lambda}{b} + v\right) = h. \qquad (5.5) \end{aligned}$$

Using rules of logarithms, we write

$$\log\left(\frac{\mu}{c} + u\right) = \log\left(\frac{\mu}{c}\left[1 + \frac{cu}{\mu}\right]\right) = \log\left(\frac{\mu}{c}\right) + \log\left(1 + \frac{cu}{\mu}\right).$$

We now use the quadratic Taylor approximation $\log(1+x) \approx x - \frac{x^2}{2}$ (see Appendix B) for values of x close to 0, to approximate $\log\left(1 + \frac{c}{\mu}u\right) \approx \frac{c}{\mu}u - \frac{c^2}{2\mu^2}u^2$ for values of u close to 0 — that is, for (y,z) close to $(\frac{\mu}{c}, \frac{\lambda}{b})$, or values of $V(y,z) = h$ close to h_0. This gives us

$$\log\left(\frac{\mu}{c} + u\right) \approx \log\left(\frac{\mu}{c}\right) + \frac{c}{\mu}u - \frac{c^2}{2\mu^2}u^2,$$

and similarly we may approximate $\log\left(\frac{\lambda}{b} + v\right)$ by $\log\left(\frac{\lambda}{b}\right) + \frac{b}{\lambda}v - \frac{b^2}{2\lambda^2}v^2$. (This is the same approximation process we would use in linearizing about the equilibrium, as introduced in Section 4.1, but here we are keeping the quadratic as well as the linear terms.) When we substitute these approximations into (5.5) we obtain the curve

$$-\mu \log\left(\frac{\mu}{c}\right) - cu + \frac{c^2}{2\mu}u^2 + \mu + cu - \lambda \log\left(\frac{\lambda}{b}\right) - bv + \frac{b^2}{2\lambda}v^2 + \lambda + bv = h$$

or

$$\frac{c^2}{2\mu}u^2 + \frac{b^2}{2\lambda}v^2 = h + \mu \log \frac{\mu}{c} - \mu + \lambda \log \frac{\lambda}{b} - \lambda = h - h_0$$

as an approximation to the curve (5.5). This curve, whose equation in the $y - z$ plane is

$$\frac{c^2}{2\mu}\left(y - \frac{\mu}{c}\right)^2 + \frac{b^2}{2\lambda}\left(z - \frac{\lambda}{b}\right)^2 = h - h_0, \tag{5.6}$$

represents an ellipse with center at $(\frac{\mu}{c}, \frac{\lambda}{b})$ as long as $h > h_0$.

This shows that the curve (5.6) is a closed curve around the equilibrium $(\frac{\mu}{c}, \frac{\lambda}{b})$ if $h - h_0$ is small and positive. Since the solution runs around a closed curve, it must repeat itself and is therefore periodic. Thus the Lotka-Volterra model predicts periodic fluctuations as had been observed in the shark and fish populations. It is possible to prove that the *period of oscillation*, or time for a cycle to repeat itself, is approximately $2\pi/\sqrt{\lambda\mu}$. The *phase portrait* (Figure 5.2), or sketch of orbits in the phase plane, shows that, because the orbit is traversed in a counterclockwise direction (see Exercise 7 below), the maximum prey population (reached at the rightmost end of each orbit) occurs one quarter of a cycle before the maximum predator population (reached at the top of each orbit).

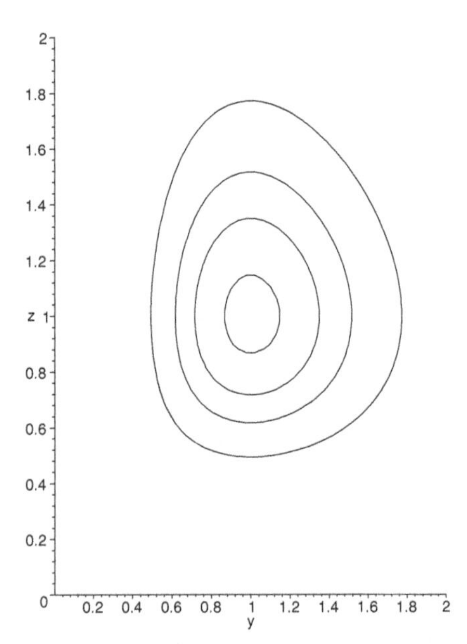

FIGURE 5.2: Phase portrait for the Lotka-Volterra model.

Now that we have seen an example, let us restate more generally the terminology we shall use in this chapter. The Lotka-Volterra system is an example of a *two-dimensional autonomous system of first-order differential equations,*

$$\begin{aligned} y' &= F(y, z), \\ z' &= G(y, z). \end{aligned} \tag{5.7}$$

The *phase plane* for such a system is the $y - z$ plane.

An *equilibrium* of the system (5.7) is a solution (y_∞, z_∞) of the pair of equations

$$F(y, z) = 0, \quad G(y, z) = 0.$$

Geometrically, an equilibrium is a point in the phase plane. In terms of the system (5.7), an equilibrium gives a constant solution $y = y_\infty$, $z = z_\infty$ of the system. This definition is completely analogous to the definition of an equilibrium given for a difference equation in Section 2.3, and for a first-order differential equation in Section 4.1.

The *orbit* of a solution $y = y(t)$, $z = z(t)$ of the system (5.7) is the curve in the $y - z$ phase plane consisting of all points $(y(t), z(t))$ for $0 \le t < \infty$. A closed orbit corresponds to a periodic solution.

There is a geometric interpretation of orbits which is analogous to the interpretation given for solutions of first-order differential equations in Section 3.3. Just as the curve $y = y(t)$ has slope $y' = f(t, y)$ at each point (t, y) along its length, an orbit of (5.7) (considering z as an implicit function of y) has slope

$$\frac{dz}{dy} = \frac{z'}{y'} = \frac{G(y, z)}{F(y, z)}$$

at each point of the orbit. The phase plane *direction field* for a two-dimensional autonomous system is a collection of line segments with this slope at each point (y, z), and an orbit must be a curve which is tangent to the direction field at each point of the curve. Computer algebra systems such as Maple, MATLAB and Mathematica may be used to draw the direction field for a given system. For example, Figure 5.3 gives a direction field drawn using Maple for the Lotka-Volterra system (5.1) with $\lambda = 1$, $\mu = 1$, $b = 1$, $c = 1$, with the orbits from Figure 5.2 superimposed on it.

EXAMPLE 1.

Describe the orbits of the system

$$y' = z, \quad z' = -y.$$

Solution: If we consider z as a function of y, we have $\frac{dz}{dy} = z'/y' = -\frac{y}{z}$. Solution by separation of variables gives

$$\int z\,dz = -\int y\,dy,$$

and integration gives $\frac{z^2}{2} = -\frac{y^2}{2} + c$. Thus every orbit is a circle $y^2 + z^2 = 2c$ with center at the origin, and every solution is periodic. □

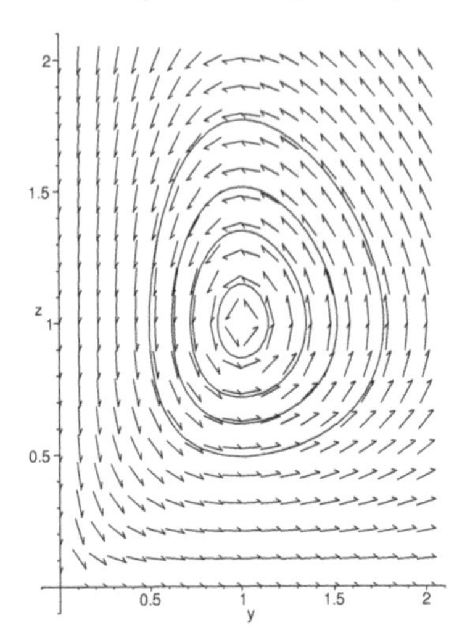

FIGURE 5.3: Direction field and phase portrait for the Lotka-Volterra model.

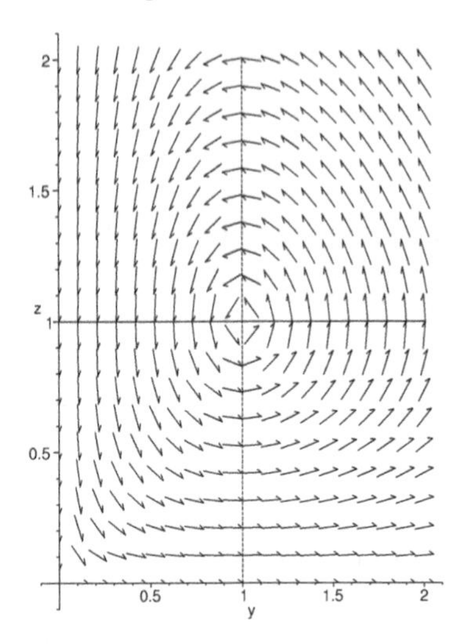

FIGURE 5.4: Direction field and nullclines for the Lotka-Volterra model (the coordinate axes are also nullclines).

To find equilibria of a system (5.7), it is helpful to draw the *nullclines*, the curves $F(y, z) = 0$ on which $y' = 0$, and $G(y, z) = 0$ on which $z' = 0$. An equilibrium is an intersection of these two curves. In order to distinguish between the two nullclines, we will use a solid curve for the y-nullcline and a dotted curve for the z-nullcline. Thus for the Lotka-Volterra system (5.1) we may add the y-nullclines $y = 0$ and $z = \frac{\lambda}{b}$, and the z-nullclines $z = 0$ and $y = \frac{\mu}{c}$ to the direction field of Figure 5.3 to obtain Figure 5.4.

Exercises

In each of Exercises 1–6, describe the orbits of the given system.

1. $y' = yz^2$, $z' = zy^2$
2. $y' = e^z$, $z' = e^{-y}$
3. $y' = \cos z$, $z' = y$
4. $y' = ye^z$, $z' = yze^y$
5. $y' = -\beta yz$, $z' = \beta yz$ (see Exercise 7, Section 3.4)
6. $y' = -\alpha y$, $z' = -\beta yz$ (see Exercise 8, Section 3.4)
7. * Show that the orbits of the Lotka-Volterra system (5.1) are traversed in a counterclockwise direction as t increases. [Hint: For a point on the orbit with $z = \frac{\lambda}{b}$ and $y > \frac{\mu}{c}$, $y' = 0$ and $z' > 0$; thus at this point y is a maximum and z is increasing].
8. * Find in which direction the system in Example 1 traverses its orbits.

5.2 Linearization of a system at an equilibrium

Sometimes it is possible to find the orbits in the phase plane of a system of differential equations, as we were able to do for the Lotka-Volterra system in the preceding section, but it is rarely possible to solve a system of differential equations analytically. For this reason, our study of systems will concentrate on qualitative properties. The linearization of a system of differential equations at an equilibrium is a linear system with constant coefficients, whose solutions approximate the solutions of the original system near the equilibrium. In this section we shall see how to find the linearization of a system. In the next section we shall see how to solve linear systems with constant coefficients, and this will enable us to understand much of the behavior of solutions of a system near an equilibrium.

Let (y_∞, z_∞) be an equilibrium of a system

$$\begin{aligned} y' &= F(y, z), \\ z' &= G(y, z), \end{aligned} \tag{5.8}$$

that is, a point in the phase plane such that

$$F(y_\infty, z_\infty) = 0, \quad G(y_\infty, z_\infty) = 0. \tag{5.9}$$

We will assume that the equilibrium is *isolated*, that is, that there is a circle centered around (y_∞, z_∞) which does not contain any other equilibrium. We shift the origin to the equilibrium by letting $y = y_\infty + u$, $z = z_\infty + v$, and then use Taylor's theorem for two variables (see Appendix B) to approximate $F(y_\infty + u, z_\infty + v)$ and $G(y_\infty + u, z_\infty + v)$. The difference here between a one-dimensional system and a two-dimensional one is that the approximation via Taylor's theorem in two or more dimensions uses partial derivatives. Our approximations are

$$\begin{aligned} F(y_\infty + u, z_\infty + v) &= F(y_\infty, z_\infty) + F_y(y_\infty, z_\infty)u + F_z(y_\infty, z_\infty)v + h_1, \\ G(y_\infty + u, z_\infty + v) &= G(y_\infty, z_\infty) + G_y(y_\infty, z_\infty)u + G_z(y_\infty, z_\infty)v + h_2, \end{aligned} \tag{5.10}$$

where h_1 and h_2 are functions which are "quadratic" in u and v, in the sense that they are negligible relative to the linear terms in (5.10) when u and v are small (i.e., close to the equilibrium).

The linearization of the system (5.8) at the equilibrium (y_∞, z_∞) is defined to be the linear system with constant coefficients

$$\begin{aligned} u' &= F_y(y_\infty, z_\infty)u + F_z(y_\infty, z_\infty)v, \\ v' &= G_y(y_\infty, z_\infty)u + G_z(y_\infty, z_\infty)v. \end{aligned} \tag{5.11}$$

To obtain it, we first note that $y' = u'$, $z' = v'$, and then substitute (5.10) into (5.8). By (5.9) the constant terms are zero, and for the linearization we neglect the higher-order terms h_1 and h_2. The *coefficient matrix* of the linear system (5.11) is the matrix of constants

$$\begin{bmatrix} F_y(y_\infty, z_\infty) & F_z(y_\infty, z_\infty) \\ G_y(y_\infty, z_\infty) & G_z(y_\infty, z_\infty) \end{bmatrix}.$$

In population models, this matrix is often called the *community matrix* of the system at equilibrium. Its entries describe the effect of a change in each variable on the growth rates of the two variables.

EXAMPLE 1.
Find the linearization at each equilibrium of the Lotka-Volterra system

$$y' = y(\lambda - bz), \; z' = z(-\mu + cy).$$

Solution: As noted in the previous section, the equilibria are the solutions of $y(\lambda - bz) = 0$, $z(-\mu + cy) = 0$. One solution is the origin, $y = 0$, $z = 0$, representing extinction of both fish and sharks, and a second is $y = \frac{\mu}{c}$, $z = \frac{\lambda}{b}$, representing coexistence. Because the partial derivatives of the functions on the right side of the system are, respectively,

$$\frac{\partial}{\partial y}\left[y(\lambda - bz)\right] = \lambda - bz, \qquad \frac{\partial}{\partial z}\left[y(\lambda - bz)\right] = -by,$$
$$\frac{\partial}{\partial y}\left[z(-\mu + cy)\right] = cz, \qquad \frac{\partial}{\partial z}\left[z(-\mu + cy)\right] = -\mu + cy,$$

the linearization at an equilibrium (y_∞, z_∞) is

$$\begin{aligned} u' &= (\lambda - bz_\infty)u \quad -by_\infty v, \\ v' &= cz_\infty u \quad +(-\mu + cy_\infty)v. \end{aligned}$$

Thus the linearization at (0,0) is

$$u' = \lambda u, \; v' = -\mu v,$$

and the linearization at $(\frac{\mu}{c}, \frac{\lambda}{b})$ is

$$u' = -\frac{b\mu}{c}v, \; v' = \frac{c\lambda}{b}u.$$

(What these linearized systems tell us about the predator-prey system's behavior is investigated later in this section, cf. Example 3). □

EXAMPLE 2.
Find the linearization at each equilibrium of the system

$$y' = -\beta yz, \; z' = \beta yz - \gamma z.$$

This corresponds to the classical Kermack-McKendrick SIR model for an epidemic, which we shall study in Section 6.1. Here y corresponds to the number of susceptible, uninfected individuals, z to the number of infected, infective individuals, and β and γ to the infection and recovery rates, respectively.

Solution: The equilibria are the solutions of $-\beta yz = 0$, $\beta yz - \gamma z = 0$. To satisfy both these equations, we must have $z = 0$, but there is no restriction on y (so we have an entire line of equilibria $(y, 0)$). Since

$$\frac{\partial}{\partial y}[-\beta yz] = -\beta z, \qquad \frac{\partial}{\partial z}[-\beta yz] = -\beta y,$$
$$\frac{\partial}{\partial y}[\beta yz - \gamma z] = \beta z, \quad \frac{\partial}{\partial z}[\beta yz - \gamma z] = \beta z - \gamma,$$

the linearization at an equilibrium (y_∞, z_∞) is

$$\begin{aligned} u' &= -\beta z_\infty u \quad && -\beta y_\infty v, \\ v' &= \beta z_\infty u && +(\beta y_\infty - \gamma)v. \end{aligned}$$

The equilibria are the points $(y_\infty, 0)$ with arbitrary y_∞, and the corresponding linearization is

$$u' = -\beta y_\infty v, \; v' = (\beta y_\infty - \gamma)v.$$

If we examine the second of these equations, we see that $v(t)$ experiences either exponential growth or exponential decay, depending on the sign of the coefficient $(\beta y_\infty - \gamma)$. If $y_\infty > \gamma/\beta$, this coefficient is positive and solutions to the linearized system exhibit exponential growth in v, which corresponds to growth in z, the number of infected individuals. Only if $y_\infty < \gamma/\beta$ does v decay, corresponding to a drop in the number of infectives. As we will discover in Section 6.1, this comes from the modeling assumption that infectious contacts, like the predator-prey contacts in the Lotka-Volterra model, occur at a rate proportional to the product of both populations, yz. Infection control might therefore concentrate on reducing y, the number of susceptibles who make potentially infectious contacts; two common ways to do so, illustrated in Figure 5.5, are vaccination and protection from infectious contact. □

An equilibrium of the system (5.8) with the property that every orbit with initial value sufficiently close to the equilibrium remains close to the equilibrium for all $t \geq 0$, and approaches the equilibrium as $t \to \infty$, is said to be *asymptotically stable*. An equilibrium of (5.8) with the property that some solutions starting arbitrarily close to the equilibrium move away from it is said to be *unstable*. These definitions are completely analogous to those given in Section 2.4 for difference equations and Section 4.1 for first-order differential equations.

The fundamental property of the linearization which we will use to study stability of equilibria is the following result, which we state without proof. The proof may be found in any text which covers the qualitative study of nonlinear

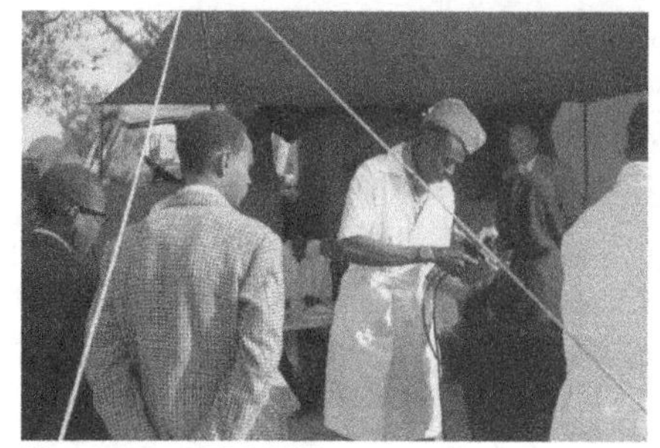

Photo courtesy CDC PHIL

Photo by Eneas De Troya

FIGURE 5.5: Two common strategies for reducing the number of individuals at risk of infection in an epidemic are vaccination and protection from contact. At left, a man vaccinates a child (in his mother's arms) against smallpox in the Republic of Chad during a worldwide vaccination program that began in 1967 and lasted over a decade. At right, riders on a Mexico City subway train wear breathing masks during the 2009 H1N1 influenza epidemic.

differential equations. Here we suppose F and G to be twice differentiable — that is, smooth enough for the linearization to give a correct picture.

LINEARIZATION THEOREM: If (y_∞, z_∞) is an equilibrium of the system

$$y' = F(y,z), \quad z' = G(y,z),$$

and if every solution of the linearization at this equilibrium approaches zero as $t \to \infty$, then the equilibrium (y_∞, z_∞) is asymptotically stable. If the linearization has unbounded solutions, then the equilibrium (y_∞, z_∞) is unstable.

For a first-order differential equation $y' = g(y)$ at an equilibrium y_∞, the linearization is the first-order linear differential equation $u' = g'(y_\infty)u$. We may solve this differential equation by separation of variables and see that all solutions approach zero if $g'(y_\infty) < 0$, and there are unbounded solutions if $g'(y_\infty) > 0$. We have seen in Section 4.1, without recourse to the linearization, that the equilibrium is asymptotically stable if $g'(y_\infty) < 0$ and unstable if $g'(y_\infty) > 0$. The linearization theorem is valid for systems of any dimension and is the approach needed for the study of stability of equilibria for systems of dimension higher than 1.

Note that there is a case where the theorem above does not draw any conclusions: the case where the linearization about the equilibrium is neither asymptotically stable nor unstable. In this case, the equilibrium of the original (nonlinear) system may be asymptotically stable, unstable, or neither. To see that this case may indeed occur, recall the Lotka-Volterra model we studied in the previous section, where periodic solutions arose.

EXAMPLE 3.
Show that the equilibrium (0,0) of the Lotka-Volterra system

$$y' = y(\lambda - bz), \; z' = z(-\mu + cy)$$

is unstable.

Solution: As we have seen in Example 1, the linearization of the system at the equilibrium (0,0) is $u' = \lambda u$, $v' = -\mu v$. These two equations can be solved separately by separation of variables, and every solution of the linearization has the form $u = c_1 e^{\lambda t}$, $v = c_2 e^{-\mu t}$. As every solution with $c_1 \neq 0$ is unbounded, the linearization theorem shows that the equilibrium (0,0) is unstable.

If we try to apply the theorem to the interior equilibrium $\left(\frac{\mu}{c}, \frac{\lambda}{b}\right)$, we must consider the linearization $u' = -\frac{b\mu}{c}v$, $v' = \frac{c\lambda}{b}u$. Although we have not yet discussed any way to analyze such linearizations in general, methods we will discuss later in this chapter will enable us to show that all solutions of this linearization are periodic — namely, combinations of $\sin\sqrt{\lambda\mu}\,t$ and $\cos\sqrt{\lambda\mu}\,t$. Since the linearization's solutions neither approach zero nor become unbounded, the linearization theorem above does not allow us to draw any conclusions about the stability of this equilibrium; instead, we must turn to other methods, such as the phase plane approach we took in the last section.
□

EXAMPLE 4.
For each equilibrium of the system

$$y' = z, \; z' = -2(y^2 - 1)z - y,$$

determine whether the equilibrium is asymptotically stable or unstable.

Solution: The equilibria are the solutions of $z = 0$, $-2(y^2 - 1)z - y = 0$, and thus the only equilibrium is (0,0). Since $\frac{\partial}{\partial y}[z] = 0$, $\frac{\partial}{\partial z}[z] = 1$, and

$$\frac{\partial}{\partial y}[-2(y^2 - 1)z - y] = -4yz - 1, \; \frac{\partial}{\partial z}[-2(y^2 - 1)z - y] = -2(y^2 - 1),$$

the linearization at (0,0) is $u' = 0u + 1v = v$, $v' = -u + 2v$. We can actually solve this system of equations outright by using a clever trick[4] to reduce it

[4]It has been said that the best way to solve a differential equation is by correctly guessing the solution. It has also been said that a technique is a trick that can be used more than once. Many techniques (and tricks) that have been developed to solve differential equations take advantage of some peculiarity of the particular equations under consideration. The trick used above certainly falls into that category. However, as our approach in this text concentrates on qualitative methods, we shall not attempt to provide a list of such tricks, referring the interested reader instead to a more general introductory text on solving differential equations.

to a single equation. We subtract the first equation from the second to give $(v-u)' = (v-u)$. This is a first-order differential equation for $(v-u)$, whose solution is $v - u = c_1 e^t$. This partial result is already enough to tell us that the equilibrium is unstable, as since the difference between u and v grows exponentially, at least one of them must therefore grow exponentially as well. However, to give a more explicit conclusion we shall complete the solution: Substitution of $v = u + c_1 e^t$ into $u' = v$ gives another first-order linear equation $u' - u = c_1 e^t$. The solution of this equation, obtained by the method of Section 3.5, is $u = (c_1 t + c_2)e^t$, and therefore $v = (c_1 t + c_1 + c_2)e^t$. As this has unbounded solutions, we (again) conclude that the equilibrium (0,0) is unstable. □

Exercises

In Exercises 1–8, find the linearization of the given system at each equilibrium.

1. $y' = y + z - 2$, $z' = z - y$
2. $y' = y + z - 1$, $z' = y$
3. $y' = y + z^2$, $z' = y + 1$
4. $y' = z - 2$, $z' = y^2 - 8z$
5. $y' = e^z$, $z' = e^{-y}$
6. $y' = z$, $z' = \sin y$
7. $y' = y(\lambda - ay - bz)$,
   $z' = z(\mu - cy - dz)$, $a, b, c, d > 0$
8. $y' = y(\lambda - ay + bz)$,
   $z' = z(\mu + cy - dz)$, $a, b, c, d > 0$

In each of Exercises 9–12, for each equilibrium of the given system determine whether the equilibrium is asymptotically stable or unstable.

9. $y' = -2y$, $z' = -z$
10. $y' = y$, $z' = -z$
11. $y' = -y$, $z' = y^2 - z$
12. $y' = y + z$, $z' = z - 1$

13. Show that the equilibrium (0,0) of the system $y' = y(\lambda - ay - bz)$, $z' = z(\mu - cy - dz)$ is unstable (see Exercise 7).
14. Show that the equilibrium (0,0) of the system $y' = y(\lambda - ay + bz)$, $z' = z(\mu + cy - dz)$ is unstable (see Exercise 8).
15. How would the epidemic model of Example 2 have to change if
    (a) the whole population were partially protected from infection?
    (b) part of the population were born [permanently] partially protected from infection?
    (c) part of the population were initially temporarily protected?
    (d) the outbreak were extended enough to warrant including demographic renewal (births and deaths) in the model?

    In each case write the new model.

5.3 Linear systems with constant coefficients

We have seen in the preceding section that the stability of an equilibrium of a system of differential equations is determined by the behavior of solutions of the system's linearization at the equilibrium. This linearization is a linear system with constant coefficients (recall that a linear system has right-hand sides linear in the state variables, here y and z). Thus, in order to be able to decide whether an equilibrium is asymptotically stable, we need to be able to solve linear systems with constant coefficients. We were able to do this in the examples of the preceding section because the linearizations took a simple form with one of the equations of the system containing only a single variable. In this section we shall develop a more general technique.

The problem we wish to solve is a general two-dimensional linear system with constant coefficients,

$$\begin{aligned} y' &= ay + bz, \\ z' &= cy + dz, \end{aligned} \tag{5.12}$$

where a, b, c, and d are constants. We look for solutions of the form

$$y = Ye^{\lambda t}, \quad z = Ze^{\lambda t}, \tag{5.13}$$

where λ, Y, and Z are constants to be determined, with Y and Z not both zero. When we substitute the form (5.13) into the system (5.12), using $y' = \lambda Y e^{\lambda t}$, $z' = \lambda Z e^{\lambda t}$, we obtain two conditions

$$\begin{aligned} \lambda Y e^{\lambda t} &= aYe^{\lambda t} + bZe^{\lambda t}, \\ \lambda Z e^{\lambda t} &= cYe^{\lambda t} + dZe^{\lambda t}, \end{aligned}$$

which must be satisfied for all t. Because $e^{\lambda t} \neq 0$ for all t, we may divide these equations by $e^{\lambda t}$ to obtain a system of two equations which do not depend on t, namely

$$\begin{aligned} \lambda Y &= aY + bZ, \\ \lambda Z &= cY + dZ \end{aligned}$$

or

$$\begin{aligned} (a - \lambda)Y + bZ &= 0, \\ cY + (d - \lambda)Z &= 0. \end{aligned} \tag{5.14}$$

The pair of equations (5.14) is a system of two homogeneous linear algebraic equations for the unknowns Y and Z. For certain values of the parameter λ this system will have a solution other than the obvious solution $Y = 0$, $Z = 0$. In order that the system (5.14) have a non-trivial solution for Y and

Z, it is necessary that the determinant of the coefficient matrix, which is $(a-\lambda)(d-\lambda)-bc$, be equal to zero. This gives a quadratic equation, called the *characteristic equation*, of the system (5.12) for λ. We may rewrite the characteristic equation as

$$\lambda^2-(a+d)\lambda+(ad-bc)=0. \tag{5.15}$$

We will assume that $ad-bc\neq 0$, which implies that $\lambda=0$ is not a root of (5.15). Our reason for this assumption is the following: If $\lambda=0$ is a root (or, equivalently, $ad-bc=0$), then equations (5.14) become the same as the equilibrium conditions for (5.12), obtained by setting the right-hand sides of (5.12) to zero. With $ad-bc=0$, the equilibrium conditions will reduce to a single equation: for instance, substituting $d=bc/a$ into the second equation and then multiplying by a/c will result in the first equation. Consequently, there will be a line of non-isolated equilibria. We do not wish to explore this problem (but see Exercise 19 below), in part because the treatment of non-isolated equilibria is complicated, and in part because it is difficult to apply the linearization theorem of the previous section when the linearization of a system about an equilibrium is of this form.

With $ad-bc\neq 0$, the characteristic equation has two roots λ_1 and λ_2, which may be real and distinct, real and equal, or complex conjugates. (Students of linear algebra will take note that these roots are in fact the eigenvalues of the community matrix; the characteristic equation of the system is also the characteristic equation of this matrix.) If λ_1 and λ_2 are the roots of (5.15), then there are a solution (Y_1,Z_1) of (5.14) corresponding to the root λ_1, and a solution (Y_2,Z_2) of (5.14) corresponding to the root λ_2. These, in turn, give us two solutions

$$y=Y_1e^{\lambda_1 t},\ z=Z_1e^{\lambda_1 t}\ \text{ and }\ \mathrm{y}=\mathrm{Y}_2\mathrm{e}^{\lambda_2 \mathrm{t}},\ \mathrm{z}=\mathrm{Z}_2\mathrm{e}^{\lambda_2 \mathrm{t}}$$

to the system (5.12). We note that if λ is a root of (5.15), then equations (5.14) reduce to a single equation, so that we may always choose a value for one of Y, Z, with the other then determined by (5.14). If (5.14) sets one of Y, Z to zero, then we may choose a value for the other.

Because system (5.12) is linear, it is possible to show that if we have two different solutions of the system (5.12), then every solution of (5.12) is a constant multiple of the first solution plus a constant multiple of the second solution. By "different" we mean that neither solution is a constant multiple of the other. Here, if the roots λ_1 and λ_2 of the characteristic equation (5.15) are distinct, we do have two different solutions of (5.12), and then every solution of system (5.12) has the form

$$\begin{aligned} y&=K_1Y_1e^{\lambda_1 t}+K_2Y_2e^{\lambda_2 t},\\ z&=K_1Z_1e^{\lambda_1 t}+K_2Z_2e^{\lambda_2 t}\end{aligned} \tag{5.16}$$

for some constants K_1 and K_2. The form (5.16) with two arbitrary constants

K_1 and K_2 is called the *general solution* of the system (5.12). If initial values $y(0)$ and $z(0)$ are specified, these two initial values may be used to determine values for the constants K_1 and K_2, and thus to obtain a *particular solution* in the family (5.16).

EXAMPLE 1.
Find the general solution of the system

$$y' = -y - 2z, \quad z' = y - 4z$$

and also the solution such that $y(0) = 3$, $z(0) = 1$.

Solution: Here $a = -1$, $b = -2$, $c = 1$, $d = -4$, so that $a+d = -5$, $ad-bc = 6$, and the characteristic equation is $\lambda^2+5\lambda+6 = 0$, with roots $\lambda_1 = -2$, $\lambda_2 = -3$. With $\lambda = -2$, both equations of the algebraic system (5.14) are $Y - 2Z = 0$, and we may take $Y = 2$, $Z = 1$. The resulting solution of (5.12) is $y = 2e^{-2t}$, $z = e^{-2t}$. With $\lambda = -3$, both equations of the algebraic system (5.14) are $Y - Z = 0$, and we may take $Y = 1$, $Z = 1$. The resulting solution of (5.12) is $y = e^{-3t}$, $z = e^{-3t}$. Thus the general solution of the system is

$$y = 2K_1e^{-2t} + K_2e^{-3t}, \quad z = K_1e^{-2t} + K_2e^{-3t}.$$

To satisfy the initial conditions, we substitute $t = 0$, $y = 3$, $z = 1$ into this form, obtaining a pair of equations $2K_1 + K_2 = 3$, $K_1 + K_2 = 1$. We may subtract the second of these from the first to give $K_2 = 2$, and then $K_1 = -1$. This gives, as the solution of the initial value problem,

$$y = 4e^{-2t} - e^{-3t}, \quad z = 2e^{-2t} - e^{-3t}. \ \Box$$

If the characteristic equation (5.15) has a double root, this method gives only one solution of the system (5.12), and we need to find a second solution in order to form the general solution. The characteristic equation has a double root if the *discriminant* $(a+d)^2-4(ad-bc) = (a-d)^2+4bc = 0$. It is possible to show (and the reader can verify) that if λ is a double root of (5.15), $\lambda = \frac{a+d}{2}$, then in addition to the solution $y = Y_1e^{\lambda t}$, $z = Z_1e^{\lambda t}$ of (5.12) there is a second solution, of the form

$$y = (Y_2 + Y_1t)e^{\lambda t}, \quad z = (Z_2 + Z_1t)e^{\lambda t},$$

where Y_1, Z_1 are as in (5.14) and Y_2, Z_2 are given by

$$(a - \lambda)Y_2 + bZ_2 = Y_1,$$
$$cY_2 + (d - \lambda)Z_2 = Z_1.$$

Thus the general solution of the system (5.12) in the case of equal roots is

$$y = (K_1Y_1 + K_2Y_2)e^{\lambda t} + K_2Y_1te^{\lambda t}, \tag{5.17}$$
$$z = (K_1Z_1 + K_2Z_2)e^{\lambda t} + K_2Z_1te^{\lambda t}.$$

Note that if (5.15) has a single root and $b = 0$, then $\lambda = a = d$, and we have $Y_1 = 0$, $Z_1 = cY_2$, so that the general solution becomes

$$y = K_3 e^{\lambda t}, \; z = K_4 e^{\lambda t} + cK_3 t e^{\lambda t},$$

where $K_3 \equiv K_2 Y_2$ and $K_4 \equiv cK_1 Y_2 + K_2 Z_2$ are arbitrary constants. Likewise if (5.15) has a single root and $c = 0$, then $\lambda = a = d$, $Z_1 = 0$, $Y_1 = bZ_2$, and the general solution reduces to

$$y = K_5 e^{\lambda t} + bK_6 t e^{\lambda t}, \; z = K_6 e^{\lambda t},$$

with arbitrary constants $K_5 \equiv bK_1 Z_2 + K_2 Y_2$ and $K_6 \equiv K_2 Z_2$. (We cannot have b and c both zero here, or else the system becomes uncoupled: $y' = ay$, $z' = dz$.)

EXAMPLE 2.
Find the general solution of the system

$$y' = z, \quad z' = -y + 2z,$$

and also the solution such that $y(0) = 2$, $z(0) = 3$.

Solution: Since $a = 0$, $b = 1$, $c = -1$, $d = 2$, we have $a + d = 2$, $ad - bc = 1$. The characteristic equation is $\lambda^2 - 2\lambda + 1 = 0$, with a double root $\lambda = 1$. With $\lambda = 1$, both equations of the system (5.14) are $Y - Z = 0$, and we may take $Y = 1$, $Z = 1$ to give the solution $y = e^t$, $z = e^t$ of (5.12). Substituting these values into the equations for Y_2, Z_2, we find that they reduce to the single equation $-Y_2 + Z_2 = 1$, so we may take $Y_2 = 0$, $Z_2 = 1$. Equations (5.17) now give the general solution $y = K_1 e^t + K_2 t e^t$, $z = (K_1 + K_2)e^t + K_2 t e^t$. To find the solution with $y(0) = 2$, $z(0) = 3$, we substitute $t = 0$, $y = 2$, $z = 3$ into this form, obtaining the pair of equations $K_1 = 2$, $K_1 + K_2 = 3$. Then $K_2 = 1$, and the solution satisfying the initial conditions is $y = 2e^t + te^t$, $z = 3e^t + te^t$.
□

Another complication arises if the characteristic equation (5.15) has complex roots. While the general solution of (5.12) is still given by (5.16), in this case the constants λ_1 and λ_2 are complex, and the solution is in terms of complex functions. Complex exponentials, however, can be defined with the aid of trigonometric functions ($e^{i\theta} \equiv \cos\theta + i\sin\theta$ for real θ), so it is still possible to give the solution of (5.12) in terms of real exponential and trigonometric functions. In this case, if the characteristic equation (5.15) has conjugate complex roots $\lambda = \alpha \pm i\beta$, where α and β are real and $\beta > 0$, equations (5.16) become

$$\begin{aligned} y &= (K_1 Y_1 + K_2 Y_2)e^{\alpha t}\cos\beta t + i(K_1 Y_1 - K_2 Y_2)e^{\alpha t}\sin\beta t, \\ z &= (K_1 Z_1 + K_2 Z_2)e^{\alpha t}\cos\beta t + i(K_1 Z_1 - K_2 Z_2)e^{\alpha t}\sin\beta t. \end{aligned}$$

We can eliminate the imaginary coefficients by defining $Q_1 \equiv i(K_1 - K_2)$,

$Q_2 \equiv K_1 + K_2$, and taking $(a - \lambda_i)Y_i + bZ_i = 0$ for $i = 1, 2$ from (5.16) to arrive at the form

$$\begin{aligned} y &= Q_1 b e^{\alpha t} \sin \beta t + Q_2 b e^{\alpha t} \cos \beta t, \\ z &= -[Q_1(a-\alpha) - Q_2\beta] e^{\alpha t} \sin \beta t + [Q_1\beta - Q_2(a-\alpha)] e^{\alpha t} \cos \beta t. \end{aligned}$$

EXAMPLE 3.

Find the general solution of the system

$$y' = -2z, \quad z' = y + 2z,$$

and also the solution with $y(0) = -2$, $z(0) = 0$.

Solution: We have $a = 0$, $b = -2$, $c = 1$, $d = 2$, and the characteristic equation is $\lambda^2 - 2\lambda + 2 = 0$, with roots $\lambda = 1 \pm i$. The general solution is then

$$y = -2K_1 e^t \sin t - 2K_2 e^t \cos t, \quad z = (K_1 - K_2)e^t \sin t + (K_1 + K_2)e^t \cos t.$$

To find the solution with $y(0) = -2$, $z(0) = 0$, we substitute $t = 0$, $y = -2$, $z = 0$ into this form, obtaining $-2K_2 = -2$, $K_1 + K_2 = 0$, whose solution is $K_1 = -1$, $K_2 = 1$. This gives the particular solution

$$y = 2e^t \sin t - 2e^t \cos t, \quad z = -2e^t \sin t. \ \square$$

In many applications, especially in analyzing stability of an equilibrium, the precise form of the solution of a linear system is less important to us than the qualitative behavior of solutions. It will turn out that often the crucial question is whether all solutions of a linear system approach zero as $t \to \infty$. In the next section we will give an algebraic criterion to answer this question, but we may also infer the answer from a look at the direction field near the origin. It requires more effort, but a phase portrait consisting of several orbits may be even more helpful.

For example, the direction field and phase portrait for the system $y' = -y - 2z$, $z' = y - 4z$, solved explicitly in Example 1 above, are shown in Figure 5.6. The direction field suggests that every orbit approaches the origin with a fixed limiting direction, and this is correct. The origin is an equilibrium for the linear system, and an equilibrium with this behavior for orbits is called a *node*. Nodes occur when the roots of the corresponding characteristic equation are real and of the same sign (here both negative).

The direction field and phase portrait for the system $y' = y + 2z$, $z' = -z$ are shown in Figure 5.7. Orbits starting on the line $z = -y$ approach the origin, but all other orbits are repelled from the origin. Such an equilibrium is called a *saddle point*. Saddle points occur when the roots of the corresponding characteristic equation are real but have different signs.

The direction field and phase portrait for the system $y' = z$, $z' = -2y - 2z$

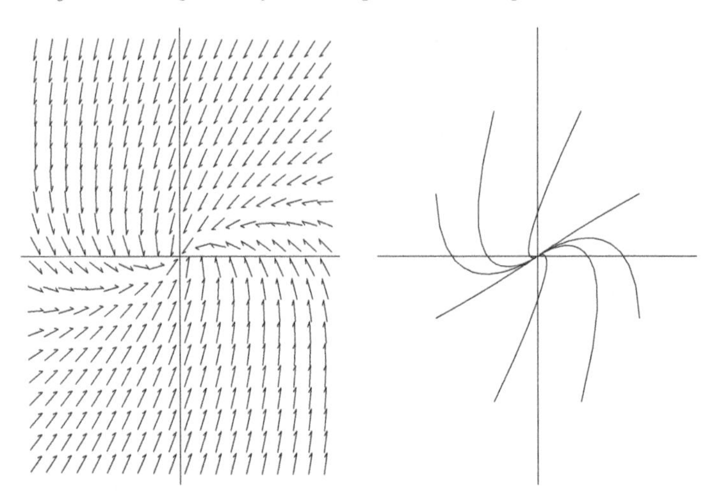

FIGURE 5.6: Direction field and phase portrait showing a stable node.

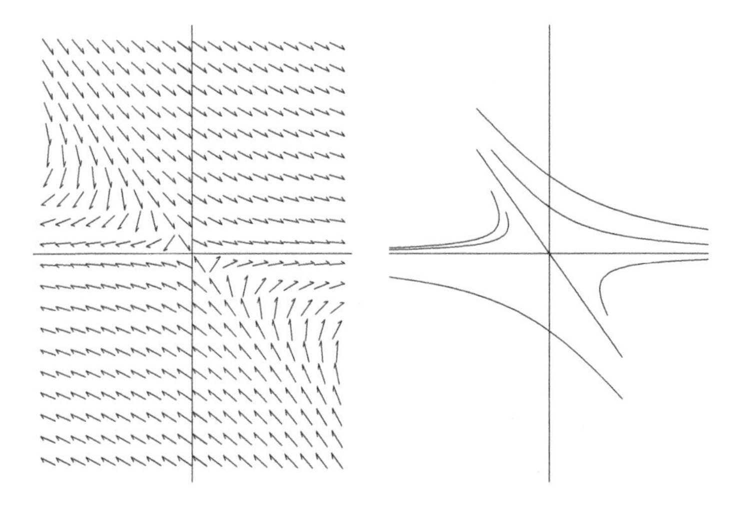

FIGURE 5.7: Direction field and phase portrait showing a saddle point.

are shown in Figure 5.8. Every orbit approaches the origin, but the approach is by an inward spiral. Such an equilibrium is called a *spiral point*. Here the roots of the characteristic equation are complex.

The direction field and phase portrait for the system $y' = z$, $z' = -y$ are shown in Figure 5.9. Every orbit is a closed orbit around the origin, and such an equilibrium is called a *center*. In this case the roots of the characteristic equation are purely imaginary.

In the above examples, every orbit approaches the origin for the node and spiral point, but we could also give examples in which the directions

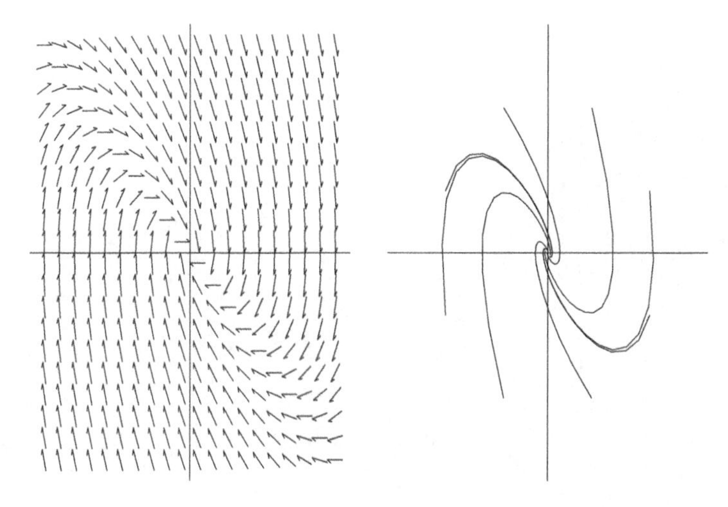

FIGURE 5.8: Direction field and phase portrait showing a stable spiral point.

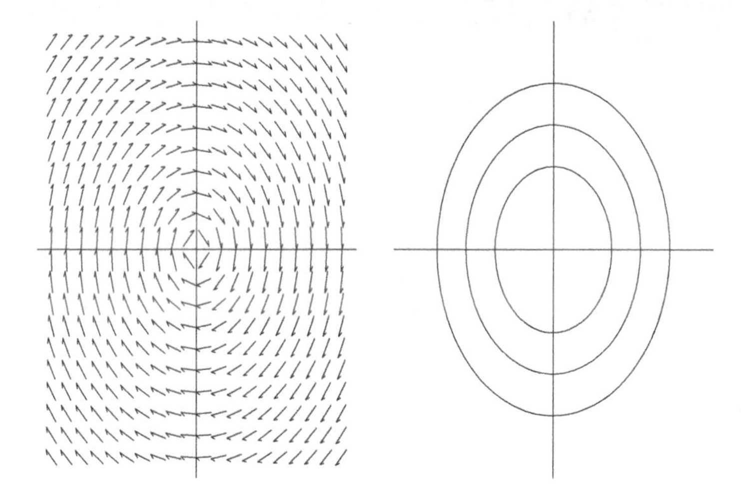

FIGURE 5.9: Direction field and phase portrait showing a center.

are reversed and all orbits are unbounded. For a center, no orbits approach the origin, and for a saddle point there are orbits which approach the origin and also unbounded orbits. Every equilibrium point for a linear autonomous system is of one of these four types.

As noted above, the nature of the origin as an equilibrium of the linear system (5.12) depends on the roots of the characteristic equation (5.15). If both roots of (5.15) are real and of the same sign, then the solutions of (5.12) are combinations of either positive exponentials or negative exponentials. This

Photo by Steve Hillebrand, US FWS

FIGURE 5.10: A system of interconnected ponds acts like a mixing system.

implies that the slope of an orbit, which is $\frac{z'(t)}{y'(t)}$, must approach a limit as $t \to \infty$, and thus that the origin is a node. If the roots of (5.15) are real and of opposite sign, then there are solutions which are positive exponentials and solutions which are negative exponentials. Thus there are solutions approaching the origin and solutions moving away from the origin, and the origin is a saddle point. If the roots of (5.15) are complex, the solutions contain trigonometric functions and the orbits oscillate. If the real parts of the roots are zero, the orbits are periodic, and the origin is a center. If the real parts are different from zero, the orbits spiral, and the origin is a spiral point.

In addition to its use in equilibrium stability analysis, the solution of linear systems with constant coefficients has some direct applications. We now conclude this section with a direct application to mixing problems. In Section 3.6 we described some mixing problems with a single compartment. Mixing problems with two compartments, whether natural (such as a system of interconnected ponds, see Figure 5.10) or artificial (connected tanks, such as in Example 4 below), may be described by characterizing the rate of change in each compartment to give a system of two differential equations. If the mixture flows from one compartment to the other, the concentrations in the two compartments are linked, and we would expect a coupled system of differential equations.

EXAMPLE 4.
A tank contains 100 liters of water and 10 kg. of salt, thoroughly mixed. Pure water is added at the rate of 5 liters/minute, and the mixture is poured off at the rate of 5 liters/minute into a second tank which initially contains 80 liters of water. The mixture in the second tank is then poured out as waste, at the same rate. Formulate and solve a model to describe the weight of salt in each tank as a function of time.

Solution: Let $y(t)$ denote the weight of salt in the first tank and $z(t)$ the weight of salt in the second tank at time t. Then $y(0) = 10$ and $z(0) = 0$. The concentration of salt in the first tank is $\frac{y}{100}$ kg./liter, and the weight of salt poured into the second tank per minute is $5\,\frac{y}{100} = \frac{y}{20}$ kg. The concentration of salt in the second tank is $z/80$ kg./liter, and the weight of salt poured out is $5\,\frac{z}{80} = \frac{z}{16}$ kg. Thus

$$y' = -y/20, \quad z' = y/20 - z/16.$$

In addition to this system of differential equations, we must impose the initial conditions $y(0) = 10$, $z(0) = 0$.

To solve this problem, we take $a = -\frac{1}{20}$, $b = 0$, $c = \frac{1}{20}$, $d = -\frac{1}{16}$. This gives the characteristic equation $\lambda^2 + \frac{1}{16}\lambda + \frac{1}{1600} = 0$, with roots $\lambda_1 = -\frac{1}{20}$ and $\lambda_2 = -\frac{1}{80}$. With $\lambda = -\frac{1}{20}$, the system (5.14) reduces to $4Y_1 + 3Z_1 = 0$, so we can take $Y_1 = 3$, $Z_1 = -4$, giving a solution $y = 3e^{-t/20}$, $z = -4e^{-t/20}$ of the system. With $\lambda = -\frac{1}{80}$, the system reduces to $Y_2 = 0$, so we can take $Z_2 = 1$, giving a solution $y = 0$, $z = e^{-t/80}$ of the system. Thus the general solution is $y = 3K_1e^{-t/20}$, $z = -4K_1e^{-t/20} + K_2e^{-t/80}$. To satisfy the initial conditions $y(0) = 10$, $z(0) = 0$, we substitute $t = 0$, $y = 10$, $z = 0$, and obtain the equations $10 = 3K_1$, $0 = -4K_1 + K_2$, with solution $K_1 = \frac{10}{3}$, $K_2 = \frac{40}{3}$. Thus the solution of the initial value problem is $y = 10e^{-t/20}$, $z = -\frac{40}{3}e^{-t/20} + \frac{40}{3}e^{-t/80}$. We note that since the solution is a sum of negative exponentials, it approaches zero as $t \to \infty$; that is, the amount of salt in each tank approaches zero asymptotically as time goes on. □

5.3.1 A liver chemistry example

The chemical bromosulfophthalene (BSP) is used to measure hepatic metabolism. BSP is administered directly into the blood via an intravenous (IV) injection, and then BSP levels in the blood are recorded periodically in order to measure the liver's ability to clear toxins, chemicals and other foreign substances from the blood. When blood flows through the liver, there is an exchange of these substances between the blood and the liver, at rates dependent upon the concentration of the given substance in the system of origin (blood or liver). At the same time, the liver eliminates excess amounts of the substance via other pathways (such as the digestive system). This exchange is described by Jolivet[5] in a simple linear two-compartment model, describing the three exchanges from blood to liver, from liver to blood, and from liver to elimination from the body. If we denote the blood with a subscript 1, the liver with a subscript 2, and elimination with a subscript 0, then we can denote the three exchange rates, respectively, as k_{21}, k_{12}, and k_{02}. Each rate can be interpreted as the reciprocal of the average time required for each exchange to occur.

Figure 5.11 provides a stylized flow chart for this exchange process. If we define $q_1(t)$ to be the quantity (or, alternatively, the concentration) of BSP in the blood and $q_2(t)$ the quantity (or concentration) in the liver, then we can describe the rates of change of q_1 and q_2 via the following two differential equations:

$$\begin{aligned} \frac{dq_1}{dt} &= -k_{21}q_1 \quad +k_{12}q_2 \\ \frac{dq_2}{dt} &= k_{21}q_1 \quad -(k_{12}+k_{02})q_2 \end{aligned} \tag{5.18}$$

where the first term in each equation represents flow of BSP from the blood to the liver, the second term in each equation represents the flow in the reverse direction, and the remaining term in the second equation represents elimination of BSP from the liver. To complete the statement of an initial value problem, we add the initial conditions $q_1(0) = Q$, $q_2(0) = 0$ representing the injection at time 0 of a quantity (or concentration) Q of BSP into the blood, at which point the liver has no BSP in it.

Since the system (5.18) is a linear system with constant coefficients, we assume that there are exponential solutions of the form $q = ae^{\lambda t}$, so that $q' = \lambda q$; then the equations (5.18) become

$$\begin{aligned} \lambda q_1(t) &= -k_{21}q_1(t) \quad +k_{12}q_2(t), \\ \lambda q_2(t) &= k_{21}q_1(t) \quad -(k_{12}+k_{02})q_2(t). \end{aligned}$$

Solutions λ are eigenvalues of the coefficient matrix $\begin{bmatrix} -k_{21} & k_{12} \\ k_{21} & -(k_{12}+k_{02}) \end{bmatrix}$, as straightforward substitution leads to elimination of q_1 and q_2, leaving the characteristic equation

$$\lambda^2 + (k_{21}+k_{12}+k_{02})\lambda + (k_{21}k_{02}) = 0.$$

The two roots are

$$\lambda_{1,2} = -\frac{1}{2}\left(k_{21}+k_{12}+k_{02} \pm \sqrt{\Delta}\right), \quad \text{where } \Delta = (k_{21}+k_{12}+k_{02})^2 - 4k_{21}k_{02} \tag{5.19}$$

is the discriminant of the quadratic characteristic equation. Some algebra

[5]Emmanuel Jolivet, *Introduction aux modèles mathématiques en biologie*, INRA/Masson, Paris, 1983. Chapter 3, pp. 37–52.

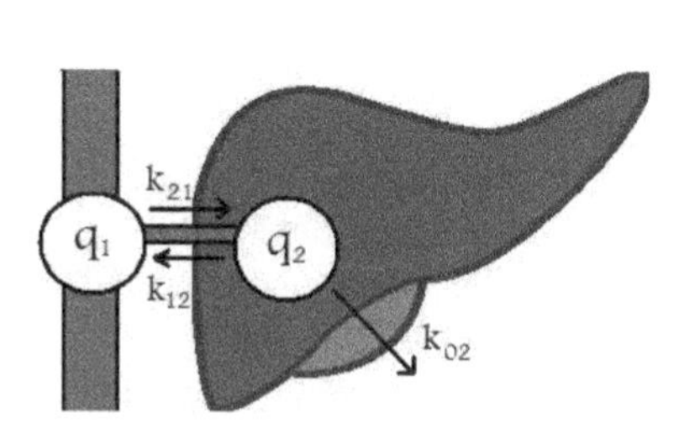

FIGURE 5.11: A sketch of a flow chart depicting the exchanges of the chemical BSP between the blood and the liver.

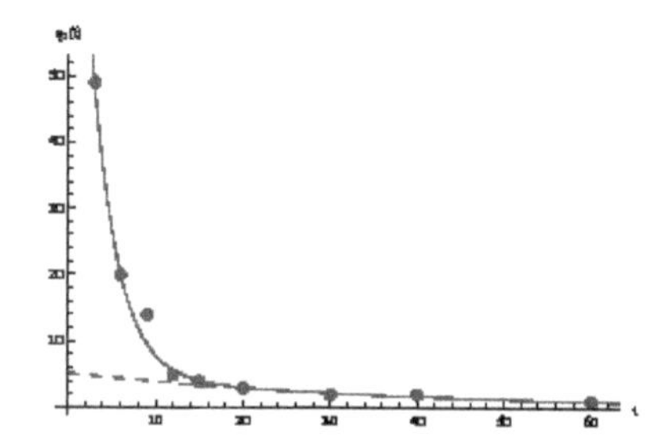

FIGURE 5.12: A comparison of the liver chemistry data (dots), the slow-decaying component of the solution (dashed curve), and the full solution (solid curve).

shows that $\Delta = (k_{21} - k_{02})^2 + k_{12}^2 + 2k_{12}(k_{02} + k_{21}) > 0$, so both roots are real and (since $\sqrt{\Delta} < k_{21}+k_{12}+k_{02}$) negative, corresponding to exponential decay. In particular, if we take $\lambda_1 > \lambda_2$, then λ_1 describes a slower rate of elimination of BSP from the body, while λ_2 describes a portion which is eliminated more quickly. Overall, then, solutions to the system of differential equations have the form

$$q_1(t) = a_1 e^{\lambda_1 t} + a_2 e^{\lambda_2 t}, \tag{5.20}$$

$$q_2(t) = b_1 e^{\lambda_1 t} + b_2 e^{\lambda_2 t}, \tag{5.21}$$

where the constants a_1, a_2, b_1 and b_2 are determined by initial conditions.

Knowing that the amounts (or concentrations) of BSP in the blood and liver can be described by these functions, we can use measurements of BSP concentration in the blood over time to calculate first the coefficients in the above equations, and then the original exchange rates, which give an understanding of how quickly the liver is able to clear BSP from the body. As an example of how this model might be used in practice, Jolivet provides the data in Table 5.1 from Feldmann and Schneider,[6] in which 0.8 mg of BSP was introduced via IV injection.

TABLE 5.1: Data from Feldmann and Schneider [24] showing the concentration q_1 of BSP in the blood as a function of time t since introduction.

t (min)	3	6	9	12	15	20	30	40	60
$q_1(t)$ (μg/L)	49	20	14	5	4	3	2	2	1

[6] U. Feldmann and B. Schneider, A general approach to multicompartment analysis and models for the pharmacodynamics, in J. Berger et al. (eds.), *Mathematical models in medicine: workshop, Mainz, March 1976* (Berlin/New York: Springer-Verlag), Lecture Notes in Biomathematics 11: 243–277, 1976.

As always, real data contains "noise," so will not fit any theoretical form (i.e., (5.20)) perfectly. There are many sophisticated numerical and statistical methods for deriving the best possible estimate for the coefficients in the equation, but for the sake of simplicity here we will use only some of these data and derive simple estimates (which nevertheless compare well with those Jolivet obtained, q.v.). First, we assume that the difference in decay rates between λ_1 and λ_2 is great enough that after enough time, most of the fast-decaying $e^{\lambda_2 t}$ component has dwindled to a negligible level, leaving primarily the slower-decaying $e^{\lambda_1 t}$ component: $q_1(t) \approx a_1 e^{\lambda_1 t}$. If we assume that this is true for the last half of the data, then we can use the concentrations after 20 and 60 minutes to calculate the coefficients a_1 and λ_1:

$$\begin{aligned} q_1(20) = 3 &\approx a_1 e^{20\lambda_1}, & q_1(60) = 1 &\approx a_1 e^{60\lambda_1}, \\ \log 3 &\approx \log a_1 + 20\lambda_1, & \log 1 = 0 &\approx \log a_1 + 60\lambda_1, \end{aligned}$$

from which $\log a_1 \approx -60\lambda_1$, and, substituting, $\log 3 \approx -40\lambda_1$, making $\lambda_1 \approx (\log 3)/40\text{min} = -0.0275/\text{min}$ and thus $a_1 \approx e^{-60\lambda_1} = 5.2\mu\text{g/L}$ (note all logarithms here are natural, not base ten).

To recover a_2 and λ_2 we return to the earliest data points (3 and 6 minutes after injection), where we assume the fast-decaying component can still be detected:

$$\begin{aligned} q_1(3) = 49 &= a_1 e^{3\lambda_1} + a_2 e^{3\lambda_2}, & q_1(6) = 20 &= a_1 e^{6\lambda_1} + a_2 e^{6\lambda_2}, \\ 49 &= 5.2e^{3(-0.0275)} + a_2 e^{3\lambda_2}, & 20 &= 5.2e^{6(-0.0275)} + a_2 e^{6\lambda_2}, \\ 49 &= 4.788 + a_2 e^{3\lambda_2}, & 20 &= 4.41 + a_2 e^{6\lambda_2}, \\ \log 44.212 = 3.789 &= \log a_2 + 3\lambda_2, & \log 15.59 = 2.747 &= \log a_2 + 6\lambda_2, \end{aligned}$$

so $\log a_2 = 2.747 - 6\lambda_2$, making $3.789 = 2.747 - 3\lambda_2$, and thus $\lambda_2 = -(3.789 - 2.747)/3\text{min} = -0.347/\text{min}$. Then, finally, $a_2 = e^{2.747-6(-0.347)} \approx 125\mu\text{g/L}$. This gives the concentration of BSP in the blood of

$$q_1(t) = 5.2e^{-0.0275t} + 125e^{-0.347t}. \tag{5.22}$$

From this we see that most of the BSP is removed fairly quickly from the blood (after an average of $1/\lambda_2 = 2.88\text{min}$), while the rest takes longer to be eliminated (an average of $1/\lambda_1 = 36.4\text{min}$). Figure 5.12 superimposes plots of the original data, the solution, and the slow-decaying component of the solution.

To recover the original exchange rates we use a somewhat *ad hoc* process, first observing that $q_1(0) = 5.2 + 125 \approx 130\mu\text{g/L}$, $q_2(0) = 0$ so that (from (5.18)

$$\frac{dq_1}{dt}(0) = -k_{21}q_1(0) + k_{12}q_2(0) = -(130\mu\text{g/L})k_{21},$$

while at the same time (from (5.22))

$$\frac{dq_1}{dt}(0) = 5.2(-0.0275) + 125(-0.347) = -43.518(\mu\text{g/L})/\text{min},$$

making $k_{21} = -43.518/(-130)\text{min} = 0.334/\text{min}$. Then, to recover the other two rates, we observe that (from (5.19), and after a little algebra)

$$\lambda_1\lambda_2 = k_{02}k_{21}, \quad \lambda_1 + \lambda_2 = -(k_{21} + k_{12} + k_{02}).$$

The first of these two equations becomes $(-0.0275)(-0.347) = 0.334k_{02}$, making $k_{02} = 0.0286/\text{min}$, and then the latter equation becomes $-0.0275-0.347 = -(0.334 + k_{12} + 0.0286)$, so that $k_{12} = 0.0119/\text{min}$. The reciprocals of these rates give the corresponding average passage times.

Exercises

In each of Exercises 1–14, find the general solution of the given system by analytic solution, use a computer algebra system to examine the behavior of solutions, and classify the origin as a node, saddle point, center, or spiral point.

1. $y' = y + 5z$, $z' = y - 3z$
2. $y' = y - z$, $z' = z$
3. $y' = 2y + z$, $z' = z$
4. $y' = -y$, $z' = y - z$
5. $y' = 4y$, $z' = 2y + 4z$
6. $y' = z$, $z' = -y + 2z$
7. $y' = z$, $z' = -2y - 3z$
8. $y' = y - z$, $z' = 4y - 3z$
9. $y' = y + 2z$, $z' = -3y + 6z$
10. $y' = 3y - 4z$, $z' = y - 2z$
11. $y' = 3y + 5z$, $z' = -5y + 3z$
12. $y' = z$, $z' = -y$
13. $y' = y + z$, $z' = z$
14. $y' = 5y + z$, $z' = 5z$
15. A tank contains 1000 liters of water and a salt solution containing 10 kg./liter is pumped into it at a rate of 200 liters /minute. The mixture is led to a second tank containing 1000 liters of water at a rate of 200 liters/minute, and the mixture is pumped out of the second tank at the same rate. What is the concentration of salt in the second tank after one hour?

16. Two tanks begin with 10 kg. of salt dissolved in 100 liters of water. Water is pumped into the first tank at a rate of 10 liters/minute, the mixture is pumped from the first tank to the second tank at a rate of 10 liters/minute, and the mixture is pumped out of the second tank at a rate of 10 liters/minute. What is the amount of salt in each tank after one hour, and what is the amount of salt in each tank after a very long time?

17. Obtain the general solution of the system $y' = ay$, $z' = cy + dz$ if $a \neq d$ by solving $y' = ay$, substituting the result into $z' = cy + dz$ and solving.

18. Obtain the general solution of the system $y' = ay$, $z' = cy + az$ by solving $y' = ay$, substituting the result into $z' = cy + az$ and solving.

19. * Consider the system

$$y' = y - z,$$
$$z' = 3y - 3z,$$

for which the condition $ad - bc \neq 0$ is violated.

(a) Show that every point of the line $z = y$ is an equilibrium.

(b) Use the fact that $z' = 3y'$ to deduce that $z = 3y + c_1$ for some constant c_1.

(c) Use the result of part (b) to eliminate z from the system and obtain a first-order linear differential equation for y.

(d) Solve for y and obtain the solution $y = -\frac{c_1}{2} + c_2e^{-2t}$, $z = -\frac{c_1}{2} + 3c_2e^{-2t}$.

(e) Show that as $t \to \infty$ every orbit approaches the point $(-\frac{c_1}{2}, -\frac{c_1}{2})$ on the line of equilibria, and that the slope of the line joining this point to any point on the orbit is the constant 3.

(f) Use the information obtained to sketch the phase portrait of the system.

5.4 Qualitative analysis of systems

In order to apply the linearization theorem of Section 5.2 to questions of stability of an equilibrium, we must determine conditions under which all solutions of a linear system with constant coefficients

$$\begin{aligned} y' &= ay + bz, \\ z' &= cy + dz \end{aligned} \tag{5.23}$$

approach zero as $t \to \infty$. As we saw in Section 5.3, the nature of the solutions of (5.23) is determined by the roots of the characteristic equation

$$\lambda^2 - (a+d)\lambda + (ad - bc) = 0. \tag{5.24}$$

If the roots λ_1 and λ_2 of (5.24) are real, then the solutions of (5.23) are made up of terms $e^{\lambda_1 t}$ and $e^{\lambda_2 t}$, or $e^{\lambda_1 t}$ and $te^{\lambda_1 t}$ if the roots are equal. In order that all solutions of (5.23) approach zero, we require $\lambda_1 < 0$ and $\lambda_2 < 0$, so that the terms will be negative exponentials. If the roots are complex conjugates, $\lambda = \alpha \pm i\beta$, then in order that all solutions of (5.23) approach zero, we require $\alpha < 0$. Thus if the roots of the characteristic equation have negative real part, all solutions of the system (5.23) approach zero as t$\to \infty$. In a similar manner, we may see that if a root of the characteristic equation has positive real part, then (5.23) has unbounded solutions.

It turns out, however, that it is not necessary to solve the characteristic equation in order to determine whether all solutions of (5.23) approach zero, as there is a useful criterion in terms of the coefficients of the characteristic equation. The basic result, whose proof may be found in Appendix C, is that the roots of a quadratic equation $\lambda^2 + a_1\lambda + a_2 = 0$ have negative real part if and only if $a_1 > 0$ and $a_2 > 0$. Applying this to the characteristic equation (5.24) and the system (5.23), we obtain the following result for linear systems with constant coefficients:

STABILITY THEOREM FOR LINEAR SYSTEMS: Every solution of the linear system with constant coefficients (5.23)

$$y' = ay + bz, \quad z' = cy + dz$$

approaches zero as $t \to \infty$ if and only if the *trace* $a + d$ of the coefficient matrix of the system is negative and the *determinant* $ad - bc$ of the system's coefficient matrix is positive. If either the trace is positive or the determinant is negative, there is at least one unbounded solution.

EXAMPLE 1.
Determine whether all solutions tend to zero or whether there are unbounded solutions for each of the following systems:

$$\begin{array}{ll} (i) & u' = -u - 2v, \quad v' = u - 4v \\ (ii) & u' = v, \quad v' = -u - 2v \\ (iii) & u' = -2v, \quad v' = u + 2v \end{array}$$

Solution: (i) The characteristic equation is $\lambda^2 + 5\lambda + 6 = 0$, with roots $\lambda = -2$, $\lambda = -3$. Thus all solutions tend to zero. Alternatively, since the trace of the coefficient matrix is $-5 < 0$ and the determinant is $6 > 0$, the stability theorem gives the same conclusion. For (ii), the characteristic equation is $\lambda^2 + 2\lambda + 1 = 0$ with a double root $\lambda = -1$, and thus all solutions tend to zero. For (iii), the characteristic equation is $\lambda^2 - 2\lambda + 2 = 0$, and since the trace is positive, there are unbounded solutions. As we indicated in the previous section, we could also have drawn this conclusion from a phase portrait. □

If we apply the stability theorem for linear systems to the linearization

$$\begin{aligned} u' &= F_y(y_\infty, z_\infty)u + F_z(y_\infty, z_\infty)v, \\ v' &= G_y(y_\infty, z_\infty)u + G_z(y_\infty, z_\infty)v \end{aligned} \tag{5.25}$$

of a system

$$y' = F(y, z), \quad z' = G(y, z) \tag{5.26}$$

at an equilibrium (y_∞, z_∞), we obtain the following result.

EQUILIBRIUM STABILITY THEOREM: Let (y_∞, z_∞) be an equilibrium of a system $y' = F(y, z)$, $z' = G(y, z)$, with F and G twice differentiable. Then if

$$F_y(y_\infty, z_\infty) + G_z(y_\infty, z_\infty) < 0 \tag{5.27}$$

and

$$F_y(y_\infty, z_\infty)G_z(y_\infty, z_\infty) - F_z(y_\infty, z_\infty)G_y(y_\infty, z_\infty) > 0, \tag{5.28}$$

the equilibrium (y_∞, z_∞) is asymptotically stable. If either

$$F_y(y_\infty, z_\infty) + G_z(y_\infty, z_\infty) > 0$$

or

$$F_y(y_\infty, z_\infty)G_z(y_\infty, z_\infty) - F_z(y_\infty, z_\infty)G_y(y_\infty, z_\infty) < 0$$

the equilibrium (y_∞, z_∞) is unstable.

EXAMPLE 2.
Determine whether each equilibrium of the system

$$y' = z, \quad z' = 2(y^2 - 1)z - y$$

is asymptotically stable or unstable.

Solution: The equilibria are the solutions of $z = 0$, $2(y^2 - 1)z - y = 0$, and thus the only equilibrium is (0,0). Here $F(y, z) = z$, with partial derivatives 0 and 1, respectively, and $G(y, z) = 2(y^2 - 1)z - y$, with partial derivatives $4yz - 1$ and $2(y^2 - 1)$ respectively. Therefore the community matrix at the equilibrium is

$$\begin{bmatrix} 0 & 1 \\ -1 & -2 \end{bmatrix}$$

with trace -2 and determinant 1, as in Example 1(ii). Thus the equilibrium (0,0) is asymptotically stable. □

EXAMPLE 3.
Determine whether each equilibrium of the system

$$\begin{aligned} y' &= y(1 - 2y - z) \\ z' &= z(1 - y - 2z) \end{aligned}$$

is asymptotically stable or unstable. This system could model a population of two competing species, say two species of fish sharing the same habitat (see Figure 5.13): each species lives according to a logistic rule in the absence of the other, with a carrying capacity of 1/2 unit (measured in thousands or millions of individuals), $y' = y(1 - 2y)$ or $z' = z(1 - 2z)$, and both species suffer from the presence of each other since they compete for the same resources (food, shelter, etc.), with the negative effect given by the same "bilinear" term (yz) which we have seen used to model encounters in predator-prey and epidemic models. We shall study populations in competition in more detail in Section 6.2.1; this is a special case where the two competitors are essentially evenly matched (same carrying capacity, same competitive disadvantage).

Solution: The equilibria are the solutions of $y(1-2y-z) = 0$, $z(1-y-2z) = 0$. One solution is (0,0), representing extinction of both species; a second is the solution of $y = 0$, $1 - y - 2z = 0$, which is $(0,\frac{1}{2})$ [the second species displaces the first]; a third is the solution of $z = 0$, $1 - 2y - z = 0$, which is $(\frac{1}{2},0)$ [the first species displaces the second]; and a fourth is the solution of $1-2y-z = 0$, $1-y-2z = 0$, which is $(\frac{1}{3},\frac{1}{3})$, representing coexistence. The community matrix at an equilibrium (y_∞, z_∞) is

$$\begin{bmatrix} 1 - 4y_\infty - z_\infty & -y_\infty \\ -z_\infty & 1 - y_\infty - 4z_\infty \end{bmatrix}.$$

At (0,0), this matrix has trace 1 and determinant 1, and thus the equilibrium is

Photo by David Burdick, courtesy NOAA

FIGURE 5.13: Yellow tang (*Zebrasoma flavescens*), black Achilles tang (*Acanthurus achilles*), and chub swim near a coral reef off the coast of Maui, Hawaii. Fish which share the same habitat may compete for food, shelter, and other resources.

unstable. At $(0,\frac{1}{2})$, this matrix has trace $-\frac{1}{2}$ and determinant $-\frac{1}{2}$, and thus the equilibrium is unstable. At $(\frac{1}{2},0)$, this matrix has trace $-\frac{1}{2}$ and determinant $-\frac{1}{2}$, and thus the equilibrium is unstable. At $(\frac{1}{3},\frac{1}{3})$, this matrix has trace $-\frac{4}{3}$ and determinant $\frac{1}{3}$, and thus this equilibrium is asymptotically stable. Thus the model suggests that evenly matched competitors can coexist. □

The careful reader will have noticed that, like the equilibrium stability theorem of Section 4.1, the equilibrium stability theorem given above has a hole of sorts in its result, in that the theorem says nothing about the stability of equilibria for which the trace and determinant lie on the boundary of conditions (5.27) and (5.28) — in other words, for which the linearization has solutions which do not approach zero as $t \to \infty$ but stay bounded. The reason for this "hole" is that in such cases, the linearization does not give enough information to determine stability. The following example recalls such a case, from the application with which we began this chapter.

EXAMPLE 4.
Determine the asymptotic stability or instability of each equilibrium of the Lotka-Volterra system

$$y' = y(\lambda - bz),$$
$$z' = z(-\mu + cy).$$

Solution: We showed in Example 3, Section 5.2 that the equilibrium (0,0) is unstable. Thus we need only examine the equilibrium (y_∞, z_∞) with $y_\infty = \frac{\mu}{c}$, $z_\infty = \frac{\lambda}{b}$. By the computation carried out in Example 1, Section 5.2, the community matrix at this equilibrium is

$$\begin{bmatrix} 0 & -\frac{b\mu}{c} \\ \frac{c\lambda}{b} & 0 \end{bmatrix}.$$

This matrix has a positive determinant, but the trace is zero. In this case, the stability theorem does not give any information. However, as we saw in Section 5.1, the orbits of the system neither tend to the equilibrium nor move away from the equilibrium. Thus the equilibrium is neither asymptotically stable nor unstable, but behaves like a center (cf. Figures 5.2, 5.9). □

If all orbits beginning near an equilibrium remain near the equilibrium for $t \geq 0$, but some orbits do not approach the equilibrium as $t \to \infty$, the equilibrium is said to be *stable*, or sometimes *neutrally stable.* If the origin is neutrally stable for the linearization at an equilibrium, then the equilibrium may also be neutrally stable for the nonlinear system, as for a Lotka-Volterra system. However, it is also possible for the origin to be neutrally stable for the linearization at an equilibrium, while the equilibrium is asymptotically stable or unstable. Thus neutral stability of the origin for a linearization at an equilibrium gives no information about the stability of the equilibrium.

We have seen in Section 4.1 that a solution of an autonomous first-order differential equation is either unbounded or approaches a limit as $t \to \infty$. For an autonomous system of two first-order differential equations, these same two possibilities exist. In addition, however, there is the possibility of an orbit which is a closed curve, corresponding to a periodic solution. Such an orbit is called a *periodic orbit* because it is traversed repeatedly.

There is a remarkable result which says essentially that these are the only possibilities.

POINCARÉ-BENDIXSON THEOREM: A bounded orbit of a system of two first-order differential equations which does not approach an equilibrium as $t \to \infty$ either is a periodic orbit or approaches a periodic orbit as $t \to \infty$.

Standard techniques from the study of complex variables can be used to show that a periodic orbit must enclose an equilibrium point in its interior. In many examples, there is an unstable equilibrium, and orbits beginning near this equilibrium spiral out toward a periodic orbit. A periodic orbit which is approached by other (non-periodic) orbits is called a *limit cycle*. One example of a limit cycle involves the system

$$y' = y(1 - y^2 - z^2) - z,$$
$$z' = z(1 - y^2 - z^2) + y,$$

which has its only equilibrium at the origin. The equilibrium is unstable, and Figure 5.14 illustrates the fact that all orbits not beginning at the origin spiral counterclockwise [in or out] toward the unit circle $y^2 + z^2 = 1$.

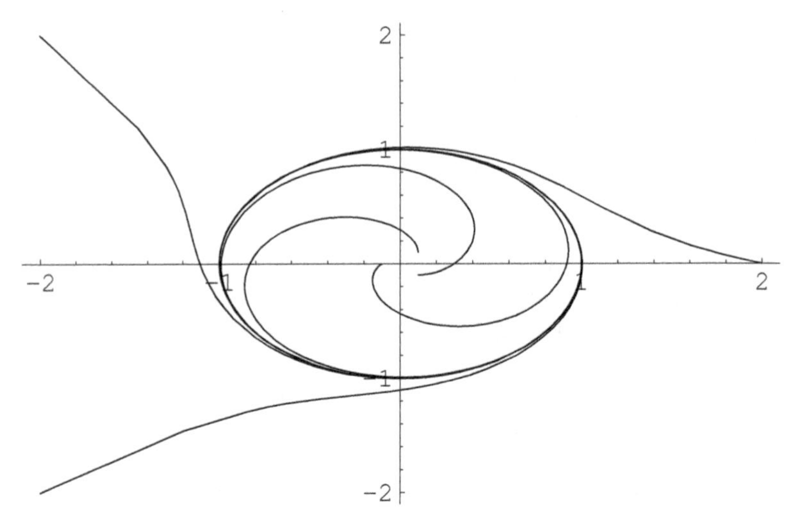

FIGURE 5.14: Trajectories approaching a limit cycle.

In many applications the functions $y(t)$ and $z(t)$ are restricted by the nature of the problem to non-negative values. For example, this is the case if $y(t)$ and $z(t)$ are population sizes. In such a case, only the first quadrant $y \geq 0$, $z \geq 0$ of the phase plane is of interest. For a system

$$y' = F(y, z), \quad z' = G(y, z)$$

which has $F(0, z) \geq 0$ for $z \geq 0$ and $G(y, 0) \geq 0$ for $y \geq 0$, then since $y' \geq 0$ along the positive z-axis (where $y = 0$) and $z' \geq 0$ along the positive y-axis (where $z = 0$), no orbit can leave the first quadrant by crossing one of the axes. The Poincaré-Bendixson theorem may then be applied to orbits in the first quadrant. In such a case, the first quadrant is called an *invariant set*, a region with the property that orbits must remain in the region.

If instead F and G are identically zero along the respective [half-]axes, then $y' = 0$ for $\{y = 0,\ z \geq 0\}$, and $z' = 0$ for $\{y \geq 0,\ z = 0\}$. In this case, orbits which begin on an axis must remain on that axis, and orbits beginning in the interior of the first quadrant (with $y(0) > 0$, $z(0) > 0$) must remain in the interior of the first quadrant (i.e., $y(t) > 0$ and $z(t) > 0$ for $t \geq 0$). If there is no equilibrium in the first quadrant, there cannot be a periodic orbit, because a periodic orbit *must* enclose an equilibrium. Thus, if there is no equilibrium in the first quadrant every orbit must be unbounded.

EXAMPLE 5.
Show that every orbit in the region $y > 0$, $z > 0$ of the system

$$\begin{aligned} y' &= y(2-y) - \frac{yz}{y+1}, \\ z' &= 4\frac{yz}{y+1} - z \end{aligned}$$

approaches a periodic orbit as $t \to \infty$.

Solution: We have $y' = 0$ when $y = 0$, and $z' = 0$ when $z = 0$, so orbits starting in the first quadrant remain in the first quadrant. Equilibria are the solutions of either $y = 0$ or $2 - y = \frac{z}{y+1}$, *and* either $z = 0$ or $\frac{4y}{y+1} = 1$. If $y = 0$, we must also have $z = 0$. If $2 - y = \frac{z}{y+1}$, we could have $z = 0$, which implies $y = 2$, or $y = \frac{1}{3}$, which implies $z = \frac{20}{9}$. Thus there are three equilibria, namely (0,0), (2,0), and $(\frac{1}{3}, \frac{20}{9})$. By checking the values of the trace and determinant of the community matrix, which is

$$\begin{bmatrix} 2 - 2y_\infty - \frac{z_\infty}{(1+y_\infty)^2} & -\frac{y_\infty}{y_\infty+1} \\ \frac{4z_\infty}{(y_\infty+1)^2} & \frac{4y_\infty}{y_\infty+1} - 1 \end{bmatrix},$$

we may see that each of the three equilibria is unstable. In order to apply the Poincaré-Bendixson theorem, we must show that every orbit starting in the first quadrant of the phase plane is bounded.

To show this, we might like to show that y' and z' are negative when y and/or z are sufficiently large, but a glance at the equations tells us this isn't necessarily so. Therefore we instead consider some positive combination of y and z whose time derivative does become negative far enough from the origin. In particular, consider the function $V(y, z) = 4y + z$. If an orbit is unbounded, then along this orbit the function $V(y, z)$ must also be unbounded. The derivative of $V(y, z)$ along an orbit is

$$\frac{d}{dt}V[y(t), z(t)] = 4y'(t) + z'(t) = 4y(2-y) - z.$$

This is negative except in the bounded region defined by the inequality $z < 4y(2-y)$. Therefore the function $V(y, z)$ cannot become unbounded, because it is decreasing ($dV/dt < 0$) whenever it becomes large ($z > 4y(2-y)$, which is true, for example, whenever $V > 9$). This proves that all orbits of the system are bounded. Now we may apply the Poincaré-Bendixson theorem to see that every orbit approaches a limit cycle. □

For autonomous systems of more than two differential equations there is no result analogous to the Poincaré-Bendixson theorem. Orbits of such systems may behave in very strange ways, and it is not possible to give a description of all the possibilities. One possibility is chaotic behavior similar to what we saw in Section 2.6 for difference equations. However, the analogue of the linearization theorem is valid, and it is still possible to decide whether an equilibrium is asymptotically stable.

Exercises

In Exercises 1–6, for each equilibrium of the given system determine whether the equilibrium is asymptotically stable or unstable.

1. $y' = y + z - 2$, $z' = z - y$ (cf. Exercise 1, Section 5.2)
2. $y' = y + z - 1$, $z' = y$ (cf. Exercise 2, Section 5.2)
3. $y' = y + z^2$, $z' = y + 1$ (cf. Exercise 3, Section 5.2)
4. $y' = z - 2$, $z' = y^2 - 8z$ (cf. Exercise 4, Section 5.2)
5. $y' = e^z$, $z' = e^{-y}$ (cf. Exercise 5, Section 5.2)
6. $y' = z$, $z' = \sin y$ (cf. Exercise 6, Section 5.2)
7. * Determine the behavior of orbits of the system
$$\begin{aligned} y' &= y(1 - y - 2z), \\ z' &= z(1 - 2y - z). \end{aligned}$$
8. * Determine the behavior of orbits of the system
$$\begin{aligned} y' &= y(2 - y) - \tfrac{yz}{y+1}, \\ z' &= 2\,\tfrac{yz}{y+1} - z. \end{aligned}$$
9. * Determine the behavior of orbits of the system
$$y' = y(\lambda - ay - bz), \quad z' = z(-\mu + cy).$$
10. * Determine the behavior of orbits of the system
$$y' = ry\left(1 - \frac{y}{K + az}\right), \quad z' = sz\left(1 - \frac{z}{M + by}\right).$$

11. * Consider the system

$$y' = z, \quad z' = -y - z^3.$$

(a) Show that (0,0) is the only equilibrium.

(b) Show that the linearization of the system at the origin has a center.

(c) Show that the non-negative function $V(y, z) = y^2 + z^2$ decreases along every orbit of the system and tends to zero, which implies that the origin must be a spiral point of the system.

12. * Consider the system in polar coordinates (r, θ)

$$r' = r(1 - r), \quad \theta' = 1.$$

(a) Show that r approaches 1 as $t \to \infty$, and θ increases unboundedly.

(b) Deduce that the circle $r = 1$ is a limit cycle which every orbit except the constant solution $r = 0$ approaches.

13. * The system in Example 5 has a predator-prey structure, with the term common to both equations representing the interaction rate between predators and prey.

(a) Which variable represents the predators, and which the prey?

(b) What is the significance of the 2 in the dy/dt equation?

(c) The interaction rate saturates as which of the two populations grows?

Miscellaneous exercises

In each of Exercises 1–2, describe the orbits of the given system.

1. $y' = y^2 z$, $z' = zy^2$

2. $y' = e^{-z}$, $z' = e^y$

In each of Exercises 3–6, find the linearization of the given system at each equilibrium.

3. $y' = y + z - 4$, $z' = y - z$
4. $y' = z - 1$, $z' = y$
5. $y' = y^2 z$, $z' = zy^2$
6. $y' = e^{-z}$, $z' = e^y$

In each of Exercises 7–8, for each equilibrium of the given system determine whether the equilibrium is asymptotically stable or unstable.

7. $y' = y^2 z$, $z' = zy^2$
8. $y' = e^{-z}$, $z' = e^y$

In each of Exercises 9–16, find the general solution of the given system by analytic solution, use a computer algebra system to examine the behavior of solutions, and classify the origin as a node, saddle point, center, or spiral point.

9. $y' = z - 3y$, $z' = 5y + z$
10. $y' = y$, $z' = y + z$
11. $y' = 2y + z$, $z' = z$
12. $y' = -y$, $z' = y - z$
13. $y' = -z$, $z' = y$
14. $y' = y - z$, $z' = y + z$
15. $y' = y + z$, $z' = y + 2z$
16. $y' = y + z$, $z' = y + z$

17. A tank contains 1000 liters of water and a salt solution containing 20 kg./liter is pumped into it at a rate of 100 liters /minute. The mixture is led to a second tank containing 1000 liters of water at a rate of 100 liters/minute, and the mixture is pumped out of the second tank at the same rate. What is the concentration of salt in the second tank after one hour?

18. Two tanks begin with 10 kg. of salt dissolved in 100 liters of water. Water is pumped into the first tank at a rate of 20 liters/minute, the mixture is pumped from the first tank to the second tank at a rate of 20 liters/minute, and the mixture is pumped out of the second tank at a rate of 20 liters/minute. What is the amount of salt in each tank after one hour, and what is the amount of salt in each tank after a very long time?

In Exercises 19–22, for each equilibrium of the given system determine whether the equilibrium is asymptotically stable or unstable.

19. $y' = y + z - 4$, $z' = y - z$
20. $y' = z - 1$, $z' = y$
21. $y' = y^2 z$, $z' = zy^2$
22. $y' = e^{-z}$, $z' = e^y$

Chapter 6

Topics in Modeling Systems of Populations

There are many questions involving the interaction of two different populations. These different populations may be members of a single population but distinguished by gender, age, or their infection status with respect to a disease present in the population, or they may be members of two quite different species, cooperating, competing for a common resource, or in a predator-prey relationship. The modeling of such questions leads naturally to systems of differential equations, typically with each differential equation describing one of the interacting populations. This chapter is devoted to some examples, describing applications of the general theory of systems of differential equations developed in the previous chapter.

6.1 Epidemiology: Compartmental models

6.1.1 An epidemic model

In Sections 3.2.3 and 4.3.2 we considered models for the spread of an infectious disease in which a population was divided into susceptibles and infectives; the underlying assumptions were that there was a rate of contracting the infection which depended on the number of susceptibles and the number of infectives, that there was a rate of recovery depending on the number of infectives, and that on recovery infectives returned to the susceptible class. In other words, it was assumed that there was no immunity against re-infection after recovery from the infection.

In this section, we shall consider some models for the spread of infectious diseases which include a third class, of removed members. Many diseases, especially diseases caused by viral agents, including smallpox, measles, and rubella (German measles), provide immunity against re-infection.

We let $S(t)$ denote the number of susceptibles, $I(t)$ the number of infectives, and $R(t)$ the number of removed members. We assume that the population has constant total size K, so that $S(t) + I(t) + R(t) = K$. We will derive differential equations expressing the rate of change of the size of each of the three classes, but in each case one of the equations may be eliminated since we can use the above relation to find S, I or R in terms of the other two. Thus we will obtain a system of two differential equations to describe the spread of diseases for which there is a removed class.

A model was proposed by W. O. Kermack and A. G. McKendrick[1] to explain the rapid rise and fall of cases frequently observed in epidemics, including the Great Plague of 1665–66 in England, cholera in London in 1865, and plague in Bombay in 1906. This model is

$$\begin{aligned} S' &= -\beta SI, \\ I' &= \beta SI - \gamma I, \\ R' &= \gamma I. \end{aligned} \tag{6.1}$$

The only difference from the model of Section 3.2.3 is that the term γI now represents a rate of transition from the class I to the class R, instead of a rate of return to the class S. The rate of recoveries in unit time is γI, and the rate of transmission of infection from infectives to susceptibles is βSI. Note that this model is only appropriate if the duration of an outbreak is short enough that demographics (natural births and deaths) can be ignored.

We consider the model as a system of two equations, viewing R as determined by S and I, $R = K - S - I$, since the first two equations do not involve R:

$$\begin{aligned} S' &= -\beta SI, \\ I' &= \beta SI - \gamma I. \end{aligned} \tag{6.2}$$

The equilibria of the system (6.2) (seen in slightly different form in Section 5.2, Example 2) are the solutions of the pair of equations $\beta SI = 0$, $\beta SI - \gamma I = 0$. The first of these implies that either $S = 0$ or $I = 0$. If $S = 0$, the second equation is satisfied only if $I = 0$, while if $I = 0$ the second equation is satisfied for every S. Thus there is a line of equilibria $(S_\infty, 0)$ with S_∞ arbitrary, $0 \leq S_\infty \leq K$. If we compute the linearization of the system (6.2) at an equilibrium $(S_\infty, 0)$, we obtain

$$\begin{aligned} u' &= -\beta S_\infty v, \\ v' &= (\beta S_\infty - \gamma)\, v, \end{aligned}$$

and the linearization theorem of Section 5.2 cannot be applied.

In order to obtain an understanding of the qualitative behavior of solutions of the system (6.2), we observe from (6.2) that $S' < 0$ whenever $S > 0$, $I > 0$. This means that the function $S(t)$ decreases for all t. In addition, $I' < 0$ whenever $I > 0$, $\beta S < \gamma$, while $I' < 0$ if $I < 0$, $\beta S > \gamma$. Thus if $S(0) < \frac{\gamma}{\beta}$, $S(t)$ remains less than $\frac{\gamma}{\beta}$ for all t, and $I(t)$ decreases to zero as t increases.

[1]W.O. Kermack and A.G. McKendrick, A contribution to the mathematical theory of epidemics, *Proc. Roy. Soc. London* 115 (1927), 700–721.

However, if $S(0) > \frac{\gamma}{\beta}$, then $I(t)$ increases so long as $\beta S > \gamma$, and thus $I(t)$ increases initially before decreasing to zero. We think of introducing a small number of infectives into a susceptible population so that $I(0) = \epsilon > 0$, $S(0) = K - \epsilon$. Then if $\beta K/\gamma < 1$, $I(t)$ decreases monotonically to zero and the infection dies out. On the other hand, if $\beta K/\gamma > 1$, an epidemic occurs, as $S(0) > \frac{\gamma}{\beta}$ (for ϵ small), so $I(t)$ increases to a maximum and then decreases to zero. This is another threshold theorem of Kermack and McKendrick with the threshold quantity $\beta K/\gamma$. This threshold quantity distinguishes between two possible behaviors just like the threshold quantity in Section 3.2.3, but the possible behaviors are not the same as in Section 3.2.3.

One might suppose that the reason for the eventual disappearance of the infection in the epidemic case is that all susceptibles become infected, but observations of epidemics indicate that this is not the case. The model (6.2) agrees with observation in that it implies that the limiting value $S(\infty) = \lim_{t\to\infty} S(t)$ of every solution of the system (6.2) obeys $S(\infty) > 0$. We may see this by calculating

$$\begin{aligned}\frac{d}{dt}\left[S(t) + I(t) - \frac{\gamma}{\beta}\log S(t)\right] &= S'(t) + I'(t) - \frac{\gamma}{\beta}\frac{S'(t)}{S(t)} \\ &= -\beta SI + [\beta SI - \gamma I] - \frac{\gamma}{\beta}(-\beta I) = 0\end{aligned}$$

(motivated by observing that $S' + I' = -\gamma I$, and finding a function of S and I whose time derivative is γI). Thus $S(t) + I(t) - \frac{\gamma}{\beta}\log S(t)$ is a constant. Since $I(\infty) = 0$ and $I(0) \approx 0$, this gives

$$S(\infty) - \frac{\gamma}{\beta}\log S(\infty) = S(0) - \frac{\gamma}{\beta}\log S(0).$$

It follows that

$$S(0) - S(\infty) = \frac{\gamma}{\beta}\left[\log S(0) - \log S(\infty)\right] = \frac{\gamma}{\beta}\log\frac{S(0)}{S(\infty)},$$

and therefore

$$\beta/\gamma = \frac{\log\left[\frac{S(0)}{S(\infty)}\right]}{S(0) - S(\infty)}. \tag{6.3}$$

The quantity β/γ is known as the *contact number*. Not only does (6.3) imply $S(\infty) > 0$ (because if S_∞ were zero, the right side of (6.3) would be infinite but the left side is finite), but it also gives a means of estimating the contact rate β, which generally cannot be measured directly. By making a serological survey (testing for immune responses in the blood) in the population before and after an epidemic, one may estimate $S(0)$ and $S(\infty)$, and then (6.3) gives β/γ. If the mean infective period $1/\gamma$ is known as well, then β can be calculated. The contact rate β depends on the disease as well as on other factors such as the rate of mixing in the population.

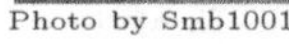

FIGURE 6.1: During the Great Plague of 1665–66, the village of Eyam in England voluntarily quarantined itself in hopes of preventing the plague from spreading to neighboring villages. Inhabitants of the neighboring populations left food and other supplies for Eyam residents at the Boundary Stone (pictured) just outside the village.

FIGURE 6.2: A phase portrait for model (6.2).

For example, the village of Eyam in England maintained isolation from other villages during the Great Plague of 1665–66 (see Figure 6.1), and its population decreased from 350 to 83 during the course of the epidemic. There is reason to believe that there were actually two separate epidemics, the first of which reduced the susceptible population to 254. By substituting $S(0) = 254$, $S(\infty) = 83$ into (6.3), we obtain

$$\beta/\gamma = \frac{\log \frac{254}{83}}{254 - 83} = 6.54 \times 10^{-3}.$$

The infective period was 11 days, or 0.3667 months. Using a month as the unit of time we obtain the estimate $\beta = 0.0178$. This data, with 7 initial infectives, gives the phase portrait of Figure 6.2, traversed from right to left as time progressed and the number of susceptibles decreased. Note that in this case infected individuals were removed to the R class through death for the most part, rather than recovery with immunity. Our simple model (6.2) still describes this process, however different the interpretation may be, as the R class of the model simply includes individuals no longer involved in the spread of the disease.

The criterion $\beta K/\gamma > 1$ for the establishment of a disease can also be expressed as the requirement that the susceptible population density exceeds a certain critical value $\frac{\gamma}{\beta}$. For fox rabies in Europe, observations indicate a critical population density of approximately 1 fox/km^2; rabies dies out in regions which are more sparsely populated. This data together with the average life expectancy of 5 days for a rabid fox gives the estimate $\beta \approx 72$ km^2/fox year.

In order to avoid an epidemic, it is necessary to reduce below 1 the quantity $\beta K/\gamma$, which is called the *basic reproductive number* and often denoted by $\mathcal{R}_0$. This may sometimes be achieved by immunization, which has the effect of transferring members from the susceptible class to the removed class and thus reducing $S(0)$. (This idea was first mentioned briefly in Section 5.2, Example 2.) If we immunize a fraction p of the susceptible population, we would replace K by $(1-p)K$, and this would give a basic reproductive number $\beta K(1-p)/\gamma$. In order to make this basic reproductive number less than 1, we require $\beta K(1-p)/\gamma < 1$. This is equivalent to $1-p < \frac{\gamma}{\beta K}$, or $p > 1 - \frac{\gamma}{\beta K}$.

A population is said to have *herd immunity* if a sufficiently large fraction has been immunized to reduce the basic reproductive number below 1 and thus assure that the disease will not spread if an infective is introduced into the population (the term suggests that one can be protected from infection if enough of one's neighbors are immune, see Figure 6.3). The only infectious disease for which this has actually been achieved worldwide is smallpox. For measles, epidemiological data in the U.S. indicates a basic reproductive number ranging from 5.4 to 6.3 in rural areas, requiring vaccination of 81.5% to 84.1% of the population to achieve herd immunity. In urban areas, the basic reproductive number ranges from 8.3 to 13.0, requiring vaccination of 88.0% to 92.3% of the population. As measles vaccination is only about 95% effective (for vaccination at age 15 months) and not all people are willing to allow vaccination, it is impossible in practice to achieve herd immunity. For smallpox, the basic reproductive number is about 5, requiring 80% vaccination to achieve herd immunity. This is feasible because the consequences of smallpox are dire enough to encourage immunization.

EXAMPLE 1.

A survey of freshman students at Yale University[2] found that 25% were susceptible to rubella at the beginning of the year and 9.65% were susceptible at the end of the year. What fraction would have had to be immunized to avoid the spread of rubella?

Solution: Using $S(0) = 0.25$, $S(\infty) = 0.0965$ and substituting in (6.2), we obtain

$$\beta/\gamma = \frac{\log \frac{0.25}{0.0965}}{0.25 - 0.0965} = 6.20.$$

In order to avoid the spread of rubella, the requirement is $S(0) < \frac{\gamma}{\beta} = 0.16$.

[2]A.S. Evans, *Viral Infections of Humans*, 2nd ed., Plenum Press, New York (1982), reported by H. W. Hethcote, Three basic epidemiological models, *Applied Mathematical Ecology*, S. A. Levin, T. G. Hallam, and L. J. Gross (eds), Biomathematics 18, Springer-Verlag, New York-Heidelberg (1989), 119–144.

Photo by AlMare

FIGURE 6.3: The notion of herd immunity comes from the fact that members of a homogeneously mixing population (a "herd") making potentially infectious contacts with neighbors can be protected from infection if enough of those neighbors are immune. Imagine here that the individuals in the car in the background are in close contact with the neighboring members of the herd surrounding them (here, a herd of goats on a road in Greece). If enough of those neighbors are vaccinated against infection, then by the time the index individual comes into contact with an unvaccinated individual, the latter may no longer be infected. With the infection rate reduced so drastically, the infection dies out in the population.

This could be achieved by immunizing an additional 9% of the class, or $\frac{0.25-0.16}{0.25} = 36\%$ of the susceptible students. □

EXAMPLE 2.
In Example 2, Section 3.2.3, a disease was described spreading in a population of 1200 members. Suppose the disease, with $\beta = \frac{1}{3000}$, $1/\gamma = 6$ days, had conferred immunity on recovered infectives. How many members would have had to have been immunized to avoid an epidemic?

Solution: The basic reproductive number is $\beta K/\gamma = \frac{1}{3000} \times 6 \times 1200 = 2.4$. Thus herd immunity, $p > 1 - \frac{\gamma}{\beta K}$, would require immunization of a fraction $1 - \frac{1}{2.4} = 0.5833$, or 700 members. □

It may be important to know the maximum number of infectives at any given time, for example to be able to arrange enough facilities for isolation and treatment. From the model (6.2), we know that the maximum of I occurs when $S = \frac{\gamma}{\beta}$, and also that the quantity $S + I - \frac{\gamma}{\beta} \log S$ is constant. Thus if

t^* is the time when $S = \frac{\gamma}{\beta}$, we have

$$S(t^*) + I(t^*) - \frac{\gamma}{\beta}\log S(t^*) = S(0) + I(0) - \frac{\gamma}{\beta}\log S(0)$$

and

$$\frac{\gamma}{\beta} + I(t^*) - \frac{\gamma}{\beta}\log\frac{\gamma}{\beta} = S(0) - \frac{\gamma}{\beta}\log S(0).$$

From this we conclude that $I(t^*)$, the maximum number of infectives, is given by

$$I(t^*) = S(0) - \frac{\gamma}{\beta}\log S(0) + \frac{\gamma}{\beta}\log\frac{\gamma}{\beta} - \frac{\gamma}{\beta} = S(0) - \frac{\gamma}{\beta} - \frac{\gamma}{\beta}\frac{\log \beta S(0)}{\gamma}. \quad (6.4)$$

For the Great Plague in Eyam, this gives a maximum infective population of 30.4, confirmed by the phase portrait of Figure 6.2.

EXAMPLE 3.
What is the maximum number of infectives in the rubella epidemic of Example 1?

Solution: In Example 1 we were given $S(0) = 0.25$, and we calculated $\beta/\gamma = 6.20$. Then (6.4) gives

$$I(t^*) = 0.25 - \frac{1}{6.20} - \frac{1}{6.20}\log(6.20)(0.25) = 0.018.$$

Thus at most 1.8% of the population is infected at any one time. □

6.1.2 A model for endemic situations

The model (6.2) studied earlier in this section is appropriate for describing an epidemic, a single outbreak of a disease. Many diseases are always present, especially in less-developed countries, and this is described by saying that they are *endemic*. In order to model a long-term endemic situation, we must add demographics—births and natural deaths—to the model (6.2). If we assume a proportional death rate μ in each compartment and a birth rate μK (to keep the total population size constant), we would have a model

$$\begin{aligned} S' &= \mu K - \beta S I - \mu S \\ I' &= \beta S I - (\gamma + \mu) I. \end{aligned} \quad (6.5)$$

We analyze this model by finding equilibria and checking their asymptotic stability. Equilibria (S, I) are solutions of the pair of equations

$$\beta S I = \mu(K - S), \quad \beta S I = (\gamma + \mu) I.$$

The second of these equations factors into two possibilities, namely $I = 0$

(disease-free equilibrium) and $\beta S = \gamma + \mu$ (endemic equilibrium). For a meaningful endemic equilibrium, we must have $S \leq K$, or

$$\mathcal{R}_0 = \frac{\beta K}{\gamma + \mu} \geq 1.$$

This quantity $\mathcal{R}_0$ has an epidemiological meaning, because a single infective in a wholly susceptible population would make βK contacts in unit time, all of which would produce new infections, and would continue to transmit infection for a mean infective period (corrected for natural deaths) of $1/(\gamma + \mu)$. Thus $\mathcal{R}_0$ is the total number of secondary infections caused by a single infective inserted into a wholly susceptible population.

At the disease-free equilibrium, $S = K, I = 0$. At an endemic equilibrium, if there is one (that is, if $\mathcal{R}_0 > 1$), it is easy to calculate that

$$S = \frac{\gamma + \mu}{\beta}, \quad I = \frac{\mu}{\gamma + \mu}\left(1 - \frac{1}{\mathcal{R}_0}\right) K.$$

The matrix of the linearization of (6.5) at an equilibrium (S, I) is

$$\begin{bmatrix} -(\beta I + \mu) & -\beta S \\ \beta I & \beta S - (\gamma + \mu) \end{bmatrix}.$$

At the disease-free equilibrium, this matrix is

$$\begin{bmatrix} -\mu & -\beta K \\ 0 & \beta K - (\gamma + \mu) \end{bmatrix},$$

and it is easy to see that this has negative eigenvalues if and only if $\mathcal{R}_0 < 1$. Thus the disease-free equilibrium is asymptotically stable if and only if $\mathcal{R}_0 < 1$.

At the endemic equilibrium (S, I) the matrix is

$$\begin{bmatrix} -(\beta I + \mu) & -\beta S \\ \beta I & 0 \end{bmatrix}.$$

Since this matrix has negative trace and positive determinant, the endemic equilibrium is always asymptotically stable if it exists. We may summarize the analysis by saying that there is always a single asymptotically stable equilibrium, the disease-free equilibrium if $\mathcal{R}_0 < 1$ and the endemic equilibrium if $\mathcal{R}_0 > 1$.

For most human diseases, the mean infective period $1/\gamma$ is less than one month, much smaller than the mean life span $1/\mu$, which is onthe order of 70 years. Thus an endemic equilibrium, with

$$I = \frac{\mu}{\gamma + \mu}\left(1 - \frac{1}{\mathcal{R}_0}\right) K,$$

the number of infectives is a tiny fraction of the carrying capacity. In a population of moderate size, say 1000, the number of infectives at an endemic equilibrium might be less than 1, and small random effects might be enough to wipe out the infective population. A numerical simulation of an epidemic model (6.2) and an endemic model (6.5) would appear indistinguishable. However, in a large population of, say 1,000,000, there might be 1,000 infectives at an endemic equilibrium, and this is a number of disease cases large enough to be significant.

For animal diseases, where the life span may be much shorter and the disease infective period may be much longer, the differences between an epidemic and an endemic situation may be much more readily noticed. An extreme example would be a disease with no recovery, such as rinderpest (a cattle disease of ancient origin which has quite recently become the second disease, after smallpox, to be eliminated). Such a disease would be described by an SI model, like (6.5) but with $\gamma = 0$,

$$\begin{aligned} S' &= \mu K - \beta S I - \mu S, \\ I' &= \beta S I - \mu I, \end{aligned} \tag{6.6}$$

having

$$\mathcal{R}_0 = \frac{\beta K}{\mu},$$

and an endemic equilibrium

$$S = \frac{\mu}{\beta}, \quad I = \left(1 - \frac{1}{\mathcal{R}_0}\right) K.$$

Exercises

1. The same survey of Yale students described in Example 1 reported that 91.1% were susceptible to influenza at the beginning of the year, and 51.4% were susceptible at the end of the year. Estimate the contact number β/γ and decide whether there was an epidemic.

2. An influenza epidemic was reported at an English boarding school in 1970 which spread to 512 of the 673 students. Estimate the contact number β/γ.

3. What fraction of the Yale students of Exercise 1 would have had to be immunized to prevent an epidemic?

4. What fraction of the boarding school students of Exercise 2 would have had to be immunized to prevent an epidemic?

5. What was the maximum number of Yale students of Exercises 1 and 3 missing classes because of influenza at any given time?

6. What was the maximum number of boarding school students of Exercises 2 and 4 suffering from influenza at any given time?

7. * If some members of a population susceptible to a disease which provides immunity against re-infection moves out of the region of the epidemic, at a per capita rate λ, the situation may be modeled by a system

$$S' = -\beta SI - \lambda S, \quad I' = \beta SI - \gamma I.$$

Show that both S and I approach zero as $t \to \infty$.

8. * Consider a model for a disease which confers only temporary immunity after recovery, so that recovered individuals lose their immunity at a per capita rate of c (per time unit).

(a) Reformulate model (6.1) to reflect the new structure for the disease cycle.

(b) Reduce the model to a two-dimensional system as done for (6.1), by eliminating R.

(c) Identify the equilibria.

(d) Describe the qualitative behavior of the system. Is there a threshold condition for this SIRS model, as there was with the SIR model?

9. Perform numerical simulations for the models (6.2) and (6.5) with parameter values $K = 1000, \beta = 1/2000, \gamma = 1/4, \mu = 1/25,000$ and a variety of initial values.

10. Perform numerical simulations for the SI model (6.6) with parameter values $A = 100, \beta = 1/2000, \mu = 1/10$ and a variety of initial values. Calculate the basic reproduction number and equilibrium susceptible and infective population sizes. Does the model approach an endemic equilibrium directly or does it go through an epidemic-like behavior first like the SIR model?

11. Suppose that a vaccine is developed and approved for an infection which confers immunity upon recovery, and that susceptible individuals get vaccinated at a per capita rate ϕ. How would the SIR model systems with (6.5) and without (6.2) demographics change to incorporate this vaccination program, if vaccinated individuals have the same immunity (assumed lifelong) as recovered individuals?

12. * Identify the equilibria of the model in the previous exercise for the system with demographics, and analyze their stability. From the results, deduce the expression for the *control reproductive number*, analogous to R_0, and interpret the difference from the expression given in the text for R_0.

13. Rewrite the SI model (6.6) to incorporate *vertical transmission*, in which a proportion p of offspring born to infected mothers are born infected. State any additional assumptions.

6.2 Population biology: Interacting species

The Lotka-Volterra system considered in Section 5.1 is an example of a model for the sizes of two interacting populations. The formulation of models for two interacting species depends on the nature of the interaction as well as on the assumptions about the behavior of each population in the absence of the other population. In this section, we shall examine some models for two different kinds of interaction of two species, namely species in competition, species in a predator-prey relation, and symbiotic relations. These are the three possible relationships between two species: both species may be damaged (competition), one species may profit at the expense of the other species (predator-prey relationship), or each species may benefit from the presence of the other species (symbiosis).

6.2.1 Species in competition

Let us consider two species whose population sizes at time t are $y(t)$ and $z(t)$, respectively. Suppose that each species would grow according to a logistic law if there were no interaction with the other species. Suppose also that the two species are competing for resources, and that the effect of this competition is to decrease the per capita growth rate of each species by an amount proportional to the population size of the other species. These assumptions lead to a model of the form

$$\begin{aligned} y' &= y(\lambda - ay - bz) \\ z' &= z(\mu - cy - dz) \end{aligned} \tag{6.7}$$

with λ, μ, a, b, c, and d positive constants. The carrying capacities of the two species are, respectively, $\frac{\lambda}{a}$ and $\frac{\mu}{d}$. It is easy to see that (0,0), $(\frac{\lambda}{a},0)$, and $(0,\frac{\mu}{d})$ are equilibria of the system (6.7). In addition, there may be an equilibrium which we will call (y_∞,z_∞) with $y_\infty > 0$, $z_\infty > 0$, if the lines $ay + bz = \lambda$ and $cy + dz = \mu$ intersect in the first quadrant. Their point of intersection,

$$y_\infty = \frac{d\lambda - b\mu}{ad - bc}, \; z_\infty = \frac{a\mu - c\lambda}{ad - bc}, \tag{6.8}$$

which exists as long as $ad \neq bc$, lies in the first quadrant when the numerators and denominator in (6.8) are either all positive or all negative. Some algebra shows that this happens when the relative growth ratio λ/μ (of y to z) lies between a/c and b/d.

Because $F(y,z) = \lambda y - ay^2 - byz$ and $G(y,z) = \mu z - cyz - dz^2$, the community matrix at an equilibrium (y,z) is

$$\begin{bmatrix} \lambda - 2ay - bz & -by \\ -cz & \mu - cy - 2dz \end{bmatrix}.$$

Thus at the equilibrium (0,0) this matrix has trace $\lambda+\mu > 0$ and determinant $\lambda\mu > 0$, and the equilibrium is unstable. At the equilibrium $(\frac{\lambda}{a},0)$, the matrix has trace $-\lambda+\frac{a\mu-c\lambda}{b}$ and determinant $-\lambda(\frac{a\mu-c\lambda}{a})$, so the trace is negative and the determinant positive if $a\mu - c\lambda < 0$. A similar argument shows that the equilibrium $(0,\frac{\mu}{d})$ is asymptotically stable if $d\lambda - b\mu < 0$. For the equilibrium (y_∞,z_∞), if there is one, the community matrix is

$$\begin{bmatrix} -ay_\infty & -by_\infty \\ -cz_\infty & -dz_\infty \end{bmatrix}.$$

Thus the trace is negative, and the determinant is positive if and only if $ad - bc > 0$ (or $a/c > b/d$).

Therefore the equilibrium (0,0) is always unstable, and we may summarize the results for the other three equilibria as follows:

I. If $b/d < \lambda/\mu < a/c$, there is an equilibrium (y_∞,z_∞) which is asymptotically stable, and the other three equilibria are unstable.

II. If $a/c < \lambda/\mu < b/d$, there are an unstable equilibrium (y_∞,z_∞) and two locally asymptotically stable equilibria $(\frac{\lambda}{a},0)$ and $(0,\frac{\mu}{d})$.

III. If $\lambda/\mu < a/c, b/d$, the equilibrium $(\frac{\lambda}{a},0)$ is unstable and the equilibrium $(0,\frac{\mu}{d})$ is asymptotically stable.

IV. If $\lambda/\mu > a/c, b/d$, the equilibrium $(\frac{\lambda}{a},0)$ is asymptotically stable and the equilibrium $(0,\frac{\mu}{d})$ is unstable.

Writing the possibilities in terms of the above ratios allows us to consider the relative importances of the natural growth rates (λ/μ) and the limiting effects of species y (a/c) and species z (b/d). Thus in case I the relative growth and self-limiting ratios for each species dominate the relative effects of competition — for species y relative to species z, this is $b/d < \lambda/\mu < a/c$ as written above (competition for species y means the effects of species z), while for species z relative to species y this is $c/a < \mu/\lambda < d/b$, with c/a representing the relative effects of competition with species y. Here competition is not severe enough to prevent coexistence of the two species. Mathematically speaking, in this case there is one asymptotically stable equilibrium with both species surviving, and every orbit in the first quadrant approaches this equilibrium.

In case II, the reverse of the above is true: the relative effects of competition dominate both the relative natural growth and self-limiting ratios, $a/c < \lambda/\mu < b/d$ as given above for species y relative to z, or $d/b < \mu/\lambda < c/a$ for species z relative to y. Here the level of competition is so high that the two species cannot coexist. Therefore there are two locally asymptotically stable equilibria, each corresponding to survival of one species and extinction of the other. Which species wins the competition depends on the initial values.

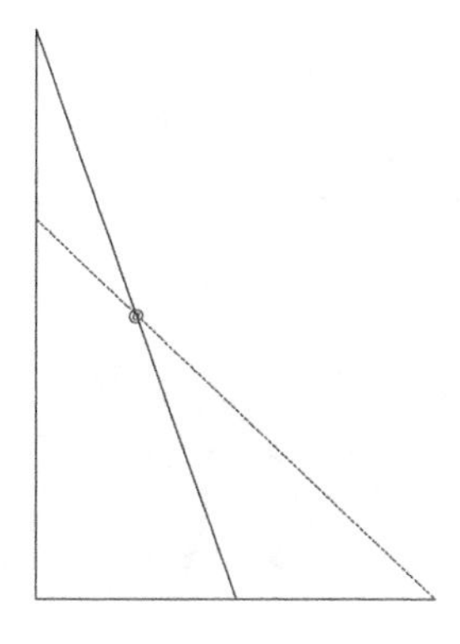

FIGURE 6.4: Case I: Coexistence.

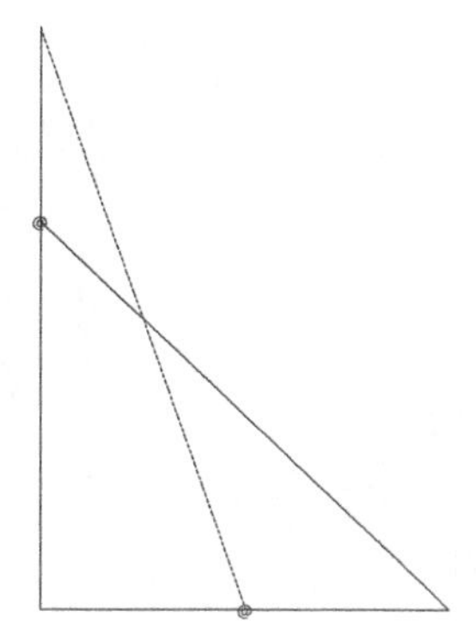

FIGURE 6.5: Case II: Competitive exclusion.

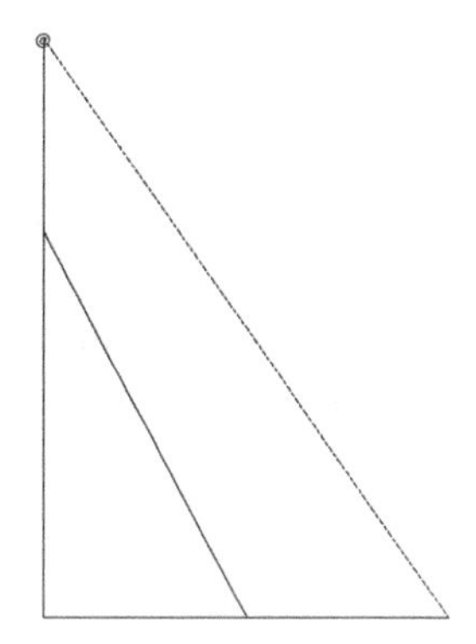

FIGURE 6.6: Case III: z survives.

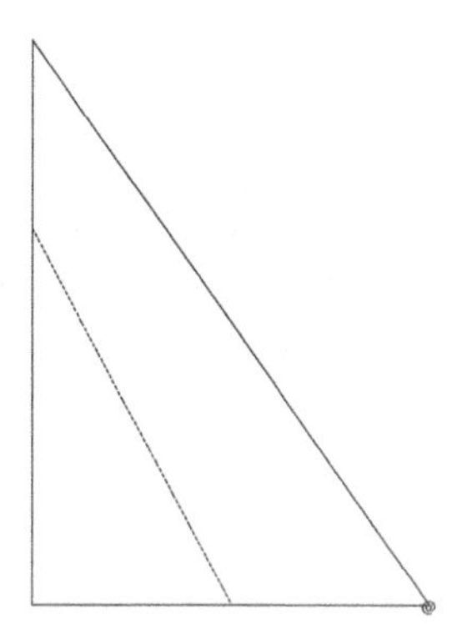

FIGURE 6.7: Case IV: y survives.

In case III, the relative growth ratio for species y (relative to z) is dominated by both of the limiting ratios, $\lambda/\mu < a/c, b/d$, while the relative growth ratio for species z (relative to y) exceeds the two limiting ratios, $\mu/\lambda > c/a, d/b$. Therefore the more robust z-species survives, and the y-species is wiped out.

In case IV, the tables are turned: the relative growth ratio for y (to z) exceeds the two limiting ratios, $\lambda/\mu > a/c, b/d$, while the relative growth ratio for z (to y) is less than the two limiting ratios, $\mu/\lambda < c/a, d/b$. Thus the y-species survives, and the z-species is wiped out.

The four cases may also be distinguished by the locations of the isoclines, which are all straight lines, as shown in Figures 6.4–6.7 for the four cases, respectively. In these figures, each asymptotically stable equilibrium has been marked @. By drawing the two nullclines for any competitive system and noting their relative positions, we may identify which of the four cases describes the system.

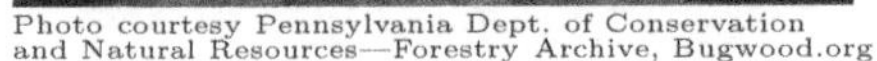
Photo courtesy Pennsylvania Dept. of Conservation and Natural Resources—Forestry Archive, Bugwood.org

Photo by Greg Webster

FIGURE 6.8: Competition occurs in southeastern Canada and the northeastern United States between the gypsy moth, *Lymantria dispar* (left), and the northern tiger swallowtail, *Papilio canadensis* (right).

Suppose now that we have two species of moth competing for the same food supply, with respective initial per capita reproduction rates of 100 and 60 (in per capita per time units), and carrying capacities on a given patch of habitat (such as a tree or field) of 25 and 30. (Figure 6.8 illustrates one such competition, between the gypsy moth and the northern tiger swallowtail butterfly.) In the next two examples we shall see that the extent to which competition adversely affects each species determines whether the two can coexist, or, if not, which one survives.

EXAMPLE 1.
Determine the outcome of a competition modeled by the system

$$y' = y(100 - 4y - 4z),\ z' = z(60 - y - 2z).$$

Solution: A coexistence equilibrium is found by solving the system of algebraic equations $4y + 4z = 100$, $y + 2z = 60$. If we subtract double the second equation from the first we obtain $2y = -20$, and thus there is no coexistence equilibrium. Alternatively, we note that $\lambda/\mu = 5/3$ is not between $a/c = 4$ and $b/d = 2$.

The two equilibria with one species surviving are (25,0) and (0,60). We know that every orbit must approach one of these two equilibria. In order to decide which, we note that as $(y, z) \to (0, 60)$, $\frac{y'}{y} = 100 - 4y - 4z \to -240 < 0$, while as $(y, z) \to (25, 0)$, $\frac{z'}{z} = 60 - y - 2z \to 35 > 0$. This shows that (y, z) must approach the equilibrium (0,60), because in order to approach (25,0) it would be necessary to have $z' < 0$. Alternatively, $\lambda/\mu < a/c, b/d$, so we are in case III above. We now see that the z-species wins the competition. □

EXAMPLE 2.
Now suppose instead that species y is only affected one quarter as much by the presence of species z — that is, that each z individual reduces species y's per capita growth rate by one instead of four. Determine the outcome of the competition modeled by the resultant system

$$y' = y(100 - 4y - z), \; z' = z(60 - y - 2z).$$

Solution: The system of equilibrium conditions $4y + z = 100$, $y + 2z = 60$ can be solved by eliminating one of the variables to obtain the coexistence equilibrium (20,20).To determine this equilibrium's asymptotic stability, we use the fact that the community matrix at the equilibrium (y_∞, z_∞) is

$$\begin{bmatrix} -ay_\infty & -by_\infty \\ -cz_\infty & -dz_\infty \end{bmatrix} = \begin{bmatrix} -80 & -20 \\ -20 & -40 \end{bmatrix}$$

with negative trace and positive determinant. Thus the coexistence equilibrium is asymptotically stable, and every orbit approaches it. Alternatively, we note that λ/μ now lies between the other two ratios, $b/d = 1 < \lambda/\mu = 5/3 < a/c = 4$, as in case I. In this case, the level of competition (coefficients b and c) is small enough for both species that they can coexist. □

Finally, we revisit the model for plant competition first explored in Section 3.2.2. Equation (3.17) gave the general equation for a ranked system of competitors, in which the proportion p_i of habitat occupied by species i changes according to four processes: colonizing uninhabited patches, natural mortality, displacing inferior competitors (with index $j > i$) and being displaced by superior competitors (with index $j < i$). Analysis showed that each species persisted at a positive equilibrium p_i^* if and only if its colonization rate c_i exceeded its own mortality rate m_i and the superior competitors left some space available for colonization, $\sum_{j=1}^{i-1} p_j < 1$. However, Bampfylde observed that displacement does not occur in rainforests; a competition model including only colonization of open spaces ($1 - \sum_{j=1}^{n} p_j$ for n species) and natural mortality, but no displacement, becomes (cf. (2.18))

$$\frac{dp_i}{dt} = c_i p_i \left(1 - \sum_{j=1}^{n} p_j\right) - m_i p_i \tag{6.9}$$

for $i = 1, 2, ..., n$.

We analyze system (6.9) through its equilibria, found by setting $dp_i/dt = 0$. For each i this yields either $p_i^* = 0$ or $1 - \sum_{j=1}^{n} p_j^* = m_i/c_i$. In general the values of m_i and c_i are independent from one species to another, so it is unlikely to have $m_i/c_i = m_j/c_j$ for any $i \neq j$; this then makes it impossible for the proportion of free sites $(1 - \sum_{j=1}^{n} p_j)$ to match simultaneously more than one m_i/c_i as in the equilibrium conditions. Thus no more than one species can have a nonzero equilibrium value—that is, any equilibrium represents a scenario where all species but one die out, an illustration of the so-called *principle of competitive exclusion* in ecology, which holds that only one species or population can occupy a given ecological niche (food, habitat, etc.) at a given time.

An example may help to illustrate. Consider a system with three such competing species. The equilibrium conditions are ($dp_1/dt = 0$) either $p_1^* = 0$ or $1 - \sum_{j=1}^{3} p_j^* = m_1/c_1$, ($dp_2/dt = 0$) either $p_2^* = 0$ or $1 - \sum_{j=1}^{3} p_j^* = m_2/c_2$, *and* ($dp_3/dt = 0$) either $p_3^* = 0$ or $1 - \sum_{j=1}^{3} p_j^* = m_3/c_3$. If we assume that m_1/c_1, m_2/c_2, and m_3/c_3 all have different values, then at most one of them can be equal to $1 - \sum_{j=1}^{3} p_j^*$. Thus the equilibria are $(0, 0, 0)$ [extinction of all 3 species], $(1 - \frac{m_1}{c_1}, 0, 0)$ [only species 1 persists], $(0, 1 - \frac{m_2}{c_2}, 0)$ [only species 2 persists], and $(0, 0, 1 - \frac{m_3}{c_3})$ [only species 3 persists]. We can determine when each equilibrium is asymptotically stable using the community matrix, which calculations show to be

$$\begin{bmatrix} c_1(g - p_1) - m_1 & -c_1 p_1 & -c_1 p_1 \\ -c_2 p_2 & c_2(g - p_2) - m_2 & -c_2 p_2 \\ -c_3 p_3 & -c_3 p_3 & c_3(g - p_3) - m_3 \end{bmatrix},$$

where $g = 1 - p_1 - p_2 - p_3$ is the gap (the proportion of unoccupied sites). At the first equilibrium this simplifies to

$$\begin{bmatrix} c_1 - m_1 & 0 & 0 \\ 0 & c_2 - m_2 & 0 \\ 0 & 0 & c_3 - m_3 \end{bmatrix},$$

which readers familiar with linear algebra will observe has the eigenvalues (solutions to the characteristic equation) $c_1 - m_1$, $c_2 - m_2$ and $c_3 - m_3$. Thus the extinction equilibrium is asymptotically stable if and only if $c_i < m_i$ for all 3 species (that is, each species' mortality rate outstrips its ability to colonize new territory), just as for the displacement model of Chapter 3.

At the second equilibrium, in which species 1 wins the competition, the community matrix simplifies to

$$\begin{bmatrix} m_1 - c_1 & -c_1 p_1^* & -c_1 p_1^* \\ 0 & c_2(m_1/c_1) - m_2 & 0 \\ 0 & 0 & c_3(m_1/c_1) - m_3 \end{bmatrix};$$

once again the eigenvalues are the diagonal entries, and this equilibrium is asymptotically stable if they are all negative, i.e.,

$$\frac{c_1}{m_1} > \max\left(1, \frac{c_2}{m_2}, \frac{c_3}{m_3}\right).$$

We can extrapolate similar conditions for the asymptotic stability of the other single-survivor equilibria, so that in the end the species with the highest c_i/m_i

ratio wins the competition. (We omit here the discussion of issues related to global stability which would be required for a rigorous proof.)

This result suggesting competitive exclusion may appear counter-intuitive when the dominant competitor will not (at equilibrium) occupy the entire rainforest: the model which allows displacement predicts coexistence as long as there is enough habitat for all, and without displacement the inferior competitors would appear to be at an advantage relative to the same scenario with displacement. However, what constitutes a superior competitor is quite different in this new model (the species with the greatest c_i/m_i ratio) than in the one studied in Chapter 3 (where species are ranked entirely independently of their c_i/m_i ratios), and a little algebra shows that any "inferior competitor" in this new model (say $c_2/m_2 < c_1/m_1$) would also approach a zero equilibrium value in the prior model.

In the boundary case, mentioned as unlikely above, of a "tie" for highest c_i/m_i ratio, there are an infinite number of [non-isolated] equilibria such that the "tied" species' proportions sum to their common m_i/c_i ratio.

6.2.2 Predator-prey systems

The Lotka-Volterra system

$$\begin{aligned} y' &= y(\lambda - bz) \\ z' &= z(-\mu + cy) \end{aligned} \tag{6.10}$$

was our first example of a model for a predator-prey system, formulated under the assumptions that the presence of the z-species (predators) reduces the growth rate of the y-species (prey), that presence of the prey increases the growth rate of the predators, and that by themselves the prey species would grow exponentially while the predator species would die out exponentially.

If we assume instead that the prey species obeys a logistic law, we replace the Lotka-Volterra system (6.10) by

$$\begin{aligned} y' &= y(\lambda - ay - bz), \\ z' &= z(-\mu + cy). \end{aligned} \tag{6.11}$$

The system (6.11) obviously has an equilibrium at (0,0). There is a second equilibrium with $z = 0$ and $ay = \lambda$, or $y = \frac{\lambda}{a}$ (prey but no predators). A coexistence equilibrium (y_∞, z_∞), in the interior of the first quadrant, is found by solving the pair of equations

$$\begin{aligned} ay + bz &= \lambda, \\ cz &= \mu, \end{aligned}$$

obtaining $y_\infty = \frac{\mu}{c}$, $z_\infty = \frac{c\lambda - a\mu}{c}$. This equilibrium is relevant only if $z_\infty > 0$, or $a\mu - c\lambda < 0$. In this case the community matrix at (y_∞, z_∞) is

$$\begin{bmatrix} -ay_\infty & -by_\infty \\ cz_\infty & 0 \end{bmatrix},$$

calculated much as in Example 4, Section 5.4. As the conditions for stability of this equilibrium, namely $-ay_\infty < 0$ and $bcy_\infty z_\infty > 0$, are satisfied automatically, the equilibrium (y_∞, z_∞) is asymptotically stable when it exists. We can show similarly that if $a\mu - c\lambda < 0$ the equilibrium $(\frac{\lambda}{a},0)$ is unstable, while if $a\mu - c\lambda > 0$ this equilibrium is asymptotically stable. The equilibrium (0,0) is always unstable. Thus the model (6.11) always has exactly one asymptotically stable equilibrium, which every orbit approaches. If $a\mu - c\lambda < 0$ $(\lambda/\mu > a/c)$ this equilibrium is $(\frac{\lambda}{a},0)$, corresponding to extinction of the predators. In neither case can there be a periodic orbit, and thus the model (6.11) is less "realistic" than the original Lotka-Volterra model. The self-limiting property of the prey dynamics prevents sustained oscillations, and as a increases can even prevent damped ones.

EXAMPLE 3.
Determine the behavior of a predator-prey system (for example, moths and birds) modeled by the system

$$y' = y(180 - y - z), \; z' = z(-500 + 10y).$$

Solution: An interior equilibrium is a solution of the system

$$y + z = 180, \; 10y = 500.$$

As this system has the solution (50,130) in the first quadrant, there is an equilibrium (50,130) which (from the above discussion) all orbits approach. The two species co-exist, with the prey at less than 1/3 its natural carrying capacity of 180. □

In the model (6.11) the term $-byz$ in the equation for y' represents the rate at which predators consume prey. Thus it is assumed that each predator's consumption is proportional to the prey population size. Biologically, it is more plausible to assume that the rate of prey consumption per predator increases with prey population size but is bounded as the prey population becomes unbounded, that is, that there is a maximum rate of consumption per predator no matter how plentiful the food supply. Beyond a certain point, the prey population no longer limits the resources of the predators. For example, we may assume that the rate of consumption of prey per predator has the form $\frac{qy}{y+A}$ where q and A are positive constants.[3] Instead of the term cyz in the equation for z', we incorporate a term proportional to $\frac{yz}{z+A}$, representing the conversion of food (prey) into predator biomass. This leads us to a model of the form

$$y' = ry\left(1 - \frac{y}{K}\right) - \frac{ayz}{y + A}, \tag{6.12}$$
$$z' = sz\left(\frac{y}{y + A} - \frac{J}{J + A}\right).$$

[3]See Section 2.1.3 for an interpretation of this Verhulst-type expression, including the significance of A.

Photo courtesy Florida Keys National Marine Sanctuary

FIGURE 6.9: A barracuda eats another fish off the Florida keys.

Here, we have also changed the names of some of the parameters in (6.11), replacing λ by r, a by $\frac{r}{K}$, and μ by $\frac{sJ}{J+A}$. The term $\frac{qyz}{y+A}$ (replacing byz) in the first equation of (6.12) is called the *predator functional response*, and the term $\frac{syz}{y+A}$ (replacing cyz) in the second equation of (6.12) is called the *predator numerical response*; the constant $\frac{s}{q}$ is the conversion efficiency of prey into predators. The model (6.12) assumes that the prey population would obey a logistic law in the absence of predators, and that the predator population would die out exponentially in the absence of prey.

Here we have rewritten the natural decay rate of the predator population in terms of J, which we can see from (6.12) is the minimum prey population required to sustain the predator population ($z' \geq 0$). As J decreases, so does the rate at which the predator population would die out in the absence of the prey. In the following two examples, we shall see that the parameter J (or, equivalently, μ) is capable of changing the nature of the system's behavior.

EXAMPLE 4.
Determine the qualitative behavior of a predator-prey system (imagine this time large and small fish in a pond, cf. Figure 6.9) modeled by the differential equations

$$y' = y\left(1 - \frac{y}{30}\right) - \frac{yz}{y+10},$$
$$z' = z\left(\frac{y}{y+10} - \frac{3}{5}\right).$$

Solution: Equilibria are solutions of the pair of equations

$$y\left(1-\frac{y}{30}-\frac{z}{y+10}\right)=0, \qquad (6.13)$$
$$z\left(\frac{y}{y+10}-\frac{3}{5}\right)=0.$$

One equilibrium is given by $y=0$, $z=0$. If $z=0$, (6.13) implies

$$y\left(1-\frac{y}{30}\right)=0,$$

and thus another equilibrium is given by $y=30$, $z=0$. An equilibrium with y and z both positive satisfies

$$1-\frac{y}{30}-\frac{z}{y+10}=0, \quad \frac{y}{y+10}=\frac{3}{5}.$$

The second of these equations is $5y=3y+30$, or $y=15$, and substitution into the first equation gives $1-\frac{1}{2}-\frac{z}{25}=0$, or $z=12.5$. Thus a third equilibrium is given by $y=15$, $z=12.5$.

We now use the equilibrium stability theorem of Section 5.4 with

$$F(y,z)=y\left(1-\frac{y}{30}\right)-\frac{yz}{z+10}, \quad G(y,z)=z\left(\frac{y}{y+10}-\frac{3}{5}\right).$$

Then the partial derivatives are

$$\begin{aligned} F_y(y,z) &= 1-\frac{y}{15}-\frac{10z}{(y+10)^2}, & F_z(y,z) &= -\frac{y}{y+10} \\ G_y(y,z) &= \frac{10z}{(y+10)^2}, & G_z(y,z) &= \frac{y}{y+10}-\frac{3}{5}. \end{aligned}$$

At the equilibrium (0,0), the community matrix is

$$\begin{bmatrix} 1 & 0 \\ 0 & -\frac{3}{5} \end{bmatrix},$$

and thus the equilibrium is unstable. At the equilibrium (30,0), the community matrix is

$$\begin{bmatrix} -1 & -\frac{3}{4} \\ 0 & \frac{3}{20} \end{bmatrix},$$

and since its determinant is $-\frac{3}{20}<0$ this equilibrium is also unstable. At the equilibrium (15,12.5), the community matrix is

$$\begin{bmatrix} -\frac{1}{5} & -\frac{3}{5} \\ \frac{1}{5} & 0 \end{bmatrix},$$

and since this has trace $-\frac{1}{5}<0$ and determinant $\frac{3}{25}>0$, this equilibrium is asymptotically stable. It is possible to show that every orbit with $y(0) > 0$ and $z(0) > 0$ — not just those which start close to (15,12.5) — approaches this equilibrium. Thus predator and prey coexist here, with prey at half their natural carrying capacity. □

EXAMPLE 5.

Suppose now that the environment of the pond in Example 4 improves in such a way that the predator fish tend to live longer (or die off more slowly), so that the natural per capita mortality rate drops from $\frac{3}{5}$ (in per time units) to $\frac{1}{3}$. Determine the qualitative behavior of this new predator-prey system, modeled by the differential equations

$$y' = y\left(1 - \frac{y}{30}\right) - \frac{yz}{y+10},$$
$$z' = z\left(\frac{y}{y+10} - \frac{1}{3}\right).$$

Solution: Equilibria are solutions of the pair of equations

$$y\left(1 - \frac{y}{30} - \frac{z}{y+10}\right) = 0,$$
$$z\left(\frac{y}{y+10} - \frac{1}{3}\right) = 0.$$

As in Example 4, there is an equilibrium at (0,0) and a second, predator-free ($z = 0$) equilibrium with $y = 30$. An equilibrium with y and z both positive satisfies

$$1 - \frac{y}{30} - \frac{z}{y+10} = 0, \quad \frac{y}{y+10} = \frac{1}{3}.$$

The second of these equations is $3y = y + 10$, or $y = 5$, and substitution into the first equation gives $1 - \frac{5}{30} - \frac{z}{15} = 0$, or $z = 12.5$. Thus a third equilibrium, representing coexistence, is given by $y = 5$, $z = 12.5$.

The community matrix is the same as in Example 4, except that $\frac{3}{5}$ is replaced by $\frac{1}{3}$ in $G(y, z)$. Thus the community matrix at (0,0) is

$$\begin{bmatrix} 1 & 0 \\ 0 & -\frac{1}{3} \end{bmatrix},$$

and this equilibrium is unstable. The community matrix at (30,0) is

$$\begin{bmatrix} -1 & -\frac{3}{4} \\ 0 & \frac{5}{12} \end{bmatrix},$$

and since this has determinant $-\frac{5}{12} < 0$, this equilibrium is unstable. The community matrix at (5,12.5) is

$$\begin{bmatrix} \frac{1}{9} & -\frac{1}{3} \\ \frac{5}{9} & 0 \end{bmatrix}.$$

Since this matrix has positive trace, the equilibrium (5,12.5) is also unstable, and the system has no asymptotically stable equilibrium. In order to show that all orbits in the first quadrant are bounded, we apply the technique introduced in Example 5, Section 5.4, adding the two equations of the model to obtain

$$(y+z)' = y\left(1 - \frac{y}{30}\right) - \frac{z}{3}.$$

Thus $y + z$ is decreasing except in the bounded region defined by $\frac{z}{3} < y\left(1 - \frac{y}{30}\right)$. In order for an orbit to be unbounded, $y + z$ must be unbounded, and, as in Example 5, Section 5.4, this is impossible since $y + z$ is decreasing whenever $y + z$ is large. Thus all orbits in the first quadrant are bounded, and the Poincaré-Bendixson theorem may be applied to show that there must be a limit cycle (with the equilibrium (5,12.5) in its interior) to which every orbit tends. Thus the two species co-exist, but their population sizes fluctuate periodically. Some orbits are shown in Figure 6.10. □

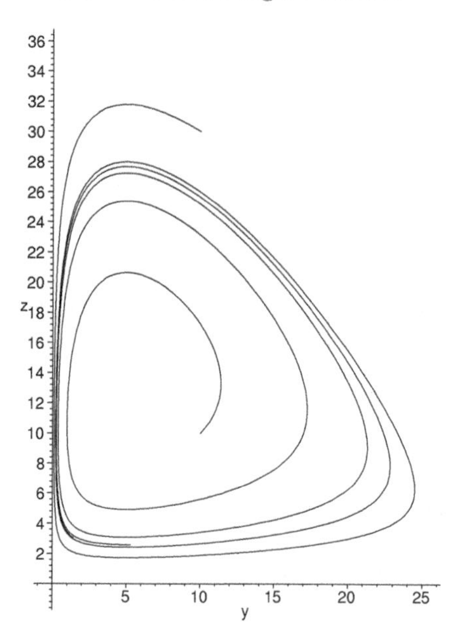

FIGURE 6.10: Some orbits of the system in Example 5.

Examples 4 and 5 show that for a model of the form (6.12) there may be periodic orbits, periodic orbit] or every orbit may approach an equilibrium. Which behavior occurs depends on the values of the parameters in the model rather than on the form of the model. In fact, there is a class of models considerably more general than (6.12) exhibiting the same two possible behaviors.

We shall now consider models of the general form

$$\begin{aligned} y' &= yf(y) - yz\phi(y), \\ z' &= z[sy\phi(y) - c]. \end{aligned} \tag{6.14}$$

In this model, as before, $y(t)$ is the prey population size and $z(t)$ is the predator population size. Here the term $yf(y)$ represents the prey population growth rate in the absence of predators. We assume that the per capita growth rate of the prey decreases as prey population size increases, and that there is a prey carrying capacity K, so that

$$f'(y) < 0 \; [y \geq 0], \; f(K) = 0. \tag{6.15}$$

The term $yz\phi(y)$ represents the predator functional response, with consumption of $y\phi(y)$ prey per predator in unit time. We assume that consumption of prey is positive, and that the prey consumption per predator $y\phi(y)$ increases with prey population size, but that the fraction $\phi(y)$ of prey population consumed per predator decreases:

$$\phi(y) > 0, \; [y\phi(y)] > 0, \; \phi'(y) \leq 0. \tag{6.16}$$

The term $syz\phi(y)$ is the predator numerical response, and cz is predator mortality. Because of (6.16), the predator per capita growth rate $sy\phi(y) - c$ increases with prey population size and is positive for prey population size above some minimum J, $sJ\phi(J) = c$. Normally, it is assumed that the minimum prey population size for predator survival J is less than the prey carrying capacity K, as otherwise it is clear that the predator population cannot survive.

Let us try to analyze the behavior of solutions of a system (6.14) under the assumptions (6.15) and (6.16). The equilibria of (6.14) are the solutions of

$$y[f(y) - z\phi(y)] = 0, \tag{6.17}$$
$$syz\phi(y) = cz. \tag{6.18}$$

If $y = 0$, then (6.17) is satisfied. In order to satisfy (6.18) as well we must take $z = 0$. Thus one equilibrium is (0,0). If $z = 0$, then (6.18) is satisfied and (6.17) reduces to $yf(y) = 0$, which implies that either $y = 0$ or $y = K$. Thus (K,0) is a second equilibrium, as we should expect. If y and z are both positive, (6.17) becomes $f(y) - z\phi(y) = 0$, and (6.18) becomes $sy\phi(y) = c$, which implies $y = J$. If $y = J$, (6.17) implies $z = \frac{f(J)}{\phi(J)}$. Since $\phi(J) > 0$, then $z > 0$ provided $f(J) > 0$, i.e., provided $J < K$. Thus if $J < K$, there is a third equilibrium (y_∞,z_∞) with $y_\infty > 0$, $z_\infty > 0$ given by

$$y_\infty = J, \; z_\infty = \frac{f(J)}{\phi(J)}.$$

In order to calculate the community matrix at each equilibrium of (6.14), we take $F(y, z) = yf(y) - yz\phi(y)$, $G(y, z) = syz\phi(y) - cz$. Then

$$\begin{aligned} F_y(y, z) &= yf'(y) + f(y) - z[\phi(y) + y\phi'(y)], \quad & F_z(y, z) = -y\phi(y) \\ G_y(y, z) &= sz[\phi(y) + y\phi'(y)], & G_z(y, z) = sy\phi(y) - c. \end{aligned}$$

Thus the community matrix at the equilibrium (0,0) is

$$\begin{bmatrix} f(0) & 0 \\ 0 & -c \end{bmatrix}.$$

Since $f(0) > 0$ and $-c < 0$, this equilibrium is unstable (by applying the equilibrium stability theorem of the previous section).

The community matrix at the equilibrium $(K, 0)$ is

$$\begin{bmatrix} Kf'(K) & -K\phi(K) \\ 0 & sK\phi(K) - c \end{bmatrix}.$$

Because $f'(K) < 0$ and $sK\phi(K) - c > 0$ (if $K > J$), we see that the equilibrium $(K, 0)$ is unstable if $K > J$. If instead $K < J$, then $sK\phi(K) - c < 0$, and in this case the equilibrium $(K, 0)$ is asymptotically stable.

If $K > J$, there is a third equilibrium (y_∞, z_∞) with community matrix

$$\begin{bmatrix} cc[yf(y)]'_{y_\infty} - z_\infty[y\phi(y)]'_{y_\infty} & -y_\infty\phi(y_\infty) \\ sz_\infty[y\phi(y)]'_{y_\infty} & 0 \end{bmatrix}.$$

The conditions for asymptotic stability are

$$[yf(y)]'_{y_\infty} - z_\infty[y\phi(y)]'_{y_\infty} < 0 \tag{6.19}$$

and

$$y_\infty\phi(y_\infty)\, sz_\infty[y\phi(y)]'_{y_\infty} > 0. \tag{6.20}$$

Because $\phi(y_\infty) > 0$, and $[y\phi(y)]'_{y_\infty} > 0$ from (6.16), the condition (6.20) is satisfied, and the equilibrium (y_∞, z_∞) is asymptotically stable if and only if (6.19) is satisfied. The condition (6.19) is equivalent to

$$y_\infty f'(y_\infty) + f(y_\infty) - y_\infty z_\infty \phi'(y_\infty) - z_\infty\phi(y_\infty) < 0,$$

and since $z_\infty = \frac{f(y_\infty)}{\phi(y_\infty)}$, this is the same (after multiplying by $\phi(y_\infty)/y_\infty$) as

$$f'(y_\infty)\phi(y_\infty) - f(y_\infty)\phi'(y_\infty) < 0. \tag{6.21}$$

The equation of the prey nullcline is $z = \frac{f(y)}{\phi(y)}$, and differentiation shows that the slope of the prey nullcline is

$$\frac{dz}{dy} = \frac{\phi(y)f'(y) - f(y)\phi'(y)}{[\phi(y)]^2}. \tag{6.22}$$

Comparing (6.22) and (6.21), we obtain a simple geometric characterization of stability of the equilibrium (y_∞, z_∞):

The equilibrium (y_∞, z_∞) of the system (6.14) with $y_\infty > 0$, $z_\infty > 0$ under the hypotheses (6.15) and (6.16) is asymptotically stable if and only if the prey nullcline has negative slope at the equilibrium.

Because of the hypotheses (6.15) and (6.16), the slope of the prey nullcline at its y-intercept $(K, 0)$ is $\frac{f'(K)}{\phi(K)} < 0$. The slope of the prey nullcline at $y = 0$ may be either positive or negative, depending on the functions $f(y)$ and $\phi(y)$ and the parameter values. For example, if $f(y) = r\left(1 - \frac{y}{K}\right)$ and $\phi(y) = \frac{q}{y+A}$, the prey nullcline has positive slope at $y = 0$ if $A < K$. Thus there are cases in which the prey nullcline increases to a maximum and then decreases to zero as y increases. In such cases the equilibrium (y_∞, z_∞) is unstable if it is on the increasing portion of the prey nullcline, and asymptotically stable if it is on the decreasing portion of the prey nullcline.

Increasing the carrying capacity of the prey species amounts to increasing the food supply for the predators. Geometrically, this would have the effect of moving the prey nullcline upward and to the right. This could move the equilibrium from the decreasing portion of the prey nullcline to the increasing portion of the prey nullcline and thus change it from an asymptotically stable equilibrium to an unstable equilibrium. This possibility, that increasing the food supply for the predators could destabilize the population system, has been called the "paradox of enrichment."

If the equilibrium (y_∞, z_∞) is unstable, the system (6.14) has no asymptotically stable equilibrium. It is possible to show that every solution of the system (6.14) in the first quadrant is bounded as $t \to \infty$ (see Exercise 13 below). Then by the Poincaré-Bendixson theorem (Section 5.4), every orbit must approach a limit cycle, and the prey and predator populations oscillate as in Example 5. It is possible that an orbit could come very close to one of the axes, where a small perturbing force, say an environmental change, could wipe out one of the populations and lead to the collapse of the population system. Thus the oscillations caused by enrichment could turn out to be very harmful to the predator species.

EXAMPLE 6.
Describe the long-term behavior of a predator-prey system modeled by the system

$$y' = y(10 - y) - yz, \; z' = z(y - 5).$$

Solution: This system is of the form (6.14) with $f(y) = 10 - y$, so that $K = 10$, and $\phi(y) = y$, so that $J = 5$. The hypotheses (6.15) and (6.16) are satisfied. The prey nullcline is the line $z = 10 - y$ with negative slope, and thus the coexistence equilibrium given by $y = 5$, $z = 10 - y$, or $z = 5$, is asymptotically stable. Every orbit approaches this equilibrium, so the species will coexist. □

6.2.3 Symbiosis

There are situations in which the interaction of two species is mutually beneficial, for example, plant-pollinator systems (Figure 6.11 gives another example). Such an interaction is called mutualistic or symbiotic. The interaction may be *facultative*, meaning that the two species could survive separately,

Photo by Jan Derk

FIGURE 6.11: Two goby fish and a shrimp in a symbiotic relationship: the shrimp maintains a burrow in the sand on the sea floor which also serves as a safe refuge for the goby and their eggs, while the goby alert the shrimp (which has poor eyesight) if danger approaches.

or *obligatory*, meaning that each species would become extinct without the assistance of the other.

If we model a symbiotic system by a pair of differential equations with linear per capita growth rates

$$\begin{aligned} y' &= y(\lambda - ay + bz), \\ z' &= z(\mu + cy - dz), \end{aligned} \tag{6.23}$$

the mutualism of the interaction is modeled by the positive nature of the interaction terms cy and bz. In a facultative interaction, the constants λ and μ are positive, while in an obligatory relation the constants λ and μ are negative. In each type of interaction there are two possibilities, depending on the relationship between the slope a/b of the y isocline and the slope c/d of the z-isocline. If $ad > bc$, the mutualistic effects are smaller than the self-limiting terms in the per capita growth rates, and the slope of the y-isocline is greater than the slope of the z-isocline.

In both facultative and obligatory interactions, if $ad < bc$ there is a region of the phase plane in which solutions become unbounded, and this suggests that either we must restrict models of this form by requiring $ad > bc$, or we must consider models with nonlinear per capita growth rates.

For models with linear per capita growth rates and $ad > bc$, the only asymptotically stable equilibrium in the facultative case is the intersection (y_∞, z_∞) of the lines $ay - bz = \lambda$, $-cy + dz = \mu$ with $y_\infty > 0, z_\infty > 0$, and every orbit tends to this equilibrium. To see this, we calculate the equilibrium

$$y_\infty = \frac{d\lambda + b\mu}{ad - bc} > 0, z_\infty = \frac{c\lambda + a\mu}{ad - bc} > 0.$$

The community matrix at (y_∞, z_∞) is

$$\begin{bmatrix} \lambda - 2ay_\infty + bz_\infty & by_\infty \\ cz_\infty & \mu - 2dz_\infty + cy_\infty \end{bmatrix} = \begin{bmatrix} -ay_\infty & by_\infty \\ cz_\infty & -dz_\infty \end{bmatrix}$$

and since this matrix has negative trace and positive determinant the equilibrium (y_∞, z_∞) is asymptotically stable. It is easy to verify that the other equilibria $(0, 0), (\lambda/a, 0), (0, \mu/d)$ are unstable.

In the obligatory case with $ad > bc$, since $\lambda < 0$, $\mu < 0$ the only equilibrium in the first quadrant is $(0, 0)$, asymptotically stable since the community matrix is

$$\begin{bmatrix} \lambda & 0 \\ 0 & \mu \end{bmatrix},$$

and asymptotic stability follows from $\lambda < 0$, $\mu < 0$. Thus the only asymptotically stable equilibrium is the origin, and every orbit tends to the origin. In the obligatory case neither species survives. While our model may be acceptable in the facultative case, it is clear that the possibility of obligatory mutualism is not described by this model. If we consider the obligatory case with $ad < bc$, there is an equilibrium (x_∞, y_∞) with $x_\infty > 0$, $y_\infty > 0$, which may be shown to be a saddle point whose stable separatrices separate the phase plane into a region of mutual extinction and a region of unbounded growth. Such a separation is plausible biologically, but it would be necessary to alter the model so as to rule out the possibility of unbounded growth in order to give a more realistic model.

Exercises

In each of Exercises 1–4, determine the outcome of the competition modeled by the given system.

1. $y' = y(120 - 3y - z)$, $z' = z(80 - y - z)$
2. $y' = y(75 - y - 4z)$, $z' = z(60 - y - 3z)$
3. $y' = y(90 - 2y - z)$, $z' = z(40 - y - z)$
4. $y' = y(80 - y - z)$, $z' = z(80 - 2y - 3z)$

In each of Exercises 5–10, determine the qualitative behavior of the predator-prey system modeled by the given system.

5. $y' = y\left(1 - \frac{y}{20}\right) - \frac{yz}{y+10}$, $z' = 3z\left(\frac{y}{y+10} - \frac{1}{2}\right)$

6. $y' = 3y\left(1 - \frac{y}{40}\right) - \frac{2yz}{y+15},\ z' = z\left(\frac{2y}{y+15} - \frac{6}{5}\right)$

7. $y' = y\left(1 - \frac{y}{20}\right) - \frac{yz}{y+10},\ z' = 3z\left(\frac{y}{y+10} - \frac{1}{6}\right)$

8. $y' = 3y\left(1 - \frac{y}{40}\right) - \frac{2yz}{y+15},\ z' = z\left(\frac{2y}{y+15} - \frac{1}{2}\right)$

9. $y' = y\left(1 - \frac{y}{20}\right) - \frac{yz}{y+20},\ z' = 3z\left(\frac{y}{y+10} - \frac{3}{4}\right)$

10. $y' = 3y\left(1 - \frac{y}{40}\right) - \frac{2yz}{y+15},\ z' = z\left(\frac{2y}{y+15} - \frac{8}{5}\right)$

11. * Show that the coexistence equilibrium (y_∞, z_∞) of the system

$$y' = ry\left(1 - \frac{y}{K}\right) - \frac{ayz}{y+A},\quad z' = sz\left(\frac{ay}{y+A} - \frac{aJ}{J+A}\right)$$

is unstable if $K > A + 2J$ and asymptotically stable if $J < K < A + 2J$.

12. * Determine the behavior as $t \to \infty$ of solutions of the system

$$y' = ry\left(1 - \frac{y}{K + az}\right),\quad z' = sz\left(1 - \frac{z}{M + by}\right)$$

with a, b, K, M, r, s positive constants. [Warning: The behavior if $ab < 1$ is different from the behavior if $ab > 1$.]

13. * Show that every first-quadrant orbit of the system $y' = yf(y) - yz\phi(y)$, $z' = z[sy\phi(y) - c]$, under the assumptions $f'(y) < 0$ $[y \geq 0]$, $f(K) = 0$, and $\phi(y) > 0$, $\phi'(y) \leq 0$, $[y\phi(y)]' \geq 0$, is bounded, by showing that the function $sy + z$ is decreasing except in a bounded set, much as in Example 5.

In Exercises 14 through 17, find all equilibria of the given mutualistic system and determine their stability. Are there any unbounded orbits?

14. $y' = y(-50 + z - y), z' = z(-20 - z + 2y)$

15. $y' = y(-50 + z - y), z' = z(-20 - z + 2y)$

16. $y' = y(10 - y + z/(1 + y)), z' = z(y/(1 + z) - 1)$

17. $y' = y(1 - y/20), z' = z(1 - z/10) \cdot ((2y + 1)/(1 + y))$

18. For a system of the form (6.23) with $\lambda > 0, \mu < 0$, show that both species survive, even though the z-species would go extinct in the absence of the y-species.

19. * Models for cell growth often assume that a cell can be in various states, with switching (transfer) between one state an another.[4] Assume that there are two states, in only one of which there is proliferation (cell division). If $P(t)$ and $Q(t)$ represent the concentrations of cells in the two states, the following equations can be assumed, with all Greek letters representing positive constants

$$P' = (\gamma - \delta P)P - \alpha P + \beta Q, \quad Q' = \alpha P - (\lambda + \beta)Q. \tag{6.24}$$

(a) Is P or Q the proliferating state? Why?

(b) Find all biologically meaningful steady states of (6.24) and determine their stability. You should distinguish the cases $(\alpha - \gamma)/\alpha < \beta/(\lambda + \beta)$ and $(\alpha - \gamma)/\alpha > \beta/(\lambda + \beta)$.

For further reading about this type of model, see Eisen and Schiller [20].

20. * This problem[5] is concerned with a theory of liver regeneration due to Bard [2, 3]. Normally, the rate of cell division in the liver is very low, but if up to two thirds of the liver of a rat is removed, then the liver grows back to its original size in about a week. Bard discusses two theories. One theory, based on the assumed existence of a growth stimulator, predicts that the liver volume V will overshoot its normal value before finally settling down to a steady state. Such an overshoot has not been observed. Here we will show something of how an alternative inhibitor model can account for the facts. Bard assumes that liver cells are prevented from dividing by an inhibitor of short half-life. The inhibitor is synthesized by the liver at a rate proportional to its size and is secreted into the blood, where its concentration is the same as in the liver. Let $V(t)$ be the volume of the liver and $S(t)$ the concentration of the inhibitor. Bard postulates the equations

$$V' = V[f(S) - r], \quad S' = \frac{pV}{W + V} - qS, \tag{6.25}$$

with W (blood volume), r, p, q constants. (a) Why should the function f be assumed to have a negative derivative?

(b) Show that the system (6.25) has a unique positive equilibrium, and that the linearization of (6.25) at this equilibrium has the form

$$u' = -\gamma v, \quad v' = \alpha u - qv,$$

with positive constants γ, α.

---

[4]This problem is taken from Lee A. Segel, *Modeling dynamic phenomena in molecular and cellular biology*, Cambridge University Press, Cambridge (1984), p. 267.

[5]Taken from Lee A. Segel, *Modeling dynamic phenomena in molecular and cellular biology*, Cambridge University Press, Cambridge (1984), p. 269.

21. * [6] Consider the system

$$A' = \alpha A - a_1 A^3 - a_2 a B^2, \quad B' = \beta B - b_1 B A^2 - b_2 B^3.$$

All coefficients are assumed to be positive., and we are interested only in non-negative values of $A(t), B(t)$.

(a) Discuss the suitability of this system as a model for the interaction of two bacteria populations.

(b) Show that there are four possible equilibria,

(i) $\mathrm{I} = (0, 0)$,
(ii) $\mathrm{II} = (0, \sqrt{\beta/b_2})$,
(iii) $\mathrm{III} = (\alpha/a_1, 0)$,
(iv) $\mathrm{IV} = (\sqrt{(a_2\beta - \alpha b_2)/(a_2 b_1 - a_1 b_2)}, \sqrt{(\alpha b_1 - \beta a_1)/(a_2 b_1 - a_1 b_2)})$.

(c) Examine the stability of the equilibria and show that there are four possibilities, namely
(i) II and III are unstable and if IV exists it is stable.
(ii) II is stable, III is unstable, and IV cannot exist.
(iii) II is unstable, III is stable, and IV cannot exist.
(iv) II and III are stable and IV exists but is unstable.

22. Consider a simple age-structured model in which individuals are classified as juveniles J or adults A (discrete age structure applies naturally to species with distinct life stages, or more arbitrarily to species with continuous development for whom a maturation age is defined). We assume that juveniles are born at a rate dependent on the number of adults but limited (logistically) by the resources used by juveniles and adults; that juveniles mature at a certain rate; and that juvenile and adult mortality rates may differ. This leads to the model

$$J' = bA\left(1 - \frac{\epsilon J + A}{K}\right) - (\mu + \gamma)J, \quad A' = \gamma J - dA,$$

for constant rates b, γ, μ, d, carrying capacity K, and a dimensionless constant ϵ which represents how many resources a juvenile uses relative to an adult (we might then assume $0 \le \epsilon \le 1$).

Perform a complete qualitative analysis for this model, identifying any equilibria and the conditions for their stability. As with many population models, the long-term survival of the population can be determined by a demographic reproductive number R_d similar to the basic reproductive number used in epidemiological models. Derive and interpret biologically an expression for R_d.

[6]This problem is taken from Lee A. Segel, *Modeling dynamic phenomena in molecular and cellular biology*, Cambridge University Press, Cambridge (1984), p. 270.

6.3 Numerical approximation to solutions of systems

Just as for single differential equations, the numerical approximation of solutions of a system of differential equations is often a useful approach. The qualitative methods on which we have concentrated in this chapter are directed at obtaining information about the behavior of solutions for large t; numerical approximation methods may be the only way to obtain information about the behavior of solutions on a finite interval. There are many computer programs available commercially, either as part of a computer algebra system or as stand-alone programs. These programs can generate both numerical data and a graphic portrayal of solutions either by showing both y and z as functions of t or by showing orbits in the phase plane. We will not discuss the implementation of approximation methods in detail but will confine ourselves to brief descriptions of the analogues of the methods described in Section 4.5 for first-order differential equations.

Our goal is to estimate the values $y(T)$ and $z(T)$ for $t = T$ of the solution $[y(t), z(t)]$ of an initial value problem

$$\begin{aligned} y' &= F(y, z), \quad y(0) = y_0, \\ z' &= G(y, z), \quad z(0) = z_0. \end{aligned} \tag{6.26}$$

We divide the interval $0 \le t \le T$ into N subintervals of length h by defining $t_1 = h$, $t_2 = 2h$, ..., $t_N = Nh$. The number N of subintervals and the length h of each subinterval are related by $T = Nh$. Just as for first-order equations, the fundamental idea is to replace the system of differential equations (6.26) by a system of difference equations which can be solved recursively. In the *Euler method* we achieve this by replacing y' and z' by the respective difference quotients

$$\frac{y(t+h) - y(t)}{h} \quad \text{and} \quad \frac{z(t+h) - z(t)}{h}$$

to obtain the approximation scheme

$$\begin{aligned} y_{k+1} &= y_k + hF(y_k, z_k), \\ z_{k+1} &= z_k + hG(y_k, z_k), \quad [k = 0, \ldots, N-1]. \end{aligned} \tag{6.27}$$

As noted in Section 4.5, it can be shown via Taylor approximations that the truncation error of the Euler method at each step is no greater than a constant multiple of h^2, and that the accumulated truncation error is no more than a constant multiple of h.

The *modified Euler method* is given by

$$\begin{aligned} y_{k+2} &= y_k + 2h\,F(y_{k+1}, z_{k+1}), \\ z_{k+2} &= z_k + 2h\,G(y_{k+1}, z_{k+1}), \quad [k = 0, \ldots, N-1]. \end{aligned} \tag{6.28}$$

As in the first-order case, it requires the use of some starting method to give y_1 and z_1. It can be shown that the truncation error of the modified Euler method in each step is no more than a constant multiple of h^3, and that the accumulated truncation error is no more than a constant multiple of h^2. This represents a considerable improvement over the Euler method.

The *Runge-Kutta method* is still more accurate, having a truncation error in each step of no more than a constant multiple of h^5 and an accumulated truncation error of no more than a constant multiple of h^4. It is given by

$$\begin{aligned} y_{k+1} &= y_k + \tfrac{h}{6}[K_1 + 2K_2 + 2K_3 + K_4], \\ z_{k+1} &= z_k + \tfrac{h}{6}[L_1 + 2L_2 + 2L_3 + L_4], \end{aligned} \tag{6.29}$$

where

$$\begin{aligned} K_1 &= F(y_k, z_k), & L_1 &= G(y_k, z_k), \\ K_2 &= F(y_k + \frac{hK_1}{2}, z_k + \frac{hL_1}{2}), & L_2 &= G(y_k + \frac{hK_1}{2}, z_k + \frac{hL_1}{2}), \\ K_3 &= F(y_k + \frac{hK_2}{2}, z_k + \frac{hL_2}{2}), & L_3 &= G(y_k + \frac{hK_2}{2}, z_k + \frac{hL_2}{2}), \\ K_4 &= F(y_k + hK_3, z_k + hL_3), & L_4 &= G(y_k + hK_3, z_k + hL_3). \end{aligned}$$

The warnings in Section 4.5 about the proper choice of step size h are also relevant here. Some commercial software uses a variable step size to minimize problems.

6.3.1 Example: A two-sex model

The population models studied up to this point have ignored the fact that humans and most other animal species reproduce sexually, requiring both males and females. Ignoring gender and treating the population as a single group is a modeling assumption made to simplify aspects of the biological system which do not directly address the research question under investigation. However, there are certainly many questions which do require separating a population into males and females. We shall now develop a reasonably simple two-sex model, and quickly see that even simple systems may be too complex to complete qualitative analysis, in which case numerical analysis can provide helpful insights.

Let us denote the numbers of females and males in a population by $F(t)$ and $M(t)$, respectively. If we assume (i) that the total birth rate is a function $r(F, M)$ of F and M, (ii) that the proportions of those births which are female or male, denoted f and m (so that $f + m = 1$), are independent of population size, and (iii) that per capita death rates μ_f and μ_m may differ by gender but are also independent of population density, then we can write the system

$$F' = f\,r(F,M) - \mu_f\,F, \quad M' = m\,r(F,M) - \mu_m\,M. \tag{6.30}$$

One idea for describing the reproduction rate is to consider it a contact rate between females and males, and to base it on the law of mass action, $r(F,M) = cFM$ for some contact rate c. However, such a model turns out (see Exercise 5) to have solutions (for any parameter values) which become infinite in finite time. The problem is that the growth rates for such a model are not only unconstrained by resource limitations, but increase too quickly even for small populations: this choice of r makes growth increase quadratically, rather than linearly, in population size (recall from Chapter 3 that growth rates linear in population size, leading to exponential population growth over time, do a good job of describing observed initial population growth). Therefore, a more realistic reproduction rate should increase roughly linearly in $N = F + M$.

A ratio-dependent approach can provide a roughly linear increase in reproduction. If we suppose females' maximum reproduction rate is c_f while males' is c_m, the function

$$r(F,M) = \frac{(c_f F)(c_m M)}{(c_f F) + (c_m M)}$$

gives a rate which is bounded above by both $c_f F$ and $c_m M$: that is,

$$r = (c_f F)\,\frac{c_m M}{c_f F + c_m M} < (c_f F) \text{ and } r = (c_m M)\,\frac{c_f F}{c_f F + c_m M} < (c_m M).$$

This choice of r does still allow unbounded growth, like a simple linear ODE, but the population does not become infinite in finite time. Substituting this function into (6.30) and solving the equilibrium conditions yields only the extinction equilibrium (0,0). However, the usual qualitative analysis techniques fail to provide conditions for its stability, because the Jacobian matrix is indeterminate at (0,0):

$$J = \begin{bmatrix} c_f f(1-x)^2 - \mu_f & c_m f x^2 \\ c_f m(1-x)^2 & c_m m x^2 - \mu_m \end{bmatrix}, \text{ where } x = \frac{c_f F^*}{c_f F^* + c_m M^*},$$

but the quantity x can take on any value from 0 to 1 as $(F,M) \to (0,0)$, depending upon the path taken to approach the origin (as students of multivariable calculus will recall). [The equilibrium stability theorem of Chapter 5 does not apply here since the growth rates are not differentiable at (0,0).]

Therefore, in order to explore what determines the long-term behavior of this system, we turn to numerical analysis. Computer systems such as Mathematica, Maple, and MATLAB can perform symbolic calculations in some cases (such as computing the Jacobian matrix above), but here we illustrate their ability to compute and graph numerical solutions (approximations) for systems. Since exploration is inherently *ad hoc* rather than formulaic, we here present a sample of commands and results representative of the problem solving process. Using Mathematica syntax, we may first define the equations and their numerical solution, as well as a specialized command to plot them (both variables $F(t)$ and $M(t)$ on the same graph, with $F(t)$ in thick light gray and $M(t)$ dashed to distinguish them) as follows, with input commands given in **bold**):

```
In[1]:= twosexeqns :=
   {F'[t] == f*cf*cm*M[t]*F[t]/(cm*M[t]+cf*F[t]) - muf*F[t],
    M'[t] == m*cm*cf*M[t]*F[t]/(cm*M[t]+cf*F[t]) - mum*M[t],
    F[0] == F0, M[0] == M0}
```

```
In[2]:= twosexsoln := NDSolve[twosexeqns, {F[t],M[t]}, {t,0,tf}][[1]]
```

```
In[3]:= PopPlot[x_] := Plot[{F[t]/.x, M[t]/.x}, {t,0,10}, AxesLabel→
   {"t", "P(t)"}, PlotStyle→{{Thickness[0.02], LightGray},
   {Thickness[0.01], Dashing[0.1], Black}}]
```

We must now set parameter values (here arbitrarily setting all parameters to 1 for simplicity, except that since $f + m = 1$ by definition, we set them both equal to 1/2):

```
In[4]:= cf=1; cm=1; f=1/2; m=1/2; muf=1; mum=1; K=1;
```

...as well as the initial conditions and final time:

```
In[5]:= F0 = 1; M0 = 1; tf = 100;
```

We now request a solution using the given values, and plot it:

```
In[6]:= set1 = twosexsoln
Out[6]= {F[t] → InterpolatingFunction[{{0.,100.}},<>][t],
         M[t] → InterpolatingFunction[{{0.,100.}},<>][t]}
```

```
In[7]:= plot1 = PopPlot[set1]
```

Out[7]=
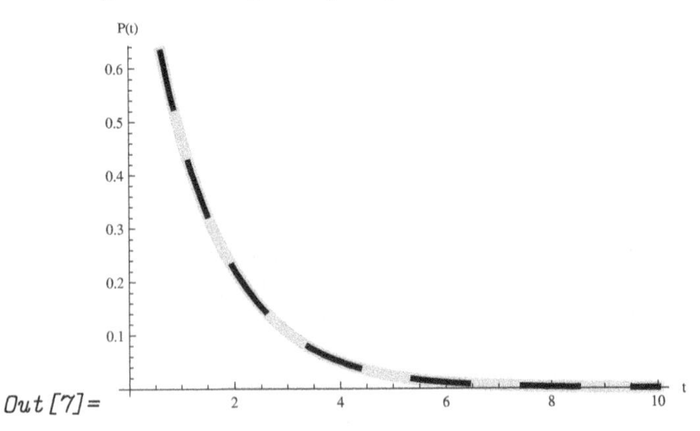

It should not be surprising that $F(t)$ and $M(t)$ are equal (their graphs superimposed) since we have given symmetric parameter values. However, we see that for the given values, both populations die out quickly. Perhaps larger initial population sizes would make a difference?

```
In[8]:= F0=10; M0=10;
```

```
In[9]:= set2=twosexsoln;
```

```
In[10]:= plot2 = PopPlot[set2]
```

Out[10]=

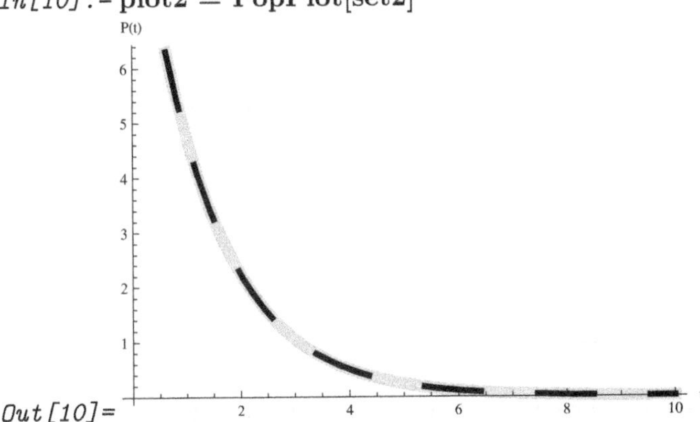

This graph is identical to the previous one except for scale (the numbers are 10 times as great). So initial conditions may not matter. A partial phase portrait may make this clearer, showing that trajectories from the two different sets of initial conditions lead to the same equilibrium: the origin.

```
In[11]:= pp1=ParametricPlot[{F[t]/.set1, M[t]/.set1}, {t,0,tf},
    AxesLabel→{"F", "M"}, PlotStyle→Thickness[0.01], PlotRange→All];
```

```
In[12]:= pp2=ParametricPlot[{F[t]/.set2, M[t]/.set2}, {t,0,tf},
    AxesLabel→{"F", "M"}, PlotStyle→Dashing[0.05]];
```

```
In[13]:= Show[pp1,pp2]
```

Out[13]=

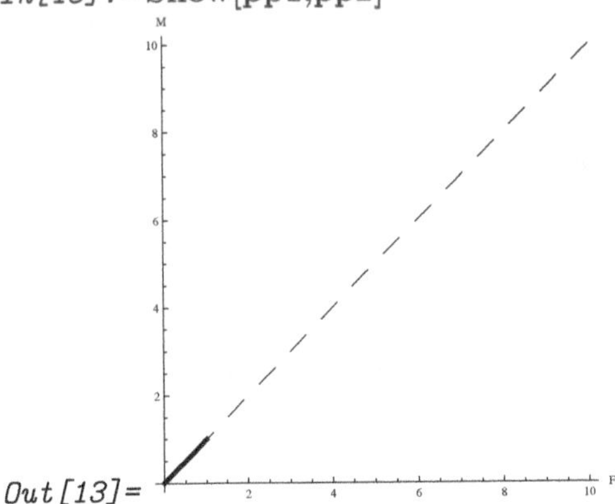

(The trajectories are superimposed since they both fall on the line $F = M$.)

Now what if we increase the reproduction rates, say by an order of magnitude?

```
In[14]:= cf=10; cm=10;
```

In[15]:= **set3=twosexsoln;**

In[16]:= **plot3 = PopPlot[set3]**

P(t)
6×10^7
5×10^7
4×10^7
3×10^7
2×10^7
1×10^7

Out[16]= 2 4 6 8 10 t

The outcome has now changed qualitatively: the population grows exponentially instead of dying out exponentially. What if we now reduce one of the reproduction rates, creating an asymmetry?

In[17]:= **cf=3;**

In[18]:= **set4=twosexsoln;**

In[19]:= **plot4 = PopPlot[set4]**

Out[19]=

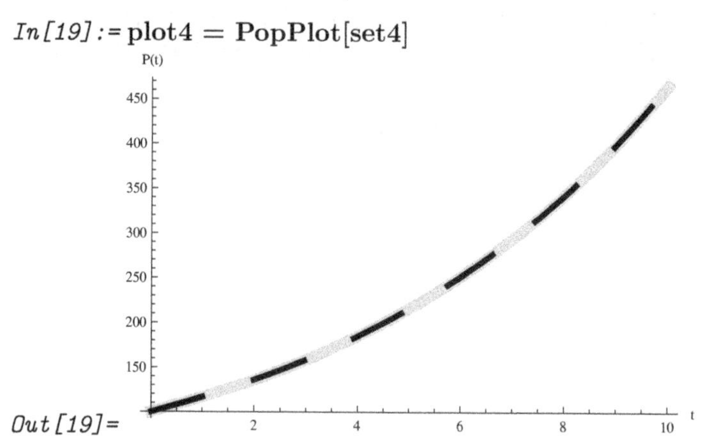

The overall growth rate is certainly slowed, but the population still grows, and both populations remain matched. What if we force an asymmetry in the population values by changing one of the initial conditions?

In[20]:= **F0=50;**

In[21]:= **set5=twosexsoln;**

In[22]:= **plot5 = PopPlot[set5]**

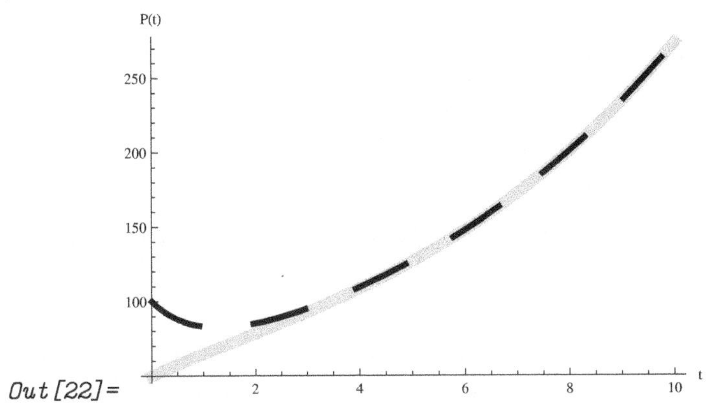

Despite the initial disparity in sizes, the two sexes' populations quickly return to even before growing, because all new births are evenly divided between females and males.

Here we have used arbitrary values rather than realistic estimates because the goal is to uncover qualitative insights about the model's behavior. The original question of the precise threshold condition between the two possible behaviors (growth and decay) has not yet been answered precisely, although a few data points have been established. A more methodical approach would fix all but one or two key parameters (say c_f and c_m), vary those independently over a regular grid (say each taking on values between 1 and 10), then mark on the grid which combinations led to extinction and which to growth, and finally abstract from the results an overall pattern describing the influence of these two parameters. We invite the reader to try such an experiment.

As one final note in this example, we observe that the unbounded growth of the model above can be eliminated by introducing a logistic term in each equation, that is, by multiplying each sex's growth rate by a factor of $\left(1 - \frac{F+M}{K}\right)$ for some carrying capacity K. While the resulting model is more complicated to analyze (and still has the problem of not being linearizable at the origin), even a partial analysis (see Exercise 6) sheds some light on the threshold between survival and extinction for both that model and the one explored above.

Exercises

Use a computer and available software to approximate the solution of each of the following initial value problems on the given interval.

1. The two-species [moth] competition model $y' = y(100 - 4y - z)$, $z' = z(60 - y - 2z)$ of Example 1, Section 6.2, with symmetric initial conditions $y(0) = 10$, $z(0) = 10$, over the time range $0 \le t \le 10$.

2. The [pond] predator-prey system $y' = y(1 - \frac{y}{30}) - \frac{yz}{y+10}$, $z' = z(\frac{y}{y+10} - \frac{3}{5})$

of Example 4, Section 6.2, with initial conditions $y(0) = 40$, $z(0) = 10$, over the time range $0 \leq t \leq 10$.

3. The [pond] predator-prey system $y' = y(1 - \frac{y}{30}) - \frac{yz}{y+10}$, $z' = z(\frac{y}{y+10} - \frac{1}{3})$ of Example 5, Section 6.2, with initial conditions $y(0) = 40$, $z(0) = 10$, over the time range $0 \leq t \leq 10$.

4. The SIR disease model $S' = -\frac{1}{200}SI$, $I' = \frac{1}{200}SI - \frac{1}{6}I$ of Section 6.1, with initial conditions $S(0) = 1190$, $I(0) = 10$, over the time range $0 \leq t \leq 20$.

In each case, compare the numerical approximation with the qualitative insights obtained in Sections 6.1 and 6.2 where these systems are studied as examples.

5. Analyze the [discarded] mass-action two-sex model described in the text, substituting $r(F, M) = cFM$ into (6.30), as follows:

   (a) Identify all equilibria for the system.

   (b) Analyze each equilibrium's local stability using the Jacobian matrix.

   (c) Draw a phase portrait, using a computer and available software as necessary. Include some sample solutions, showing that some lead to extinction while others lead to unbounded growth.

6. Analyze the logistic two-sex model

$$\begin{aligned} F' &= f\,\frac{(c_f F)(c_m M)}{(c_f F) + (c_m M)}\left(1 - \frac{F+M}{K}\right) - \mu_f F, \\ M' &= m\,\frac{(c_f F)(c_m M)}{(c_f F) + (c_m M)}\left(1 - \frac{F+M}{K}\right) - \mu_m M \end{aligned}$$

   (with $f + m = 1$), as follows:

   (a) Identify analytically all equilibria of the system. Rewrite all nonzero expressions in terms of the dimensionless reproductive quantities $R_f = c_f f/\mu_f$, $R_m = c_m m/\mu_m$ and the ratio c_f/c_m.

   (b) Since the Jacobian matrix is indeterminate at the origin, use the condition that the nonzero equilibrium be positive to conjecture stability conditions for the equilibria.

   (c) Explore these stability conditions by graphing (with the aid of a computer and appropriate software) numerical solutions for several sets of parameter values that obey each condition.

Chapter 7

Systems with Sustained Oscillations and Singularities

In this chapter we discuss some processes in which oscillations occur naturally as well as some situations in which singularities and multiple time scales arise. These lead to some new methods of analysis more advanced than those of previous chapters and some new biological fields of study.

7.1 Oscillations in neural activity

Neurons are cells in the body which transmit information to the brain and the body by amplifying an incoming stimulus (electrical charge input) and transmitting it to neighboring neurons and then turning off to be ready for the next stimulus. Neurons have fast and slow mechanisms to open ion channels in response to electrical charges. The key quantities are the concentration of sodium ions and potassium ions (both positively charged). A resting neuron has an excess of potassium and a deficit of sodium, and a negative resting potential (an excess of negative ions). Neurons use exchanges of sodium and potassium ions across the cell membrane to amplify and transmit information. There are *voltage-gated channels* for each kind of ion which open and close in response to voltage differences and which are closed in a resting neuron. When a burst of positive charge enters the cell, making the potential less negative, the voltage gated sodium channels open. Since there is an excess of sodium outside the cell, more sodium ions enter, increasing the potential until it eventually becomes positive. Next, a slow mechanism acts to block the voltage-gated sodium channels and another slow mechanism begins to open voltage-gated potassium channels. Both of these diminish the buildup of positive charge by blocking sodium ions from entering and by allowing excess potassium ions to leave. When the potential decreases to or below the resting potential, these slow mechanisms turn off, and then the process can start over. If the electrical excitation reaches a sufficiently high level, called an *action potential*, the neuron fires, transmitting the excitation to other neurons.

7.1.1 The Fitzhugh-Nagumo equations

In order to describe a simple model for this neuron firing process, we denote the potential by v, scaled so that $v = 0$ is the resting potential. We let $v = a$ be the potential above which the neuron fires and $v = 1$ the potential with sodium channels open ($0 < a < 1$). A model of the form

$$v' = -v(v-a)(v-1),$$

which has[1] asymptotically stable equilibria at 0 and 1 and an unstable equilibrium at a, would explain part of the observed behavior. If the initial potential is above a, the potential increases to 1, and if the initial potential is below a, the potential decreases to 0. Thus the model allows the signal amplification of the neuron, but stops at $v = 1$.

We must also build in a blocking mechanism, to lower the voltage by switching the ion channels. Let w denote the strength of this blocking mechanism, with $w = 0$ (turned off) when $v = 0$. As $v \to 1$ the mechanism gets stronger but remains bounded, and we assume an equation of the form

$$w' = \epsilon(v - \gamma w),$$

with a limiting value for w of $\frac{v}{\gamma}$ if v is fixed. If $v = 0$, then $w \to 0$, and if $v = 1$, then $w \to \frac{1}{\gamma}$ (the maximum strength of the blocking mechanism). The scaling parameter ϵ does not affect w's equilibrium value but does influence its rate of approach to equilibrium. A small value of ϵ indicates a slow-acting mechanism (relative to the potential v).

Finally, to incorporate the effect of the blocking mechanism on v, we add a term $-w$ to the rate of change of v. The model we shall examine, known as the *Fitzhugh-Nagumo system*,[2] is the two-dimensional system

$$\begin{aligned} v' &= -v(v-a)(v-1) - w, \\ w' &= \epsilon(v - \gamma w). \end{aligned} \tag{7.1}$$

Alternatively, to develop this model in electrical terms, we may think of the cell membrane as consisting of three components: a capacitor representing the membrane capacity, a nonlinear current voltage device for the fast current, and a resistor, inductor and battery in series for the recovery current. Using Kirchhoff's laws, we may write equations for this membrane circuit, namely

$$\begin{aligned} C_m \frac{dV}{d\tau} &= -F(V) - i, \\ L\frac{di}{d\tau} &= -Ri + V - V_0. \end{aligned} \tag{7.2}$$

[1] We leave the straightforward analysis of this solo equation as an exercise for the reader; see Exercise 12.

[2] R. Fitzhugh, Impulses and physiological states in theoretical models of nerve membrane, *Biophy. J.* 1 (1961), 445–466; J. S. Nagumo, S. Arimoto, and S. Yoshizawa, An active pulse transmission line simulating nerve axon, *Proc. Inst. Radio Engineers* 50 (1962), 2061–2071.

Here, i is the current through the resistor and inductance, $V = V_i - V_e$ is the membrane potential, and V_0 is the potential gain across the battery. The variable τ is used to represent time, because we will presently want to use t as a dimensionless time variable. The first equation gives the fast current across the membrane; the second gives the voltage drops across the inductor and across the resistor and battery (completing the slow current series circuit).

The function $F(V)$ which describes the current through the voltage device is assumed to have three zeros, with the smallest, $V = V_0$, and the largest, $V = V_0 + V_1$, being asymptotically stable equilibria of the differential equation $dV/d\tau = -F(V)$. We take R_1 to be the "passive" resistance of the nonlinear current voltage device, $R_1 = 1/F'(V_0)$.

To simplify the model, we reformulate the system (7.2) in dimensionless variables, by letting

$$v = \frac{V - V_0}{V_1}, \quad w = \frac{R_1 i}{V_1}, \quad f(v) = \frac{R_1 F(V_1 v)}{V_1}, \quad t = \frac{L\tau}{R_1}.$$

This transforms the system (7.2) to

$$\begin{aligned} \epsilon \frac{dv}{dt} &= -f(v) - w, \\ \frac{dw}{dt} &= v - \gamma w \end{aligned} \tag{7.3}$$

with

$$\epsilon = \frac{R_1^2 C_m}{L}, \quad v_o = \frac{V_0}{V_1}, \quad \gamma = \frac{R}{R_1}.$$

The simplest choice for the function $f(v)$ with three zeros is the cubic polynomial

$$f(v) = v(v-1)(v-a)$$

and this gives the Fitzhugh-Nagumo model.

The Fitzhugh-Nagumo model is a simplification of the four-dimensional *Hodgkin-Huxley model* proposed in the early 1950's by Sir Alan Hodgkin and Sir Andrew Huxley,[3] for which they received the Nobel Prize in Physiology and Medicine in 1963 and which is still used now for studying neurons and other kinds of cells. The model describes an excitable system, and these occur in a large variety of biological systems.

If γ is not too large, the only equilibrium of the system (7.1) is $v = 0$, $w = 0$, and the matrix of the linearization at this equilibrium is

$$\begin{bmatrix} -a & -1 \\ \epsilon & -\epsilon\gamma \end{bmatrix};$$

as this matrix has negative trace and positive determinant, the equilibrium

[3] A.L. Hodgkin and A.F. Huxley, A quantitative description of membrane current and its application to conduction and excitation in nerve, *J. Physiology* 117 (1952), 500–544.

(0,0) is asymptotically stable. However, it is interesting to see *how* a solution which starts at a point $(v_0,0)$, $v_0 > a$, approaches the equilibrium. We may do so by examining the nullclines, but it is perhaps more convincing, if less rigorous, to use a computer algebra system to draw this orbit. Figure 7.1 shows this orbit, traversed counterclockwise, with $a = 0.3$, $\gamma = 1$, $\epsilon = 0.01$, $v_0 = 0.4$.

Thus v amplifies quickly and then returns to zero. An unexpected property of this orbit is that after the neuron fires and the potential drops, it overshoots zero. This property can be derived from the nullcline analysis, and is also observed in practice.

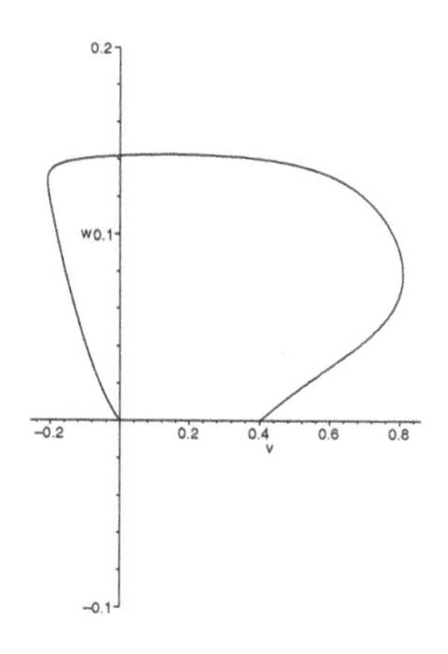

FIGURE 7.1: One orbit of system (7.1).

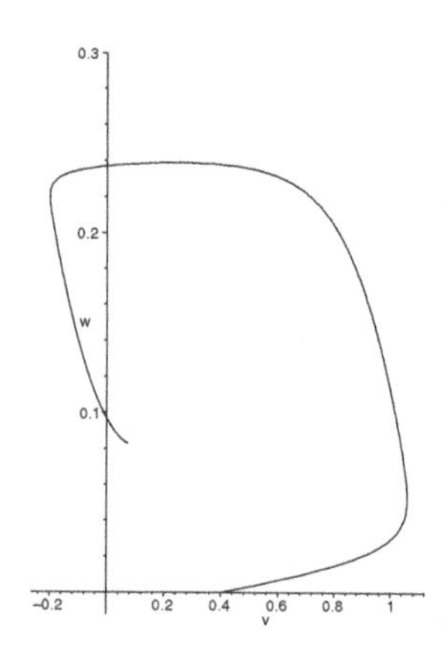

FIGURE 7.2: An orbit of system (7.4) with stable equilibrium.

Another experiment gives the cell a constant input of positive ions instead of a single pulse. If we apply a constant current J, we add J to the rate of change of potential (assuming that 1 unit of current raises the potential by 1 unit in unit time), so that the modification of (7.1) is

$$\begin{aligned} v' &= -v(v-a)(v-1) - w + J, \\ w' &= \epsilon(v - \gamma w). \end{aligned} \tag{7.4}$$

This moves the equilibrium (0,0) into the first quadrant. For small values of J this equilibrium is asymptotically stable (Figure 7.2). For larger inputs it becomes unstable and a periodic orbit is set up (Figure 7.3).

Thus a steady input leads to a periodic solution. Examination of v as a function of t (Figure 7.4) shows a "bursting" behavior, similar to what is observed in real neurons, with the potential rising close to 1 and then dropping below zero.

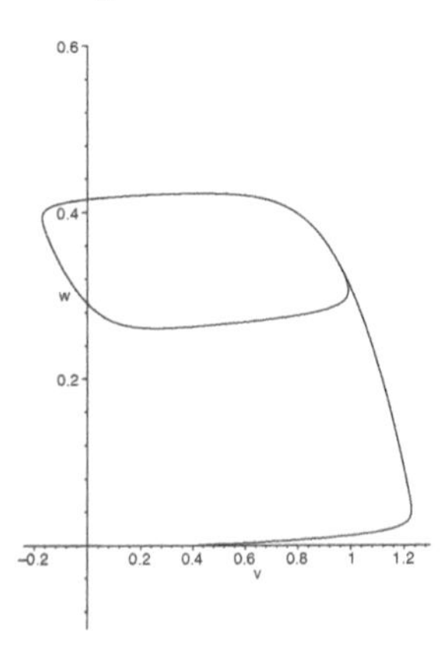

FIGURE 7.3: An orbit of system (7.4) with stable limit cycle.

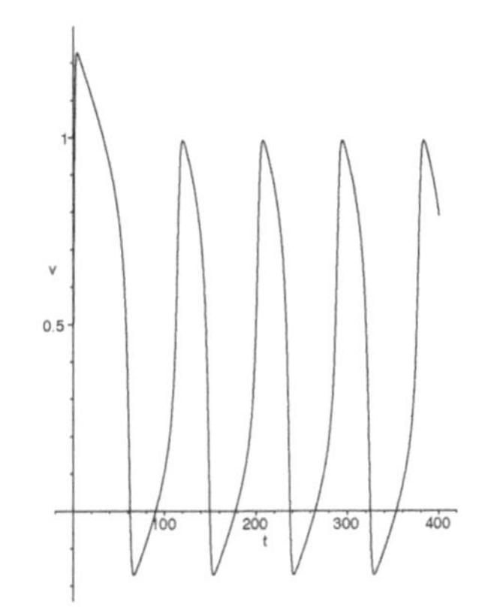

FIGURE 7.4: Periodic "bursting" behavior of system (7.4) for large J.

7.1.2 A model for cat neurons

A model by McCarley and Hobson[4] seeks to explain regular oscillations in cat neuron activity associated with the sleep cycle (in particular the deepest part of sleep, called REM for rapid eye movement), measured in number of electrical discharges per second. The model assumes the sleep cycle can be described primarily in terms of two cell groups, aminergic and cholinergic cells, with activity levels (signal strengths) $x(t)$ and $y(t)$. Aminergic neurons fire steadily during waking, inhibiting the activity of REM-activating cholinergic cells, but during sleep the aminergic neurons decrease their firing, and the cholinergic neurons activate. This causes REM episodes but also begins to excite the aminergic cells, until eventually they resume firing, once more inhibiting the cholinergic cells and ending the REM period. During sleep this cycle repeats several times (cf. Figure 7.5).

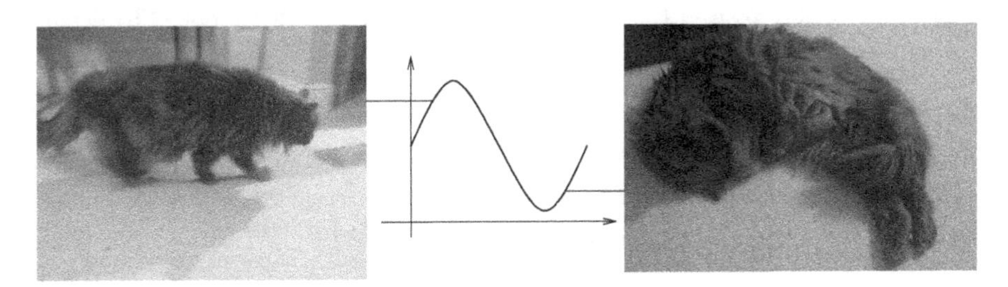

FIGURE 7.5: Aminergic cell activity is high while cats are awake but low during REM sleep.

[4]R.W. McCarley and J.A. Hobson, Neuronal excitability modulation over the sleep cycle: A structured and mathematical model, *Science* 189: 58–60 (1975).

The rate of change of neuron activity can thus be described as proportional to the current level of activity, with nonlinear interaction between cell groups. The model suggested is

$$\begin{aligned} x' &= ax - bxy, \\ y' &= -cy + dxy, \end{aligned} \tag{7.5}$$

where a, b, c, d are positive constants. The equilibrium conditions are

$$x(a - by) = 0, \quad y(dx - c) = 0.$$

These imply that there are two equilibria: $(0, 0)$ and $(c/d, a/b)$. The matrix of the linearization at an equilibrium (x, y) is

$$\begin{bmatrix} a - by & -bx \\ dy & dx - c \end{bmatrix}.$$

At the equilibrium $(0, 0)$ this matrix is

$$\begin{bmatrix} a & 0 \\ 0 & -c \end{bmatrix},$$

and it is clear that since the eigenvalues of this matrix are $a, -c$, this equilibrium is unstable.

At the equilibrium $(c/d, a/b)$ the matrix of the linearization is

$$\begin{bmatrix} 0 & -bc/d \\ ad/b & 0 \end{bmatrix}.$$

Since the eigenvalues of this matrix are complex conjugates with real part zero, $\pm i\sqrt{ac}$, the linearization at this equilibrium has periodic solutions. In fact, the model (7.5) is the same as the model (5.1) for the Lotka-Volterra system studied in Section 5.1, where it is shown that all solutions are given by periodic orbits around the equilibrium. Thus the simple model (7.5) predicts regular oscillations as observed.

Exercises

In each of Exercises 1–9, use a computer algebra system to display the behavior of solutions of the Fitzhugh-Nagumo system (7.4) with the given values of the parameters.

1. $a = 0.3,\ v_0 = 0.2,\ \gamma = 1,\ \epsilon = 0.01,\ J = 0$
2. $a = 0.3,\ v_0 = 0.4,\ \gamma = 1,\ \epsilon = 0.01,\ J = 0$
3. $a = 0.3,\ v_0 = 0.4,\ \gamma = 1,\ \epsilon = 1,\ J = 0$
4. $a = 0.3,\ v_0 = 0.2,\ \gamma = 1,\ \epsilon = 0.01,\ J = 0.3$

5. $a = 0.3$, $v_0 = 0.2$, $\gamma = 1$, $\epsilon = 1$, $J = 0.3$

6. $a = 0.3$, $v_0 = 0.4$, $\gamma = 1$, $\epsilon = 1$, $J = 0.3$

7. $a = 0.3$, $v_0 = 0.2$, $\gamma = 10$, $\epsilon = 0.01$, $J = 0$

8. $a = 0.3$, $v_0 = 0.2$, $\gamma = 10$, $\epsilon = 0.01$, $J = 0.3$

9. $a = 0.3$, $v_0 = 0.4$, $\gamma = 10$, $\epsilon = 1$, $J = 0.3$

10. Summarize the results of the above exercises: which cases showed periodic "bursting" behavior? Which parameter(s) appear to be most important in determining the qualitative behavior of the system?

11. Show that the Fitzhugh-Nagumo system (7.1) has positive equilibria if and only if $\gamma(1-a)^2 \geq 4$. Use a computer algebra system to display some solutions of the systems (7.1) and (7.4) with $a = 0.3, \gamma = 20, \epsilon = 0.01$ and $J = 0, J = 0.1, J = 0.3$.

12. (a) Use a phase line or linearization to analyze the behavior of the solo equation $v' = -v(v-a)(v-1)$, which represents neuron voltage in the absence of a blocking mechanism.

    *(b) Analyze the behavior of the following model, in which the blocking mechanism w exists but fails to act on the voltage: $v' = -v(v-a)(v-1)$, $w' = \epsilon(v - \gamma w)$.

7.2 Singular perturbations and enzyme kinetics

For differential equations or systems of differential equations which depend on a parameter there is a general theorem to the effect that solutions are continuous functions of the parameter on any finite interval. However, if a derivative is multiplied by a parameter which may be allowed to tend to zero, this is not necessarily true. Such a situation is called a *singular perturbation.*

The mathematical analysis of singular perturbation problems was developed long after the idea appeared in applications in the physical and biological sciences, where it manifests in phenomena which act on multiple spatial or time scales. In the analysis of the flow of a fluid with small viscosity there are boundary layer effects. These are large changes in the behavior of the flow in a small spatial region at the boundary of the region of flow, first studied in 1905. In such cases it may be helpful to study the flow on two different *spatial* scales: for instance, fluid passing through a tube may flow freely in the center of the tube but not close to the edges of the tube. Many problems in the biological sciences involve actions on very different *time* scales, and these may lead to a rapid change in some of the variables on a very short initial time interval while other variables act more slowly (in fact, we described this kind of behavior in the last section, in developing the Fitzhugh-Nagumo model). One early example was the study of enzyme kinetics, where the chemical reactions which occur take place at vastly different rates. In this section we shall study a typical enzyme kinetics problem, and revisit models for neuron bursting, through the lens of singular perturbations, which allows us to isolate the two (or more) scales on which the process under study operates and deal with them one at a time.

Singular perturbation problems arise in models (systems of differential equations) containing a small parameter ϵ, of the form

$$\begin{aligned} \epsilon y' &= f(y, z, \epsilon), \quad y(0) = y_0 \\ z' &= g(y, z, \epsilon), \quad z(0) = z_0 \end{aligned} \tag{7.6}$$

with solution $(y(t, \epsilon), z(t, \epsilon))$. There is a corresponding reduced system obtained by setting $\epsilon = 0$,

$$\begin{aligned} f(y, z, 0) &= 0 \\ z' &= g(y, z, 0), \quad z(0) = z_0 \end{aligned} \tag{7.7}$$

with solution $(y_0(t), z_0(t))$.

Since ϵ is assumed to be small, the form (7.6) suggests that the y reaction time is much faster than the z reaction time. Thus y goes to its equilibrium value rapidly, and at its equilibrium $f(y, z, 0) = 0$. Then we might expect that the reduced problem (7.7) is a good approximation to the full problem (7.6) *after* a short initial time interval near $t = 0$ during which y moves to its equilibrium value. Because (7.7) is a first-order differential equation (which requires one initial condition to identify a unique solution) and (7.6) is a two dimensional system (requiring two initial conditions for a unique solution) we must expect to lose an initial condition in the reduction, and this suggests that the solutions of (7.7) and (7.6) (each derived on a different time scale) may not agree close to $t = 0$.

If the partial derivative $f_y(y, z, 0) \neq 0$ we may solve the equation $f(y, z, 0) = 0$ for y as a function of z, $y = \phi(z)$. Thus the reduced system (7.7) is equivalent to the first-order initial value problem

$$z' = g(\phi(z), z, 0), \quad z(0) = z_0. \tag{7.8}$$

Then we have the solution of (7.7) with $z_0(t)$ the solution of (7.8) and $y_0(t) = \phi(z_0(t))$. Then $y_0(0) = \phi(z_0)$. If $y_0 \neq \phi(z_0)$ it is not possible for the solution of the reduced problem (7.7) to satisfy the two initial conditions of the full problem (7.6). The solution $(y_0(t), z_0(t))$ of the reduced problem is called the *outer solution.* In order to use the solution of the reduced problem (7.7) as an approximation to the solution of the full problem (7.6), we would need a result to the effect that for each t away from $t = 0$ the solution of the reduced problem (7.7) is the limit as $\epsilon \to 0$ of the solution of the full problem (7.6). There is such a result, and we will state it shortly.

In applications one often makes a *quasi-steady-state hypothesis*, that y remains almost constant, so that $y' \approx 0$. This hypothesis is expressed as $f(y, z, 0) = 0$; in singular perturbation language the hypothesis is just that the full problem is approximated by the reduced problem.

Of course, because $y(t, \epsilon)$ (the solution to the full problem) and $y_0(t)$ (the solution to the reduced problem) do not match at $t = 0$, we should expect that $y(t, \epsilon)$ changes rapidly for t close to 0. To analyze this, we change the time scale by making the change of independent variable $t = \epsilon s$ (making s "fast" time, relative to "slow" time t). Then for any function u we have, by the Chain Rule of calculus,

$$\frac{du}{dt} = \frac{du}{ds} \cdot \frac{ds}{dt} = \frac{1}{\epsilon} \cdot \frac{du}{ds}$$

and the system (7.6) is transformed to

$$\begin{aligned} \frac{dy}{ds} &= f(y, z, \epsilon), \quad y(0) = y_0; \\ \frac{dz}{ds} &= \epsilon g(y, z, \epsilon), \quad z(0) = z_0. \end{aligned} \tag{7.9}$$

A very small t-interval, called a *boundary layer*, corresponds to a much longer s-interval. Since ϵ is small, the second equation of (7.9) says that the second variable z remains almost constant initially and may be replaced by z_0. This suggests solving the initial value problem

$$\frac{dy}{ds} = f(y, z_0, 0), \quad y(0) = y_0, \tag{7.10}$$

called the boundary layer system, to give an approximation, called the *inner solution* to the behavior of the system near $t = 0$. Away from $t = 0$ we hope that the solution of the full problem is approximated well by the outer solution.

Sometimes a problem is given in the form (7.9) from the start. The underlying idea in a singular perturbation problem is that there are two different time scales inherent in the problem, and this makes it possible to analyze the problem separately on each time scale. The reduction in dimension because of this separation simplifies the analysis.

The mathematical treatment of singular perturbations began in the 1940's from the perspective of asymptotic expansions. A few years later the qualitative result which justifies the use of the reduced system as an approximation to the full system was obtained independently in the U.S.A. and the Soviet Union:[5]

THEOREM (Levinson-Tihonov): Suppose that

1. f, g are smooth functions,
2. the equation $f(y, z, 0) = 0$ can be solved for y as a smooth function of z, $y = \phi(z)$,
3. the reduced system (7.7) has a solution on an interval $0 \leq t \leq T$,
4. the boundary layer system (7.10) has an asymptotically stable equilibrium.

Then $y(t, \epsilon) \to y_0(t)$, $z(t, \epsilon) \to z_0(t)$ as $\epsilon \to 0+$ for $0 < t \leq T$. The convergence of y is non-uniform at $t = 0$.

There is an extension of this result to infinite time intervals.[6]

THEOREM: Suppose, in addition to the hypotheses of the Levinson-Tihonov theorem, that the reduced system (7.7) has a solution which is asymptotically stable and that the boundary layer system (7.10) has a solution which is asymptotically stable uniformly in z_0. Then the convergence is uniform on closed subsets of $0 < t < \infty$.

The essential content of these results is that if ϵ is sufficiently small the solution of the reduced system is a good approximation to the solution of the singularly perturbed system except very close to $t = 0$. The relation

[5] N. Levinson, Perturbations of discontinuous solutions of nonlinear systems of differential equations, *Acta Math.* 82 (1950), 71–106; A.N. Tihonov, On the dependence of the solutions of differential equations on a small parameter, *Mat. Sbornik NS* 22 (1948), 193–204.

[6] F.C. Hoppensteadt, Singular perturbations on the infinite interval, *Trans. Amer. Math. Soc.* 123 (1966), 521–535.

$f(y, z, 0) = 0$ is called the *quasi-steady-state hypothesis.* Close to $t = 0$ the solution of the boundary layer system (7.10) describes the behavior of solutions. Thus the analysis of a singular perturbation problem can be decomposed into the analysis of two simpler problems, namely the boundary layer system and the reduced problem. Curiously, in fluid dynamic applications the focus of attention has been on the boundary layer system, whereas in most biological applications the primary interest has been in the long-term behavior, that is, the reduced problem.

EXAMPLE 1.
Describe the solution of the first-order differential equation

$$\epsilon y' = -y, \quad y(0) = 1.$$

Solution: The solution of this initial value problem may be obtained easily by separation of variables and is

$$y(t, \epsilon) = e^{-t/\epsilon}.$$

We may calculate

$$\lim_{\epsilon \to 0} y(t, \epsilon) = \begin{cases} 1 & \text{for } t = 0 \\ 0 & \text{for } t > 0. \end{cases}$$

This limit is discontinuous at $t = 0$. When $\epsilon = 0$ the problem is no longer an initial value problem but is just the relation $y = 0$ together with the (incompatible) initial condition $y(0) = 1$. This indicates that the solution for $\epsilon > 0$ begins with the value 1 at $t = 0$ and then decreases rapidly to 0. A graph of the solution for a small value of ϵ would indicate this boundary layer (see Exercise 3 below).

Another way to approach this problem is to change the time scale by making the change of independent variable $t = \epsilon s$ to transform the problem to

$$\frac{dy}{ds} = -y, \quad y(0) = 1,$$

whose solution is $y(s) = e^{-s} = e^{-t/\epsilon}$.□

EXAMPLE 2.
Describe the behavior of solutions of the initial value problem

$$\begin{aligned} \epsilon y' &= -y, \quad y(0) = 1, \\ z' &= -yz, \quad z(0) = 1. \end{aligned} \tag{7.11}$$

Solution: We let $(y(t, \epsilon), z(t, \epsilon))$ be the solution of this problem and we let $(y_0(t), z_0(t))$ be the solution of the *reduced* problem

$$\begin{aligned} y &= 0, \\ z' &= -yz, \quad z(0) = 1, \end{aligned} \tag{7.12}$$

which is obtained by setting $\epsilon = 0$ in (7.11). It is easy to see that the solution of the reduced problem (7.12) is given by $y_0(t) = 0$, $z_0(t) = 1$. In order to solve the full problem (7.11) we may solve its first equation as in Example 1 to obtain $y(t,\epsilon) = e^{-t/\epsilon}$ and then substitute this solution into the second equation to obtain $z' = -e^{-t/\epsilon}z$, $z(0) = 1$. The solution of this is

$$z(t,\epsilon) = e^{-\epsilon(1-e^{-t/\epsilon})}$$

(see Exercise 1 below). Since $\lim_{\epsilon\to 0} z(t,\epsilon) = 1$ for all $t \geq 0$ (see Exercise 2 below) we see that $z(t,\epsilon) \to y_0(t)$ as $\epsilon \to 0$ for all $t \geq 0$. However, as we have seen in Example 1,

$$\lim_{\epsilon\to 0} y(t,\epsilon) = \begin{cases} 1 & \text{for } t = 0, \\ 0 & \text{for } t > 0, \end{cases}$$

and

$$\lim_{\epsilon\to 0} y(t,\epsilon) \neq y_0(t) \equiv 0$$

for $t = 0$. It is possible, just as in Example 1, to stretch the time variable by setting $t = \epsilon s$ to give the system

$$\begin{aligned} \frac{dy}{ds} &= -y, \quad y(0) = 1 \\ \frac{dz}{ds} &= -\epsilon yz, \quad z(0) = 1 \end{aligned}$$

and to solve this version of the problem (see Exercise 4 below). □

EXAMPLE 3: THE FITZHUGH-NAGUMO MODEL.
In the previous section we described the Fitzhugh-Nagumo model, and remarked that the two variables in this system operated on different time scales. However, we did not make any use of this in our description of the behavior of the model. Now we revisit this model thinking of ϵ as a small parameter and view it as a singular perturbation problem.

Solution: We recall the Fitzhugh-Nagumo model

$$\begin{aligned} \epsilon\frac{dv}{dt} &= -f(v) - w, \\ \frac{dw}{dt} &= v - \gamma w \end{aligned} \tag{7.13}$$

with

$$f(v) = v(v-1)(v-a).$$

Then v is a fast variable and w is a slow variable if ϵ is small, with the quasi-steady-state ($dv/dt = 0$) given by the cubic

$$w = -f(v) = -v(v-a)(v-1).$$

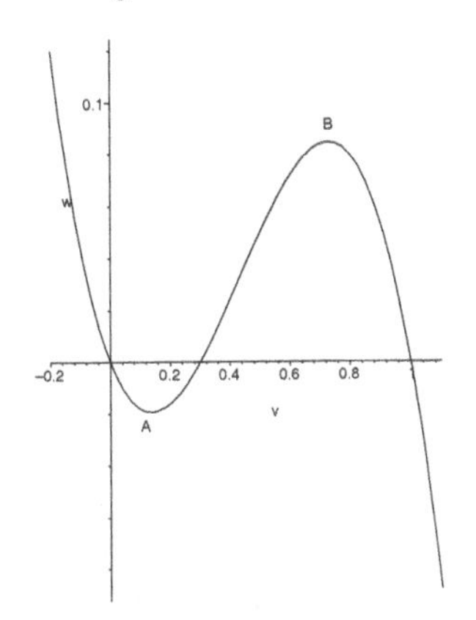

FIGURE 7.6: Quasi-steady state for the Fitzhugh-Nagumo system.

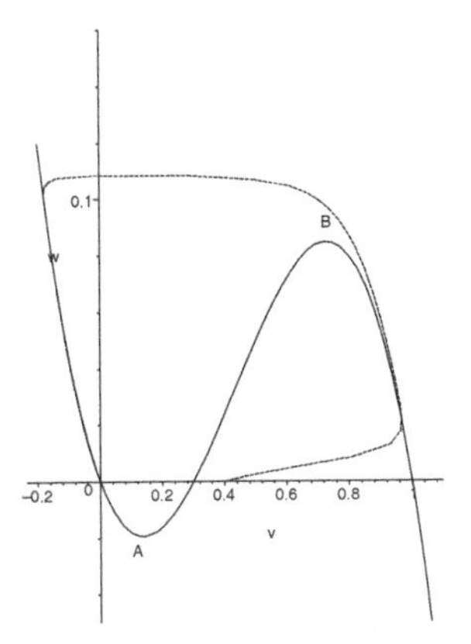

FIGURE 7.7: An orbit of the Fitzhugh-Nagumo system.

If we represent the system graphically, the quasi-steady state (QSS) is a curve in the (v, w) plane with three monotone branches which we denote

$$v = v_1(w), \quad v = v_2(w), \quad v = v_3(w),$$

seen in Figure 7.6 as the branches to the left of the point A, between the points A and B, and to the right of the point B, respectively. The fast-time dynamics in v make solutions move (horizontally) toward [the nearest stable branch of] the QSS, and then the slow-time dynamics in w send solutions up or down the QSS curve toward the w-coordinate v/γ. To see this, one can first analyze the behavior of the "fast" dynamics in v (for fixed w, see Exercise 11) and then the behavior of the "slow" dynamics in w (see Exercise 12). Analysis of the v equation shows that the branch v_2 is unstable while v_1 and v_3 are locally asymptotically stable (in the v direction), so if we take $w(0) = 0$, then $v = a$, where a is the middle intersection of the quasi-steady-state curve (the unstable v_2) with the v axis, separates solution behaviors into two distinct groups, depending on the initial value of v. If $v(0) < a$, then v goes directly to the equilibrium $(0, 0)$. If $v(0) > a$, then v goes rapidly to the right branch $v = v_3(w)$ while w remains constant. Then, in the slow time, the orbit follows the curve $v = v_3(w)$ until it comes near the point B; this is the excited phase of the motion. If γ is small enough, say $\gamma < 4.5$ (this seems arbitrary but it's not, see Exercise 13), then near B we have $dw/dt = v - \gamma w > 0$, indicating a continued upward trajectory, but at B the graph of the quasi-steady-state $w = -f(v)$ turns downward. The orbit thus cannot continue to follow the quasi-steady-state curve past B; it is then governed by a fast system

$$\epsilon v' = -f(v) - w, \quad w' = 0$$

and moves rapidly horizontally until it reaches the branch $v = v_1(w)$. It then follows this branch of the quasi-steady-state curve to the equilibrium $(0, 0)$.

This behavior is illustrated in Figure 7.7, which shows an orbit for the system (7.13) with parameter values $a = 0.3, \gamma = 1, \epsilon = 0.002$ along with the quasi-steady-state curve.

Now suppose a constant current is applied, changing the model to

$$\epsilon \frac{dv}{dt} = -f(v) - w + J, \tag{7.14}$$
$$\frac{dw}{dt} = v - \gamma w.$$

Then the effect is to move the quasi-steady-state curve upward (it is now $w = -f(v) + J$) and the equilibrium into the first quadrant (since the line $v = \gamma w$ where $dw/dt = 0$ only passes through the first and third quadrants). It is possible to show that when J is large enough for the equilibrium to reach the portion $v = v_2(w)$ of the quasi-steady-state curve (at the point labeled A) the equilibrium becomes unstable and the orbit becomes periodic, oscillating between the branches $v = v_1(w)$ and $v = v_3(w)$. □

7.2.1 Bursting

In the Fitzhugh-Nagumo model, and also in the four-dimensional Hodgkin-Huxley model, which models the behavior of a neuron somewhat more closely, neurons may fire periodically if stimulated by a constant applied current. Many cells exhibit a more complicated behavior called bursting, in which periods during which the potential changes slowly alternate with periods of rapid oscillation. Such behavior is observed, for example, in groups of pancreatic β-cells. Action potential bursting in these cells, clustered together in the pancreas in groups called the islets of Langerhans (see Figure 7.8), plays a critical role in secreting the hormone insulin, used for maintaining glucose levels in the body. (Type II diabetes, hyperglycemia, develops when this bursting behavior does not compensate correctly in response to glucose levels in blood plasma.) Here we will not explore the rather complicated models which have been formulated to attempt to explain bursting phenomena,[7] but we will sketch out a way to use multiple timescales and singular perturbations to design a model that exhibits bursting.

The basic idea is to build a model which can exhibit either oscillation or quiescence (a stable equilibrium), and then use a slow timescale to switch back and forth between the two behaviors, to replicate bursting. In order to exhibit oscillation within the fast timescale, we will need at least two dimensions there, so we suggest a three-dimensional model with two fast variables and

[7]One review of models for pancreatic β-cell bursting is Arthur Sherman and Richard Bertram, Integrative modeling of the pancreatic β-cell, in *Wiley Interscience Encyclopedia of Genetics, Genomics, Proteomics, and Bioinformatics*, Part 3 Proteomics, M. Dunn, ed., Section 3.8 Systems Biology, R. L. Winslow, ed., John Wiley and Sons, Ltd., 2005. DOI: 10.1002/047001153X.g308213.

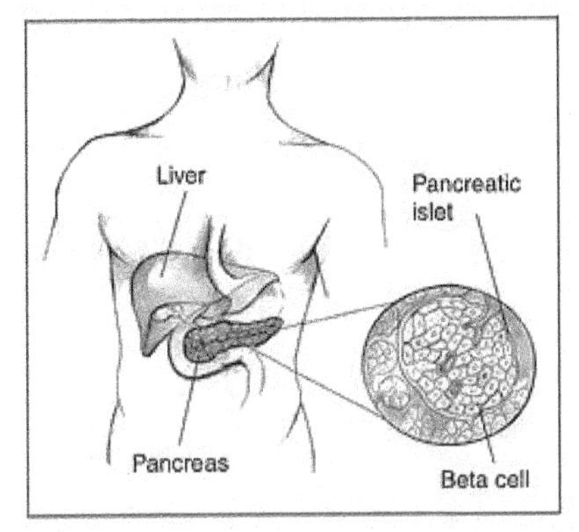

Image courtesy National Diabetes Information Clearinghouse

FIGURE 7.8: β-cells in the pancreas, clustered in islets, exhibit bursting behavior central to insulin secretion.

one slow variable, say

$$\begin{aligned} \epsilon v' &= f_1(v, w, z), \\ \epsilon w' &= f_2(v, w, z), \\ z' &= g(v, w, z). \end{aligned} \tag{7.15}$$

We would like to arrange the fast variable system to have three equilibria: an asymptotically stable node, a saddle point, and an unstable node with an asymptotically stable limit cycle around it. In this way the unstable equilibrium (saddle point) serves to separate the fast variable state space into two different regions, one where solutions approach the stable equilibrium and one where solutions approach the limit cycle; the dividing curve (or surface) that extends from the unstable equilibrium to do this is called a *separatrix*, formed by the *stable manifold* of the saddle point, separating the *domain of attraction* of the asymptotically stable node from the domain of attraction of the limit cycle. (The unstable equilibrium branch v_2 in the fast dynamics of the Fitzhugh-Nagumo example serves similarly to separate the domains of attraction for the stable branches v_1 and v_3.)

The slow variable z then follows the quasi-steady state curve

$$f_1(v, w, z) = 0, \quad f_2(v, w, z) = 0.$$

We would like to arrange for the slow variable to move solutions back and forth across the separatrix. A move into the domain of attraction of the limit cycle will trigger a rapid oscillation, and then a move back into the domain of attraction of the asymptotically stable node will lead to a quiescent period. This cycle will be repeated to give bursting.

An (admittedly artificial) example is a fast system

$$\begin{aligned} v' &= w - v^3 + 3v^2, \\ w' &= 1 - 5v^2 - w \end{aligned} \tag{7.16}$$

coupled with a slow variable z and an applied voltage J to give the system

$$\begin{aligned} v' &= w - v^3 + 3v^2 + J - z, \\ w' &= 1 - 5v^2 - w, \\ z' &= \epsilon[s(v - v_1) - z] \end{aligned} \tag{7.17}$$

where $v_1 = (\sqrt{5} - 1)/2$. In the slow dynamics, when v is large, z increases and becomes large, but then v' is decreased by the larger z, until eventually v decreases, and then so does z. With the parameter values $J = 2, \epsilon = 0.002, s = 4$ the graph of v for this system is shown in Figure 7.9.

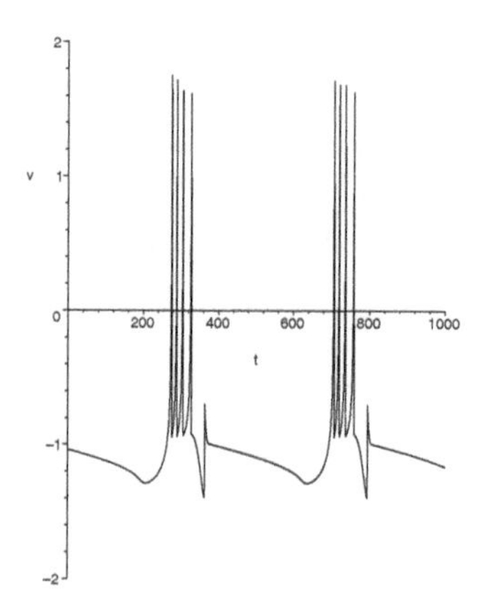

FIGURE 7.9: Bursting.

The phenomenon of bursting arises in many real cell situations, and the above example is an extremely simplified version of one of the types of model that displays bursting. Since study of the behavior of such models requires very detailed examination of the phase portrait of the fast subsystem, we will not go further into this subject.

7.2.2 An example from enzyme kinetics

Most biochemical reactions in living organisms involve proteins, called *enzymes*, which act as catalysts. Enzymes react on certain compounds, called *substrates*. For example, hemoglobin in red blood cells is an enzyme, and the oxygen with which it combines is a substrate. A large part of enzyme kinetics is concerned with the study of rates of reaction and the behavior of the various reactants.

A basic enzymatic reaction, proposed by Michaelis and Menten,[8] involves a substrate S reacting with an enzyme E to form a complex SE which is converted into a product P and the enzyme. We let s, e, c, p denote the concentrations of substrate, enzyme, complex, and product, respectively. We assume that the substrate and enzyme combine according to a mass action law with rate $k_1 se$, and that the complex decomposes into substrate and enzyme at a rate $k_{-1}c$ and converts to product at a rate $k_2 c$. This leads to the system of differential equations

$$\begin{aligned} s' &= -k_1 se + k_{-1}c, \\ e' &= -k_1 se + (k_{-1} + k_2)c, \\ c' &= k_1 se - (k_{-1} + k_2)c, \\ p' &= k_2 c, \end{aligned} \tag{7.18}$$

with initial conditions at the beginning of the process

$$s(0) = s_0, \quad e(0) = e_0, \quad c(0) = 0, \quad p(0) = 0.$$

Since $(c+e)' = 0, c+e$ is a constant e_0, and we may replace e by $e_0 - c$ in (7.18). Also, if s and c are known, then e is determined, and we may calculate p by integration. Thus the system (7.18) may be reduced to the two-dimensional system

$$\begin{aligned} s' &= -k_1(e_0 - c)s + k_{-1}c, \\ c' &= k_1(e_0 - c)s - (k_{-1} + k_2)c \end{aligned} \tag{7.19}$$

with initial conditions $s(0) = s_0, c(0) = 0$.

It is easy to verify (see Exercise 9 below) that the only equilibrium of (7.19) is $s = 0, c = 0$ and that this equilibrium is asymptotically stable. We are interested in more detailed qualitative information. At $t = 0$, since $c = 0, s = s_0$, we have $s' = -k_1 e_0 s_0 < 0, c' = k_1 e_0 s_0 > 0$. Thus initially s decreases from s_0 while c increases from zero, and c continues to increase until $c' = 0$, or $k_1(e_0 - c)s - (k_{-1} + k_2)c = 0$. Thus c increases until

$$c = \frac{k_1 e_0 s}{k_{-1} + k_2 + k_1 s}. \tag{7.20}$$

Similarly, we may see that s decreases until

$$c = \frac{k_1 e_0 s}{k_{-1} + k_1 s}. \tag{7.21}$$

Since the value of c defined by (7.21) is greater than the value defined by (7.20), s continues to decrease for all t (since c never reaches that higher threshold) while c increases from zero to a maximum and then decreases.

---

[8] L. Michaelis and M.I. Menten, Die Kinetik der Invertinwirkung, *Biochem. Z.* 49 (1913), 333–369.

The hallmark of a singular perturbation problem is that it contains differential equations with very different reaction rates. In order to identify (7.19) as a singular perturbation problem we put it into dimensionless form. Because each term in (7.19) must have the same dimension we see that k_1 has dimension 1/(cell concentration)(time) and k_{-1}, k_2 have dimension 1/(time). We let

$$y = \frac{c}{e_0}, \qquad z = \frac{s}{s_0},$$

both dimensionless, and then define the dimensionless time $\tau = k_1 e_0 t$. Then

$$\begin{aligned} \frac{dc}{dt} &= e_0 \frac{dy}{dt} &= e_0 \frac{dy}{d\tau}\frac{d\tau}{dt} &= k_1 e_0^2 \frac{dy}{d\tau} \\ \frac{ds}{dt} &= s_0 \frac{dz}{dt} &= s_0 \frac{dz}{d\tau}\frac{d\tau}{dt} &= k_1 e_0 s_0 \frac{dz}{d\tau} \end{aligned}$$

and the system (7.18) is transformed to

$$\begin{aligned} k_1 e_0^2 \frac{dy}{d\tau} &= k_1 e_0 (1-y) s_0 z - (k_{-1} + k_2) e_0 y, \\ k_1 e_0 s_0 \frac{dz}{d\tau} &= -k_1 e_0 (1-y) s_0 z + k_{-1} e_0 y. \end{aligned} \tag{7.22}$$

We now define the parameters

$$\lambda = \frac{k_2}{k_1 s_0}, \qquad K = \frac{k_{-1} + k_2}{k_1 s_0}, \qquad \epsilon = \frac{e_0}{s_0}$$

so that $K - \lambda > 0$. Then the system (7.22) becomes

$$\begin{aligned} \epsilon \frac{dy}{d\tau} &= z(1-y) + Ky, \quad y(0) = 0, \\ \frac{dz}{d\tau} &= -z(1-y) + (K-\lambda) y, \quad z(0) = 1. \end{aligned} \tag{7.23}$$

In many enzyme reactions the enzymes are very effective catalysts and the concentration of enzyme needed is very small compared to the substrate concentration. This means that e_0 is much smaller than s_0 and thus that ϵ is very small. Often ϵ is between 10^{-2} and 10^{-7}. Thus we may view (7.23) as a singular perturbation problem of the form (7.6) with

$$f(y,z) = z(1-y) - Ky, \quad y_0 = 0, \quad g(y,z) = -z(1-y), \quad z_0 = 1.$$

This means that except in the boundary layer very close to $t = 0$, where c may change rapidly, we may approximate the solution of (7.19) by the solution of the reduced problem

$$\begin{aligned} s' &= -k_1 (e_0 - c) s + k_{-1} c, \\ 0 &= k_1 (e_0 - c) s - (k_{-1} + k_2) c. \end{aligned} \tag{7.24}$$

This gives

$$c = \frac{k_1 e_0 s}{k_1 s + k_{-1} + k_2},$$

and then (7.20) reduces to the single differential equation

$$\begin{aligned} s' &= -k_1 e_0 s + c(k_1 s + k_{-1}) \\ &= -k_1 e_0 s + \frac{k_1 e_0 s(k_1 s + k_{-1}}{k_1 s + k_{-1} + k_2} \\ &= -\frac{k_2 e_0 s}{s + \frac{k_{-1}+k_2}{k_1}}. \end{aligned}$$

If we let $K_m = \frac{k_{-1}+k_2}{k_1}, Q = k_2 e_0$, this becomes

$$s' = -\frac{Qs}{s + K_m}. \tag{7.25}$$

This substrate uptake rate is called a Michaelis-Menten uptake, and reaction rates of this form occur in other problems. For example, the predator functional response in predator-prey models is often assumed to be of this form. We recall that in the original model (7.18) the rate of formation of product is given by $p' = k_2 c$. For the reduced problem we may replace c in this expression by (7.20), and then we obtain $p' = -s'$. Thus the rate of reaction is given by (7.25).

Exercises

1. Show that the solution of $z' = -e^{-t/\epsilon} z, z(0) = 1$ is $z(t, \epsilon) = e^{-\epsilon(1-e^{-t/\epsilon})}$.

2. Show that $\lim_{\epsilon\to 0} e^{-\epsilon(1-e^{-t/\epsilon})} = 1$ for all $t \geq 0$. [Hint: Show that $\lim_{\epsilon\to 0} \epsilon(1 - e^{-t/\epsilon}) = 0$.]

3. Graph the solution of the initial value problem in Example 1 with the values $\epsilon = 0.2, \epsilon = 0.1, \epsilon = 0.01$.

4. Find the solution of the stretched version of the initial value problem of Example 2.

5. Solve the initial value problem $\epsilon y' = y - y^3, y(0) = 1/2$ and compare the solution with the solution of the corresponding reduced problem.

6. Solve the initial value problem

$$\begin{aligned} \epsilon y' &= y - y^3, \quad y(0) = 1/2, \\ z' &= yz, \quad z(0) = 1. \end{aligned}$$

7. Find the equilibria and analyze their stability for the system (7.16).

8. Use a computer algebra system to sketch the graph of v as a function of t for the system (7.17) with $J = 1.5, J = 2$, and $J = 2.5$.

9. Show that the only equilibrium of (7.19) is $s = 0, c = 0$ and that this equilibrium is asymptotically stable.

10. Sketch in the phase plane the orbit with initial value $s = s_0, c = 0$ and identify the portion of the orbit which corresponds to the initial fast reaction.

11. Complete a qualitative analysis of the behavior of solutions of the "fast" dynamics of the Fitzhugh-Nagumo model

$$\frac{dv}{dt} = -v(v-a)(v-1) - w,$$

considering w as a constant. In particular, without attempting to solve explicitly for equilibrium values v^*, find the points A and B shown in Figure 7.6 and discussed in Example 3, and show that any equilibrium located between them must be unstable, while all other equilibria outside $[A, B]$ must be locally asymptotically stable.

12. Complete a qualitative analysis of the behavior of solutions of the "slow" dynamics in the Fitzhugh-Nagumo model, $\frac{dw}{dt} = v - \gamma w$, considering v as a constant (even though it's really not constant, this can give an indication of the direction of change of w at any point in time).

13. * Use the results of the previous two exercises (in particular, the expression for B in terms of a, and the fact that $dw/dt > 0$ when $v > \gamma w$) to find a condition on γ that makes $dw/dt > 0$ near the point $(B, -f(B))$ on the QSS curve. [Hint: first write the condition in terms of a and B, then just in terms of a, and finally consider the range of values of the bound on γ as a varies from 0 to 1.]

14. Use a computer and software such as *Maple, Mathematica* or *MATLAB* to verify numerically the claim made at the end of Example 3 regarding the behavior of the Fitzhugh-Nagumo model as a current J is applied across the cell membrane. For fixed values of the parameters a, γ and ϵ (such as those used to generate Figure 7.7), graph the two nullclines of system (7.14) and the solution to the system, for several values of J beginning at 0. How does the behavior of the solutions change as J increases? (There is a second change as J continues increasing, beyond the change mentioned in the text.)

15. To review the section, list terms associated with fast vs. slow systems.

7.3 HIV: An example from immunology

The human immunodeficiency virus (HIV) is the virus which causes Acquired Immune Deficiency Syndrome (AIDS). Since the early 1980s there has been a major effort to combat this disease, and one part of this effort has been to understand the immunology of HIV as a pathogen.

Immunology is the study of the body's immune system, which protects the body from external attacks, such as by pathogens. When a pathogen is introduced into the body, there is an immune response which attempts to clear it. Pathogens are captured by macrophages. These are cells which present digested pieces of the pathogen, called antigens, to the CD4 positive lymphocytes ($CD4^+T$ cells). In response these T cells reproduce by subdivision and activate a second type of T cell, the CD8 positive T lymphocytes ($CD8^+T$ cells) which can seek out and destroy cells which are infected with pathogens. A second type of response is the antibody response, in which the $CD4^+T$ cells signal the B lymphocyte cells in the blood which produce antibodies to destroy the pathogen. In the remainder of this section we will use the term T cell to mean a $CD4^+T$ cell unless otherwise specified.

HIV and AIDS both refer to immunodeficiency because HIV attacks the immune system, infecting and destroying the T cells, removing the body's ability to defend itself against not only HIV but other, more common pathogens. Progression to AIDS is typically defined in terms of a patient's T cell count. For these reasons this section will focus on describing and explaining the long-term interaction between HIV and T cells within an average individual.

When HIV infects the body (see Figure 7.10) it attacks the T cells. It causes an infected host cell to produce new virus particles, either slowly during the lifetime of the cell or (eventually) in a rapid burst which also kills the cell (lysis). Immediately after infection the amount of virus detected in the blood rises rapidly. This acute stage is accompanied by flu-like symptoms. After a few weeks the virus concentration decreases and the symptoms disappear (the infection is then said to be clinically latent). An immune response to the virus occurs, and antibodies against the virus can be detected in the blood. If these antibodies are detected in a person's blood, that person is said to be HIV-positive. The level to which the virus falls after the primary infection is called the *set point*. The virus concentration then remains almost constant for several years, and although there are no disease symptoms, there is a gradual decrease in the T cell count. This asymptomatic period may last as long as 10 years. This is followed by a passage to AIDS, in which the T cell count drops to a very low level and the body is unable to fight off infections. For an uninfected individual the T cell count is approximately $1000/mm^3$. Progression to AIDS is usually measured by a drop below $200/mm^3$ in the T cell count. Figure 7.11 shows the decrease in T cell count over ten years for a typical patient, superimposed on a graph of HIV viral load over the same time frame.

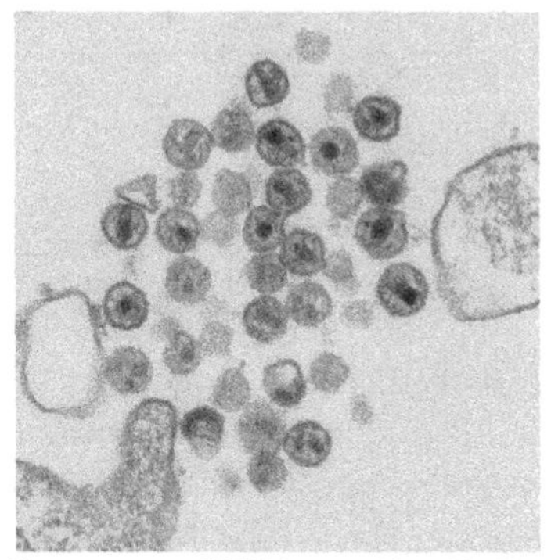

Photo courtesy CDC/Maureen Metcalfe, Tom Hodge

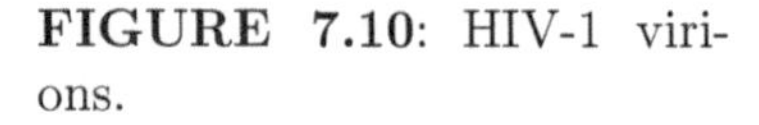

FIGURE 7.10: HIV-1 virions.

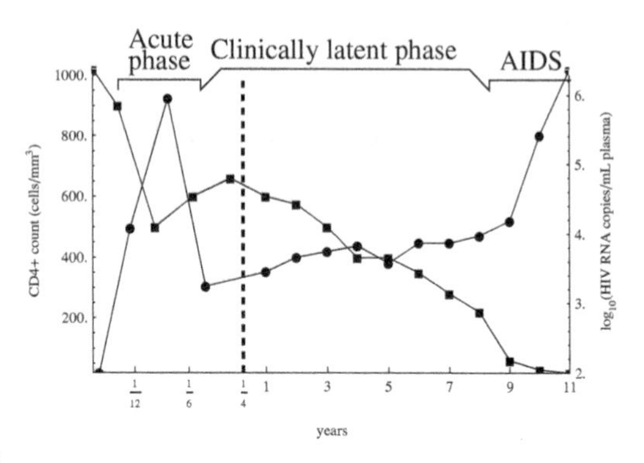

FIGURE 7.11: This graph, adapted from widely circulated ones in Pantaleo et al. (1993) and Fauci et al. (1996), shows the progression of within-host viral load (circles) and corresponding CD4+ T cell count (squares) [vs. time since infection] in a typical HIV patient, over the initial acute phase, the clinically latent phase, and eventual progression to AIDS. Note the change in timescale after 3 months.

In modeling the interaction between T cells and HIV, one must make decisions about which processes and approaches to include. Must all the immune responses be modeled separately? Infected T cells spawn HIV virions internally during a quiescent or latent period, followed by lysis and the sudden release of free virions into the system; since much of the viral load comes from other types of infected cells, need each of these stages (or indeed the infected T cells at all) be represented explicitly? Is the initial acute phase important in determining the long-term outcome? Do viral and immune system dynamics take place on different enough timescales that a singular perturbation approach would help? Some answers are offered by the specificity of one's hypothesis (which often claims one mechanism to be most important in explaining observed behavior), while others can only be determined by trying several models and/or analysis approaches, and then selecting the simplest one which explains observations well. We shall now employ both ways, via multiple tries beginning with a simple model for a healthy immune system.[9]

[9]Much of the pioneering work in modeling HIV immunology builds on work by Perelsen and colleagues; one set of models whose structure parallels that developed in this section is discussed in Alan S. Perelson and Patrick W. Nelson, Mathematical analysis of HIV-1 dynamics in vivo, *SIAM Review* 41(1): 3–44, March 1999. Another helpful reference for this topic is M.A. Nowak and R.M. May, *Virus dynamics: mathematical principles of immunology and virology*, Princeton University Press, 2003.

7.3.1 A basic model

T cells are produced mainly in the thymus. The production is not well understood, but it is plausible to assume a constant production rate in an uninfected person, and a proportional death rate. (The T cell production rate is also known to be a decreasing function of viral load.) We let T denote the concentration of T cells and assume a production rate s_1 and a proportional death rate μ in the absence of infection (making the average cell lifetime $1/\mu$). Then the T cell concentration in a healthy individual is described by the initial value problem

$$T' = s_1 - \mu T, \quad T(0) = T_0$$

which has an asymptotically stable equilibrium s_1/μ.

We attempt to model the immune system's interaction with HIV in terms of the concentrations of healthy T cells, infected T cells (denoted by I), and free virus (denoted by V). We begin with a relatively simple model which adds only the presence and direct influence of free virions V, incorporating just four of the forces described earlier. The virus's presence decreases the production of T cells, altering the production function; if we assume that the maximal decrease still leaves the net production rate positive (regardless of viral load), we can describe it via a Michaelis-Menten type saturation term:

$$s_1 - \frac{s_2 V}{a + V}$$

with $s_1 > s_2$ (and some positive half-saturation constant a). We also assume a similar form for the baseline growth rate of the virus due to growth from other infected cells such as macrophages and infected thymocytes,

$$\frac{gV}{b + V},$$

with maximum growth rate g and half-saturation constant b. (We justify the dependence on V by taking it as an indicator of the overall number of infected cells producing virions, which does not explicitly appear in the model.) Finally, we assume the encounter rate between the immune system and free virus has a mass action form TV; such encounters result in the virus infecting T cells, say at a rate $k_1 TV$, and (taking T as an indicator of the immune system's capacity to eliminate the virus) in clearance of free virus, say at a rate $k_2 TV$. Since this latter term includes the loss of free virus in the process of infecting T cells, we take $k_2 > k_1$. (Since the T cell level is not directly responsible for clearance of free virus, an alternative assumption would be a proportional clearing rate of free virus, but without this interaction term the virus is completely unaffected by the immune system in the model.) These assumptions yield the two-dimensional system

$$\begin{aligned} T' &= s_1 - \frac{s_2 V}{a + V} - \mu T - k_1 TV, \quad T(0) = T_0, \\ V' &= \frac{gV}{b + V} - k_2 TV, \quad V(0) = V_0. \end{aligned} \tag{7.26}$$

A reasonable initial condition for T is the stable equilibrium of the healthy T cell model, $T_0 = s_1/\mu$.

Note that (7.26) includes the healthy T cell model as a special case where $V_0 = 0$, and therefore has the virus-free equilibrium $T = T_0$, $V = 0$. The stability of the virus-free equilibrium $(T_0, 0)$ is straightforward to determine. The matrix of the linearization of system (7.26) at $(T_0, 0)$ is

$$\begin{bmatrix} -\mu & -\frac{s_2}{a} - k_1 T_0 \\ 0 & \frac{g}{b} - k_2 T_0 \end{bmatrix},$$

which has eigenvalues $-\mu$ and $\frac{g}{b} - k_2 T_0$. Thus this equilibrium is asymptotically stable if and only if $g < k_2 T_0 b$ (the maximum virus growth rate is less than the clearance rate when $T = T_0$, $V = b$).

The existence of any endemic equilibrium is more complicated to determine analytically (the condition is quadratic, so there are 0, 1, or 2 of them) and best considered graphically, using a phase portrait. It is straightforward to solve both equilibrium conditions for T as functions of V, so we write the nullclines as

$$T = F_T(V) = \frac{(s_1 - s_2)V + a s_1}{(V + a)(k_1 V + \mu)}, \quad T = F_V(V) = \frac{g}{k_2(V + b)}.$$

Both of these are positive, decreasing, concave up functions of V (for $V \geq 0$). Examining the T nullcline ($dT/dt = 0$), we see $F_T(0) = \frac{s_1}{\mu} = T_0$ and $\lim_{V\to\infty} F_T(V) = 0$; for the V nullcline ($dV/dt = 0$), $F_V(0) = \frac{g}{k_2 b}$ and $\lim_{V\to\infty} F_V(V) = 0$. Comparing the two values at $V = 0$, we see their order is determined by the stability condition for the virus-free equilibrium: $F_V(0) < F_T(0)$ if and only if $g < k_2 T_0 b$. Since both functions approach zero as $V \to \infty$, to determine their order as $V \to \infty$ we instead compare $\lim_{V\to\infty} V\,F(V)$. We find $\lim_{V\to\infty} V\,F_T(V) = (s_1 - s_2)/k_1$ while $\lim_{V\to\infty} V\,F_V(V) = g/k_2$. Thus $F_V(V) < F_T(V)$ as $V \to \infty$ if and only if $\frac{s_1 - s_2}{k_1} > \frac{g}{k_2}$. There are four combinations of these two conditions, and therefore at least as many different possible phase portraits for the system. Here we shall consider only two of them (see Exercise 2 for the full set).

If $g > bk_2T_0$, so that the virus-free equilibrium is unstable, and in addition $\frac{s_1 - s_2}{k_1} > \frac{g}{k_2}$, then F_V begins above F_T (at $V = 0$) but ends up beneath it (as $V \to \infty$), so the two nullclines must cross, with an endemic steady state at their intersection. The phase portrait (Figure 7.12) shows this equilibrium to be [globally] asymptotically stable. (This proves that these two conditions are sufficient to bring about a globally stable endemic state; in theory it might be possible for one to exist some other way, but later in this section we will show that these conditions are indeed necessary as well.)

If instead $g > bk_2T_0$ and $\frac{g}{k_2} > \frac{s_1 - s_2}{k_1}$, then F_V begins and ends above F_T. The simplest possible behavior here is for the two nullclines never to meet, with the V nullcline lying above the T nullcline for $0 \leq V < \infty$. In this case all crossings of the V nullcline are downward, and crossings of the T nullcline

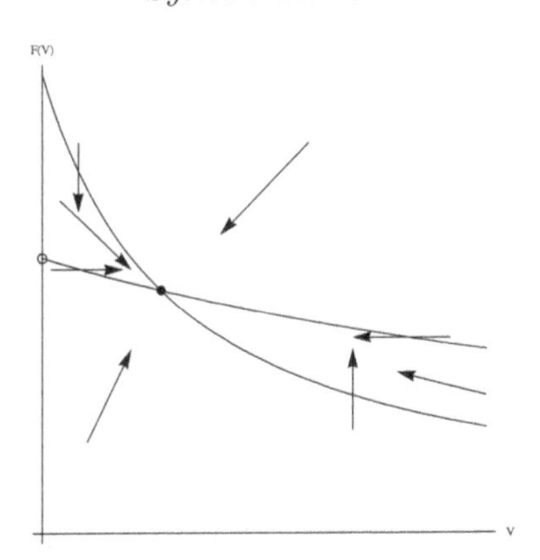

FIGURE 7.12: Nullclines of the system (7.26) in the V-T plane, in the case where $\frac{g}{k_2} > bT_0$ and $\frac{g}{k_2} < \frac{s_1 - s_2}{k_1}$.

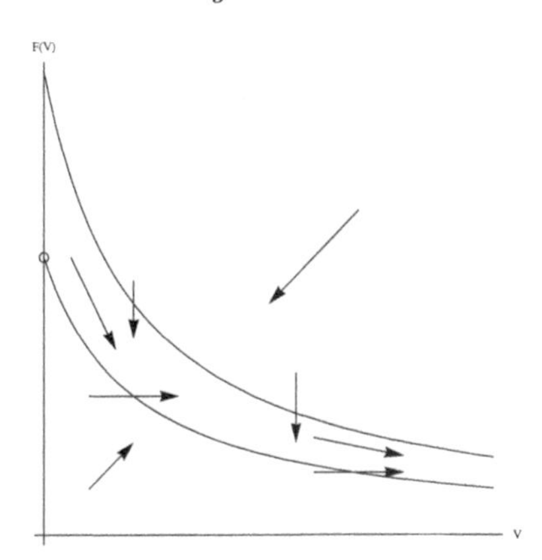

FIGURE 7.13: Nullclines of the system (7.26) in the V-T plane, in the case where $\frac{g}{k_2} > bT_0$, $\frac{g}{k_2} > \frac{s_1 - s_2}{k_1}$ and the nullclines do not cross.

are to the right (Figure 7.13). Thus solutions will enter the region between the curves, either from above or from below; any solution lying between the two nullclines must remain between them for all further time (moving down and to the right). Thus T will decrease and V will increase for all t, with $V \to \infty$ and $T \to 0$, indicating progression to AIDS.

Given the relatively complicated analysis, a simplification technique may help us better understand the model system's behavior. Since the changes in V take place on a much faster time scale (a few weeks) than the changes in T (several years), it is reasonable to consider rewriting (7.26) as a singular perturbation problem of the form

$$\begin{aligned} T' &= c_T \left[s_1 - \frac{s_2 V}{a+V} - \mu T - k_1 T V \right], \\ \epsilon V' &= c_V \left[\frac{gV}{b+V} - k_2 T V \right], \end{aligned}$$

with proportionality coefficients c_T and c_V, allowing us to rescale the time variable to study either fast (inner system) or slow (outer system) time. Let us suppose that the values of the coefficients in (7.26) support this notion. Then our analysis of (7.26) can be decomposed into the analysis of the inner system with T constant, $T = T_0$:

$$\frac{dV}{d\tau} = \frac{gV}{b+V} - k_2 T_0 V, \quad V(0) = V_0 \geq 0, \tag{7.27}$$

in the boundary layer near $t = 0$ (where τ is the rescaled, "fast" time), and an outer system for T (with V given in terms of T).

For the inner system, the first-order equation (7.27) can easily be shown to have a unique asymptotically stable equilibrium: $V = 0$ if $g < bk_2T_0$, and $V = \frac{g}{k_2T_0} - b > 0$ if $g > bk_2T_0$ (see Exercise 1 below). Thus if $g < bk_2T_0$ the virus is eradicated quickly while if $g > bk_2T_0$ the viral load has a positive limit.

The outer system, meanwhile, is

$$T' = s_1 - \frac{s_2V}{a+V} - \mu T - k_1TV, \quad T(0) = T_0, \tag{7.28}$$

with

$$\frac{gV}{b+V} - k_2TV = 0. \tag{7.29}$$

In order to analyze the outer system we first solve (7.29) for V as a function of T, obtaining either $V = 0$ or (if $T < g/k_2b$)

$$V = \frac{g}{k_2T} - b, \tag{7.30}$$

next substitute this solution into (7.28), and then analyze the resulting first-order equation. If $V = 0$, the outer system is the healthy T cell equation $T' = s_1 - \mu T$ for which $T(t) \to T_0$ as $t \to \infty$. This is the case in which the virus is eliminated rapidly. Note that if, at any point in time, $T(t) > g/k_2b$, then $V = 0$ is the only nonnegative solution for V, and (revisiting the inner system with this value of T) V then quickly (i.e., in fast time) approaches 0, leaving T to approach T_0.

To treat the case $V = \frac{g}{k_2T} - b$, we substitute (7.30) for V into (7.28), which yields, after some simplification, the outer system

$$T' = \mathcal{F}(T) = s_1 - s_2\frac{g - k_2bT}{g - k_2(b-a)T} - (\mu - k_1b)T - k_1\frac{g}{k_2}, \quad T(0) = T_0, \tag{7.31}$$

with $g > k_2bT_0$ and, since the virus terms should make $\mathcal{F}(T) < s_1 - \mu T$ (and thus $T(t) \le T_0$), $T(t) < g/k_2b$ more broadly. Thus in analyzing the outer system we restrict our attention to the interval $[0, g/k_2b]$. Given the complexity of $\mathcal{F}$, we proceed using basic properties of $\mathcal{F}$ rather than finding its roots. We observe

$$\mathcal{F}(0) = k_1\left(\frac{s_1 - s_2}{k_1} - \frac{g}{k_2}\right), \quad \mathcal{F}(g/k_2b) = \mu\left(T_0 - \frac{g}{k_2b}\right) < 0,$$

and calculate $\mathcal{F}'(T) = k_1b - \mu + \dfrac{s_2agk_2}{[g + (a-b)k_2T]^2}$,

$$\mathcal{F}'(0) = k_1b - \mu + \frac{s_2ak_2}{g}, \quad \mathcal{F}'(g/k_2b) = k_1b - \mu + \frac{s_2k_2b^2}{ga}.$$

Both $\mathcal{F}$ and $\mathcal{F}'$ have a *pole* (i.e., a vertical asymptote) at $T = \frac{g}{k_2(b-a)}$, but it does not fall within $[0, g/k_2b]$: if $a > b$ the pole is negative and $\mathcal{F}'$ (which decreases away from the pole) is then monotone decreasing on $[0, g/k_2 b]$, while if $a < b$ some algebra shows that the pole is greater than $g/k_2 b$, making $\mathcal{F}'$ monotone increasing on $[0, g/k_2 b]$. In either case $\mathcal{F}'$ is monotone, so $\mathcal{F}'$ has at most one root ($\mathcal{F}' = 0$) in $[0, g/k_2 b]$, meaning $\mathcal{F}$ can "turn around" at most once.

Thus if $\mathcal{F}(0) > 0$ there is a unique equilibrium in $(0, g/k_2 b)$ (since $\mathcal{F}(g/k_2 b) < 0$), and since $\mathcal{F}$ crosses from positive to negative there, $\mathcal{F}' < 0$ there, making the equilibrium asymptotically stable (locally, and also globally since there are no other stable equilibria). This indicates an endemic state.

If $a > b$ (making $\mathcal{F}'$ monotone decreasing and thus $\mathcal{F}$ concave down) and $\mathcal{F}'$ does have a root, T_c, in $[0, g/k_2 b]$, then T_c is a local maximum for $\mathcal{F}$, and it is possible to have $\mathcal{F}(0) < 0$ but $\mathcal{F}(T_c) > 0$. In this case there are two equilibria: one in $(0, T_c)$ (call it E_0) which is unstable (since $\mathcal{F}$ crosses from negative to positive there, making $\mathcal{F}' > 0$), and one in $(T_c, g/k_2 b)$ (call it E_1), asymptotically stable (since $\mathcal{F}$ crosses from positive to negative there, making $\mathcal{F}' < 0$ as before). Here the dynamics are more complex, as the unstable E_0 acts as a separatrix: If $T_0 < E_0$, then $T \to 0$ (and then $V \to \infty$) and the infection progresses to AIDS, but if $T_0 > E_0$, then $T \to E_1$, and the infection persists in an endemic state.

If neither of the two sets of conditions above hold (i.e., $\mathcal{F}(0) < 0$ and either T_c does not exist in $[0, g/k_2 b]$ or $\mathcal{F}(T_c) < 0$), then $\mathcal{F} < 0$ on all of $[0, g/k_2 b]$, making $T' < 0$ so that $T \to 0, V \to \infty$ and the infection progresses to AIDS regardless of initial conditions.

The condition $\mathcal{F}(0) > 0$ (for a single, globally stable endemic equilibrium) is equivalent to

$$\frac{g}{k_2} < \frac{s_1 - s_2}{k_1}, \tag{7.32}$$

seen in the analysis of the original (full) system (7.26). The more complicated conditions that lead to two equilibria can be condensed (see Exercise 3) to

$$\frac{b}{a}\sqrt{\frac{k_2}{g}} < \sqrt{\frac{\mu - k_1 b}{s_2 a}} < \sqrt{\frac{k_2}{g}} - \sqrt{\frac{k_1}{s_2}\left(1 - \frac{b}{a}\right)\left(1 - \frac{s_1 - s_2}{k_1}\frac{k_2}{g}\right)}. \tag{7.33}$$

By viewing (7.26) as a singular perturbation problem we have been able to obtain some information that we could not obtain directly. In the first place, we can see that if the virus is eradicated, this takes place in the inner system and thus happens very quickly. Second, the analysis of the outer system shows that the conditions $g > bk_2 T_0$ and $\frac{g}{k_2} < \frac{s_1 - s_2}{k_1}$ obtained graphically using nullclines are not only sufficient for the existence of a unique, globally stable endemic equilibrium but necessary as well. Also, the relation (7.29) between T and V is useful information obtained from the outer system.

The information that we have obtained about the model (7.26) is that if $g < bk_2 T_0$ the virus is eradicated quickly, if $g > bk_2 T_0$ and $\frac{g}{k_2} < \frac{s_1 - s_2}{k_1}$ there is an asymptotically stable infected steady state without progression to AIDS (as these two inequalities may be contradictory for some parameter

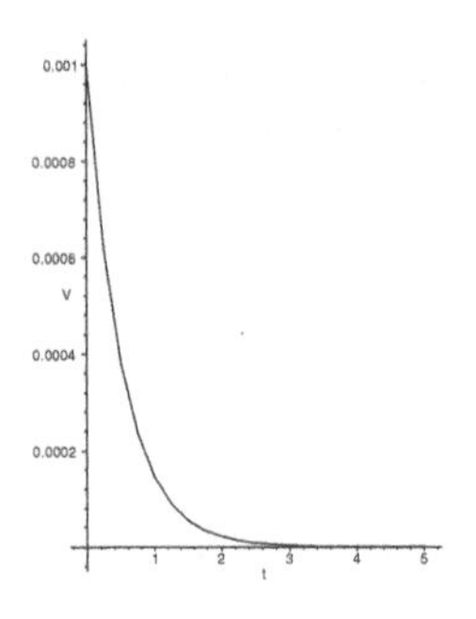

FIGURE 7.14: $V(t)$ for $0 \leq t \leq 5$, $g = 1$: the virus is cleared quickly.

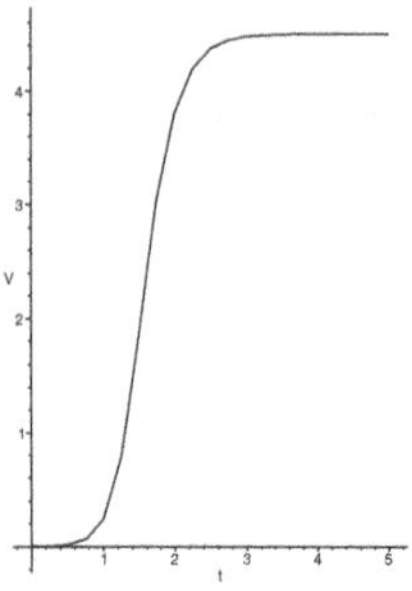

FIGURE 7.15: $V(t)$ for $0 \leq t \leq 5$, $g = 125$: the virus quickly reaches a plateau.

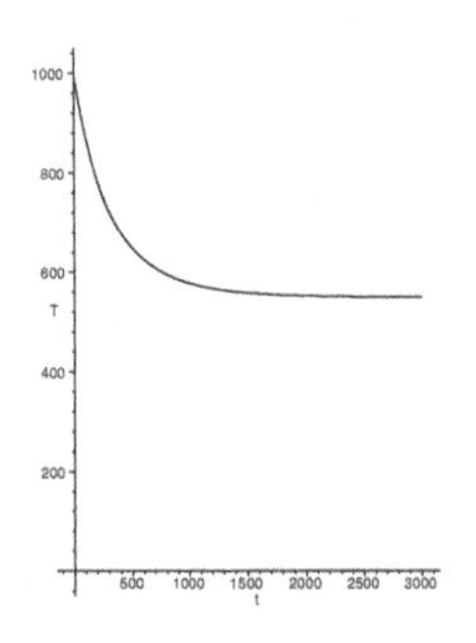

FIGURE 7.16: $T(t)$ for $0 \leq t \leq 3000$, $g = 100$: a long-term endemic state.

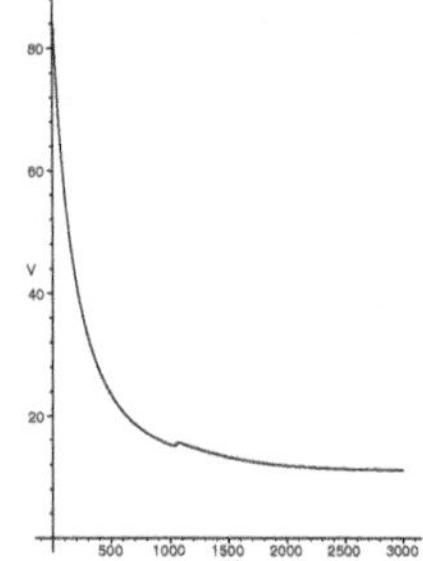

FIGURE 7.17: $V(t)$ for $0 \leq t \leq 3000$, $g = 100$: a long-term endemic state.

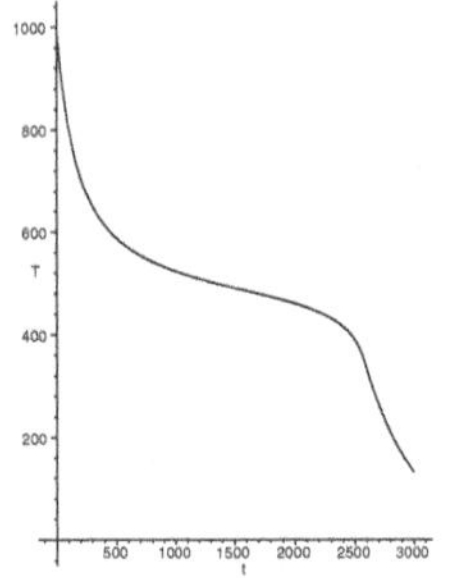

FIGURE 7.18: $T(t)$ for $0 \leq t \leq 3000$, $g = 125$: long-term progression to AIDS.

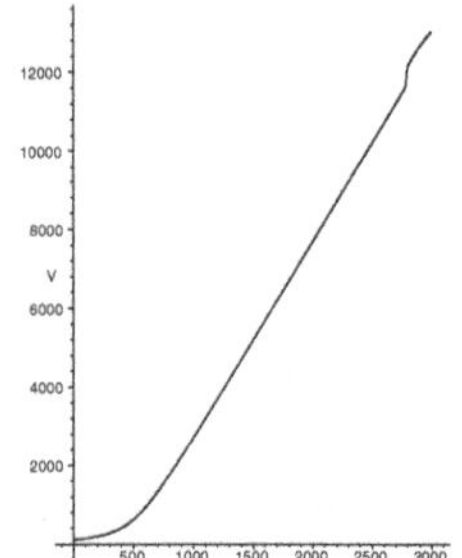

FIGURE 7.19: $V(t)$ for $0 \leq t \leq 3000$, $g = 125$: long-term progression to AIDS.

values, this possibility may not arise), and under most other conditions there is progression to AIDS.

For numerical simulations with given sets of parameter values, the rapid changes in V near $t = 0$ may cause difficulties in the approximation. It is advisable to use the equilibrium of the inner system as an initial value. This is essentially equivalent to finding a solution of the outer problem and matching it to the solution of the inner problem, thus using the singular perturbation approach at least implicitly. We will give the results of some numerical simulations, using the parameter values

$$s_1 = 10\,\frac{cells}{mm^3}/day,\ \ s_2 = 7\,\frac{cells}{mm^3}/day,\ \ a = 12\,\frac{virus}{mm^3},\ \ b = 8\,\frac{virus}{mm^3},\tag{7.34}$$
$$k_1 = 2.5\times 10^{-4}\left(\frac{virus}{mm^3}\,day\right)^{-1},\ \ k_2 = 0.01\left(\frac{cells}{mm^3}\,day\right)^{-1},\ \ \mu = 0.01 day^{-1}.$$

The values for s_1, s_2, μ, k_1, k_2 have been derived from experimental data. The graphs are very sensitive to changes in the parameters a and b. We use different values of g to illustrate the possibilities of eradication of the virus, an infected steady state, and progression to AIDS. Figures 7.14 and 7.15 show the short term behavior with $g = 1$ and $g = 125$, respectively (units for g in all cases are $\frac{virus}{mm^3}/day$ but will hereinafter be omitted for space constraints). Figures 7.16 and 7.17 show the long term behavior of T and V for $g = 100$, illustrating an asymptotically stable infected steady state, while Figures 7.18 and 7.19 show the long term behavior for $g = 125$, illustrating progression to AIDS.

7.3.2 Including infected cells

The system (7.26) neglects the impact of infected T cells. Since the fraction of T cells which are infected is very small, perhaps of the order of 10^{-4} or 10^{-5}, this may be a plausible simplification. We can determine the extent to which this simplification affects the model's ability to describe the immune system–virus interaction by developing and analyzing a model which includes infected T cells. Infected cells I are created by the infection encounters already described as occurring at a rate k_1TV; we now also assume that they die (through lysis) at a proportional death rate δ (the reciprocal of how long an average infected T cell takes to rupture). Finally, we assume that, when lysis occurs, on average each infected T cell produces N virions in its lifetime. These virions are released when the cell dies, adding to the concentration of free virus. These assumptions lead to the three-dimensional model

$$\begin{aligned} T' &= s_1 - \frac{s_2 V}{a+V} - \mu T - k_1 TV, \quad T(0) = T_0, \\ I' &= k_1 TV - \delta I, \quad I(0) = 0, \\ V' &= N\delta I + \frac{gV}{b+V} - k_2 TV, \quad V(0) = V_0. \end{aligned}\tag{7.35}$$

The analysis of this more complicated system follows the same lines as that of (7.26). Like the simpler system, (7.35) has a virus-free equilibrium $(T_0, 0, 0)$, and the matrix of the linearization at this equilibrium is

$$\begin{bmatrix} -\mu & 0 & -\frac{s_2}{a} - k_1 T_0 \\ 0 & -\delta & k_1 T_0 \\ 0 & \delta N & \frac{g}{b} - k_2 T_0 \end{bmatrix},$$

whose eigenvalues are $-\mu$ and the eigenvaluesof the 2×2 matrix

$$\begin{bmatrix} -\delta & k_1 T_0 \\ \delta N & \frac{g}{b} - k_2 T_0 \end{bmatrix},$$

which we denote by B. Thus the virus-free equilibrium is asymptotically stable if and only if

$$\begin{aligned} \mathrm{tr} B &= \frac{g}{b} - k_2 T_0 - \delta < 0, \\ \det B &= \delta \left(-\frac{g}{b} + k_2 T_0 - k_1 N T_0 \right) > 0. \end{aligned}$$

Since the determinant condition is stronger than the trace condition, the equilibrium is asymptotically stable if and only if the determinant is positive, or

$$g < b T_0 (k_2 - N k_1). \tag{7.36}$$

In particular, stability requires that $k_2 > N k_1$. The interpretation is again that the maximum virus growth rate g must be outweighed by the net clearance rate [when $T = T_0, V = b$], here $b T_0 (k_2 - N k_1)$, in order for the infection to die out; the clearance rate due to T cell–virus encounters is reduced by the indirect virus production that also results from such encounters (each of which eventually produces N virions).

It is possible to show by an elementary but technically complicated argument that if (7.36) is not satisfied there is an asymptotically stable infected equilibrium for g in a suitable range. We will omit the argument, however, since once again we will be able to obtain a stronger result by viewing the problem as a singular perturbation.

We recast (7.35) as a singular perturbation with I and V as fast variables and T as a slow variable. Then the singular perturbation approach would give a two-dimensional system for I and V as an inner system and a first-order equation for T as an a outer problem. For this model the mean number N of virions produced by an infected T cell turns out to be significant in the long-term behavior of the model.

The inner problem is the system

$$\begin{aligned} \frac{dI}{d\tau} &= k_1 T_0 V - \delta I, \\ \frac{dV}{d\tau} &= N \delta I + \frac{gV}{b+V} - k_2 T_0 V \end{aligned} \tag{7.37}$$

(where τ again denotes rescaled, fast time). The Jacobian matrix of this system at the equilibrium $I = 0, V = 0$ is just the matrix B that we obtained in analyzing the full three-dimensional model (7.35), and thus this equilibrium is asymptotically stable if and only (7.36) is satisfied. It is not difficult to show that the system (7.37) has another equilibrium, which is positive and asymptotically stable, if (7.36) is not satisfied but $0 < k_2 - Nk_1 < \frac{g}{T_0 b}$, given by

$$I = \frac{k_1 T_0}{\delta}\left(\frac{g}{(k_2 - Nk_1)T_0} - b\right), \qquad V = \frac{g}{(k_2 - Nk_1)T_0} - b.$$

If, finally, $k_2 < Nk_1$, so that the virus clearance rate is outweighed by the indirect virus creation rate from infected T cells alone, then one can show (Exercise 6) that $I, V \to \infty$, corresponding to AIDS.

The outer problem is

$$T' = s_1 - \frac{s_2 V}{a + V} - \mu T - k_1 T V$$

with V the appropriate limiting value of the inner system. If $k_2 - Nk_1 > g/T_0 b$ (so $V = 0$), which is the case of virus eradication, then the outer system reduces to the healthy T cell equation and $T \to T_0$. If $k_2 < Nk_1$ (so $V \to \infty$), then $T \to 0$ and the infection progresses to AIDS and death. If $0 < k_2 - Nk_1 < g/T_0 b$, so that

$$V(T) = \frac{g}{(k_2 - Nk_1)T} - b,$$

then we may write the outer problem in the form (7.31) with k_2 replaced by $k_2 - Nk_1$ in the expression for $\mathcal{F}(T)$. The argument used in analyzing (7.31) shows there is a positive asymptotically stable equilibrium if and only if

$$\frac{g}{k_2 - Nk_1} < \frac{s_1 - s_2}{k_1} \tag{7.38}$$

(analogous to (7.32)). For larger values of g ($\frac{g}{k_2 - Nk_1} > \frac{s_1 - s_2}{k_1}$) two endemic equilibria may exist if (analogous to (7.33))

$$\frac{b}{a}\sqrt{\frac{k_2 - Nk_1}{g}} < \sqrt{\frac{\mu - k_1 b}{s_2 a}} < \sqrt{\frac{k_2 - Nk_1}{g}} - \sqrt{\frac{k_1}{s_2}\left(1 - \frac{b}{a}\right)\left(1 - \frac{s_1 - s_2}{k_1}\frac{k_2 - Nk_1}{g}\right)}.$$

Otherwise $f(T) < 0$ for all T, and once again the solutions of the outer problem are monotone decreasing, and without a positive equilibrium every solution must approach zero, so that we have progression to AIDS.

For numerical simulations, it is reasonable to use the parameter values of (7.34) together with $\delta = 0.5/day$, $N = 10$ virions per infected T cell. However, the value of N is not well determined experimentally, and different values may give quite different behavior.

Another way to improve the model (7.35) would be to incorporate the fact that T cells do not become infective immediately on being infected but go first into a latent stage. This can be modeled most simply by adding a class L of latent T cells which die at the same rate as uninfected T cells and move into the infected class at a proportional rate k_3, corresponding to an exponential distribution of latent periods with mean $1/k_3$. This would give a four-dimensional model

$$\begin{aligned} T' &= s_1 - \frac{s_2 V}{a+V} - \mu T - k_1 T V, \quad T(0) = T_0, \\ L' &= k_1 T V - \mu_T L - k_3 L, \quad L(0) = 0, \\ I' &= k_3 L - \delta I, \quad I(0) = 0, \\ V' &= N \mu_I I - k_2 T V + \frac{g V}{b+V}, \quad V(0) = V_0. \end{aligned} \tag{7.39}$$

This system may be analyzed in the same way as the simpler system (7.35). The behavior of solutions is similar to that of solutions of (7.35) but agrees somewhat more closely with experimental data.

The models we have proposed in this section are radical simplifications of reality. Although the concentrations in the blood are thought to be representative, only about 2% of the T cells and virions are in the blood plasma. More accurate models would also include the concentrations in the lymphorecticular system and would distinguish between naive and memory cells, which have different roles. Nevertheless, the very simple models we have introduced do show many of the features observed in real life. One observation which is not seen in our models is that after the initial rapid increase in viral load there is a decrease with a relatively long period in which the viral load varies little before the sharp increase in progression to AIDS (Figure 7.11).

One of the main purposes of AIDS modeling is to evaluate and compare treatment strategies. The main treatments being used currently are combinations of drugs such as AZT which are inhibitors of reverse transcriptase and act to decrease k_1. Treatment models might consider at what point in the progression one should begin treatment. A problem is the development of drug resistant virus strains. Early treatment is considered to be treatment beginning before the T cell count has dropped below 500, but then drug-resistant strains of virus tend to replace drug-sensitive strains rapidly. Treatment delayed until the T cell count falls below 200 is too late to have much effect. It now appears that treatment beginning when the T cell count is between 200 and 500 may be the most effective choice with the treatments currently available. Models which incorporate treatment and drug resistance appear to corroborate this, but the question is by no means decided.

Exercises

1. Determine the behavior of solutions of the first-order equation (7.27).

2. For the system (7.26) draw all possible phase portraits for the cases:

   (a) $\frac{s_1-s_2}{k_1} < \frac{g}{k_2} < bT_0$

   (b) $bT_0 < \frac{g}{k_2} < \frac{s_1-s_2}{k_1}$

   (c) $\frac{g}{k_2} < bT_0, \frac{s_1-s_2}{k_1}$

   (d) $\frac{g}{k_2} > bT_0, \frac{s_1-s_2}{k_1}$

   Use the fact that the endemic equilibrium condition is quadratic to limit the number of crossings of the two nullclines (two of the four cases above will nevertheless have two possible portraits each).

3. Derive the inequality (7.33) from the set of conditions that lead to two endemic equilibria for the system (7.31), as follows:

   (a) Find T_c such that $\mathcal{F}'(T_c) = 0$. (The equation has 2 roots, but one is always outside the interval $[0, g/k_2b]$.) Find the condition that T_c exist (be a real number.)

   (b) Find conditions that $0 < T_c < g/k_2b$ when $a > b$. Verify that these conditions imply that $a > b$ and T_c is real. Show also that these conditions are equivalent to $\mathcal{F}'(g/k_2b) < 0 < \mathcal{F}'(0)$.

   (c) Show that $\mathcal{F}(T_c) = \mathcal{F}(0) + \left(\sqrt{s_2a} - \sqrt{(\mu - k_1b)\frac{g}{k_2}}\right)^2 /(a-b)$.

   (d) Find conditions that $\mathcal{F}(0) < 0 < \mathcal{F}(T_c)$ by solving $\mathcal{F}(T_c) > 0$ for $\frac{\mu-k_1b}{s_2a}$. Simplify to (7.33), and verify that it implies all the previous conditions (i.e., that $\mathcal{F}(0) < 0$, $0 < T_c < \frac{g}{k_2b}$, $a > b$, T_c is real).

4. Suppose it were possible to decrease k_1 to a value of $2.5 \times 10^{-5} \left(\frac{virus}{mm^3}\, day\right)^{-1}$. Use the model (7.26) to predict the values of g for which the behavior would change.

5. Would being able to decrease the value of s_2 be helpful in treatment for the model (7.26)? Explain.

6. Show that if $k_2 < Nk_1$ there is progression to AIDS ($I, V \to \infty$) for all $g \geq 0$, (a) in system (7.37), (b) in system (7.35). [Hint: Write first an inequality for V', and then an inequality for the derivative of the sum of V and a certain multiple of I.]

7. Determine conditions for the stability of the virus-free equilibrium of (7.39).

7.4 Slow selection in population genetics

In Exercise 9 of Section 2.3 we established the Hardy-Weinberg law of genetics, that the frequencies of genes (and their constituent alleles) remain constant from one generation to the next if mating is random with respect to genotype and if all genotypes have equal fitness (as measured through birth and death rates). The assumption of equal fitness may be true for some genes, such as those that determine widow's peak hairlines or attached earlobes in humans, but it is clearly not so for others, such as the gene which causes sickle-cell disease, in which the hemoglobin molecules in red blood cells are deformed and rigid (Figure 7.20). Now, we wish to investigate whether the Hardy-Weinberg principle remains true if there are differences in fitness. We shall use modeling to examine gene frequencies in a population with density-dependent population dynamics and a genotype-dependent death rate. Incorporating ongoing population dynamics shifts the context from the discrete-time models of Chapter 2, difference equations with distinct generations, to the continuous-time models provided by differential equations. In keeping with the theme of this chapter, we shall use a singular perturbation approach to simplify the eventual analysis, by assuming that the differences in death rates among different genotypes are small, and separating the time scales of the population dynamics and the genetic selection.

To establish a baseline result and distinguish the effects of continuous population dynamics from those of genotype-dependent fitness, we begin with the case in which all genotypes have equal fitness.

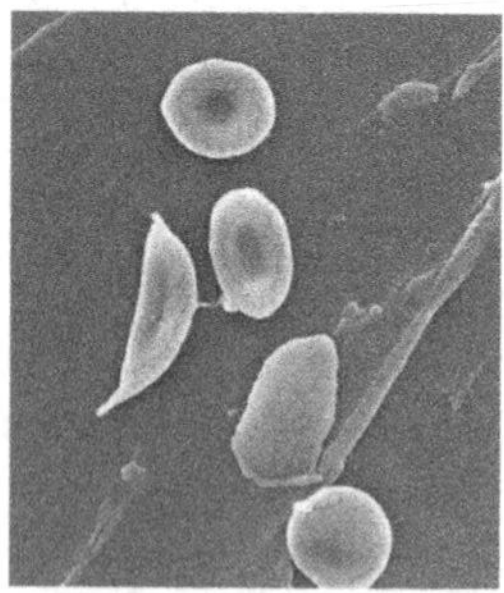

Photo by Janice Haney Carr, courtesy CDC/ Sickle Cell Foundation of Georgia: Jackie George, Beverly Sinclair

FIGURE 7.20: This electron micrograph shows a deformed red blood cell (left) alongside normal ones in a person with sickle-cell anemia.

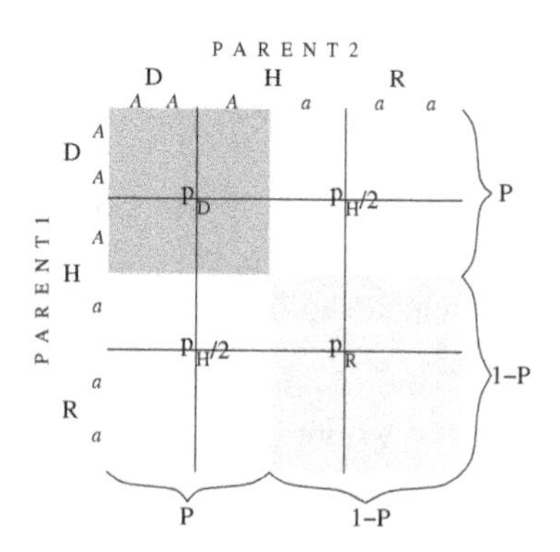

FIGURE 7.21: Proportions p_i of offspring with each genotype in a population with genotype populations y_i and genotype-independent mating.

7.4.1 Equally fit genotypes

To develop a model we must first define some notation. We consider diploid organisms with alleles A, a in a population of size $N(t)$. Let $y_1(t), y_2(t), y_3(t)$ be the number of members of genotype AA, Aa, aa, respectively, so that

$$y_1 + y_2 + y_3 = N.$$

Suppose that the density-dependent birth rate $\Lambda(N)$ is the same for all genotypes, while proportional death rates d_1, d_2, d_3 for genotypes AA, Aa, aa, respectively, may differ. Initially, we will assume that these death rates are equal, $d_1 = d_2 = d_3 = d$. We also make some assumptions on the form of $\Lambda(N)$: namely, that $\Lambda(0) = 0$ and that $\Lambda(N)$ is differentiable, with

$$\Lambda'(N) \leq \frac{\Lambda(N)}{N} \text{ for } N > 0. \tag{7.40}$$

That is, Λ increases slower than linearly as N increases. This bound prevents unbounded growth (since the total death rate does increase linearly in N) and is satisfied by most common growth functions, including logistic.

To track the quantities of interest, we define the *genotype* frequencies D (for dominant), H (for heterozygous) and R (for recessive):

$$D = \frac{y_1}{N}, \quad H = \frac{y_2}{N}, \quad R = \frac{y_3}{N},$$

so that $D + H + R = 1$, and the *allele* frequencies P for A and $1 - P$ for a. Note that $P = D + \frac{H}{2}$ since all of the alleles of the dominant (AA) genotype and half those of the heterozygous (Aa) genotype are As. Correspondingly $1 - P = R + \frac{H}{2}$.

We retain the assumption from Section 2.3 that mating is random with respect to these alleles and their associated genotypes. As noted above, we also retain the assumption of genotype-independent fecundity, so that the birth rate $\Lambda(N)$ applies to all genotypes. Then Figure 7.21 shows the possible offspring for all combination of parent genotypes, and we can write expressions for the proportions p_D, p_H, p_R of offspring of each genotype, in terms of the parent frequencies D, H, R. Note that in the figure, y_1, y_2, y_3 are deliberately shown as different in size to emphasize that variations in area within the square correspond to proportions of offspring with different genetic compositions. The dark shading shows how the proportion p_D of homozygous dominant (D) offspring can be computed as the number of offspring from two D parents, half the offspring from a D parent and an H parent, and one quarter of the offspring of two H parents, that is,

$$p_D = D^2 + 2 \cdot \frac{1}{2}DH + \frac{1}{4}H^2 = \left(D + \frac{H}{2}\right)^2 = P^2. \tag{7.41}$$

Likewise the light shading in Figure 7.21, which represents homozygous recessive (R) offspring, leads to the equation

$$p_R = R^2 + 2 \cdot \frac{1}{2} RH + \frac{1}{4} H^2 = \left(R + \frac{H}{2} \right)^2 = (1-P)^2. \tag{7.42}$$

Finally, the two unshaded areas, both equal in proportion, representing heterozygous offspring, yield

$$p_H = 2 \left(D + \frac{H}{2} \right) \left(R + \frac{H}{2} \right) = 2P(1-P). \tag{7.43}$$

(Note that thus $p_D + p_H + p_R = \left[(D + \frac{H}{2}) + (\frac{H}{2} + R) \right]^2 = P^2 + 2P(1-P) + (1-P)^2 = 1$, as it should.)

The dynamics of the frequencies D, H, R (and P) follow from those of the populations with each genotype. Under the assumptions that mating and fecundity are genotype-independent, we have

$$\begin{aligned} y_1' &= p_D \Lambda(N) - d_1 y_1, \\ y_2' &= p_H \Lambda(N) - d_2 y_2, \\ y_3' &= p_R \Lambda(N) - d_3 y_3. \end{aligned} \tag{7.44}$$

If in addition we assume genotype-independent death, $d_1 = d_2 = d_3 = d$, then the population dynamics decouple from the genetics, and taking the sum of the three equations, we have

$$N' = \Lambda(N) - dN. \tag{7.45}$$

To convert these to differential equations for the frequencies, we write $y_1 = DN$, so that $y_1' = D'N + DN'$ and $D'N = y_1' - DN'$. Into this last equation we now substitute y_1' from (7.44), $p_D = P^2$ from (7.41), and N' from (7.45):

$$\begin{aligned} D'N &= (p_D \Lambda(N) - d_1 y_1) - D(\Lambda(N) - dN) \\ &= P^2 \Lambda(N) - dDN - D\Lambda(N) + dDN = \Lambda(N)(P^2 - D), \end{aligned}$$

so that

$$D' = \frac{\Lambda(N)}{N} (P^2 - D). \tag{7.46}$$

Similarly $y_2 = HN$, so $y_2' = H'N + HN'$ and $H'N = y_2' - HN'$, from which, by (7.43),(7.44),(7.45),

$$\begin{aligned} H'N &= (p_H \Lambda(N) - d_2 y_2) - H(\Lambda(N) - dN) \\ &= 2P(1-P)\Lambda(N) - dHN - H\Lambda(N) + dHN = \Lambda(N)(2P(1-P) - H), \end{aligned}$$

so that

$$H' = \frac{\Lambda(N)}{N} (2P(1-P) - H). \tag{7.47}$$

Since by definition $D + H + R = 1$, we need not write an equation for R'; instead we may deduce $R = 1 - D - H$ from D and H. In fact, however, we may reduce the dimension of the frequency dynamics model even further, by recalling that $P = D + \frac{H}{2}$, and calculating

$$\begin{aligned} P' &= D' + \frac{H'}{2} = \frac{\Lambda(N)}{N}(P^2 - D) + \frac{\Lambda(N)}{N}(P(1-P) - H/2) \\ &= \frac{\Lambda(N)}{N}\left[P^2 + P(1-P) - \left(D + \frac{H}{2}\right)\right] = \frac{\Lambda(N)}{N}(P^2 + P - P^2 - P) = 0, \end{aligned}$$

from which we see that the frequency of allele A remains constant over time ($P' = 0$), and thus so does the frequency $(1 - P)$ of allele a. This makes sense given that all genotypes reproduce at the same rate and die at the same rate: each member of y_1 reproduces pairs of A's at the same rate as each member of y_3 reproduces pairs of a's and each member of y_2 reproduces sets of A and a.

The result that $P' = 0$ answers one of the original questions: continuous, density-dependent population dynamics does not disturb the constancy of the allele frequencies. To track the genotype frequencies over time, we need only adjoin equations (7.45),(7.46), since H can be found from D and P, and then R from D and H (or directly as $R = 1 + D - 2P$). The system for both frequency types can thus be summarized as

$$N' = \Lambda(N) - dN, \quad D' = \frac{\Lambda(N)}{N}(P^2 - D), \quad P' = 0. \tag{7.48}$$

Since the first equation, (7.45), is independent of the other two, the population dynamics can be analyzed separately. The assumptions made on the form of Λ can be used (see Exercise 1) to show that (7.45) has a unique, globally stable positive equilibrium if and only if $\Lambda'(0) > d$; otherwise the extinction equilibrium $N^* = 0$ is globally stable. In what follows, therefore, we assume that $\Lambda'(0) > d$, so that $N(t)$ approaches a stable positive limit, say K.

If we consider the two-dimensional system in N and D in (7.48) with $P = P_0$, it has the unique positive equilibrium (K, P_0^2). Linearization produces the Jacobian matrix

$$J(N, D) = \begin{bmatrix} \Lambda'(N) - d & 0 \\ \frac{\Lambda'(N)}{N} - \frac{\Lambda(N)}{N^2} & -\frac{\Lambda(N)}{N} \end{bmatrix},$$

which has, for any $N, D > 0$,

$$\mathrm{tr} J(N, D) = \Lambda'(N) - \frac{\Lambda(N)}{N} - d < -d < 0, \ \det J(N, D) = (\Lambda'(N) - d)\left(-\frac{\Lambda(N)}{N}\right),$$

and, for the equilibrium (K, P_0^2),

$$\det J(K, P_0^2) = (\Lambda'(K) - d)(-d) > 0$$

since from (7.40) $\Lambda'(K) < \Lambda(K)/K = dK/K = d$, making both factors of the determinant negative. Thus, by the equilibrium stability theorem of Section 5.4, the equilibrium (K, P_0^2) is locally asymptotically stable. In addition, since $\text{tr}J(N, D) < 0$ for all $N, D > 0$, a theorem called Bendixson's Criterion implies that the system (7.45),(7.46) does not have a periodic orbit in the first quadrant. Since $0 \leq N \leq K, 0 \leq D \leq 1$, solutions of (7.45),(7.46) are bounded, and now by the Poincaré-Bendixson theorem the equilibrium (K, P_0^2) is globally asymptotically stable (see Exercise 2 for discussion of $N^* = 0$). Since $H = P - D$ and $R = 1 + D - 2P$ the genotype frequencies D, H, and R approach $P_0^2, P_0(1 - P_0)$, and $(1 - P_0)^2$, respectively.

In the discrete generation model of Section 2.3 the genotype frequencies were constant. With nonlinear population dynamics included this is not necessarily true (unless the population starts out already at this equilibrium), but still the Hardy-Weinberg distribution is approached in the limit. This result serves as a baseline for studying the effects of varying genotype fitness.

7.4.2 Slow genetic selection

To study the consequences of genotype-dependent death rates on genotype frequencies, we maintain the hypotheses of genotype-independent mating and reproduction but assume the genotype affects survival; in particular, we assume homozygous individuals' death rates differ from those of heterozygous individuals as follows:

$$d_1 = d + \epsilon\Delta_D, \quad d_2 = d, \quad d_2 = d + \epsilon\Delta_R. \tag{7.49}$$

If $\Delta_D > 0$, dominant homozygous (AA) individuals are less fit (die sooner) than heterozygous (Aa); if $\Delta_D < 0$, they are instead more fit (die later) than heterozygous. Likewise for Δ_R and recessive homozygous (aa). Both differences are multiplied by a dimensionless parameter $\epsilon << 1$ to indicate that the differences in death rates are small compared to the overall death rates and the overall population dynamics.

Under the new set of hypotheses, system (7.44) becomes, using (7.41), (7.42), (7.43), (7.49),

$$\begin{aligned} y_1' &= P^2\Lambda(N) - (d + \epsilon\Delta_D)y_1, \\ y_2' &= 2P(1-P)\Lambda(N) - dy_2, \\ y_3' &= (1-P)^2\Lambda(N) - (d + \epsilon\Delta_R)y_3, \end{aligned} \tag{7.50}$$

with sum

$$N' = \Lambda(N) - dN - \epsilon(\Delta_D y_1 + \Delta_R y_3). \tag{7.51}$$

The equation $D'N = y_1' - DN'$ now becomes instead

$$\begin{aligned} D'N &= [P^2\Lambda(N) - (d + \epsilon\Delta_D)DN] - D[\Lambda(N) - dN - \epsilon(\Delta_D DN + \Delta_R RN)] \\ &= \Lambda(N)(P^2 - D) + \epsilon DN[\Delta_D(D-1) + \Delta_R R] \\ D' &= \frac{\Lambda(N)}{N}(P^2 - D) + \epsilon D[\Delta_D(D-1) + \Delta_R(1 + D - 2P)]. \end{aligned} \tag{7.52}$$

Similarly $H'N = y_2' - HN'$ now becomes

$$\begin{aligned} H'N &= [2P(1-P)\Lambda(N) - dHN] - H[\Lambda(N) - dN - \epsilon(\Delta_D DN + \Delta_R RN)] \\ &= \Lambda(N)[2P(1-P) - H] + \epsilon HN[\Delta_D D + \Delta_R(1+D-2P)] \\ H' &= \frac{\Lambda(N)}{N}(2P(1-P) - H) + \epsilon H[\Delta_D D + \Delta_R(1+D-2P)]. \end{aligned} \tag{7.53}$$

Again we omit an equation for R' in lieu of $D+H+R=1$ and focus instead on $P = D + \frac{H}{2}$. As before, the terms without ϵ in the equation for $P' = D' + \frac{H'}{2}$ cancel out, leaving

$$P' = \epsilon[\Delta_D D(P-1) + \Delta_R P(1+D-2P)]. \tag{7.54}$$

We collect the differential equations for population dynamics (7.51), genotype frequency (7.52), and allele frequency (7.54) in the system

$$\begin{aligned} N' &= \Lambda(N) - dN - \epsilon N(\Delta_D D + \Delta_R(1+D-2P)), \\ D' &= \frac{\Lambda(N)}{N}(P^2 - D) + \epsilon D[\Delta_D(D-1) + \Delta_R(1+D-2P)], \\ P' &= \epsilon[\Delta_D D(P-1) + \Delta_R P(1+D-2P)]. \end{aligned} \tag{7.55}$$

Note that, for small ϵ, P is a slow variable and N and D are fast variables; in particular, the inner system ($\epsilon = 0$) is given by (7.48), the system with genotype-independent death(!). As seen in the previous subsection, the inner solution approaches the Hardy-Weinberg proportions, $D \to P^2$ (with P a constant, say P_0). The outer system, which governs the long-term evolution of the slow variable P, is obtained using the slow-time variable[10] $s = \epsilon t$, so that $\frac{dP}{ds} = \frac{1}{\epsilon}\frac{dP}{dt}$, and substituting $D = P^2$ in (7.54) (see Exercise 3), to get

$$\frac{dP}{ds} = f(P) = (\Delta_D + \Delta_R)P(1-P)(P^* - P), \tag{7.56}$$

where

$$P^* = \frac{\Delta_R}{\Delta_D + \Delta_R}.$$

The equation (7.56) has three equilibria, $P = 0, P = P^*, P = 1$; the equilibrium P^* is biologically meaningful only if $0 \leq P^* \leq 1$. The question now is which of the meaningful equilibria are asymptotically stable, and this depends on the signs of Δ_D and Δ_R.

We consider four separate cases depending on these signs, cf. Figure 7.22. We sketch out the results in each case and leave the details for the reader to work through (Exercises 4–7):

[10] The variable s is here used to represent slow time relative to t, in contrast to its use in Section 7.2, where it was used as fast time.

I If $\Delta_D < 0, \Delta_R > 0$ then $d_1 < d_2 < d_3$ and the dominant allele A is clearly an advantage, since members of y_1 live longer than those in y_2, who live longer than homozygous recessive y_3. This is classic dominance; we expect natural selection to eliminate the recessive a allele over time, and model analysis shows (Exercise 4) that this is precisely what should happen: as $t \to \infty, P \to 1, D = P^2 \to 1, H = P - D \to 0, R = 1 + D - 2P \to 0$.

In practice, many mutations are of the "loss-of-function" type, harmful but recessive because one copy of the normal allele is enough to maintain some level of proper functionality (often a protein synthesis) of whatever part of the organism is affected by the given gene. Case I includes such scenarios.

II If $\Delta_D > 0, \Delta_R < 0$ then $d_1 > d_2 > d_3$ and the recessive allele a clearly confers a fitness advantage. This is the reverse situation of Case I, and consistent with our intuition, model analysis (Exercise 5) predicts fixation of the recessive allele: as $t \to \infty, P \to 0, D = P^2 \to 0, H = P - D \to 0, R = 1 + D - 2P \to 0$.

A well-known example of a recessive allele which confers a clear survival advantage, more to homozygotes than to heterozygotes, is the CCR5-Δ32 allele, which guards against both bubonic plague and HIV. CCR5-Δ32 homozygotes appear to be immune to both one type of plague[11] and one variant of HIV; heterozygotes (with one copy) have partial resistance to infection, and when infected appear to have milder or slower-progressing cases. This occurs with HIV because many types of HIV use the protein CCR5 to enter host cells, but individuals with the mutation produce smaller (defective) versions of this protein, which HIV cannot use. Over the centuries, the periodic waves of plague which ravaged Europe gradually increased the frequency of this allele (since disproportionately many of the survivors had it) to a point where it now occurs at a roughly 10% frequency in modern humans of European descent. The fact that despite conferring a clear advantage the allele only occurs at 10% frequency after many centuries underscores the difference in time scales between population dynamics and natural selection.

III If $\Delta_D > 0, \Delta_R > 0$ then both types of homozygote are disadvantaged (have higher death rates and thus shorter lifetimes) relative to heterozygotes Aa. This scenario is called *overdominance* and occurs with the aforementioned sickle-cell allele as well as that for cystic fibrosis: one copy of the [defective] recessive allele does not impede normal functionality (i.e., no disease) but confers protection against an infectious disease—malaria in the case of sickle-cell, and cholera in the case of cystic fibrosis—while two copies of the recessive allele produce a lethal

[11] It has been widely speculated that CCR5-Δ32 protected against the bubonic plague, but some studies suggest that it instead conferred protection against smallpox.

genetic disease. Natural selection cannot select for exclusively heterozygous individuals, because even if one started with such a population, half of their offspring would be homozygotes of one or the other type. Overdominance instead results in an asymptotic equilibrium distribution of all three genotypes, and analysis shows (Exercise 6) that this equilibrium is the Hardy-Weinberg distribution: as $t \to \infty$, $P \to P^*, D = P^2 \to P^{*2}, H = P - D \to P^*(1 - P^*), R = 1 + D - 2P \to (1 - P^*)^2$.

IV If $\Delta_D < 0, \Delta_R < 0$ then both homozygous genotypes AA, aa have lower death rate (greater fitness) than the heterozygous genotype Aa. This situation is referred to as *underdominance*; the selection against heterozygotes promotes survival of homozygotes only, but in eliminating heterozygotes, only one type of homozygote may survive (otherwise there will continue to be heterozygotes under random mating). Which homozygote (and allele) survive depends on both the relative advantages $-\Delta_D, -\Delta_R$ and the initial frequency of each type. Model analysis (Exercise 7) shows that P^* is an unstable equilibrium here which separates the interval (0,1) into two basins of attraction (for $P = 0$ (A) and $P = 1$ (a)): if the initial frequency of the A allele is less than P^*, it will disappear, replaced by a, but if its initial frequency is greater than P^*, then it will become fixated and the recessive allele a will gradually disappear from the population.

An example of underdominance occurs in the African butterfly *Pseudacraea eurytus*, which employs several different forms of mimicry in order to avoid predators (see Figure 7.23). Two alleles each duplicate the appearance of a different local species of butterfly, one orange and one blue, which are toxic to predators (so predators leave them alone). However, heterozygotes have an intermediate appearance which resembles nothing in particular, and thus are subject to predation. A less understood example of underdominance is the interaction of two alleles called HLA-DR3 and HLA-DR4 which increase a human's risk of developing type 1 diabetes, especially in heterozygous DR3/DR4 individuals. Despite underdominance promoting elimination of one allele (historically, diabetes was nearly universally fatal prior to the development of insulin treatments), both persist in some populations, but they are not the only alleles for the given gene.

In the first two cases the fittest genotype, a homozygote, is the only one that survives long-term. However, this is not true in the other cases. Genotype-dependent fitness (in particular, death) therefore clearly affects long-term gene frequencies, with both allele and genotype frequencies time-varying, and asymptotic to Hardy-Weinberg proportions only in the case of overdominance.

In many singular perturbation problems which model problems of viscous fluid flow, the boundary layer (inner solution, associated with fast time) has more significance. In biological models, the quasi-steady-state solution (outer solution, associated with slow time) often has more significance. The situation

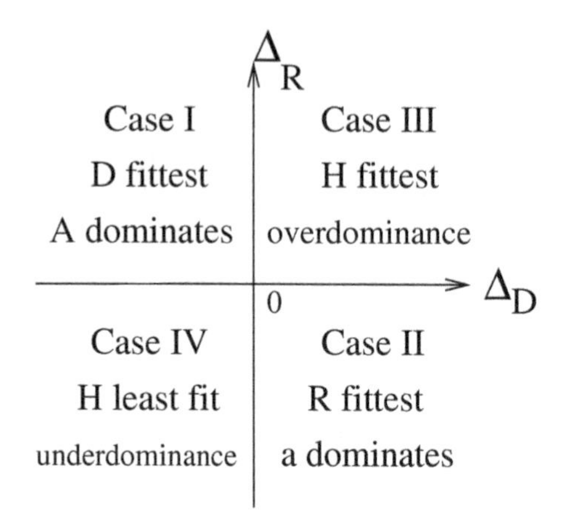

FIGURE 7.22: The slow dynamics of equation (7.56) are determined by the signs of the differential mortality rates Δ_D and Δ_R.

FIGURE 7.23: Three variants of the African mimic butterfly *Pseudacraea eurytus*, called the False Wanderer, as drawn by William Chapman Hewitson in *Illustrations of new species of exotic butterflies: selected chiefly from the collections of W. Wilson Saunders and William C. Hewitson*, Volume III. John van Voorst, London, 1868 (where it is referred to as *Diadema eurytus*).

described in this section is one where both the short-term and long-term behavior are of interest.

Exercises

1. Prove that the differential equation (7.45) has a globally asymptotically stable equilibrium which is positive if $\Lambda'(0) > d$ and zero otherwise, as follows:

   (a) Suppose that $\Lambda'(0) > d$. Explain why (7.40) implies that the graph of $y = \Lambda(x)$ will nevertheless eventually cross the line $y = d \cdot x$ (from above to below). [Recall $\Lambda(0) = 0$.]

   (b) Suppose that for some x_0 $\Lambda(x_0) < d \cdot x_0$. Explain why (7.40) implies that the graphs of $y = \Lambda(x)$ and $y = d \cdot x$ cannot intersect for $x > x_0$ (this covers both the case $\Lambda'(0) < d$ and the case where

$\Lambda'(0) > d$ but the graph of $y = \Lambda(x)$ has already crossed below the line $y = d \cdot x$).

(c) The two previous parts prove that a unique positive equilibrium exists iff $\Lambda'(0) > d$. The extinction equilibrium $N^* = 0$ of (7.45) always exists, by inspection. Use linearization to determine stability conditions for each equilibrium and complete the proof of the desired result.

2. In this section we analyzed the stability of the equilibrium (K, P_0^2) of system (7.45),(7.46). Although the right-hand side of (7.46) is undefined at $N = 0$, we can extend the system to this point by observing that $\lim_{N\to 0} \Lambda(N)/N = \Lambda'(0)$ from the definition of the derivative, and supposing that $\Lambda'(0)$ is positive (and finite). Then let $D' = h(N)(P^2 - D)$, where $h(N) = \frac{\Lambda(N)}{N}$ if $N > 0$ and $h(0) = \Lambda'(0)$. Then $(0, P_0^2)$ is an equilibrium of this system. Determine the condition for its stability by taking the limit of the expressions related to the Jacobian in the text, as $N \to 0$. How does this fit with the stability of (K, P_0^2)?

3. Derive the differential equation (7.56) from (7.54) using the change of independent variable $s = \epsilon t$ and the slow-time equilibrium condition $D = P^2$.

4. Complete the analysis of Case I for equation (7.56), in which $\Delta_D < 0, \Delta_R > 0$. You will need to consider two subcases, depending on the sign of $\Delta_D + \Delta_R$, but the result should be the same for both.

5. Complete the analysis of Case II for equation (7.56), in which $\Delta_D > 0, \Delta_R < 0$.

6. Complete the analysis of Case III for equation (7.56), in which $\Delta_D > 0, \Delta_R > 0$.

7. Complete the analysis of Case IV for equation (7.56), in which $\Delta_D < 0, \Delta_R < 0$.

8. Determine the asymptotic behavior of the system (7.56) in the case $\Delta_D = 0$, and interpret the results in terms of the genotype distribution.

9. Determine the asymptotic behavior of the system (7.56) in the case $\Delta_R = 0$, and interpret the results in terms of the genotype distribution.

10. * Rewrite system (7.48) to incorporate genotype-dependent reproduction (while maintaining genotype-independent mating preferences and death): Let AA homozygous individuals have $1 + \epsilon\Delta_D$ times as many offspring as heterozygous individuals, and aa homozygous individuals $1 + \epsilon\Delta_R$ times as many. Derive new differential equations for N, D, P.

7.5 Second-order differential equations: Acceleration

7.5.1 The harmonic oscillator

The simple harmonic oscillator, described by the second-order differential equation

$$y'' + \omega^2 y = 0,$$

is an elementary description of motion under a restoring force proportional to displacement. It arises in many physical situations such as mechanical springs,[12] and is a prototypical example of periodic motion described by its general solution

$$y = c_1 \cos \omega t + c_2 \sin \omega t.$$

Every second-order differential equation may be viewed as a system of two first-order differential equations (as explained immediately below), and thus the techniques we have developed to study first-order equations, including the machinery (singular perturbations) we have developed in this chapter, are also applicable to second-order equations. In fact, as the first-order system which corresponds to a second-order equation has a particular form, some simplifications are possible for second-order equations.

The general second-order autonomous differential equation

$$y'' = F(y, y'), \tag{7.57}$$

with F a function of two variables, may be converted to a system by defining $z = y'$. Then $y'' = z' = F(y, z)$, and the equation (7.57) is converted to the system

$$\begin{aligned} y' &= z, \\ z' &= F(y, z). \end{aligned} \tag{7.58}$$

It is easy to verify that if $(y(t), z(t))$ is a solution of (7.58), then $y(t)$ is a solution of (7.57), and vice versa: if $y(t)$ is a solution of (7.57), then $(y(t), y'(t))$ is a solution of (7.58). Thus we can switch between the second-order equation (7.57) and the system (7.58) at will, and we could develop all the theory of second-order equations as a special case of the theory for first-order systems. We shall consider below the simplest special case, the linear homogeneous equation with constant coefficients.

The general second-order linear homogeneous differential equation with constant coefficients has the form

$$y'' + py' + qy = 0, \tag{7.59}$$

[12]Consider, for example, Hooke's Law $F = -kx$ in conjunction with Newton's second law $F = ma = mx''$, which produce the equation $mx'' = -kx$ equivalent to the equation given above.

where p and q are constants. To transform this into a system, we let $y' = z$, so that $y'' = z' = -py' - qy = -pz - qy$. Thus the corresponding system is

$$\begin{aligned} y' &= z, \\ z' &= -qy - pz, \end{aligned} \tag{7.60}$$

with coefficient matrix

$$\begin{bmatrix} 0 & 1 \\ -q & -p \end{bmatrix}.$$

The trace of this matrix is $-p$, and the determinant of this matrix is q. To solve the system (7.60), we write its characteristic equation, which takes the form

$$\lambda^2 + p\lambda + q = 0. \tag{7.61}$$

In contrast to the solution of systems described in Section 5.3, now we need only the solution for y; z may be obtained from y by differentiation. Thus (paralleling the discussion in Section 5.3) if the characteristic equation (7.61) has two distinct roots λ_1 and λ_2, we obtain the solutions $y = e^{\lambda_1 t}$ and $e^{\lambda_2 t}$ of (7.59), and the general solution is $y = K_1 e^{\lambda_1 t} + K_2 e^{\lambda_2 t}$. If the characteristic equation (7.61) has a double root λ_1, the general solution is $y = K_1 e^{\lambda_1 t} + K_2 t e^{\lambda_1 t}$. If the characteristic equation (7.61) has complex conjugate roots $\alpha \pm i\beta$, the general solution of (7.59) is $y = e^{\alpha t}(K_1 \cos \beta t + K_2 \sin \beta t)$. In each case, if a solution satisfying prescribed initial conditions is sought, the constants K_1 and K_2 may be determined from the initial conditions. For systems, it is appropriate to prescribe initial values for y and z; for a second-order equation the analogue is to prescribe initial values $y(0)$ and $y'(0)$.

EXAMPLE 1.
Find the general solution of the differential equation $y'' + 4y' + 3y = 0$, and the particular solution satisfying the initial conditions $y(0) = 1$, $y'(0) = 0$.

Solution: The characteristic equation is $\lambda^2 + 4\lambda + 3 = 0$, with roots -3 and -1. Thus the general solution of the differential equation is $y = K_1 e^{-3t} + K_2 e^{-t}$. To satisfy the given initial conditions, we differentiate this general solution, obtaining $y' = -3K_1 e^{-3t} - K_2 e^{-t}$ and then substitute $t = 0$, $y = 1$, $y' = 0$. This gives the conditions $1 = K_1 + K_2$, $0 = -3K_1 - K_2$, whose solution is $K_1 = -\frac{1}{2}$, $K_2 = \frac{3}{2}$. Thus the solution of the initial value problem is $y = -\frac{1}{2}e^{-3t} + \frac{3}{2}e^{-t}$. □

EXAMPLE 2.
Find the solution of the initial value problem

$$y'' + 2y' + y = 0, \quad y(0) = 0, \ y'(0) = 1.$$

Solution: The characteristic equation $\lambda^2 + 2\lambda + 1 = 0$ has a double root $\lambda = -1$. Thus the general solution of the differential equation is $y = K_1 e^{-t} + K_2 t e^{-t}$, and $y' = -K_1e^{-t} + K_2e^{-t} - K_2te^{-t} = (K_2 - K_1)e^{-t} - K_2te^{-t}$. Substitution of $t = 0$, $y = 0$, $z = 1$ gives $K_1 = 0$, $K_2 = 1$, and the solution is $y = -te^{-t}$. □

EXAMPLE 3.

Find the general solution of the differential equation $y'' + \omega^2 y = 0$.

Solution: The characteristic equation $\lambda^2 + \omega^2 = 0$ has complex conjugate roots $\pm i\omega$, and thus the general solution of the differential equation is $y = K_1 \cos \omega t + K_2 \sin \omega t$. □

EXAMPLE 4.

We can consider the motion of the human diaphragm in breathing as that of a massive spring being moved up and down by actuating forces: muscles which pull it down during inhalation (and, to some extent, up during exhalation), the natural springlike restoring force of the diaphragm itself, and a frictionlike resistance proportional to (and opposite) the movement. If we define y to be the downward displacement of the diaphragm ($y = 0$ at rest), we can write a simple model for the net force F acting on the diaphragm:

$$F = f - ky - ry',$$

where f is the external muscular forcing, k is the proportionality constant for the restoring force, and r is the proportionality constant for the resistive force.

Using Newton's second law, $F = ma = my''$, write a second-order differential equation for y, the corresponding first-order system, and the characteristic equation for the homogeneous ($f = 0$) system.

Solution: We substitute for F in the first equation given above, and move all the terms involving y to the left side, to obtain the model

$$my'' + ry' + ky = f.$$

We can write this as the following system, defining $z = y'$:

$$\begin{aligned} y' &= z, \\ z' &= -\frac{k}{m}y - \frac{r}{m}z + \frac{f}{m}. \end{aligned}$$

Finally, the characteristic equation (which can be read from the single second-order equation with $f = 0$, cf. (7.59), (7.61)) is $m\lambda^2 + r\lambda + k = 0$.

Note, however, that because of the external forcing f, this model is not homogeneous, so finding the solution is somewhat more complicated than finding the solution to a homogeneous equation, as done in the previous examples of this section. □

7.5.2 The van der Pol oscillator

Nonlinear differential equations of second-order may also be reduced to systems, and their equilibria and stability may be studied as for systems. In the remainder of this section, we consider an equation known as the *van der Pol equation*, which arose originally as a model for cardiac oscillations, but has also had many applications in mechanical and electrical problems. The Dutch scientist Balthazar van der Pol discovered in the 1920's an electrical circuit which produced sustained oscillations, and proposed it as a model for the pacemaker of the heart.[13] In this equation, $y'' + \mu(y^2 - 1)y' + y = 0$, y originally represented the level of electric potential, with μ a scaling factor regulating the strength of the nonlinearity (note that $\mu = 0$ gives us a special case of the simple linear oscillator with which we began this section).

EXAMPLE 5.
Write the van der Pol equation as a system.

Solution: We let $z = y'$, and then $z' = y'' = -\mu(y^2-1)y' - y = -\mu(y^2-1)z - y$, and we have the system

$$\begin{aligned} y' &= z, \\ z' &= -\mu(y^2 - 1)z - y. \quad \square \end{aligned} \tag{7.62}$$

We note that because of the special form of the converted system (7.58), any equilibrium of such a system satisfies $z = 0$, and thus may be found simply by solving the equation $F(y, 0) = 0$ for y. The linearization of the system (7.58) at an equilibrium $(y_\infty, 0)$ is the system

$$\begin{aligned} u' &= v, \\ v' &= F_y(y_\infty, 0)u + F_z(y_\infty, 0)v, \end{aligned}$$

which is equivalent to the second-order differential equation

$$u'' = F_z(y_\infty, 0)u' + F_y(y_\infty, 0)u,$$

so we could write down this linearization directly, without transforming to a system and back.

EXAMPLE 6.
Find the equilibria of the van der Pol equation, the linearization at each equilibrium, and the stability of each equilibrium.

Solution: An equilibrium of (7.62) is found by solving $z = 0$, $-\mu(y^2-1)z - y =$

[13] B. van der Pol and J. van der Mark, The heartbeat considered as a relaxation oscillation, and an electrical model of the heart, *Phil. Mag.* 6 (1928), 763–775.

0, and thus the only equilibrium is $y = 0$, $z = 0$. From above, the linearization of the van der Pol equation about this equilibrium is

$$u'' = \mu u' + (-1)u,$$

or

$$u'' - \mu u' + u = 0,$$

since the linear approximation to $y^2 - 1$ at $y = 0$ is -1. The roots of the characteristic equation $r^2 - \mu r + 1 = 0$ are $r = \frac{1}{2}\left(\mu \pm \sqrt{\mu^2 - 4}\right)$. Since $\mu > 0$, these roots are both positive if $\mu > 2$, and complex conjugates with positive real part if $\mu < 2$; in either case, the origin is an unstable equilibrium. □

It is possible to show, although not without some difficulty, that every orbit of the system (7.62) is bounded, and this implies by the Poincaré-Bendixson theorem that there is a limit cycle. In fact, there is a unique limit cycle approached by every orbit, as suggested by the phase portrait in Figure 7.24 with $\mu = 1.5$.

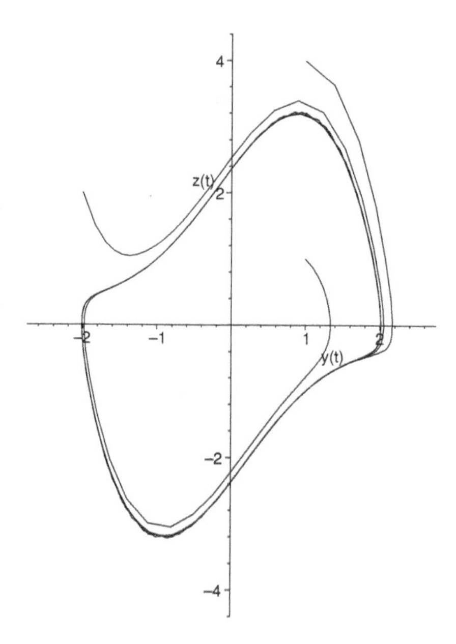

FIGURE 7.24: Phase portrait for system (7.62).

If we wish to treat the van der Pol equation as a second-order differential equation $y'' + \mu(y^2 - 1)y' + y = 0$, we would find equilibria as constant solutions (with $y' = 0$, $y'' = 0$), and thus $y = 0$ is the only equilibrium. The linearization at this equilibrium is the second-order differential equation $y'' - \mu y' + y = 0$ identified above.

The van der Pol equation is usually treated as a system, but more information can be derived by using a different system from (7.62) which allows us to see more clearly the role of the parameter μ in determining the shape of solutions (recall μ determines the strength of the nonlinear term). We first define the function $f(y) = y^2 - 1$, so that the van der Pol equation may be written as $y'' + \mu f(y)y' + y = 0$. We then define

$$F(y) = \int_0^y f(u)\,du = \frac{y^3}{3} - y, \quad w = y' + \mu F(y).$$

Then by the Chain Rule

$$w' = y'' + \mu F'(y)y' = -(\mu f(y)y' + y) + \mu f(y)y' = -y,$$

and we have a system

$$\begin{aligned} y' &= w - \mu F(y), \\ w' &= -y. \end{aligned}$$

It is also convenient to make a change of scale by letting $w = \mu z$, to give the new system

$$\begin{aligned} y' &= \mu(z - F(y)), \\ z' &= -\frac{1}{\mu}y. \end{aligned} \tag{7.63}$$

The advantage of the form (7.63), discussed in more detail below, is that for large μ the two variables operate on quite different time scales if μ is large.

It is easy to see that the only equilibrium of (7.63) is (0,0). Since both (7.62) and (7.63) are systems equivalent to the van der Pol equation, we already know that the origin must be unstable for (7.63) since we have shown that it is unstable for (7.62). It is possible to show that all orbits of (7.63) are bounded, so that we may apply the Poincaré-Bendixson theorem. The Poincaré-Bendixson theorem gives the existence of a limit cycle. A more refined argument shows that this limit cycle is unique, and is approached by every orbit except the constant orbit $y = 0$, $z = 0$.

If μ is large, we can give a more precise description of the orbit. The y-nullcline is the cubic curve $z = F(y)$, and the z-nullcline is the axis $y = 0$. Except near the y-nullcline, y' is large and z' is small, and thus the motion is essentially horizontal in the y-z plane (i.e., in the y direction). Above $z = F(y)$, the trajectory moves to the right. When it reaches the y-nullcline (and crosses it vertically), y' and z' are comparable in magnitude, and the orbit follows the nullcline to its minimum at $(1, -\frac{2}{3})$. Then it goes horizontally to the left until it reaches the nullcline at $(-2, -\frac{2}{3})$, after which it follows the nullcline up to its maximum at $(-1, -\frac{2}{3})$, and then repeats the cycle. Figure 7.25 shows this orbit, traversed clockwise, superimposed on the y-nullcline.

For large μ, the oscillations consist of time intervals in which y increases

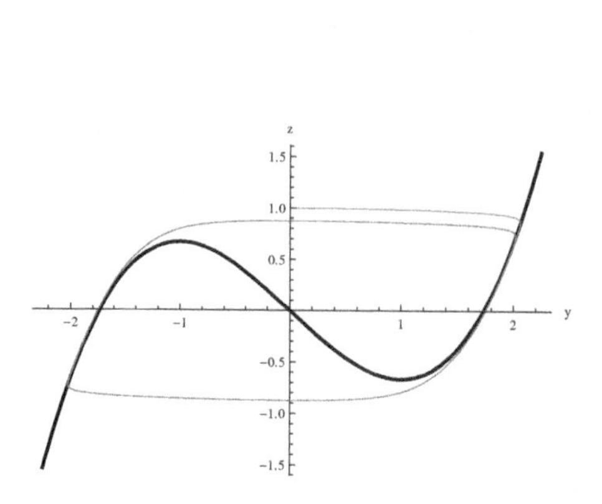

FIGURE 7.25: Representative solution (in gray) for system (7.63) beginning at (0,1) for $\mu = 5$, with y-nullcline (in black) superimposed.

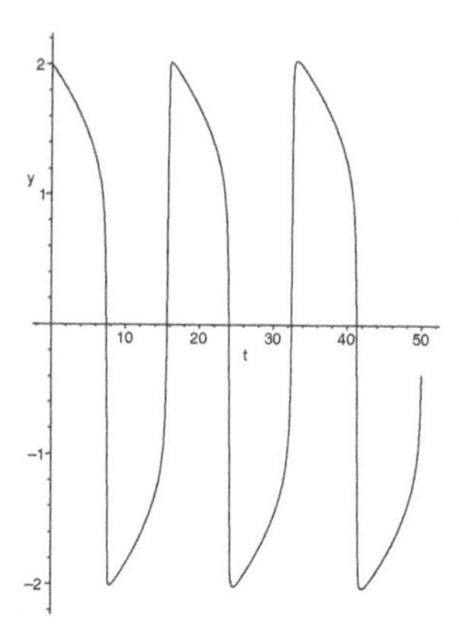

FIGURE 7.26: A solution of system (7.63) with $\mu = 8$.

or decreases rapidly (primarily horizontal movement in Figure 7.25), alternating with time intervals in which y remains almost constant (primarily vertical movement in Figure 7.25). Such oscillations are called *relaxation oscillations*; Figure 7.26 displays a solution with $\mu = 8$. Such oscillations, with relatively slow changes in y interrupted by sudden changes, are similar to cardiac electrical activity, and modifications of the van der Pol equation have been used as models for the study of oscillations in cardiac tissues.

7.5.3 A model of oxygen diffusion in muscle fibers

The muscles of the body require a continuous supply of oxygen even when at rest, due to the need to maintain certain conditions (chemical and electrical) within the cells. Most muscles are formed of long, thin fibers, which oxygen enters from the periphery (delivered via the blood system). Oxygen and oxymyoglobin (oxygenated myoglobin) diffuse throughout the fibers to transport oxygen to the cells in the (axial) centers of the fibers. We can make a simple model to describe this process in the following way.

In reality, diffusion occurs in all three spatial dimensions within the muscle fibers. However, to simplify our model, we will consider the fiber as a long, thin cylinder of uniform radius a, with all diffusion occurring radially. That is, we suppose that there is no significant axial component of diffusion (parallel to the fiber axis), and that the concentrations of oxygen and oxymyoglobin at any given point inside the fiber depend only on the distance r from the center of the fiber (or, equivalently, from the boundary). In mathematical terms, we use cylindrical coordinates (r, θ, z) to describe location inside the muscle fiber, but assume that the diffusion of oxygen and oxymyoglobin is independent of axial height z and angle θ (around the axis, see Figure 7.27). Assuming these two symmetries allows us to consider only the matter of how oxygen is transported from the boundary of the fiber (at $r = a$) inward to the center ($r = 0$).

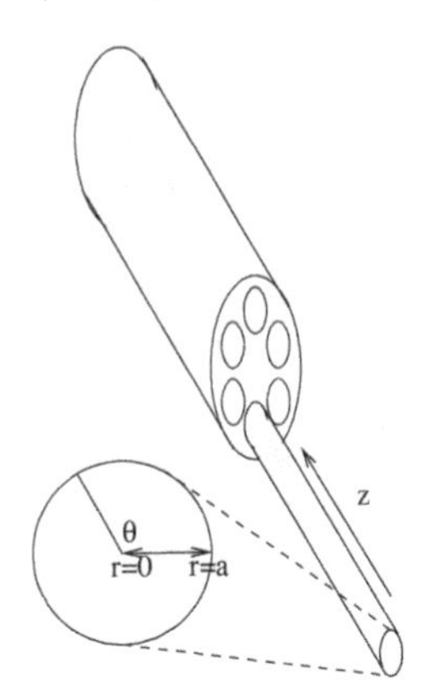

FIGURE 7.27: Viewing a single muscle fiber in a bundle as a cylinder, taking a cross-section fixes z, with $r = 0$ at the center (axis) of the fiber and $r = a$ at the edge.

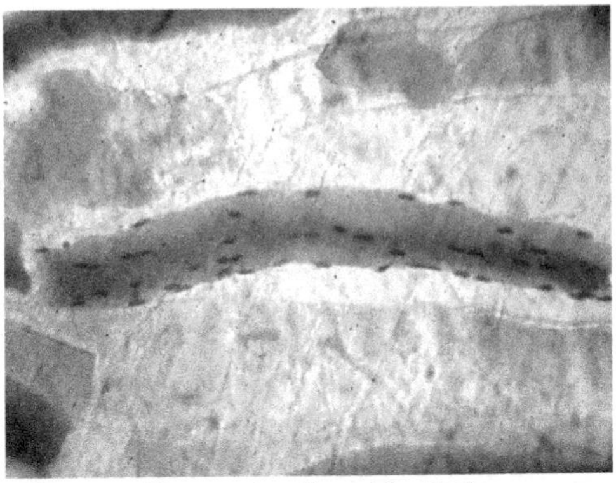

Photo courtesy Howard Swatland

FIGURE 7.28: A single muscle fiber (myofibre) is roughly cylindrical in shape. As they are formed by the fusion of myoblasts, these long cells have many nuclei (visible as dark spots on the surface).

What we need to do now is account for the radial diffusion of oxygen for $0 \leq r \leq a$. There are three factors which change the concentration of oxygen: the uptake of oxygen by myoglobin to form oxymyoglobin, the ongoing consumption of oxygen by the muscle cells, and the diffusion process. We will suppose that the oxygenization of myoglobin takes place at a rate f, and that oxygen is being consumed at a constant rate g (constant, at least, with respect to r: let us consider a muscle either at rest or at a given level of activity). This leaves the diffusion term. The function commonly used to describe the diffusion of a substance is the *Laplacian operator*, which, when converted into the cylindrical coordinates of our model, has radial component $\frac{1}{r}\frac{d}{dr}\left(r\frac{d\cdot}{dr}\right)$. Once the concentration of oxygen within the muscle reaches a steady-state distribution, these three factors (the first two negative, since oxygen is being removed via them, and the third positive) balance out, so that their sum is zero.

This gives us the model

$$K_1 \frac{1}{r}\frac{d}{dr}\left(r\frac{dy}{dr}\right) - f - g = 0 \tag{7.64}$$

for the amount y of oxygen (O_2) at a distance r from the center of the fiber, where the constant K_1 describes the rate of diffusion. Since (under our assumptions of symmetry) the diffusion at the center is even, the rate of change of the concentration should level off as one approaches the center; this gives us one boundary condition, $\frac{dy}{dr}(0) = 0$. The concentration of oxygen entering the fiber from outside (at $r = a$) is known, giving us our second boundary condition, $y(a) = Y_a$. (Since the diffusion begins outside the fiber boundaries, we cannot know that $\frac{dy}{dr}(a) = 0$.)

By completing the differentiation in (7.64) we can obtain a second-order differential equation, which can be rewritten as a first-order system and analyzed using the methods illustrated in this section. Exercises 19–22 below outline some extensions and analysis of model (7.64). The simplest model gives a quadratic distribution for $y(r)$, reflecting the higher concentration closer to the source ($r = a$). Mathematically the most interesting feature of the model and its extensions is that the nature of the first boundary condition, specifying dy/dr rather than r at 0, means there are no boundary layers for y: a slight correction (corresponding to a fast-time or inner solution) is needed only for $dy/dr(0)$, not for $y(0)$.

A more realistic model would consider the specific locations (in terms of angle θ) of the surface capillaries from which diffusion originates, and might further consider what happens when the muscle contracts, increasing the radius and changing the locations of the capillaries.

Exercises

In each of Exercises 1–12, find the general solution and the solution satisfying the initial conditions $y(0) = 1$, $y'(0) = -1$.

1. $y'' = 0$
2. $y'' - 9y = 0$
3. $y'' - 5y' + 6y = 0$
4. $y'' + 9y = 0$
5. $y'' + 10y' + 25y = 0$
6. $4y'' - y = 0$
7. $y'' + 5y' + 10y = 0$
8. $4y'' + 4y' + 13y = 0$
9. $y'' - 4y' + 13y = 0$
10. $y'' - 4y' + y = 0$
11. $y'' - 2y' + 2y = 0$
12. $\epsilon y'' + 2y' + y = 0 \quad (0 < \epsilon < 1)$

13. Find all equilibria of the differential equation $y'' + ay' + by + y^2 = 0$, where a and b are real constants and $b \neq 0$.

In each of Exercises 14–17 find all equilibria of the given differential equation, and the linearization at each equilibrium.

14. $y'' + y(1 - y^2) = 0$
15. $y'' + y' + y^3 = 0$

16. $y'' + g(y) = 0$, where $g(y)$ is a smooth function with $yg(y) > 0$ for $y \neq 0$.

17. $y'' + f(y)y' + g(y) = 0$, where $f(y)$ is a smooth function with $f(y) > 0$ for all y and $g(y)$ is a smooth function with $yg(y) > 0$ for $y \neq 0$.

18. * To illustrate the complexity of the structure when the origin is not an isolated equilibrium, consider the differential equation $y'' + y' = 0$.

    (a) Show that there is a line of equilibria.

    (b) Sketch the phase portrait in the y-y' phase plane.

19. In the model (7.64), treating the conversion rate f as a constant, complete the differentiation in the above model, and write it first as a second-order differential equation, and then as a system of two first-order equations. Can you solve the resulting equation for y'?

20. * Given that oxymyoglobin is not consumed within the muscle fiber (it acts to transport the oxygen), but that the oxygenization process mentioned above increases the amount of oxymyoglobin present, and also that the oxymyoglobin does not leave the fiber (i.e., diffusion is limited to $0 \leq r \leq a$), write an equation modeling the radial diffusion of oxymyoglobin in the muscle fiber, analogous to the one given above for oxygen, using $w(r)$ to represent the [radial] concentration of oxymyoglobin.

21. * In fact, f is not a constant, but depends on the amounts of reactants (oxygen, myoglobin and oxymyoglobin) present, as well as on some constants of proportionality. Let us take $f = K_3xy - K_4w$, where w is the concentration of oxymyoglobin, x and y are the concentrations of myoglobin and oxygen, respectively, and K_3 and K_4 are the constants of proportionality. Rewrite the second-order equations for y and w using this expression for f, and transform them into a system of four first-order differential equations.

22. * Approximate values for the parameters in this model are as follows: $a = 2.5 \times 10^{-3}$ cm, $Y_a = 3.5 \times 10^{-8}$ mol/cm^3, $g = 5 \times 10^{-8}$ mol/cm^{3}sec, $K_1 = 10^{-5}$ cm^2/sec, $K_2 = 5 \times 10^{-7}$ cm^2/sec (diffusion rate constant for oxymyoglobin), $K_3 = 2.4 \times 10^{10}$ cm^3/mol sec, $K_4 = 65$/sec, $x = 2.8 \times 10^{-7}$ mol/cm^3 (the concentration of myoglobin in muscle fibers).[14] Use a computer to find the distributions the two models predict for oxygen and myoglobin as functions of r.

[14] J. Keener and J. Sneyd, *Mathematical Physiology*, Springer-Verlag, New York (1998), p. 42.

Part III

Appendices

Appendix A

An Introduction to the Use of Maple^{TM}

Maple is a computer algebra system which can assist you in solving mathematical problems which would be quite inaccessible without its help. This introduction is not intended to make you an expert on Maple but it should be a reasonably self-contained start which will enable you to handle many problems in the analysis of ordinary differential equations and difference equations which are found in this book. Maple is intended as an aid, not as a replacement for thinking. Ideally, you will begin by working on a problem to understand what is required and will turn to Maple to provide graphic output where appropriate. While Maple has considerable capacity for calculation (good enough to pass most calculus courses), this is an aspect which we shall try not to emphasize. Our use of Maple will be concentrated on

(i) plotting graphs of functions or pairs of functions and thus finding intersections of two graphs,

(ii) graphical solution of ordinary differential equations, including first-order equations and systems of first-order differential equations,

(iii) difference equations, including the cobweb method for first-order difference equations and the graphing of solutions of difference equations of first or higher order by iterative calculations.

Maple can be run on many different kinds of computers, including Windows, Macintosh, and many UNIX systems. On many university campuses it is installed on computers in student labs. In addition, for students planning on a career which will use mathematics, the "academic version" of Maple available for sale at a moderate price in many universities is an excellent investment.

Maple can be used for many "word-processor" functions to dress up your worksheets and convert them into more formal reports. You can toggle between Maple ([>) and text (T) input on the toolbar in order to add text. Inside a Maple computation you can add comments; any Maple input beginning with # is treated as a comment and is ignored as far as calculation is concerned. Note that every time you hit the "Enter" key you begin a new input and if you want a comment to extend for more than a paragraph you need to begin each paragraph with "#".

Here are a few general instructions for using Maple. They are not meant to be complete, and a manual on the use of Maple or the online Help available in Maple contains much more useful information.

Every input for which you want Maple to do something must end with a semicolon. If you want Maple to act but not display the output you would use a colon.

The command `restart` at the beginning of a worksheet tells Maple to forget any previous instructions. If you have been working on one problem and defined a symbol and then begin working on another problem which uses the same symbol, this instruction will avoid confusion.

The command `with` tells Maple to read in a package containing some additional definitions. For example, when we solve differential equations we will always use the package DEtools and thus will begin the worksheet with the command `with(DEtools):`. If we use a semicolon rather than a colon at the end of the command, Maple will print out a list of the commands included in the package.

The syntax for defining a quantity for later use is : =. Thus the command `a: = 1` assigns the value 1 to the quantity a. This syntax is also used to define functions.

A.1 Plotting graphs of functions

The basic command to plot the graph of a given function on a given interval is

```
> plot(f(x),x=a..b);
```

Options include specification of the range of the ordinate, specification of an infinite interval, and plotting of more than one function on the same set of axes, using the formats

```
> plot(f(x),x=a..b,y=c..d);
> plot(f(x),x=0..infinity);
> plot([f(x),g(x)],x=a..b);
```

The command to plot a graph given parametrically is

```
> plot([f(t),g(t),t=a..b]);
```

The option `scaling=constrained` may be used to make the units on the two axes the same. This may give a more accurate geometric picture but may hide important information if the units are on a different scale. Figures drawn with Maple can be exported as .eps files.

We give a few examples to illustrate.

```
> restart:
> f(x):=x*exp(-x): g(x):=0.2*x:
> plot(f(x),x=0..5);
```

(see Figure A.1)

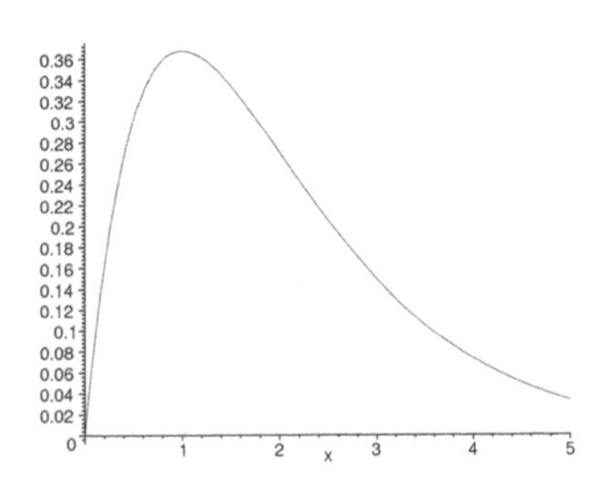

FIGURE A.1: Graph of $f(x) = xe^{-x}$.

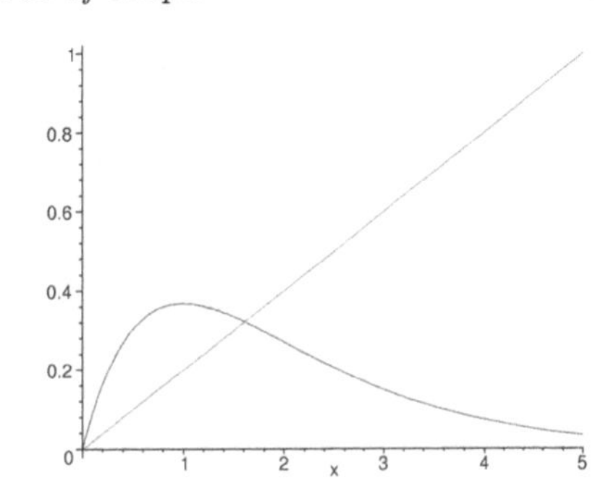

FIGURE A.2: Graph of $f(x) = xe^{-x}, g(x) = 0.2x$.

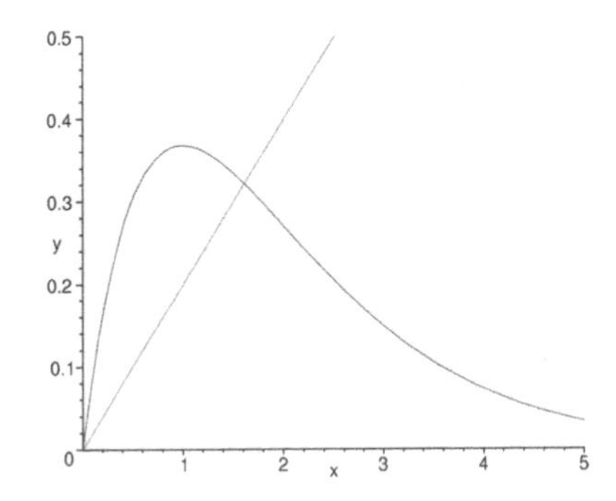

FIGURE A.3: Graph of $f(x) = xe^{-x}, g(x) = 0.2x$,y-axis specified.

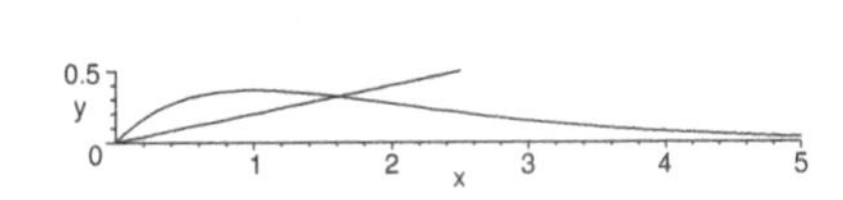

FIGURE A.4: Graph of $f(x) = xe^{-x}, g(x) = 0.2x$, scaling constrained.

```
> plot([f(x),g(x)],x=0..5);
```
(see Figure A.2)
```
> plot([f(x),g(x)],x=0..5,y=0..0.5);
```
(see Figure A.3)
```
> plot([f(x),g(x)],x=0..5,y=0..0.5,scaling=constrained):
```
(see Figure A.4)

A.2 Graphical solution of first-order differential equations

The Maple package DEtools includes commands which draw the direction field for a first-order differential equation, the direction field together with solution curves corresponding to prescribed initial values, or solution curves corresponding to prescribed initial conditions without the direction field. Each of these can be carried out using the command `DEplot`. Our description here will be restricted to its use for first-order differential equations but the command may be used for systems of differential equations and equations of order higher than one. We shall describe its use for systems later.

The command `DEplot` is followed by up to six arguments, namely the differential equation, the variable to be plotted, the range of the independent variable, the initial conditions (if solution curves are to be drawn but not if only the direction field is needed), the range of the dependent variable, and (possibly) options such as the arrows to be drawn.

A differential equation of the form $y' = f(t, y)$ may be specified by `diff(y(t),t)=f(t,y)`. There are other possible forms which may be used. However, a differential equation like $y' + y = g(t)$ may not be specified as `diff(y(t),t)+y=g(t)`; it should be written `diff(y(t),t)=-y+g(t)`.

The variable to be plotted is given as $y(t)$ and the range of the variables to be plotted is given in the form `t=a..b`, or `y=c..d`. Initial conditions are specified as lists given in the form {`[y(0)=a], [y(0) = b],...`}.

The default is to draw thin arrows. To draw solution curves without the direction field using `DEplot`, it is necessary to specify `arrows = none`. Other possible specifications for the kind of arrows include `small`, `medium`, `large`, and `line`. In the examples below, `colors` and `linecolors` have been set to `black` except where there are both arrows and solutions. This makes the printout clearer but is not necessary. Incidentally, as Maple is a system developed in Canada, it understands both the spellings "color" and "colour". We illustrate these procedures with the differential equation $y' = y(1-y)$ on the interval [0, 4], using the initial conditions $y(0) = -0.5$, $y(0) = 0$, $y(0) = 0.2$, $y(0) = 0.4$, $y(0) = 0.6$, $y(0) = 0.8$, $y(0) = 1.0$, $y(0) = 1.2$, $y(0) = 1.4$.

```
> restart: with(DEtools):
```

To draw a direction field:

```
> DEplot(diff(y(t),t)=y*(1-y), y(t), t=0..4, y=-1.0..1.4,
arrows=small,color=black);
```

(see Figure A.5)

To draw solution curves:

```
> DEplot(diff(y(t),t)=y*(1-y), y(t), t =0..4, {[y(0)=-0.5],
[y(0)=0], [y(0)=0.2], [y(0)=0.4], [y(0)=0.6], [y(0)=0.8],
[y(0)=1.0], [y(0)=1.2], [y(0)=1.4]}, y=-1.0..1.4, arrows=none,
thickness=0, linecolor=black);
```

(see Figure A.6)

To draw a direction field with solution curves:

```
> DEplot(diff(y(t),t)=y*(1-y), y(t), t =0..4, {[y(0)=-0.5],
[y(0)=0], [y(0)=0.2], [y(0)=0.4], [y(0)=0.6], [y(0)=0.8],
[y(0)=1.0], [y(0)=1.2], [y(0)=1.4]},y=-1.0..1.4, arrows=line,
thickness=0, linecolor=black);
```

(see Figure A.7)

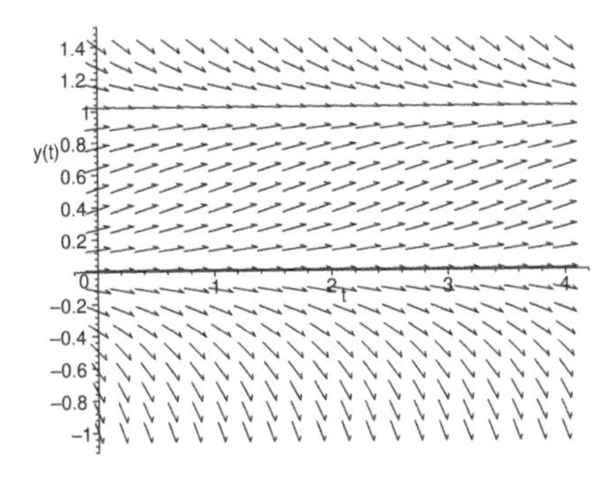

FIGURE A.5: Direction field $y' = y(1-y)$.

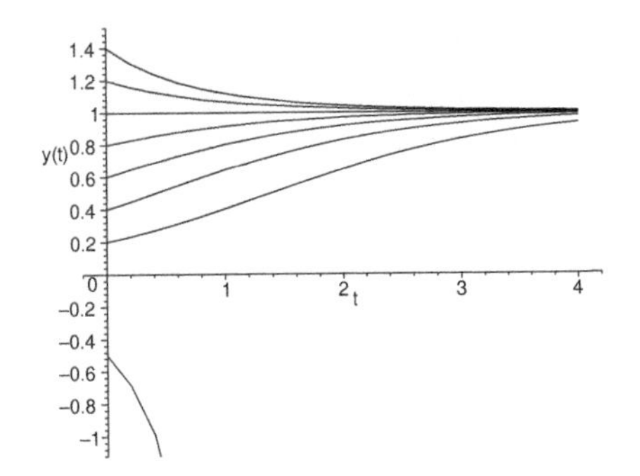

FIGURE A.6: Solution curves $y' = y(1-y)$.

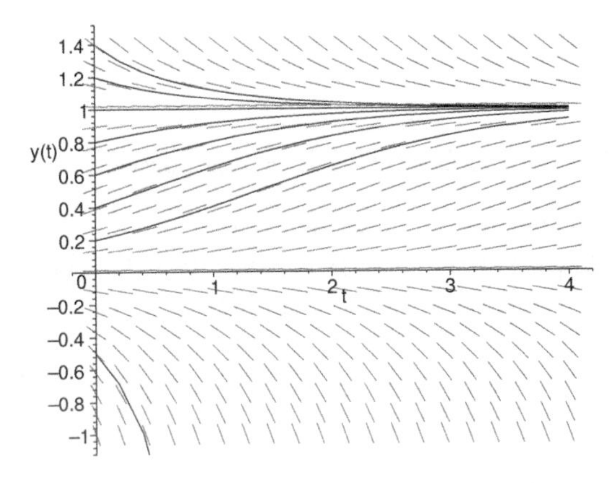

FIGURE A.7: Direction field with solution curves, $y' = y(1-y)$.

A.3 Graphical solution of systems of differential equations

The solution of an autonomous system of two first-order differential equations is similar to the procedure used for a first-order differential equation. The default is to plot the phase portrait of the two dependent variables but the "scene" option allows plotting either of the dependent variables as a function of the independent variables. The default is to show arrows in the direction field. One example should suffice to give the idea.

```
> restart: with (DEtools):
DEplot({diff(x(t),t)=1*x*(1-x/40)-x*y/(x+10),
diff(y(t),t)=1*y*10*(x-20)/(30*(x+10))-0.4},
{x(t),y(t)},t=0..25,[[x(0)=50,y(0)=10],[x(0)=50,y(0)=20],
[x(0)=50,y(0)=30],[x(0)=50,y(0)=15],[x(0)=50,y(0)=5]],
```

```
x=0..50,y=0..40,scene=[x,y],stepsize=0.05,linecolour=black,
thickness=0,arrows=none);
```

(see Figure A.8)

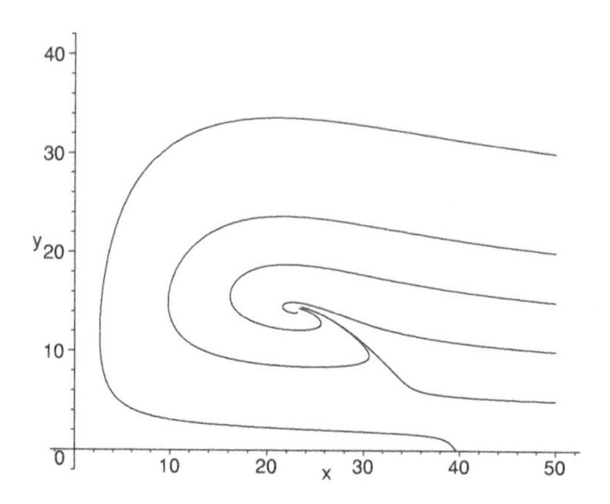

FIGURE A.8: Phase portrait.

```
> DEplot({diff(x(t),t)=1*x*(1-x/40)
-x*y/(x+10),diff(y(t),t)=1*y*10*(x-20)/(30*(x+10))-0.4},
{x(t),y(t)},t=0..25,[[x(0)=50,y(0)=10],[x(0)=50,y(0)=20],
[x(0)=50,y(0)=30],[x(0)=50,y(0)=15],[x(0)=50,y(0)=5]],
x=0..50,y=0..40,scene=[t,x],stepsize=0.05,linecolour=black,
thickness=0,arrows=none);
```

(see Figure A.9)

```
> DEplot({diff(x(t),t)=1*x*(1-x/40)
-x*y/(x+10),diff(y(t),t)=1*y*10*(x-20)/(30*(x+10))0.4},
{x(t),y(t)},t=0..25,[[x(0)=50,y(0)=10],[x(0)=50,y(0)=20],
[x(0)=50,y(0)=30],[x(0)=50,y(0)=15],[x(0)=50,y(0)=5]],
x=0..50,y=0..40,scene=[t,y],stepsize=0.05,linecolour=black,
thickness=0,arrows=none);
```

(see Figure A.10)

A.4 The cobwebbing method for graphical solution of first-order difference equations

In Section 2.2 we described the cobwebbing method for solving first-order difference equations. This graphical method requires repeated calculation of functional values and is ideally suited to execution by means of a computer program. Here we define a program called "cobweb" which needs input of a specified function (`func`), an initial value (`x0`), a number of iterations to

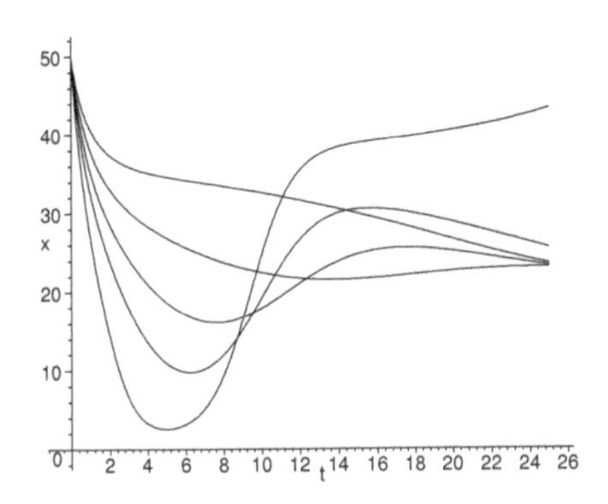

FIGURE A.9: x as function of t.

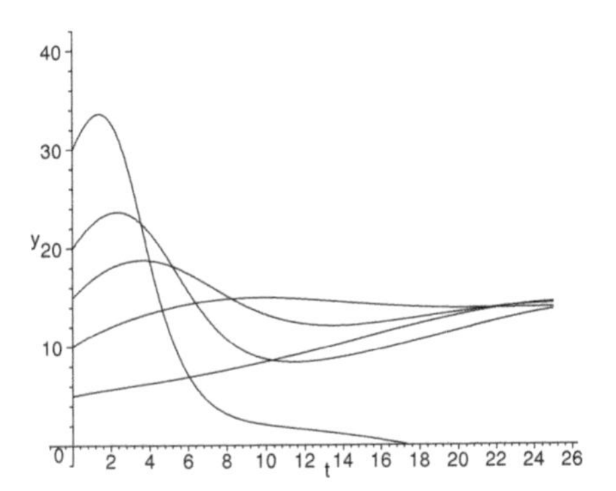

FIGURE A.10: y as function of t.

be performed (**numb**), a minimum value of the independent variable (**xmin**), and a maximum value of the independent variable (**xmax**). When these are defined and entered into the program, the cobweb method will run and plot the output.

```
> restart: with(plots): with(plottools):
> cobweb :=proc(func, x0, numb, xmin, xmax)
> local curve,diagonal, oldx, newx, lines, i, l1, l2:
> curve := plot(func(x),x=xmin..xmax):
> diagonal := line([xmin,xmin],[xmax,xmax]):
> oldx := x0: newx := func(oldx):
> lines := [line([oldx,0],[oldx, newx])]:
>   for i from 0 to numb do;
>   l1 := line([oldx, newx],[newx,newx]):
>   oldx := newx: newx := func(oldx):
>   l2 := line([oldx, oldx],[oldx,newx]):
>   lines := [op(lines),l1,l2]:
>   od:
> display(lines,curve,diagonal,scaling=
constrained, thickness=2);
> end;
```

Here is an example using the logistic difference equation. We begin by defining the function and its parameter.

```
> r:=2.9:
> logistic := x -> r*x*(1-x):
> cobweb(logistic,.45,10,0,1);
```

(see Figure A.11)

We suggest some additional examples using the logistic function all on the interval [0,1] but with different parameter values, initial points, and number of steps, namely:

1. r = 3.4, x0= 0.45, numb = 20

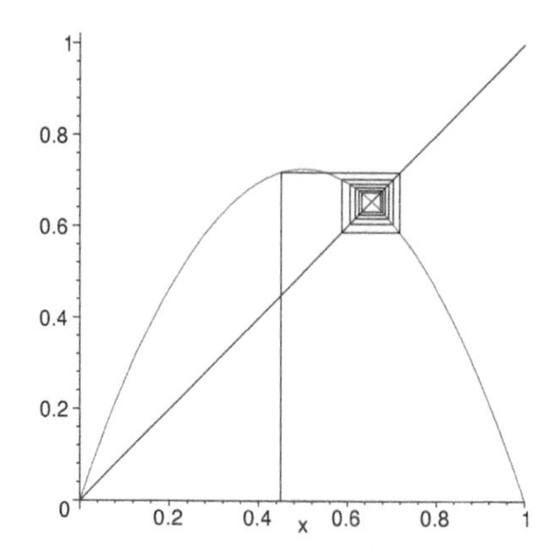

FIGURE A.11: Cobwebbing for $x_{n+1} = 3.62x_n(1 - x_n)$.

2. r = 3.5, x0 = 0.385, numb = 20
3. r = 3.58, x0 = 0.48, numb = 20
4. r = 3.8, x0 = 0.3, numb = 20
5. r = 3.8, x0 = 0.3, numb = 40
6. r = 3.8, x0 = 0.3, numb = 100
7. r = 3.8, x0 = 0.3, numb = 200.

A.5 Solution of difference equations and systems of difference equations

Here is a simple program for the iterative calculations required in solving a difference equation and plotting the resulting solution as a line segment graph.

```
> restart: with(plots):
> logistic:=y->r*y*(1-y): r:=3.62: y(0):=0.5:
>   for k from 0 to 25 do
>   y(k+1):=logistic(y(k)):
>   od:
> Q:=[seq([k,y(k)], k=0..25)]: > pointplot(Q,style=line);
```

(see Figure A.12)

The program is easily adapted to systems of difference equations, as we illustrate with an example from Section 2.8:

```
> restart: with(plots):
> r:=1: b:=0.5: p:=0.3:
```

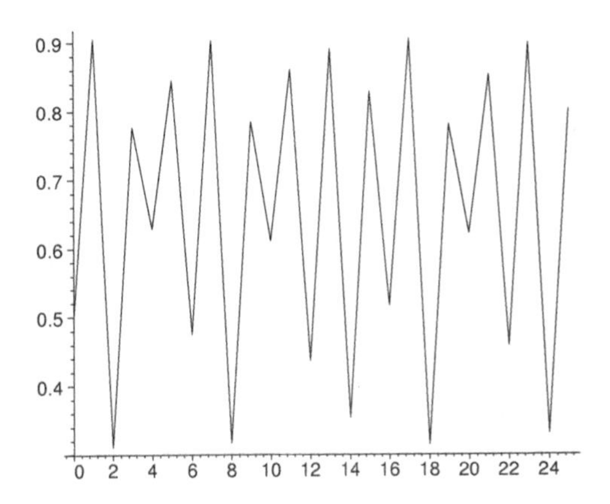

FIGURE A.12: Solution plot, $x_{n+1} = 3.62x_n(1 - x_n)$.

```
> y(0):=1: z(0):=1:
>   for k from 0 to 35 do
>   y(k+1):=r*y(k)*exp(-b*z(k))+p*y(k):
>   z(k+1):=r*y(k)*(1-exp(-b*z(k))):
>   od:
> Q:=seq([k,y(k)],k=0..35):
> R:=seq([k,z(k)],k=0..35):
> pointplot(Q,style=line);
> pointplot(R,style=line);
```

(see Figures A.13 and A.14)

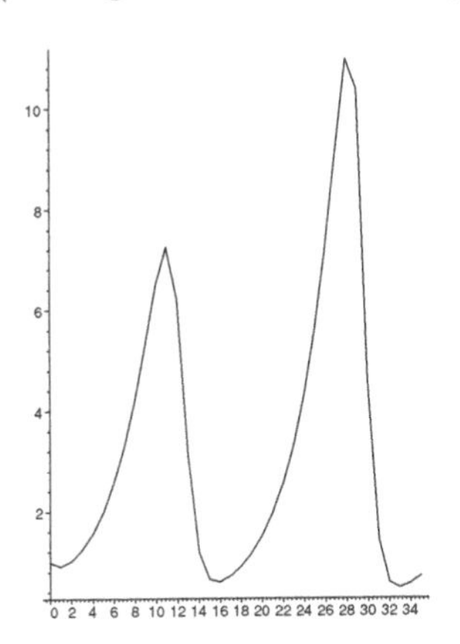

FIGURE A.13: Solution plot for y_n in system.

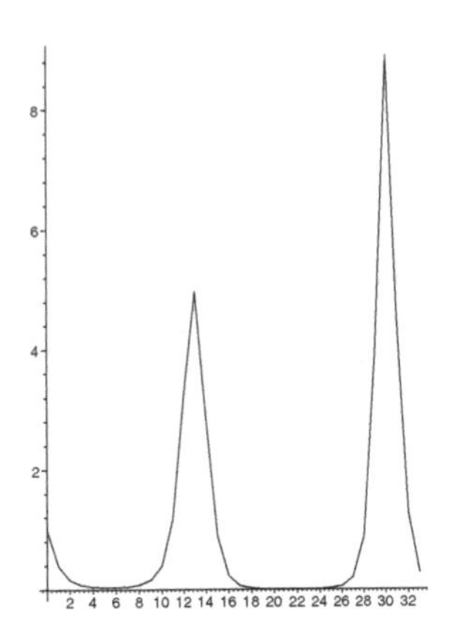

FIGURE A.14: Solution plot for z_n in system.

A.6 A bifurcation program

In Section 2.6, we described the process of bifurcation for the logistic difference equation and gave a bifurcation diagram. Here, we give a Maple program to generate a bifurcation diagram. A bifurcation diagram shows the periodic orbits for a range of values of the parameter r. The idea behind the program is to iterate the logistic function for a given value of r and to plot the values obtained after enough iterations for the orbit to have been approached. Then, for example, if two values are obtained there is a solution of period 2. The program does this for a range of values of r and plots x values as a function of r.

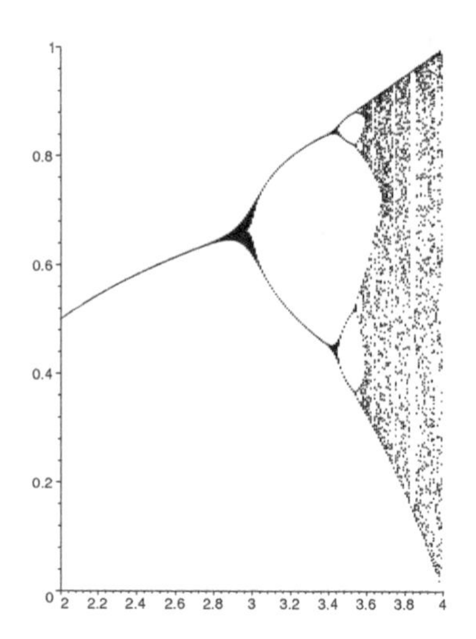

FIGURE A.15: Bifurcation diagram for the logistic difference equation.

```
> restart: with(plots):
> x(1):=0.5: r:=200:
>   for r from 200 to 400 do
>     for j from 1 to 100 do
>     x(j+1):=r*x(j)*(1-x(j))/100:
>     L(r):=seq([r/100,x(j)],j=20..100):
>     od:
>   od:
>   for r from 200 to 400 do
>   A(r):=pointplot (L(r)):
>   od:
> display (seq(A(r),r=200..400),symbol=point,view=[2..4,0..1]);
```
(see Figure A.15)

Appendix B

Taylor's Theorem and Linearization

The linearization of a dynamical, either discrete- or continuous-time, system at an equilibrium is an essential tool in the study of the equilibrium's stability. For one-dimensional systems the stability of an equilibrium can be determined without recourse to the linearization. In fact, we did not use the linearization in Section 4.1 in discussing stability for first-order differential equations, and one can give an alternative to the approach used in Section 2.3 for stability of equilibria of first-order difference equations.

If y_∞ is an equilibrium of a first-order difference equation $y_{k+1} = g(y_k)$ or a first-order differential equation $y' = g(y)$, we make the change of variable $u = y - y_\infty$, so that u represents deviation from the equilibrium. Then substitution into the difference or differential equation produces an expression $g(y_\infty + u)$. We wish to expand this expression in powers of u, because near the equilibrium u is small and higher powers of u are small compared to lower powers. Taylor's theorem provides such an expansion together with an estimate of how close the expansion is to the function being expanded. We need only the special case of the theorem covering expansions limited to first-order terms, that is, to approximate a function near a base point y_∞ by a linear expression in u.

We assume that g is a sufficiently smooth function, specifically that it is continuous with continuous first derivative and a bounded second derivative in an interval I with y_∞ in its interior. Then for any u such that $y_\infty + u$ is also in I, we may write $g(y_\infty + u)$ as the integral of its derivative,

$$g(y_\infty + u) = g(y_\infty) + \int_{y_\infty}^{y_\infty + u} g'(t)dt. \tag{B.1}$$

Likewise, we may write $g'(t)$ as the integral of its derivative,

$$g'(t) = g'(y_\infty) + \int_{y_\infty}^{t} g''(s)ds. \tag{B.2}$$

Substitution of (B.2) into (B.1) gives

$$\begin{aligned} g(y_\infty + u) &= g(y_\infty) + \int_{y_\infty}^{y_\infty + u} g'(y)dt + \int_{y_\infty}^{y_\infty + u} \left[\int_{y_\infty}^{t} g'(s)ds \right] dt \\ &= g(y_\infty) + g'(y_\infty)u + \int_{y_\infty}^{y_\infty + u} \left[\int_{y_\infty}^{t} g'(s)ds \right] dt. \end{aligned} \tag{B.3}$$

If the second derivative $g''(s)$ is bounded below by m and above by M on the interval I, then the integral $\int_{y_\infty}^{t} g'(s)ds$ is between $m(t - y_\infty)$ and $M(t - y_\infty)$ for every t in the interval, and this implies that the iterated integral (which is the error term in our expansion)

$$\int_{y_\infty}^{y_\infty + u} \left[\int_{y_\infty}^{t} g'(s)ds \right] dt$$

is between $\int_{y_\infty}^{y_\infty+u} m(t - y_\infty)dt = m\frac{u^2}{2}$ and $\int_{y_\infty}^{y_\infty+u} M(t - y_\infty)dt = M\frac{u^2}{2}$. Now we have a version of Taylor's theorem which meets our needs.

TAYLOR'S THEOREM: Let g be a function which is continuous and has a continuous first derivative and a bounded second derivative on an interval I containing a point y_∞ in its interior. Suppose the second derivative is at least m and at most M in this interval. Then the linear Taylor approximation to $g(y_\infty + u)$ is $g(y_\infty) + g'(y_\infty)u$ and the error in this approximation is between $m\frac{u^2}{2}$ and $M\frac{u^2}{2}$.

The error term can be expressed as $g''(c)\frac{u^2}{2}$, with c some (unknown) point between y_∞ and $y_\infty + u$. For our purposes, the factor u^2, which for small u makes the error term negligible compared to the linear term, is what matters.

There is a two-dimensional version of Taylor's theorem, which is needed for linearization of two-dimensional systems (Sections 2.8, 5.4). We wish to expand a function of two variables, $g(y_\infty + u, z_\infty + v)$ in powers of u and v with coefficients which depend on the value of the function g and its partial derivatives at the base point (y_∞, z_∞). In order to reduce this problem to Taylor's theorem in one variable, we define a function of one variable t,

$$G(t) = g(y_\infty + tu, z_\infty + tv).$$

Then $G(0) = g(y_\infty, z_\infty)$ and $G(1) = g(y_\infty + u, z_\infty + v)$. Taylor's theorem for one variable says that

$$G(1) = G(0) + G'(0) + R \tag{B.4}$$

where R is the error term. We may use the chain rule for partial derivatives to calculate $G'(t)$ and $G''(t)$:

$$G'(t) = ug_y(y_\infty + tu, z_\infty + tv) + vg_z(y_\infty + tu, z_\infty + tv)$$

$$G''(t) = u^2 g_{yy}(y_\infty + tu, z_\infty + tv) + 2uv g_{yz}(y_\infty + tu, z_\infty + tv) + v^2 g_{zz}(y_\infty + tu, z_\infty + tv)$$

Substitution of $t = 0$ gives

$$G'(0) = ug_y(y_\infty, z_\infty) + vg_z(y_\infty, z_\infty).$$

The expressions for $G(0)$, $G(1)$, and $G'(0)$ substituted into (B.4) give

$$g(y_\infty + u, z_\infty + v) = g(y_\infty, z_\infty) + ug_y(y_\infty, z_\infty) + vg_z(y_\infty, z_\infty) + R, \quad \text{(B.5)}$$

and it is possible to check that the error term R is quadratic in u and v, that is, it is a sum of terms which are multiples of u^2, uv, and v^2. This is the expression which is needed for linearization of a two-dimensional system at an equilibrium (y_∞, z_∞).

Appendix C

Location of Roots of Polynomial Equations

We consider the quadratic polynomial equation

$$f(\lambda) = \lambda^2 + a_1\lambda + a_2 = 0. \quad \text{(C.1)}$$

In Section 2.8, in order to determine stability of equilibria of systems of difference equations, we needed a criterion to tell when all roots of a characteristic equation (C.1) satisfy $|\lambda| < 1$. In Section 5.3, in order to determine stability of equilibria of systems of differential equations, we needed a criterion to tell when all roots of a characteristic equation (C.1) satisfy $\Re\lambda < 0$. In this appendix we shall establish both of these criteria. There are more complicated criteria of both types for higher order polynomial equations but we shall not consider these.

The equation (C.1) has two roots; if $a_1^2 - 4a_2 \geq 0$ there are two real roots λ_1 and λ_2, and if $a_1^2 - 4a_2 < 0$ the roots are complex conjugates, $\alpha \pm i\beta$. It is useful to note that in either case the sum of the roots is $-a_1$ and the product of the roots is a_2.

Our first result, needed in the study of systems of differential equations, is that the roots of (C.1) satisfy $\Re\lambda < 0$ if and only if $a_1 > 0$ and $a_2 > 0$. To see that these conditions are necessary we observe that if the roots are conjugate complex their product a_2 is automatically positive and since their sum is twice their common real part a_1 must be positive for these real parts to be negative.

If the roots are real, in order for both to be negative their product must be positive and their sum must be negative. On the other hand, the conditions $a_1 > 0$, $a_2 > 0$ are sufficient. To see this, we note that if $a_1^2 - 4a_2 \geq 0$, so that the roots are real, $a_2 > 0$ implies that they have the same sign and $a_1 > 0$ implies that both are negative. If $a_1^2 - 4a_2 < 0$, so that the roots are complex conjugates, $a_1 > 0$ implies that they have negative real part. To sum up, we have obtained the following result:

THEOREM: The roots of the quadratic equation $\lambda^2 + a_1\lambda + a_2 = 0$ have negative real part if and only if $a_1 > 0$ and $a_2 > 0$.

Our other result, needed in the study of systems of difference equations, is that the roots of (C.1) satisfy $|\lambda| < 1$. To obtain necessary conditions, we note that the product of the roots must have absolute value less than 1, so that $|a_2| < 1$. Also, we must have $f(-1) > 0$ since otherwise there would be a root less than -1, and we must have $f(1) > 0$ since otherwise there would be a root greater than 1. Thus we must have

$$f(-1) = 1 - a_1 + a_2 > 0, \ f(1) = 1 + a_1 + a_2 > 0$$

or $-(a_2 + 1) < a_1 < (a_2 + 1)$, and this is equivalent to $|a_1| < a_2 + 1$. The condition $|a_2| < 1$ may be written $-1 < a_2 < 1$, or $0 < a_2 + 1 < 2$, and we may combine our two necessary conditions into the double inequality

$$|\mathrm{a}_1| < \mathrm{a}_2 + 1 < 2. \tag{C.2}$$

These conditions are also sufficient. If (C.2) is satisfied, then $f(-1) > 0$ and $f(1) > 0$. If the roots are real, this implies that either both are less than -1 or both between -1 and $+1$, or both greater than 1. However, if both are less than -1 or greater than 1, their product is greater than 1, which contradicts $a_2 < 1$. Thus if the roots are real they must be between -1 and 1. If the roots are conjugate complex, their product a_2 is the product of their absolute values and thus less than 1. Since both roots have the same absolute value, they must both have absolute value less than 1.

To sum up, we have obtained the following result:

THEOREM (Jury criterion): The roots of the quadratic equation $\lambda^2+a_1\lambda+a_2 = 0$ satisfy $|\lambda| < 1$ if and only if

$$|\mathrm{a}_1| < \mathrm{a}_2 + 1 < 2.$$

Appendix D

Stability of Equilibrium of Difference Equations

In Section 2.3 we stated a theorem on asymptotic stability of an equilibrium of a difference equation as a consequence of the idea of linearization at an equilibrium. For a first order difference equation it is possible to establish this theorem directly, without using the linearization. The proof is an application of the mean value theorem, which is presented in most calculus courses as one of the most important results in calculus but then used mainly only to establish precisely results which might appear to be obvious.

We are concerned with the asymptotic stability of an equilibrium y_∞ of a difference equation

$$y_{k+1} = g(y_k). \tag{D.1}$$

The first result we must establish is that if a solution of (D.1) approaches a limit as $k \to \infty$, this limit must be an equilibrium of (D.1). We will not give a precise proof of this fact, but the underlying idea is that if y_k and y_{k+1} are arbitrarily close to a limit y^*, then by the continuity of the function g, $g(y_k)$ is very close to $g(y^*)$ and therefore (since $y_{k+1} = g(y_k)$ is thus close to both y^* and $g(y^*)$) we must have $g(y^*) = y^*$.

The mean value theorem states that if a function $f(x)$ is continuous and has a continuous derivative in an interval I, then for any two points a and b in this interval we have

$$f(b) - f(a) = (b - a)f'(c)$$

for some point c between a and b (actually, the requirement of continuity of f' can be relaxed somewhat). A consequence of this is that if we know that $|f'(x)| < M$ for some number M and all points in the interval, then we have the estimate

$$|f(b) - f(a)| < M|b - a| \tag{D.2}$$

for all points a and b in the interval.

Now we are ready to prove our main result, the following stability of equilibrium theorem of Section 2.3.

STABILITY OF EQUILIBRIUM THEOREM: Let y_∞ be an equilibrium of the difference equation $y_{k+1} = g(y_k)$. If $|g'(y_\infty)| < 1$, the equilibrium is asymptotically stable and if $|g'(y_\infty)| > 1$, the equilibrium is unstable.

Proof: Suppose y_∞ is an equilibrium, so that $y_\infty = g(y_\infty)$ and let $g'(y_\infty) = L$, with $|L| < 1$. We choose a number ρ with $|L| < \rho < 1$, and then there is an interval I centered at y_∞ such that $|g'(y)| < \rho$ for all points y in this interval. We take the initial value y_0 in this interval. Then since $y_1 = g(y_0)$ and $y_\infty = g(y_\infty)$ we may apply (D.2) with the function g, the points y_0 and y_∞, and $M = \rho$, and we have

$$|y_1 - y_\infty| = |g(y_0) - g(y_\infty)| < \rho|y_0 - y_\infty|.$$

Since $\rho < 1$, y_1 is closer to y_∞ than y_0 is, and therefore y_1 is also in the interval I. We may repeat the argument to give

$$|y_2 - y_\infty| = |g(y_1) - g(y_\infty)| < \rho|y_1 - y_\infty| < \rho^2|y_0 - y_\infty|$$

showing that y_2 is also in the interval I and yet closer to y_∞ than is y_1. In the same way (or by induction), we may establish that

$$|y_k - y_\infty| < \rho^k|y_0 - y_\infty|$$

for $k = 1, 2, \ldots$. This shows that each of the terms y_k is in the interval I, that the successive terms come closer to y_∞, and that y_k approaches the limit y_∞ as $k \to \infty$. Thus every solution starting close enough to y_∞ remains close to y_∞ and approaches y_∞ as $k \to \infty$, and this establishes asymptotic stability of y_∞.

If $|g'(y_\infty)| > 1$, we may pick a number $\rho > 1$ and an interval centered at y_∞ on which $|g'(y)| > \rho$ and use the Mean Value Theorem in much the same way but with the inequalities reversed to give

$$|y_k - y_\infty| > \rho^k|y_0 - y_\infty|.$$

From this we see that the terms y_k move away from y_∞ and thus that the equilibrium y_∞ is unstable. □

In the case of asymptotic stability, with $\rho < 1$, ρ gives a measure of the rapidity of approach to the equilibrium. A small value of ρ means that the equilibrium value is approached rapidly. An interesting application of this is Newton's method for approximating the solution of an equation $f(y) = 0$. The method consists of choosing a starting value y_0 sufficiently close to the desired root and iterating according to the scheme

$$y_{k+1} = y_k - \frac{f(y_k)}{f'(y_k)}.$$

Geometrically, this means drawing the tangent line to the curve $z = f(y)$ at the point y_0 and taking y_1 to be the point where this tangent line meets the y-axis (i.e., taking the tangent line as an approximation to the function near y_0 and finding where this approximation is zero). This process is then repeated.

In order to view Newton's method as a stability of equilibrium question, we define

$$g(y) = y - \frac{f(y)}{f'(y)}.$$

Then an equilibrium y_∞ of the difference equation $y_{k+1} = g(y_k)$ is a solution of the equation $f(y) = 0$ (provided $f'(y) \neq 0$). Since

$$g'(y) = 1 - \frac{[f'(y)]^2 - f(y)f''(y)}{[f'(y)]^2} = \frac{f(y)f''(y)}{[f'(y)]^2}$$

(and $f(y_\infty) = 0$) we have $g'(y_\infty) = 0$. Thus $g'(y)$ is small on an interval close to y_∞ and the approach to a root by Newton's method is very rapid once a suitable starting approximation has been found.

Answers to Selected Exercises

Chapter 2

Section 2.1

2. a.

$$w_n = \left(w_0 - \frac{\mu}{1-F}\right)F^n + \frac{\mu}{1-F}$$

   b. If $F < 1$, then eventually $w_n \to \frac{\mu}{1-F}$. This means that if the weed is less fit than its competitor, then it will approach the given equilibrium density thanks to the constant growth due to mutation.

   c. If $F > 1$, meaning the weed is more fit than its competitor, then w_n grows without bound, i.e., weed density becomes arbitrarily high.

   d. Unbounded growth in a single patch is unrealistic, so the model's operating range is limited (in w) when $F > 1$.

3. $y_1 = 3/16, y_2 = 39/256, y_3 = 8463/65536$.

5. $y_1 = 1/2, y_2 = 1/3, y_3 = 1/4$.

11. $y_k = (1.1)^k$.

13. $y_k = 1/2^k$.

15. $y_k = 1$.

17. $y_k = (1/2)^k - 1$.

19. $y_k \to 0$.

21. $y_k \to 5/4$.

Section 2.2

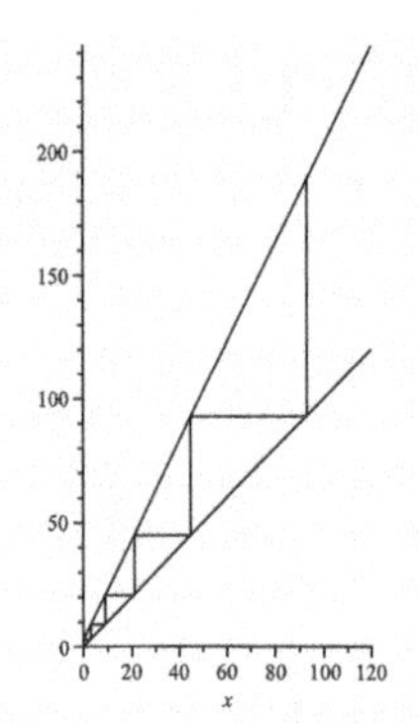

FIGURE S.1: Solution of Exercise 1

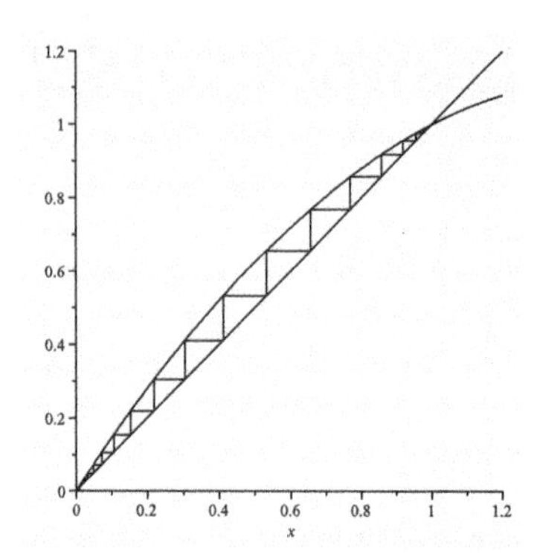

FIGURE S.2: Solution of Exercise 3

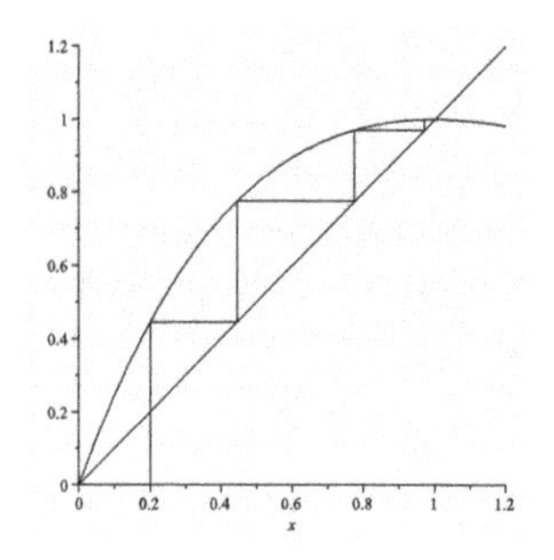

FIGURE S.3: Solution of Exercise 5

8. $y_k \to 0$.

Section 2.3

1. Equilibrium $y = 2$, unstable.

3. Equilibrium $y = \beta/(1-\alpha)$, unstable.

5. Equilibrium $y = 1 + r$ if $r > 1/e$, asymptotically stable for $r < e$.

10. a. (2.21) becomes

$$p_{n+1} = \frac{Fp_n + (1-p_n)}{Fp_n + 2(1-p_n)}.$$

The $p = 0$ equilibrium vanishes, and the coexistence eqm becomes $1/(2-F)$. The equilibrium $p = 1$ is again stable for $F > 1$, and $1/(2-F)$ is for $F < 1$.

b. Coexistence occurs only for $F < 1$, i.e., the a allele only survives if the Aa zygote is more viable than the AA zygote.

c. We can rewrite the $G = 0$ model as $p_{n+1} = \frac{K+1}{K+2}$, where $K = Fp_n/(1-p_n) > 0$. Since $\frac{K+1}{K+2} > \frac{1}{2}$ for $K > 0$, this guarantees A alleles will always be in the majority. This makes sense because all individuals who reproduce must have at least one A allele, so at least half the alleles contributed to the next generation are As.

e.

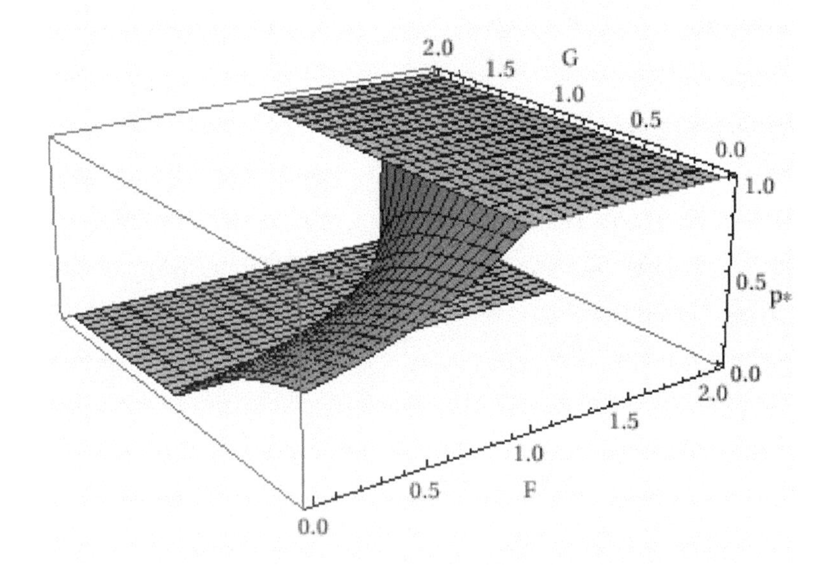

13. a.

$$\alpha_{n+1} = \frac{F\alpha_n + \frac{1}{4}(1-\alpha_n-\beta_n)}{F\alpha_n + (1-\alpha_n-\beta_n) + G\beta_n}$$
$$\beta_{n+1} = \frac{G\beta_n + \frac{1}{4}(1-\alpha_n-\beta_n)}{F\alpha_n + (1-\alpha_n-\beta_n) + G\beta_n}$$

b. We can rewrite the equations for α_{n+1} and β_{n+1} as

$$\alpha_{n+1} = \frac{\frac{1}{4}+K}{1+K+L}, \quad \beta_{n+1} = \frac{\frac{1}{4}+L}{1+K+L},$$

where $K = F\alpha_n/(1-\alpha_n-\beta_n) > 0$ and $L = G\beta_n/(1-\alpha_n-\beta_n) > 0$. This makes $(1-\alpha_n-\beta_n) = \frac{1}{2}/(1+K+L) \leq \frac{1}{2}$. This is biologically sound under 100% like-with-like mating because two homozygotes of the same type never reproduce heterozygotes, and only half of the heterozygotes' descendants will also be heterozygotes.

Section 2.4

1. Equilibrium $y = 0$ is asymptotically stable for all r.

3. $r < e$.

5. $H < (\sqrt{r} - \sqrt{A})^2$.

7. $H < 1/10$.

14. $y_{k+1} = \begin{cases} ay_k, & by_k < 1; \\ \frac{a}{2}y_k, & by_k = 1; \\ 0, & by_k > 1. \end{cases}$

Section 2.5

1. a. 233.3 (naturally occurring), 182.56 (field experiment)

   b. 158.68 (naturally occurring), 124.15 (field experiment)

   c. 154, 62.1 (naturally occurring), 104.9, 0 (field experiment)

2. a. 80.687

   b. 78.92

4. $$\frac{(\sqrt{a}-1)^2}{b} < \frac{a-1}{b} < \frac{(\sqrt{a}+1)^2}{b}$$

5. Multiply the desired double inequality by 2, subtract $(\frac{a-1}{b} - H)$, and rewrite as a single inequality

$$|H - \frac{(\sqrt{a}-1)^2}{b}| < \sqrt{\left(\frac{a-1}{b} - H\right)^2 - 4H/b}.$$

   Then square and simplify.

10. $r - 1 < E < r + 1$.

Section 2.6

1. $r = e$.

8. a. $y_{k+1} = g((1-d)y_k)$

   b. Both are the same, $r(1-d)$.

Section 2.7

1. $y = 0, z = 0$, unstable.

3. $y = 0, z = 4$, unstable.

5. $y = 0, z = 0$, unstable.

7. $y = 0$, asymptotically stable if $0 < \mu - a < 2$ and $y = \frac{a-\mu}{b\mu}$, asymptotically stable if $\mu < a$, $\frac{\mu}{a}(a - \mu) < 2$.

Section 2.8

2. The ragwort is the factor limiting the moth population size and/or reproductive ability; a certain amount of ragwort is needed to provide energy to produce each egg, and the amount available is less than the total reproductive capacity of the moths in the absence of any limitations.

3. The equilibrium $(B, 0)$ is stable if and only if $aB < 1$; the equilibrium $(\frac{1}{a}, \frac{1}{ab}\log aB)$ is stable if and only if $1 < aB < \sqrt{e}$.

5c.

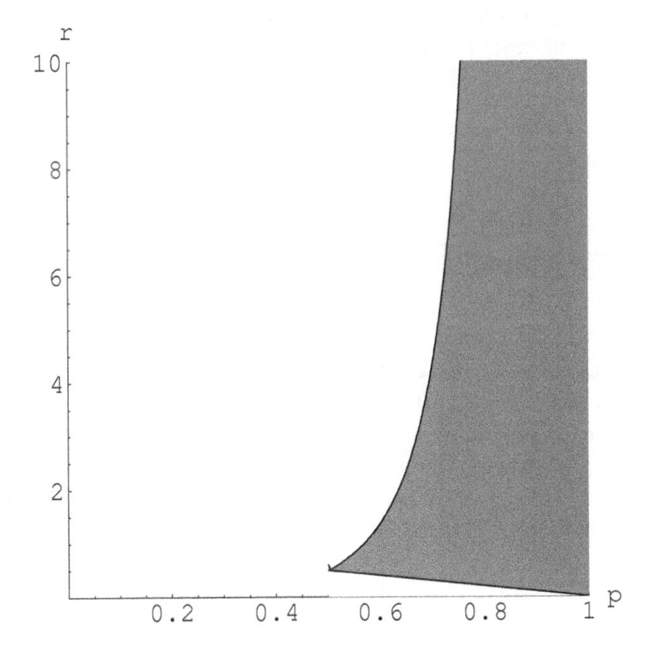

Chapter 3

Section 3.1

1. 1830.

3. 211.66

5. 17.46.

7. 199 days.

9. 4.5 billion years.

Section 3.2

3. (a) $r = 1.0668 \times 10^{-5}$, $(b)205(1990), 206(2000), 208(2010)$.

5. a. World War I and the "Spanish" flu epidemic of 1918-19

   b. The model is pretty good through 1970. Changing immigration patterns, better census counts, and better food production technology might explain changes, and changing some parameter values might give a better fit.

7. Biomass after i year is 32.5×10^6 kg. It takes 1.55 years for biomass to reach $\frac{K}{2}$.

13. $\mathcal{R}_0 = 1.4$. Reduce contact rate to 1/14 person per day.

15. a. $I' \geq 0$ with $I' > 0$ as long as $S, I > 0$ so I will continue to increase [asymptotically] toward the entire population K, slowing down only as $S \to 0$.

    b. "Eventually" here refers to the mathematical limit as time becomes infinite—that is, the conclusion in (a) deals with long-term behavior. In the long term, individuals will die, and others will be born to replace them. The model above does not include such demographic renewal. If we add a demographic renewal term to the model, the conclusion in (a) is no longer valid.

17. $y' = ry\left(1 - \frac{y}{K}\right)$.

Section 3.3

5. $y = (1 - t^2)^{-1/2}$.

8. $y = 1, y = \frac{1-e^t}{1+e^t}, y = \frac{1-e^{t-2}}{1+e^{t-2}}$

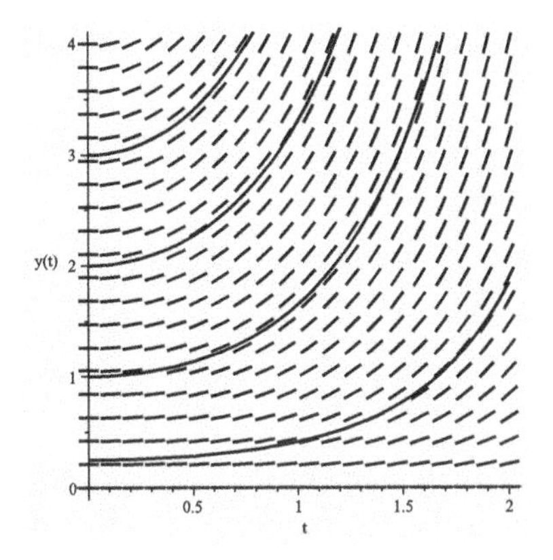

FIGURE S.4: Solution of Exercise 11

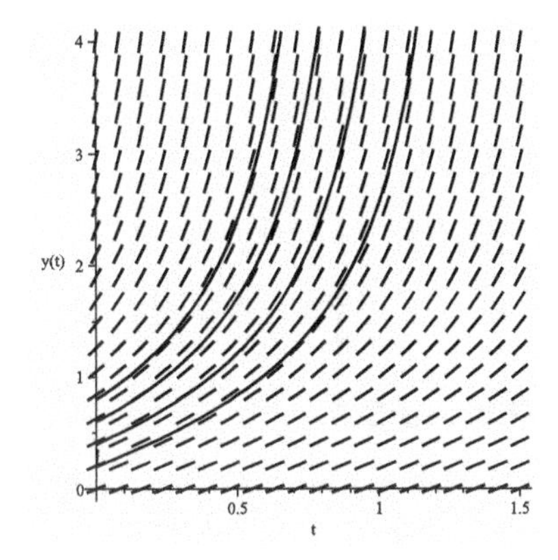

FIGURE S.5: Solution of Exercise 13

Section 3.4

1. $y = ce^{t^2}$.

3. $y = (c + 2t)^{1/2}$.

5. $y = ce^{-t^2}$.

7. $y = 4 + ce^{t^3/3}$.

9. $y = e^{(t^3-125)/3}$.

11. $y = y_0 e^{-t^2}$.

13. $y = \frac{2}{1-t^2}$.

17. a. The pacemaker signal y follows the control u, with the difference $y - u$ between them decaying over time at a rate proportional to that difference.

    b. $y(t) - u(t) = z(t) = z(0)e^{-at}$ so $y(t) = u(t) + [y(0) - u(0)]e^{-at}$.

18. a. The heart beats more slowly (or at least firing rate decreases).

    b. Firing rate decreases toward zero: no beating!

    c. Solve equation (3.29) for y_H and substitute.

19. $T_1 = T_2$

20. The solution changes only in the exponent of the exponential in the denominator of (3.37), which goes from $-rt$ to $-r_0 t + \frac{r_1}{2\pi}\cos 2\pi t$. As $t \to \infty$ the oscillation dies out since $r_0 t >> \frac{r_1}{2\pi}$, so $y \to K$ anyway.

21. a. $u' = -\mu_1 u$ leads to $u(t) = u(0)e^{-\mu_1 t}$ and $u(t_c) = ry_k e^{-\mu_1 t_c}$.

b. $u' = -\mu_2 u$ leads to $u(t) = u(t_c)e^{-\mu_2(t-t_c)} = ry_k e^{-\mu_1 t_c} e^{-\mu_2(t-t_c)}$ and $u(T) = ry_k e^{-(\mu_1-\mu_2)t_c} e^{-\mu_2 T}$

27. 99.

31. $f(t) \equiv 0$,and $f(t) = t/2$.

32. $f(t) = 0 (t < c), f(t) = (t-c)/2 (t > c)$ for every $c \geq 0$.

Section 3.5

1. Let $Q(t)$ denote the weight of salt in the tank at time t, so that $Q(0) = 3$. The rate at which salt is flowing in is 2.5 lb./minute, and the rate at which salt is flowing out is $5\left(\frac{Q}{100}\right)$, since the concentration of salt at time t is $\frac{Q}{100}$ lb./gal. and the solution is flowing out at 5 gal./minute. This leads to the initial value problem

$$Q' = 2.5 - \frac{Q}{20}, \quad Q(0) = 3,$$

which is of the form (3.46) with $k = 0.05$, $b = 2.5$, and has solution

$$Q = 50 - 47\,e^{-0.05t}.$$

In particular, $Q(60) = 50 - 47\,e^{-3} = 47.66$ lb. Also, it is clear from the expression for Q that $\lim_{t\to\infty} Q(t) = 50$.

5. The concentration is 4% times $(1 - \exp(-(0.6/1800)t))$, so set that equal to 10^{-4} and solve for t. Answer: 7.5 minutes.

7. Let $y(t)$ be the amount of drug in the bloodstream at time t. Since the drug leaves the blood at a rate proportional to $y(t)$, we have $y'(t) = -ky(t)$ for some constant $k > 0$. We are given that $y(0) = d$. The solution of the initial value problem $y' = -ky$, $y(0) = d$ is $y(t) = de^{-kt}$. We are given also that $y(1) = d/2$. Therefore $de^{-k} = d/2$, and $k = \log 2$. Now we have $y(t) = de^{-(\log 2)t} = d\,2^{-t}$. (Alternatively, $e^{-k} = \frac{1}{2}$, so $y(t) = d\left(\frac{1}{2}\right)^t$.)
   The desired time is the value of t such that $y(t) = d\,2^{-t} = d/5$. Thus $2^{-t} = 1/5$, $-t\log 2 = \log(\frac{1}{5}) = -\log 5$, or

$$t = \frac{\log 5}{\log 2} = 2.32\,\text{hours}.$$

11. 40 minutes.

13. 58.3 minutes.

Section 3.6

1. $y = 1 - e^{-2t}/2.$

3. $Y = e^{-2t} + 4e^{-3t}.$

5. $y = \frac{t^2+1}{2t}.$

7. $y = t^2/4 + 4/t^2.$

9. $y = \frac{e^t+1-e}{t}.$

11. $y = t^3 - \frac{3}{2}t^2 + \frac{3}{2}t - \frac{3}{4}[1 - e^{-2t}].$

13. $y = \sqrt{3t/2}.$

15. $Y = e^{2t(t-1)}.$

19. $y = (a+1)e^t - 1.$

24. $x(t) = 500/(t + \frac{100}{3}), x(50) = 6lb.$

25. a.
$$s(t) = 1440(25-t) - \frac{272}{25}(25-t)^{5/2}$$
b.
$$V(t) = 1000 - 40t = 40(25-t) \text{ so } \frac{s(t)}{V(t)} = 36 - 0.272(25-t)^{3/2}$$
The poor fish only have 3.56 minutes in which to be rescued.

29. The substitution $y = 1/z$, $y' = -z'/z^2$ gives
$$-\frac{1}{z^2}z' = r\,\frac{1}{z} - \frac{r}{K}\,\frac{1}{z^2},$$
$$z' + rz = \frac{r}{K}.$$
Using the integrating factor e^{rt} we obtain
$$e^{rt}z' + re^{rt}z = (e^{rt}z)' = \frac{r}{K}\,e^{rt},$$
and integration gives
$$e^{rt}z = \frac{1}{K}e^{rt} + c,$$
$$z = \frac{1}{K} + ce^{-rt}.$$
Returning to the original dependent variable y, we find the family of solutions
$$y = \frac{1}{\frac{1}{K} + ce^{-rt}} = \frac{K}{1 + cKe^{-rt}}.$$
This is the same family that we found before, in Section 3.3, except that the arbitrary constant c in Section 3.3 is replaced by the arbitrary constant cK here.

Chapter 4

Section 4.1

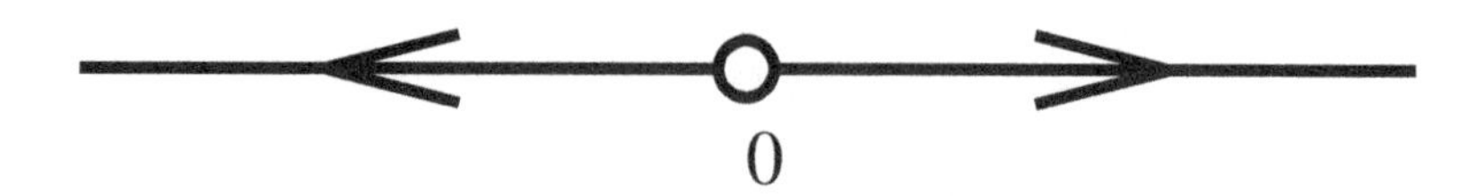

FIGURE S.6: Solution of Exercise 1

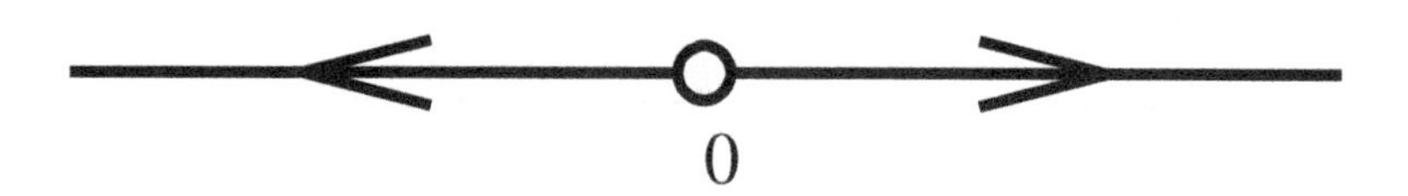

FIGURE S.7: Solution of Exercise 3

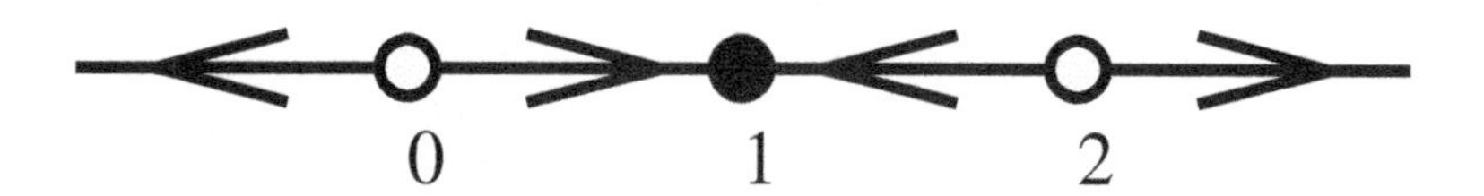

FIGURE S.8: Solution of Exercise 5

FIGURE S.9: Solution of Exercise 7

9. $y = 0$ is unstable,$y = K$ is asymptotically stable.

11. $y = 0$ is unstable,$y = K$ is asymptotically stable.

15. a.

$$\int \frac{x+K}{(\alpha+\beta x)(n-x)(x+K)-sx}\,dx = \int dt$$

In the case where $B(x)$ has three real roots, integration gives

$$A_1 \log|x-r_1| + A_2 \log|x-r_2| + A_3 \log|x-r_3| = t + c;$$

exponentiation then yields

$$(x-r_1)^{A_1}(x-r_2)^{A_2}(x-r_3)^{A_3} = ke^{-t}.$$

The only further simplification available in general is to take the A_ith root of both sides of the equation, so that the ith factor on the left is linear.

17. A single food source lasts much less than a single day for ants such as these which feed on ephemeral sources. Long-term here refers to a matter of hours or even minutes. On this scale the colony size is constant: births and deaths would accumulate significantly only over a period of days or weeks (barring catastrophic events).

18. a. $\beta = 0$

    b. $s = 0$

19. Since $0 < x/(x+K) < 1$, raising it to a higher power makes it smaller. Therefore the rate would saturate more slowly, i.e., fewer ants would be leaving the trail in unit time, though the rate still approaches s.

20. This allows the ants to optimize the trail (cut off roundabout loops), as well as find other food sources.

Section 4.2

1. In case (i), where $H < rK/4$, we can factor $dy/dt = -\frac{r}{K}(y-y_1)(y-y_2)$ which leads to

$$\frac{y(t)-y_2}{y(t)-y_1} = \frac{y(0)-y_2}{y(0)-y_1}\, exp\left(-rt\,\frac{y_2-y_1}{K}\right),$$

from which

$$y(t) = \frac{y_2 - y_1 Q \exp\left(-rt\,\frac{y_2-y_1}{K}\right)}{1 - Q\exp\left(-rt\,\frac{y_2-y_1}{K}\right)}, \quad \text{where } Q = \frac{y(0)-y_2}{y(0)-y_1}.$$

In case (ii), where $H = rK/4$, we have $dy/dt = -\frac{r}{K}(y - K/2)^2$, the solution to which is

$$y(t) = \frac{K}{2} + \frac{y(0)-\frac{K}{2}}{1+\frac{rt}{K}\left(y(0)-\frac{K}{2}\right)}.$$

Finally, in case (iii), where $H > rK/4$, there are no equilibria, so the quadratic cannot be factored over the reals. Instead we complete the square:

$$\frac{dy}{dt} = -\frac{r}{K}\left[\left(y-\frac{K}{2}\right)^2 + \left(\frac{HK}{r} - \frac{K^2}{4}\right)\right],$$

so that after integration and solving for $y(t)$ we have

$$y(t) = \frac{K}{2} + \tan\left[\tan^{-1}\left(y(0)-\frac{K}{2}\right) - \frac{rt}{K}\sqrt{\frac{HK}{r} - \frac{K^2}{4}}\right].$$

2. Factoring $dy/dt = (r/K)y(y_H^* - y)$ and the partial fraction expansion

$$\frac{1}{y(y_H^* - y)} = \frac{1}{y_H^*}\left(\frac{1}{y} - \frac{1}{(y - y_H^*)}\right)$$

lead to a solution (after integration and simplification) of

$$y(t) = \frac{y(0)y_H^*}{y(0) + (y_H^* - y(0))\exp[-(r - E)t]}.$$

4. a. From (4.12) we have $y^* = K(1 - E/r) = 3(1 - 0.023/0.5) = 3(0.954) = 2.86$ million deer. Deer-vehicle collisions (now 66,000 per year) reduce the deer population by E/r =4.6% or 140,000 deer.

   b. $dy/dt = ry(1 - y/K) - H - Ey$

   c.

$$y^* = \frac{1}{2}\left[K\left(1 - \frac{E}{r}\right) + \sqrt{\left[K\left(1 - \frac{E}{r}\right)\right]^2 - \frac{4HK}{r}}\right]$$

   gives an equilibrium value of 2.47 million deer, a much higher (17.5%) impact than the simpler estimate predicts.

5. In the absence of harvesting, we assume a logistic law, so that the population over time is given by equation (2.13). Solving this for r as we did for t in Example 4.2.1 yields the expression

$$r = \frac{1}{t_f}\ln\frac{K - y_0}{y_0\left(\frac{K}{y(t_f)} - 1\right)}.$$

For the given parameter values, we get $r = 0.097$/yr. A simple exponential (rather than logistic) estimate would give $r = \frac{1}{t_f}\ln(y_f/y_0) = 0.094$/yr, a difference of 3.2%. The critical harvest rate is $rK/4 = 5812$ whales/yr.

6. a. From (4.9), we can solve for r in terms of y_2: $r = H/y(1 - y/K) = 577/174,000(1 - 174,000/265,000) = 0.00966$/yr, an order of magnitude lower than commonly reported values for whale species. This would mean that a very weak reproductive rate is required in order for Norway's harvesting rate to hold the population so far below carrying capacity. Since the true value of r is likely much higher, the population is probably still rising back toward equilibrium.

   b. We substitute into (4.11) to find y_2 at the different r values, obtaining 244,000, 255,000, and 258,000 whales, respectively.

7. The MSY is $rK/4 = 960$ whales/yr. Estimating the population in 2000 as 11,700, there was a drop from the 1900 value of 120,000 of 108,300, or an average 1083 whales/yr. Even if whales died from no reason but whaling, the net death rate exceeds the MSY.

8. We calculate $MSY = rK/4 = (.025)24000/4 = 150$ whales/yr, well under the actual catch, which would drive even a post-recovery population extinct.

9. It would reduce it by a certain amount: calculate y_2 and subtract from 100K.

12. The logistic CYH model with $r = 0.05$/yr, $K = 10,000$ acres, $H = 100$ acres/yr has a stable equilibrium $y_2 = 7236$ acres.

14. The equilibrium size is $K(1 - E/r) = 194600(1 - 0.04/0.098) \approx 115,000$ cranes.

15. a. 12,500/(20–100) allows space for between 125 and 625 breeding pairs. We translate this to 250–1250 breeding adults, so 500–2500 cranes total; take an average and say 1500.

18. a. From the sustainability criterion $H < rK/4$ we obtain $K > 4H/r = 4(344,000)/0.07 = 19.7$ million seals as a minimum.

    b. For sustainability, the present population of 4.8 million must exceed the minimum threshold, $y(t) > y_1$. From Exercise 10, $y_1 > H/r$. Therefore a necessary condition is that $y(t) > H/r$. But here 4.8 million $< H/r = (344,000)/0.07 = 4.9$ million, so the hunt is unsustainable given the present numbers. (It is close enough that the uncertainty in the data allows for either possibility, though.)

19. a. You're always in CEH mode, below the switch.

    b. Here you begin in CYH mode, assuming the population is initially near K, so the dynamics depend on whether the stable CYH equilibrium y_2^* is greater or less than H/E. If $y_2^* > H/E$, then you remain always in CYH mode. If instead $y_2^* < H/E$, then CYH drives the population down into the CEH range, where the equilibrium population size is instead the CEH equilibrium.

20. $(1 - p)$ of the equilibrium value y_2 must exceed y_1. Thus $H < rK(1 - p)/(2 - p)^2$.

Section 4.3

1. $y = \frac{\alpha K A}{A+(K-A)e^{-\alpha t}}$.

3. $y = a - \frac{a}{\alpha a t - 1}$.

6. First solve (4.18) for K_d to get $K_d = (K_A - C^*)(K_B - C^*)/C^*$ and then substitute $K_A = A^* + C^*$, $K_B = B^* + C^*$.

8.
$$R_0 = \frac{\beta(N)}{\mu + \frac{1}{\tau}}, \quad \frac{I^*}{N} = 1 - \frac{1}{R_0},$$

and the disease persists if and only if $R_0 > 1$, that is, the disease-free equilibrium is globally asymptotically stable if $R_0 < 1$, while the endemic equilibrium is if $R_0 > 1$.

9. The model formulation would only change inasmuch as $\beta(N)$ becomes $\beta(I)$, but the equilibrium analysis would be complicated considerably with an additional function of I in the nonlinear term. If people change their behavior in response to observed prevalence, $\beta(I)$ should be a non-increasing (perhaps even sharply decreasing) function of I. If the health-care system breaks down beyond some critical level I_c, then $\beta(I)$ should be constant below I_c and increasing above it (perhaps up to some maximum).

10. The new effective population size for contacts is $S + \sigma I$, so we get

$$\frac{dI}{dt} = \beta(S + \sigma I)S\,\frac{\sigma I}{S + \sigma I} - \frac{1}{\tau}I.$$

We reduce to one variable as before with $S = N - I$, $S + \sigma I = N - (1 - \sigma)I$.

11. a. $R_0 = \sigma\beta_1 N\tau$, and $I^*/N = 1 - \frac{1}{R_0}$ as before.
    b. $R_0 = \sigma\beta_0\tau$, and $I^*/N = (R_0 - 1)/(R_0 - 1 + \sigma)$.

12. a. The recovery term disappears and $R_0 \to \infty$.
    b. The proportion of infectives at the endemic equilibrium approaches 1: everyone is infected. This is unrealistic primarily because it ignores timescale issues, such as demographic renewal.

13. Of course, dI/dt gets a new term $-\delta I$, but more significantly N is no longer constant, so we cannot represent the system with a single equation. We must include an equation for either dS/dt or dN/dt.

14. a. In addition to the DFE, endemic equilibria

$$\frac{I^*_\pm}{N} = \frac{1}{2}\left(1 \pm \sqrt{1 - \frac{4}{\beta\tau N^2}}\right)$$

exist iff $\beta\tau N^2 > 4$. To analyze stability, we calculate $f'(I) = \beta N^2[2(I/N) - 3(I/N)^2 - (1/\beta\tau N^2)]$, so that $f'(0) < 0$ and the DFE is always LAS. If the two endemic equilibria exist, then

$$f'(I^*_\pm) = -\frac{1}{2}\beta N^2\left(1 - \frac{4}{\beta\tau N^2} \pm \sqrt{1 - \frac{4}{\beta\tau N^2}}\right);$$

since $0 < 1 - \frac{4}{\beta\tau N^2} < 1$, the square root is larger than the radicand, and we conclude that $f'(I_-^*) > 0$ while $f'(I_+^*) < 0$, so that the upper endemic equilibrium is LAS and the lower one is unstable. This model behaves differently than the basic mass-action epidemic model in two important ways: first, the DFE is always LAS, which implies that $R_0 = 0$ (that is, a single infective introduced into a population of susceptibles never causes an outbreak, because infection requires many infectives (I^2)); and second, for $\beta\tau N^2 > 4$, the persistence of the infection depends on the initial number of infectives: if it is high enough ($I(0) > I_-^*$), it will establish an endemic state; otherwise it will die out quickly.

b. This model behaves mostly like the usual mass-action incidence model, with $R_0 = \sqrt{\beta\tau}$.

16. a. $dy/dt = ry(1 - y/K) - kxy$

b. $dx/dt = cxy - \mu x$

c. With y constant, the model becomes linear: $dy/dt = (cY - \mu)x$, and the predators either grow unchecked (if $cY > \mu$) or die out (if $cY < \mu$).

d. With x constant, the model becomes logistic, $dx/dt = r_2 y(1 - y/K_2)$, where $r_2 = r - kX$ and $K_2 = K(r - kX)/r$, so that the prey population x approaches a smaller carrying capacity K_2 as long as $r > kX$, and dies out if $r < kX$.

Section 4.4

1. If $r > 0$, no equilibrium. If $r = 0, y = 0$ is unstable. If $r < 0$, there are two equilibria, one unstable and one asymptotically stable.

3. $y = 0$ is asymptotically stable if and only if $r < 0$. $y = -r$ is asymptotically stable if and only if $r > 0$.

5. If $r \geq 0, y = 0$ is unstable. If $r < 0$, $y = 0$ is unstable and $y = \pm\sqrt{-1/r}$ are asymptotically stable.

7. If $-2 < r < 2$ there are no equilibria. If $r < -2$ or $r > 2$, there are two equilibria, one unstable and one asymptotically stable.

14. Let $y = x/K$ and $\tau = ts/K$. Then $dy/d\tau = a + by - cy^2 - \frac{y}{y+1}$, where $a = \alpha n/s$, $b = (\beta n - \alpha)K/s$, and $c = \beta K^2/s$.

15. a. We use A as the unit of population size and make the change of dependent variable $u = \frac{y}{A}$. This transforms the equation (4.29) to

$$Au' = rAu\left(1 - \frac{Au}{K}\right) - H\frac{u^2}{u^2 + 1}$$

or

$$\frac{A}{H}u' = r\frac{A}{H}u\left(1 - \frac{Au}{K}\right) - \frac{u^2}{u^2+1}.$$

We define the new parameters R and Q by $R = \frac{Ar}{H}$, $Q = \frac{K}{A}$ and then we have the model

$$\frac{A}{H}u' = Ru\left(1 - \frac{u}{Q}\right) - \frac{u^2}{u^2+1}. \tag{S.3}$$

We could also make the change of independent variable $s = \frac{H}{A}t$, which would replace (S.3) by

$$\frac{du}{ds} = Ru\left(1 - \frac{u}{Q}\right) - \frac{u^2}{u^2+1},$$

but we need not do this in order to carry out the equilibrium analysis. In terms of the original independent variable t we have the same equilibria and stability; the change of dependent variable changes only the exponential rate of approach to equilibrium. Effectively, we have replaced the original four parameters by the two parameters R and Q. An examination of the forest changes indicates that under change of the forest, the parameter Q remains almost constant, but the parameter R grows with the development of the forest. The process of change in the parameter in many applications is the reason for our study of bifurcations. In this case we will consider R to be our bifurcation parameter.

b. Equilibria of (S.3) are $u = 0$ and the intersections of the line $v = R\left(1 - \frac{u}{Q}\right)$ and the curve $v = \frac{u}{1+u^2}$, solutions of the equation

$$R\left(1 - \frac{u}{Q}\right) = \frac{u}{1+u^2}. \tag{S.4}$$

In order to understand the dynamics of the model, we will need to investigate how the equilibria depend on R. The curve $v = \frac{u}{1+u^2}$ starts at the origin, increases to a maximum at $(1, \frac{1}{2})$, and then decreases, approaching zero as $u \to \infty$, with an inflection point at $(\sqrt{3}, \frac{\sqrt{3}}{4})$. The line $v = R\left(1 - \frac{u}{Q}\right)$ passes through the points (0, R) and (Q, 0) in the $u - v$ plane. A typical situation is as shown in Figure S.11. There we see that, in addition to the $u = 0$ equilibrium, equation (S.4) gives three more, labelled K, L, and M, where the line and the curve intersect. We can determine their stability by applying the theorem from Section 4.1. Here (S.3) has the form $u' = uf(u)$, where $f(u) = \frac{H}{A}\left[R\left(1 - \frac{u}{Q}\right) - \frac{u}{u^2+1}\right]$ — in terms of Figure S.11, "the line minus the curve". Therefore the equilibria

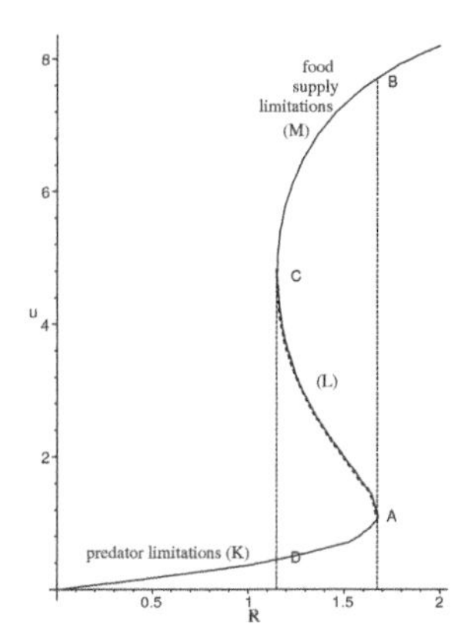

FIGURE S.10: Bifurcation diagram for equation (S.3)

will be stable which have $f'(u) < 0$ there, i.e., where the slope of the line is less than the slope of the curve. From the graph we can see that this occurs at K and M, but not at 0 or L. Therefore we have an unstable equilibrium at $u = 0$ and two asymptotically stable equilibria K and M separated by an unstable equilibrium L. The domain of attraction of the small equilibrium K is the interval [0, L), and the domain of attraction of the large equilibrium M is the interval (L, ∞).

If R (the slope of the line) increases until K and L coalesce, as in Figure S.12, then only the equilibrium M remains. Thus if a system is in equilibrium at K and R increases, there would be a jump from the small equilibrium K to the large equilibrium M. Conversely, if R is decreased until L and M coalesce, as in Figure S.13, then only the equilibrium K would remain, and a system which is in equilibrium at the large equilibrium M would crash to the small equilibrium K. These two situations therefore involve bifurcations like the saddle-node bifurcation seen earlier in this section.

c. We think of the parameter R as representing resources of the forest as food supply for the budworms, and the equilibria K and M as corresponding to budworm limitation by predators and food supply, respectively. In a young forest, R is small with only the small equilibrium K (limited by predators). As the forest develops, the food supply becomes so great that budworm growth exceeds control by predators, and R increases to the stage where there is only the large equilibrium M, signifying an outbreak of the budworm population. If this outbreak destroys the forest, the predators may regain control as R decreases, and the population may crash to the small equilibrium K. The model exhibits hysteresis; the outbreak and crash will generally occur at different population levels. This

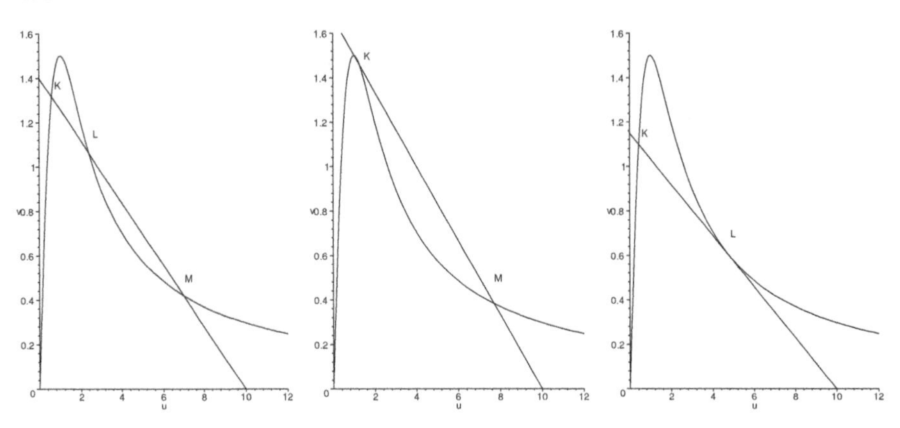

FIGURE S.11: The two sides of equation (S.4) and their intersections

FIGURE S.12: Equilibria K and L coalesce

FIGURE S.13: Equilibria L and M coalesce

may also be seen from the model (S.3) by drawing the graph of the equilibria as a function of the bifurcation parameter R. The equation (S.4) is easily solved for R as a function of y, and the bifurcation curve is just the (multi-valued) inverse function (Figure S.10). When R grows from a small value, the equilibrium moves along the bifurcation curve until it reaches the point A, and then jumps to the higher point B and continues upwards. When R decreases from a high value, the equilibrium moves down the bifurcation curve past B to C before crashing to the lower value D. Thus the discontinuity in equilibrium on the way up is at a higher value of R than the discontinuity in equilibrium on the way down.

Section 4.5

1. The modified Euler method gives 2.70796 instead of 2.5937, an error of 0.38% rather than 4.6%.

3. The Runge-Kutta approximation gives

t	0	0.2	0.4	0.6	0.8	1.0
y	1	1.177	1.368	1.566	1.765	1.956

5. The Runge-Kutta approximation gives

t	0	0.2	0.4	0.6	0.8	1.0
y	1	0.9803	0.9251	0.8441	0.7500	0.6544

8. $h < 2$.

10. 4.6 minutes

11. 8.0 minutes; there are fewer ants wandering around, so it takes longer for the pheromone trail to become established enough to attract many ants

13.
$$f_1'(P) = \frac{rm^q qP^{q-1}}{(m^q + P^q)^2},$$
$$f_1''(P) = \frac{rqm^q P^{q-2}\left[(q-1)m^q - (q+1)P^q\right]}{(P^q + m^q)^3}$$
so $f_1''(P) = 0$ when $P = 0$ or
$$P = \bar{P} = m\left(\frac{q-1}{q+1}\right)^{1/q}.$$
Now
$$f_1'(\bar{P}) = \frac{r}{m}\frac{q\left(\frac{q-1}{q+1}\right)^{\frac{q-1}{q}}}{1+\left(\frac{q-1}{q+1}\right)^2}.$$
For $q \geq 2$ this is $f_1'(\bar{P}) \approx rq/2m$ from above.

14. a. Let $p = P/m$ and $\tau = rt/m$. Then $dp/d\tau = a - bp + p^q/(1+p^q)$, where $a = L/r$ and $b = sm/r$.

b.
$$a = \frac{qp^{2q} - p^q - 1}{(1+p^q)^2}, \quad b = \frac{qp^{q-1}(1-p^q)}{(1+p^q)^2}$$

c.
$$\frac{da}{dp} = \frac{qp^{q-1}\left[(2q+1)p^{2q} + 2(q+1)p^q + 1\right]}{(1+p^q)^4},$$
$$\frac{db}{dp} = \frac{qp^{q-2}\left[p^{2q} - 3qp^q + (q-1)\right]}{(1+p^q)^3},$$
so $da/dp > 0$ for $p > 0$, i.e., a is increasing in p, with $a(0) = -1$ and $\lim_{p\to\infty} a(p) = q$, while $db/dp = 0$ at
$$p_\pm = \left(\frac{3}{2}q \pm \sqrt{\frac{9}{4}q^2 - q + 1}\right)^{1/q},$$

with $b(0) = 0$, b increasing in p for $0 < p < p_-$, decreasing on $[p_-, p_+]$, and approaching 0 again as $p \to \infty$. Meanwhile, by inspection we see that $b(p)$'s only zeroes are at $p = 0$ and 1.

d. $b(1) = 0$, while $a(1) = (q-2)/4 > 0 \Leftrightarrow q > 2$. Thus the unique x-intercept is positive iff $q > 2$. The significance of this result is that bifurcations (and hysteresis) exist only for $q > 2$. For $q > 2$, there exist three equilibria when (a, b) falls below the reverse-parametrized curve.

15. a. r up by 10, m down by 100; irreversible lake, eutrophic equilibrium

b. r down by 10, m up by 100. yields reversible lake, oligotrophic equilibrium

---

Chapter 5

Section 5.1

1. Orbits are circles with center at the origin.

3. Orbits are curves $\sin z = y^2/2 + c$.

5. Orbits are curves $z = ce^{\beta y/\alpha}$.

Section 5.2

1. Linearization at $(1, 1)$ is $u' = u + v, v' = -u + v$.

3. Linearization at $(-1, 1)$ is $u' = u + 2v, v' = u$. Linearization at $(-1, -1)$ is $u' = u - 2v, v' = u$.

5. System has no equilibria.

9. Equilibrium $(0, 0)$ is asymptotically stable.

11. Equilibrium $(0, 0)$ is asymptotically stable.

Section 5.3

1. $y = 5K_1e^{2t} + K_2e^{-4t}, z = K_1e^{2t} - K_2e^{-4t}$.

3. $y = K_2e^t, z = K_1e^{2t} - K_2e^t$.

5. $y = K_1e^{4t}, z = 2K_1te^{4t}, z = 2K_1te^{4t} + K_2e^{4t}$.

7. $y = K_1e^{-t} + K_2e^{-2t}, z = -K_1e^{-t} - 2K_2e^{-2t}$.

9. $y = K_1te^t + K_2e^t, z = K_1e^t$.

11. $y = K_1e^{3t} \sin 5t + K_2e^{3t} \cos 5t,\ y = K_2e^{3t} \sin 5t + K_1e^{3t} \cos 5t$.

13. $y = K_1te^t + K_2e^t, z = K_1e^t$.

17. $y = K_1e^{at}, z = \frac{cK_1}{a-d}e^{at} + K_2e^{at},\ (a \neq d)$.

18. $y = K_1e^{at}, z = cK_1e^{at} + K_2e^{at}$.

Section 5.4

1. $(1, 1)$ is unstable.

3. Both $(-1, 1)$ and $(-1, -1)$ are unstable.

5. System has no equilibria.

7. $(0, 0)$ and $(1/3, 1/3)$ are unstable; $(0, 1)$ and $(1, 0)$ are asymptotically stable.

9. The origin is unstable (saddle point). The equilibrium $(a/\lambda, 0)$ is asymptotically stable if and only if $\lambda c < a\mu$. If $\lambda c > a\mu$, there is an asymptotically stable equilibrium with $y > 0, z > 0$.

Chapter 6

Section 6.1

1. 1.44.

4. 17.56 %.

5. 2.83 %

9. a. New model is

$$S' = -\beta SI + cR, I' = \beta SI - \gamma I, R' = \gamma I - cR.$$

b. Reduced model is

$$S' = -\beta SI + c[N - S - I], I' = \beta SI - \gamma I.$$

c. Equilibria are $(N, 0)$ and $(\gamma/\beta, c(N - \gamma/\beta)/(\gamma + c))$.

d. Threshold condition is at $\mathcal{R}_0 = \beta N/\gamma = 1$.

Section 6.2

1. Equilibrium $(60, 20)$ is asymptotically stable; species coexist.

3. Equilibrium $(0, 45)$ asymptotically stable; y species goes extinct and z species wins competition.

5. All orbits approach the asymptotically stable equilibrium $(10, 10)$.

7. All orbits approach a limit cycle about the unstable equilibrium $(20, 10.8)$.

9. All orbits approach a limit cycle about the unstable equilibrium $(30, 25)$.

12. There are three equilibria $(0, 0)$, $(0, M)$, and $(K, 0)$. They are all unstable. If $ab \geq 1$, orbits are unbounded. If $ab < 1$, a stable equilibrium $(y_\infty, z_\infty) = \left(\frac{K+aM}{1-ab}, \frac{Mb+K}{1-ab}\right)$ appears and all trajectories approach this equilibrium.

15. $(0, 0)$ is asymptotically stable and $(70, 120)$ is unstable.

17. $(20, 10$ is asymptotically stable, and $(0, 0), (20, 0), (0, 10)$ are unstable.

22. The system has two equilibria: extinction (0,0) and persistence $\left(\frac{d}{\gamma+\epsilon d}, \frac{\gamma}{\gamma+\epsilon d}\right) \times \left(1 - \frac{1}{R_d}\right) K$; the extinction equilibrium is LAS iff $R_d < 1$, and the persistence equilibrium exists and is LAS iff $R_d > 1$. Here $R_d = \frac{b}{d} \frac{\gamma}{\mu+\gamma}$ gives the birth-to-adult-death ratio, multiplied by the proportion of juveniles who survive to maturity. Stability follows from standard methods using the Jacobian matrix and Routh-Hurwitz criteria.

Section 6.3

1. The Runge-Kutta approximation gives

t	0	0.01	0.02	0.05	1.0	10.0
y	10	14.74	18.00	20.39	20.15	20.00
z	10	12.80	14.96	18.19	19.63	20.00

3. The Runge-Kutta approximation gives

t	0	2	4	6	8	10
y	40	22.99	17.27	13.71	11.92	11.94
z	10	13.25	15.07	15.21	14.00	12.44

5. (a) $(F^*, M^*) = (0,0)$ and $(\mu_m/mc, \mu_f/fc)$

(b) $J = \begin{bmatrix} fcM^* - \mu_f & fcF^* \\ mcM^* & mcF^* - \mu_m \end{bmatrix}$, so $J(0,0) = \begin{bmatrix} -\mu_f & 0 \\ 0 & -\mu_m \end{bmatrix}$ which has both eigenvalues negative, so (0,0) is LAS, but

$$J(\mu_m/mc, \mu_f/fc) = \begin{bmatrix} 0 & \mu_m f/m \\ \mu_f m/f & 0 \end{bmatrix}$$

which has eigenvalues $\pm\sqrt{\mu_f \mu_m}$ so $(\mu_m/mc, \mu_f/fc)$ is a saddle point (hence unstable).

(c) Note one trajectory leads directly from the saddle point to the origin; perpendicular to this at the saddle point, a hyperbola-like curve (2 trajectories approaching the saddle point) divides the state space into 2 regions: below it, all solutions lead to extinction, while above it all trajectories lead to infinity.

6. (a) $(F^*, M^*) = E_0(0,0)$ or $E_1\left(\frac{R_f R_m - R_f - R_m}{R_m\left(R_f + \frac{c_f}{c_m} R_m\right)}, \frac{R_f R_m - R_f - R_m}{R_f\left(\frac{c_m}{c_f} R_f + R_m\right)}\right)$

(b) E_1 is positive iff $R_f R_m > R_f + R_m$, so conjecture E_1 is LAS when it exists, and E_0 is LAS otherwise (i.e., when $R_f R_m < R_f + R_m$). Numerical explorations should confirm this.

Chapter 7

Section 7.1

In Exercises 1–9, only Exercise 4 exhibits a limit cycle and bursting behavior. Bursting behavior requires $J > 0$ and ϵ and γ small.

Section 7.2

5. $2y\sqrt{\frac{1+y}{1-y}} = \sqrt{3}e^{t/\epsilon}$. Solution of reduced problem is $y = 0(t = 0), y = 1/2(t > 0)$.

7. Equation (6.16) has a unique equilibrium (v, w) with $1/2 < v < 2/3, w < 0$, and this equilibrium is unstable.

11. For $dv/dt = -f(v) - w$, an equilibrium v^* is LAS iff $-f'(v^*) < 0$. But this occurs precisely outside the interval $[A, B]$. We can find A and B by setting $f'(v) = 0$, which yields

$$A, B = \frac{1}{3}\left[(a+1) \pm \sqrt{a^2 - a + 1}\right].$$

12. The unique equilibrium $w^* = v/\gamma$ can be seen to be LAS since $\frac{d}{dw}\left(\frac{dw}{dt}\right) = -\gamma$ which is negative as long as $\gamma > 0$, so w always heads toward w^*.

13. $dw/dt > 0$ iff $v > \gamma w$; at $(B, -f(B))$, this yields $B > -\gamma f(B)$. Substituting for f and then B, this condition becomes

$$\gamma < \frac{1}{(B-a)(1-B)} = \frac{9}{a^2 - 4a + 1 + (a+1)\sqrt{a^2 - a + 1}}.$$

The function on the right is an increasing function of a, starting at 4.5 when $a = 0$, and increasing without bound as $a \to 1$. Thus any $\gamma < 4.5$ satisfies this criterion regardless of the value of a.

14. For parameter values with γ sufficiently low (as described in the text), when the equilibrium reaches A the orbit becomes periodic. The figure below shows the graph for $a = 0.3$, $\gamma = 1$, $\epsilon = 0.002$, and $J = 0.5$. When the equilibrium reaches B the periodic orbit collapses to a stable equilibrium again.

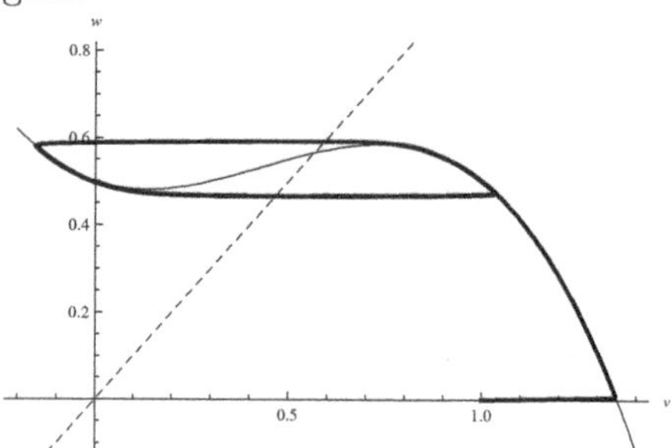

15. Fast: inner solution, boundary layer. Slow: outer solution, reduced problem, quasi-steady state.

Section 7.3

1. $V \to 0$ if $g < bK_2T_0$; $V \to g/K_2T_0 - b$ if $g > bK_2T_0$.

2. (a) Solutions approach either the virus-free equilibrium (VFE) or progression to AIDS, depending on initial conditions (an unstable endemic equilibrium provides a separatrix). (b) A unique endemic equilibrium (EE) is globally asymptotically stable (GAS). (c) Two possible portraits: the VFE is GAS if the nullcline curves do not cross, but if they do cross [twice], the VFE and an EE are both LAS (an unstable EE again providing a separatrix). (d) Two possible portraits: progression to AIDS if the nullcline curves do not cross (Figure 7.13), while if they do cross [twice] there is also a LAS EE (with an unstable EE providing a separatrix).

3. (a) $T_c = \left(\sqrt{\frac{s_2 a}{\mu - k_1 b}\frac{g}{k_2}} - \frac{g}{k_2}\right)/(a-b)$, requires $\mu > k_1 b$

   (b) $\frac{s_2 a}{\mu - k_1 b}\frac{b^2}{a^2} < \frac{g}{k_2} < \frac{s_2 a}{\mu - k_1 b}$

4. Behavior would change to progression to AIDS for $g = 1200$.

6. $V' > \frac{gV}{V+b} - NI'$, so $(V+NI)' > \frac{gV}{V+b}$ and thus increases without bound in both systems. In the system with T, the growth of V pulls down $T' < 0$ so eventually $T \to 0$.

7. The virus-free equilibrium is asymptotically stable if $g < bK_2 s_1/\mu$.

Section 7.4

1. (a) The condition (7.40) provides an upper bound on the slope of $\Lambda(N)$ which decreases as N increases: for any given N_0, $\Lambda'(N_0) < \Lambda(N_0)/N_0$, the slope of the line segment connecting (0,0) with $(N, \Lambda(N))$. Thus for any $N_1 > N_0$, $\Lambda(N_1) < (\Lambda(N_0)/N_0)N_1$, which implies that the bound on $\Lambda'(N_1)$ is even lower:

   $$\Lambda'(N_1) < \Lambda(N_1)/N_1 < \Lambda(N_0)/N_0.$$

   Thus the slope of $\Lambda(N)$ will continue to decrease until it is lower than d, and at some point the curve will cross below $d \cdot N$.

   (b) If $\Lambda(x_0) < d \cdot x_0$, then by (7.40) $\Lambda'(x_0) < d$, and as discussed in (a), since the bound decreases as x increases, the slope of the curve can never again increase past d, which would be necessary in order to cross above the line $y = d \cdot x$.

   (c) If we write (7.45) in the form $N' = g(N)$, then $g'(N) = \Lambda'(N) - d$. Thus for an equilibrium $K > 0$ we have $g'(K) = \Lambda'(K) - d < \Lambda(K)/K - d = dK/K - d = 0$ (since at an equilibrium $\Lambda(K) = dK$), and since $g'(K) < 0$ K must be asymptotically stable. Meanwhile, $g'(0) = \Lambda(0) - d$, which is asymptotically stable iff $\Lambda'(0) < d$.

2. $(0, P_0^2)$ is locally asymptotically stable iff $\Lambda'(0) < d$, consistent with the stability of (K, P_0^2) iff $\Lambda'(0) > d$.

4. For all cases we first compute $f'(P) = (\Delta_D + \Delta_R)[3P^2 - 2P(P^*+1) + P^*]$, from which $f'(0) = (\Delta_D + \Delta_R)P^*$, $f'(1) = (\Delta_D + \Delta_R)(1 - P^*)$, and $f'(P^*) = (\Delta_D + \Delta_R)P^*(P^* - 1)$.

   Now for Case I we first consider the subcase $\Delta_D + \Delta_R > 0$. Here $P^* > 1$ since $\Delta_D < 0$, so it is not of interest. If instead $\Delta_D + \Delta_R < 0$, then $P^* < 0$, so it is again not of interest. Either way, $f'(0) > 0$ and $f'(1) < 0$, so $P = 0$ is unstable and $P = 1$ is stable; thus the A allele approaches 100% frequency.

5. Case II is the reverse of Case I: if $\Delta_D + \Delta_R > 0$ then $P^* < 0$, if $\Delta_D + \Delta_R < 0$ then $P^* > 1$, and either way $f'(0) < 0$, $f'(1) > 0$, making $P = 0$ stable and $P = 1$ unstable. The a allele approaches 100% frequency.

6. With both $\Delta_D > 0, \Delta_R > 0$, $P^* \in (0,1)$, so that $f'(0) > 0$, $f'(1) > 0$, and $f'(P^*) < 0$. Thus P^* is globally stable on the interval (0,1).

7. With $\Delta_D < 0, \Delta_R < 0$, $P^* \in (0,1)$, and $f'(0) < 0$, $f'(1) < 0$, and $f'(P^*) > 0$. Thus the unstable P^* divides the interval (0,1) into two basins of attraction, with trajectories beginning on either side of it approaching the extreme on that side.

8. If $\Delta_R > 0$, $P \to 1$, and asymptotically all members are of genotype AA. If $\Delta_R < 0$, $P \to 0$, and asymptotically all members are of genotype aa.

Section 7.5

1. $y = K_1 t + K_2, y = -t + 1$.

3. $y = K_1 e^{3t} + K_2 e^{2t}, y = -3e^{3t} + 4e^{2t}$.

5. $y = K_1 e^{5t} + K_2 t e^{5t}, y = e^{5t} - 6tE^{5t}$.

7. $y = K_1 e^{-5t/2} \cos \sqrt{15}t/2 + K_2 e^{-5t/2} \sin \sqrt{15}t/2, y = e^{-5t/2} \cos \sqrt{15}/2 + \sqrt{15}/2e^{-5t/2} \sin \sqrt{15}t/2$.

9. $y = K_1 e^{2t} \cos 3t + K_2 e^{2t} \sin 3t, y = e^{2t} \cos 3t - e^{2t} \sin 3t$.

11. $y = K_1 e^t \cos t + K_2 e^t \sin t, y = e^t \cos t - 2e^t \sin t$.

13. Equilibria are $(0,0)$ and $(0, \pm\sqrt{-b})$ if $b < 0$.

15. Linearization at $(0,0)$ is $u' = v, v' = -v$.

17. Linearization at $(0,0)$ is $u' = v, v' = -g'(0) - f(0)v$.

Bibliography

Caroline Bampfylde, Modelling rainforests, M.Sc. thesis, Oxford University (1999).

J.B.I. Bard, A quantitative model of liver regeneration in the rat, *Journal of Theoretical Biology* **73**(3): 509–530 (1978).

J.B.I. Bard, A quantitative theory of of liver regeneration in the rat II: Matching an improved mitotic inhibitor model to the data, *Journal of Theoretical Biology* **79**(1): 121–136 (1979).

Madeleine Beekman, David J. T. Sumpter, and Francis L. W. Ratnieks, Phase transition between disordered and ordered foraging in Pharaoh's ants, *Proceedings of the National Academy of Sciences USA*, **98**: 9703–9706 (2001).

Hassan Benchekroun and Ngo Van Long, Transboundary fishery: a differential game model, *Economica*, **69**: 207–221 (2002).

Samuel Bernard, Jacques Bélair and Michael C. Mackey, Bifurcations in a white-blood-cell production model, *Comptes Rendus Biologies*, **327**: 201–210 (2004).

Raymond J.H. Beverton and Sidney J. Holt, *On the dynamics of exploited fish populations*, Fishery Investigations Series 2 (19), Great Britain Ministry of Agriculture, London (1954) (reprinted by Chapman & Hall (1993)).

R.J.H. Beverton and S.J. Holt, The theory of fishing, in *Sea fisheries: their investigation in the United Kingdom* (M. Graham, ed.), Edward Arnold, London, pp. 372–441 (1956).

G.E.P. Box, Science and statistics, *Journal of the American Statistical Association*, **71**: 791–799 (1976).

Fred Brauer and David A. Sánchez, Constant rate population harvesting: equilibrium and stability, *Theoretical Population Biology*, **8**: 12–30 (1975).

T. Carlson, Über Geschwindigkeit und Grösse der Hefevermehrung in Würze, *Biochemische Zeitschrift*, **57**: 313–334 (1913).

Robert J. Carman and Mary Alice Woodburn, Effects of low levels of ciprofloxacin on a chemostat model of the human colonic microflora, *Regulatory Toxicology and Pharmacology*, **33**: 276–284 (2001).

S. R. Carpenter, D. Ludwig, W. A. Brock, Management of eutrophication for lakes subject to potentially irreversible change, *Ecological Applications*, **9**: 751–771 (1999).

Hal Caswell, Masami Fujiwara, and Solange Brault, Declining survival probability threatens the north Atlantic right whale, *Proceedings of the National Academy of Sciences USA*, **96**: 3308–3313 (1999).

J. Chervinski, Salinity tolerance of young catfish, *Clarias lazera* (Burchell), *Journal of Fish Biology*, **25**: 147 (1984).

J. Davidson, On the ecology of the growth of the sheep population in South Australia, *Transactions of the Royal Society of South Australia*, **62**: 141–148 (1938).

J. Davidson, On the growth of the sheep population in Tasmania, *Transactions of the Royal Society of South Australia* **62**: 342–346 (1938).

M. Doebeli, Dispersal and dynamics, *Theoretical Population Biology*, **47**: 82–106 (1995).

J. Dumas and P. Prouzet, Variability of demographic parameters and population dynamics of Atlantic salmon *Salmo salar* L. in a south-west French river, *ICES Journal of Marine Science*, **60**: 356–370 (2003).

M. Eisen and J. Schiller, Stability analysis of normal and neoplastic growth, *Bulletin of Mathematical Biology* **39**(5): 597–605 (1977).

K.E. Emmert and L.J.S. Allen, Population persistence and extinction in a discrete-time, stage-structured epidemic model, *Journal of Difference Equations and Applications*, **10**: 1177–1199 (2004).

A.S. Evans, *Viral Infections of Humans*, 2nd ed., Plenum Press, New York (1982), reported by H.W. Hethcote, Three basic epidemiological models, *Applied Mathematical Ecology*, S.A. Levin, T.G. Hallam, and L. J. Gross (eds), Biomathematics **18** Springer-Verlag, New York-Heidelberg, pp. 119–144 (1989).

A.S. Fauci, G. Pantaleo, and D. Weissman, Immunopathogenic mechanisms of HIV infection, *Annals of Internal Medicine* **124**: 654–663 (1996).

U. Feldmann and B. Schneider, A general approach to multicompartment analysis and models for the pharmacodynamics, in J. Berger et al. (eds.), *Mathematical models in medicine: workshop, Mainz, March 1976* (Berlin/New York: Springer-Verlag), Lecture Notes in Biomathematics **11**: 243–277 (1976).

R. Fitzhugh, Impulses and physiological states in theoretical models of nerve membrane, *Biophysics Journal*, **1**: 445–466 (1961).

G.P. Garnett, An introduction to mathematical models in sexually transmitted disease epidemiology, *Sexually Transmitted Infections*, **78**: 7–12 (2002).

G.F. Gause, *The Struggle for Existence*, Williams and Wilkins, Baltimore (1934).

Tim Gerrodette, The tuna-dolphin issue, in W.F. Perrin, B. Würsig and J.G.M. Thewissen, eds. *Encyclopedia of Marine Mammals* (2nd edition). Elsevier, Amsterdam, pp. 1192–1195 (2009).

T. Gerrodette and J. Forcada, Non-recovery of two spotted and spinner dolphin populations in the eastern tropical Pacific Ocean, *Marine Ecology Progress Series*, **291**: 1–21 (2005).

Tim Gerrodette, George Watters, Wayne Perryman, Lisa Ballance, Estimates of 2006 dolphin abundance in the Eastern Tropical Pacific, with revised estimated from 1986–2003, NOAA Technical Memorandum NMFS-SWFSC-422, April 2008, available at `https://swfsc.noaa.gov/publications/TM/SWFSC/NOAA-TM-NMFS-SWFSC-422.pdf`.

L. Glass and M.C. Mackey, Pathological conditions resulting from instabilities in physiological control systems, *Annals of the N.Y. Academy of Science*, **316**: 214–235 (1979).

J. Gressel and L.A. Segel, The paucity of plants evolving genetic resistance to herbicides: possible reasons and implications, *Journal of Theoretical Biology*, **75**: 349–371 (1978).

G.H. Hardy, Mendelian proportions in a mixed population, *Science* 28(706): 49–50, July 1908 (letter to the editor).

James D. Hardy, The physical laws of heat loss from the human body, *Proceedings of the National Academy of Science USA*, **23**: 631–637 (1937).

Craig A. Harms, Mark G. Papich, M. Andrew Stamper, Patricia M. Ross, Mauricio X. Rodriguez, and Aleta A. Holm, Pharmacokinetics of oxytetracycline in loggerhead sea turtles (*Caretta caretta*) after single intravenous and intramuscular injections, *Journal of Zoo and Wildlife Medicine*, **35**: 477–488 (2004).

M.P. Hassell, Density dependence in single species populations, *Journal of Animal Ecology*, **44**: 283–295, 1975.

H.W. Hethcote and J.A. Yorke, *Gonorrhea Transmission Dynamics and Control.* Lecture Notes in Biomathematics **56**, Springer-Verlag, New York (1984).

A. L. Hodgkin and A. F. Huxley, A quantitative description of membrane current and its application to conduction and excitation in nerve, *J. Physiology*, **117**: 500–544 (1952).

F.C. Hoppensteadt, Singular perturbations on the infinite interval, *Trans. Amer. Math. Soc.*, **123**: 521–535 (1966).

G. Evelyn Hutchinson, *An introduction to population ecology*, Yale University Press, New Haven (1978).

Catherine Ryan Hyde, *Pay it forward*, Simon and Schuster, New York (1999).

International Dolphin Conservation Program Scientific Advisory Board, 7th meeting, Updated estimates of N_{MIN} and stock mortality limits, Document SAB-07-05, La Jolla, California, 30 October 2009, available at `http://www.iattc.org/PDFFiles2/SAB-07-05-Nmin-and-Stock-Mortality-Limits.pdf`.

Holger W. Jannasch, Steady state and the chemostat in ecology, *Limnology and Oceanography*, **19**: 716–720 (1974).

Emmanuel Jolivet, *Introduction aux modèles mathématiques en biologie*, INRA/Masson, Paris (1983).

D.S. Jones and B.D. Sleeman, *Differential Equations and Mathematical Biology*, CRC Press, Boca Raton, FL (2003).

Sukgeun Jung and Edward D. Houde, Recruitment and spawning-stock biomass distribution of bay anchovy (*Anchoa mitchilli*) in Chesapeake Bay, *Fishery Bulletin*, **102**: 63–77 (2004).

J. Keener and J. Sneyd, *Mathematical Physiology*, Springer-Verlag, New York (1998).

W. O. Kermack and A. G. McKendrick, A contribution to the mathematical theory of epidemics, *Proceedings of the Royal Society of London*, **115**: 700–721 (1927).

W.O. Kermack and A.G. McKendrick, Contributions to the mathematical theory of epidemics, Part II, *Proceedings of the Royal Society of London*, **138**: 55–83 (1932).

J.M. Kienzler, P.F. Dahm, W.A. Fuller, A.F. Ritter, Temperature-based estimation for the time of death in white-tailed deer, *Biometrics*, **40**: 849–854 (1984).

B.W. Knight, Dynamics of encoding in a population of neurons, *J. General Physiology*, **59**: 734–766 (1972).

G. A. Knox, The key role of krill in the ecosystem of the southern ocean with special reference to the convention on the conservation of antarctic marine living resources, *Ocean Management*, **9**: 113–156 (1984).

Mark Kot, *Elements of Mathematical Ecology*, Cambridge University Press, Cambridge (2001).

C.M. Kribs-Zaleta, To switch or taper off: the dynamics of saturation, *Math. Biosci.*, **192**: 137–152 (2004).

S.D. Lane and N.J. Mills, Intraspecific competition and density dependence in an *Ephesia kuehniella–Venturia canescens* laboratory system, *OIKOS* **101**: 578–590 (2003).

Richard C. Lathrop, Stephen R. Carpenter, Craig A. Stow, Patricia A. Soranno, and John C. Panuska, Phosphorus loading reductions needed to control blue-green algal blooms in Lake Mendota, *Canadian Journal of Fisheries and Aquatic Sciences*, **55**: 1169–1178 (1998).

R. Levins, Some demographic and genetic consequences of environmental heterogeneity for biological control, *Bulletin of the Entomological Society of America* **15**: 237–240 (1969).

N. Levinson, Perturbations of discontinuous solutions of nonlinear systems of differential equations, *Acta Mathematica*, **82**: 71–106 (1950).

D. Ludwig, D.D. Jones and C.S. Holling, Qualitative analysis of insect outbreak systems: the spruce budworm and forest, *Journal of Animal Ecology*, **47**: 315–332 (1978).

M.C. Mackey and L. Glass, Oscillation and chaos in physiological control systems, *Science*, **197**: 287–289 (1977).

T.R. Malthus, An essay on the principle of population, Harmondsworth, Middlesex (1798) (republished by Penguin, New York (1970)).

Robert M. May, Thresholds and breakpoints in ecosystems with a multiplicity of stable states, *Nature*, **269**: 471–477 (1977).

J. Maynard Smith, *Mathematical ideas in biology*, Cambridge University Press, Cambridge, 1968.

J. Maynard Smith and M. Slatkin, The stability of predator-prey systems, *Ecology*, **54**: 384–391 (1973).

R.W. McCarley and J.A. Hobson, Neuronal excitability modulation over the sleep cycle: A structured and mathematical model, *Science*, **189**: 58–60 (1975).

Curt D. Meine and George W. Archibald (Eds)., The cranes: Status survey and conservation action plan. IUCN, Gland, Switzerland, and Cambridge, U.K. 294pp. Northern Prairie Wildlife Research Center Online. `http://www.npwrc.usgs.gov/resource/birds/cranes/index.htm` (Version 02MAR98) (1996).

L. Michaelis and M.I. Menten, Die Kinetik der Invertinwirkung, *Biochem. Z.*, **49**: 333–369 (1913).

R.S. Miller and D.B. Botkin, Endangered species: models and predictions, *American Scientist*, **62**: 172–181 (1974).

Jacques Monod, La technique de culture continue: théorie et applications, *Annales de l'Institut Pasteur*, **79**: 390–410 (1950).

J.S. Nagumo, S. Arimoto, and S. Yoshizawa, An active pulse transmission line simulating nerve axon, *Proc. Inst. Radio Engineers*, **50**: 2061–2071 (1962).

M.E.J. Newman, The structure and function of complex networks, *SIAM Review*, **45**: 167–256 (2003).

A.J. Nicholson, An outline of the dynamics of animal populations, *Australian Journal of Zoology*, **3**: 9–65 (1954).

A.J. Nicholson and V.A. Bailey, The balance of animal populations, *Proc. Zoological Society of London*, **3**: 551–598 (1935).

Aaron Novick and Leo Szilard, Description of the chemostat, *Science*, **112**: 715–716 (1950).

M.A. Nowak and R.M. May, *Virus dynamics: Mathematical principles of immunology and virology*, Princeton University Press, 2003.

R. Pearl, *Introduction of medical biometry and statistics*, Saunders, Philadelphia (1930).

Alan S. Perelson and P.W. Nelson, Mathematical analysis of HIV-1 dynamics in vivo, *SIAM Review***41**: 3–44 (1999).

J.D. Reeve, D.J. Rhodes and P. Turchin, Scramble competition in the southern pine beetle, *Dendroctonus frontalis*, *Ecological Entomology*, **23**: 433–443 (1998).

R.S. Rempel and C.K. Kaufman, Spatial modeling of harvest constraints on wood supply versus wildlife habitat objectives, *Environmental Management*, **32**: 646–659 (2003). doi: 10.1007/s00267-003-0056-8.

Eric Renshaw, *Modelling biological populations in space and time*, Cambridge Studies in Mathematical Biology 11, Cambridge University Press, Cambridge (1995).

W.E. Ricker, Stock and recruitment, *J. Fisheries Research Board Canada*, **11**: 559–623 (1954).

Alex H. Ross, William S. C. Gurney, Michael R. Heath, Steven J. Hay, and Eric W. Henderson, A strategic simulation model of a fjord ecosystem, *Limnology and Oceanography*, **38**: 128–153 (1993).

W.M. Roth, Unspecified things, signs, and 'natural objects': towards a phenomenological hermeneutic of graphing, In S. B. Berenson, K. R. Dawson, M. Blanton, W. N. Coulombe, J. Kolb, K. Norwood, and L. Stiff (Eds.), *Proceedings of the Twentieth Annual Meeting of the North American Chapter of the International Group for the Psychology of Mathematics Education* (Vol. I), ERIC Clearinghouse for Science, Mathematics, and Environmental Education, Columbus, OH, pp. 291–297, (1998).

M.B. Schaefer, Some aspects of the dynamics of populations important to the management of commercial marine fisheries, *Bulletin of the Inter-Amer. Trop. Tuna Comm.* **I**: 25–56 (1954).

Russell J. Schmidt, Sally J. Holbrook, and Craig W. Osenberg, Quantifying the effects of multiple processes on local abundance: a cohort approach for open populations, *Ecology Letters*, **2**: 294–303 (1999).

M.D. Scott, S.J. Chivers, R.J. Olson, P.C. Fiedler, K. Holland, Pelagic predator associations: tuna and dolphins in the eastern tropical Pacific Ocean, *Marine Ecology Progress Series* **458**: 283–302 (2012).

Lee A. Segel, *Modeling dynamic phenomena in molecular and cellular biology*, Cambridge University Press, Cambridge (1984).

Arthur Sherman and Richard Bertram, Integrative modeling of the pancreatic β-cell, in *Wiley Interscience Encyclopedia of Genetics, Genomics, Proteomics, and Bioinformatics*, Part 3 Proteomics, M. Dunn, ed., Section 3.8 Systems Biology, R. L. Winslow, ed., John Wiley and Sons, Ltd. (2005). DOI: 10.1002/047001153X.g308213.

E. Smith, S. Haarer, J. Confrey, Seeking diversity in mathematics education: mathematical modeling in the practice of biologists and mathematicians, *Science and Education*, **6**(5): 441–472 (1997).

E.M. Stock, K.E. Emmert, Deterministic discrete-time epidemic models with applications to amphibians, masters thesis, Tarleton State University (2006).

S.H. Strogatz, Exploring complex networks, *Nature*, **410**: 268–276 (2001).

David J.T. Sumpter and Madeleine Beekman, From nonlinearity to optimality: pheromone foraging by ants, *Animal Behaviour*, **66**: 273–280 (2003).

D.J.T. Sumpter and S.C. Pratt, A modelling framework for understanding social insect foraging, *Behav. Ecol. Sociobiol.*, **53**: 131–144 (2003).

D'Arcy Wentworth Thompson, *On growth and form*, Vol. I, 2nd ed., Cambridge University Press, Cambridge (1942).

A.N. Tihonov, On the dependence of the solutions of differential equations on a small parameter, *Mat. Sbornik NS*, **22**: 193–204 (1948).

David Tilman, Competition and biodiversity in spatially structured habitats, *Ecology*, **75**: 2–16, (1994).

David Tilman, Robert M. May, Clarence M. Lehman, and Martin A. Nowak, Habitat destruction and the extinction debt, *Nature*, **371**: 65–66 (1994).

Pauline van den Driessche and James Watmough, A simple SIS epidemic model with a backward bifurcation, *Journal of Mathematical Biology*, **40**: 525–540 (2000).

E. van der Meijden, M. J. Crawley, and R. M. Nisbet, The dynamics of a herbivore-plant interaction, in *Insect Populations: in Theory and Practice* (J. P. Dempster and I. F. G. McLean, eds.), Chapman and Hall, London (1998).

B. van der Pol and J. van der Mark, The heartbeat considered as a relaxation oscillation, and an electrical model of the heart, *Phil. Mag.*, **6**: 763–775 (1928).

P.F. Verhulst, Notice sur la loi que la population suit dans son accroissement, *Corr. Math. et Phys.*, **10**: 113–121 (1838).

P.F. Verhulst, Récherches mathématiques sur la loi d'accroissement de la population, *Mem. Acad. Roy. Brussels*, **18**: 1–38 (1845).

P.R. Wade, G.W. Watters, T. Gerrodette, and S.R. Reilly, Depletion of spotted and spinner dolphins in the eastern tropical Pacific: modeling hypotheses for their lack of recovery, *Marine Ecology Progress Series*, **343**: 1–14 (2007).

Wilhelm Weinberg, Uber den Nachweis der Vererbung beim Menschen, *Jahreshefte des Vereins f ur vaterl andische Naturkunde in W urttemberg*, **64**: 368–382 (1908).

Index